PHYLOGENETIC SYSTEMATICS

Haeckel to Hennig

Species and Systematics

The *Species and Systematics* series will investigate the theory and practice of systematics, phylogenetics, and taxonomy and explore their importance to biology in a series of comprehensive volumes aimed at students and researchers in biology and in the history and philosophy of biology. The book series will examine the role of biological diversity studies at all levels of organization and focus on the philosophical and theoretical underpinnings of research in biodiversity dynamics. The philosophical consequences of classification, integrative taxonomy, and future implications of rapidly expanding data and technologies will be among the themes explored by this series. Approaches to topics in *Species and Systematics* may include detailed studies of systematic methods, empirical studies of exemplar taxonomic groups, and historical treatises on central concepts in systematics.

Editor in Chief: Kipling Will (University of California, Berkeley)

For more information visit:
www.crcpress.com/Species-and-Systematics/book-series/CRCSPEANDSYS

PHYLOGENETIC SYSTEMATICS

Haeckel to Hennig

OLIVIER RIEPPEL

CRC Press
Taylor & Francis Group
Boca Raton London New York

CRC Press is an imprint of the
Taylor & Francis Group, an **informa** business

CRC Press
Taylor & Francis Group
6000 Broken Sound Parkway NW, Suite 300
Boca Raton, FL 33487-2742

First issued in paperback 2019

© 2016 by Taylor & Francis Group, LLC
CRC Press is an imprint of Taylor & Francis Group, an Informa business

No claim to original U.S. Government works

ISBN-13: 978-0-4987-5488-0 (hbk)
ISBN-13: 978-0-367-87645-6 (pbk)

Library of Congress Cataloging-in-Publication Data

Names: Rieppel, Olivier.
Title: Phylogenetic systematics : Haeckel to Hennig / Olivier Rieppel.
Description: Boca Raton : Taylor & Francis, 2016. | Series: Species and
systematics | Includes bibliographical references and index.
Identifiers: LCCN 2016007225 | ISBN 9781498754880 (alk. paper)
Subjects: LCSH: Cladistic analysis--Germany--History. | National socialism
and science.
Classification: LCC QH83 .R54 2016 | DDC 578.01/2--dc23
LC record available at https://lccn.loc.gov/2016007225

Visit the Taylor & Francis Web site at
http://www.taylorandfrancis.com

and the CRC Press Web site at
http://www.crcpress.com

Contents

Series Preface

The Species and Systematics series is a broad-ranging venue where authors can provide the scientific community with comprehensive treatments of the history and philosophy of fundamental concepts in systematic biology, phylogenetics, and the science of taxonomy. The series also intends to connect historical perspectives to new ideas and emerging technologies that have implications for the field. The series is committed to stimulating discussion among students and researchers in biology on controversial and clarifying ideas related to the future course that we are charting in biodiversity research.

There are many approaches to the study of biological diversity and to embrace this, future volumes in the series may include detailed development and comparisons of existing and novel methods in systematics and biogeography, empirical studies that provide new insights into old questions and raise new questions to biologists and philosophers of science, and historical treatises on central and reoccurring concepts that benefit from both a retrospective and a new perspective. Some volumes will address a single important concept in great depth, giving authors the freedom to present ideas with their own slant, whereas others will be edited collections of shorter papers intended to place alternative views in sharp contrast.

For science in general and certainly for systematics, few things are more significant than placing the origin and evolution of an idea or principle in its historical context. In this book, Olivier Rieppel does a masterful job of fixing Haeckel as an anchor point for a thread that weaves through and touches virtually every important principle of phylogenetics, leading to Hennig's ideas that launched the cladistic revolution. The detailed consideration of how this thread of ideas played out in the arena of heavy biopolitics, which colors both language and intent, gives readers new insights into many concepts fundamental to current and future research. *Phylogenetic Systematics: Haeckel to Hennig* is an excellent example of scholarship and a very fine addition to the series that will make the reader pause and think and then evoke discussion, both of which are chief goals of the series.

Kipling Will
Berkeley, California

Acknowledgments

I am deeply indebted to Chuck Crumly, the editor at CRC Press, Kipling Will, the series editor, and Jennifer Ahringer, production coordinator, for their interest, help, and support in seeing this book though publication. I am equally deeply indebted to staff, who received me warmly and offered much help in the University Archives in Göttingen, Greifswald, Halle/Saale, Harvard, Tübingen, and Zurich; at the State Archives in Basel and Zurich; at the Archive of the Max Planck Society, Berlin-Dahlem; at the Bundesarchiv (Federal Archives, formerly Berlin Documentation Center), Berlin-Lichterfelde; and at the Academy of Sciences Leopoldina, Halle/Saale. Invaluable help in the search for illustrations was provided by the following individuals:

Gernot J. Abel, Copenhagen
Irmela Bauer-Klöden, Universitätsarchiv Tübingen
Tad Bennicoff, Smithsonian Institution Archives, Washington DC
Anja Bugaiski, Archive, Martin-Luther-Universität Halle-Wittenberg
Sabine Hackethal, Historische Arbeitsstelle, Museum für Naturkunde, Berlin
Heike Hartmann, ETH-Bibliothek Bildarchiv, Zurich
Gerd Hennig, Tübingen
Uwe Hoßfeld, University of Jena
Joachim W. Kadereit and Christian George, Botanical Garden, Mainz
Lynn Nyhart, University of Wisconsin, Madison
Michael Schmitt, University of Greifswald
Editha Schubert, Deutsches Entomologisches Institut, Müncheberg
Klaus Taschwer, Vienna
Susanne Uebele, Archive, Max Planck Society, Berlin-Dahlem
Renate Würsch, Universitätsbibliothek Basel
Willi Xylander, Willi-Hennig-Archive, Senckenberg Museum Görlitz
Equally invaluable help in the location of obscure literature was provided by
    Christine Giannoni, Director of the Field Museum Library

Introduction and Preview

John Lennon once controversially proclaimed the Beatles to be "more popular than Jesus." What about Charles Darwin? Whether you defend his theory of evolution, or reject it, either way you have been touched by him. It is said that Darwin, with his book *On the Origin of Species* published in 1859, revolutionized the science of biology to a degree that changed the world, or at least the Western world-view. Darwin's theory of evolution may have been revolutionary, but Darwin himself certainly was not revolutionary. Most people may have an at least vague, incomplete, perhaps sometimes bizarre understanding of evolution, but few people probably know that Darwin spent 8 years—from 1846 to 1854—studying the anatomy and classification of living and fossil barnacles. Other long-term projects that Darwin pursued aimed more in a geological direction. They concerned the formation and degradation of coral reefs through geological time, as well as the role of earthworms in the formation and continuous reworking of vegetable mould. The latter subject he first touched upon in a paper delivered to the Geological Society of London in November 1837, in which he concluded "the digestive power of animals is a geological power which acts in another region on a greater scale … a large portion of the chalk of Europe was produced from coral, by the digestive action of marine animals, in the same manner as mould has been prepared by the earth-worm on disintegrated rock" (Darwin, 1838, p. 576). The whole worm story Darwin published in the format of a best-selling book in the fall of 1881, just 6 months before he died.* In a famous passage, Darwin forcefully countered the objection that his characterization of earthworms as a geological force was way overblown:

> Here we have an instance of that inability to sum up the effects of a continually recurrent cause, which has often retarded the progress of science, as formerly in the case of geology, and more recently in that of the principle of evolution.
>
> **Darwin**
> *1881, p. 6*

Just as erosion—governed by the law of gravity—slowly but steadily carves out valleys before flattening whole mountains, so does evolution—governed by the law of variation and the principle of natural selection—slowly but continuously transform ancestral species into descendant ones. Either of these processes is too slow to be directly observable, but the effects of their continuously recurrent causes can be appreciated if enough time is allowed for them to play out. The ultimate proof lies in the Fossil Record, and the rocks that contain it.

Darwin's theory of evolution is not a revolutionary one in the sense of saltational change. Darwinian evolution plays out through geological time in a continuous, slow and gradual sequence of steps of transformation, the latter so small as to reveal themselves only in their summation. Fueled by variation and constrained by natural selection,

---

* http://darwin-online.org.uk/EditorialIntroductions/Chancellor_Earthworms.html.

Darwinian evolution makes sure that a species continuously tracks its ever-changing environment. Some have located in Darwin's principle of evolution far greater implications than mere species transformation and adaptation. The famous twentieth-century philosopher of science Sir Karl Popper (1902–1994) explained the progress made in science by the Darwinian principle of evolution. Scientific theories, in his view, behave like species. For a theory to count as scientific, Popper insisted, it must be able to generate testable predictions. Should the test show any of these predictions to be wrong, the theory must be considered falsified. The theory must consequently be dropped, or at least revised. The various predictions scientists derive from scientific theories Popper compared to variation; he compared the experimental test and potential refutation of those predictions to natural selection. Through the conjecture of theories, and the derivation of testable predictions from those, scientific research engages, according to Popper, in a process that would progressively approach true knowledge about the world without ever taking full possession of it (Popper, 1972).

This Popperian view of science and society was soon to be challenged, however, by a veritable revolution in the philosophy of science. Thomas S. Kuhn's (1922–1996) book *The Structure of Scientific Revolutions*, first published in 1962 (Kuhn, 1962), became an unlikely best seller. Never before or after has a book been published in the philosophy of science that sold as many copies, and was translated into as many languages, as Kuhn's. It was picked up and widely discussed not only by philosophers and historians of science but also by scientists and the broader public. This is all the more surprising because the essence of his argument is not at all easy to understand, as it is ultimately rooted in the philosophy of language that lies at the core of Anglo-American analytic philosophy. In general terms, what Kuhn was saying is that science does not progress slowly and steadily toward a true understanding of the world, but instead moves by leaps and bounds between changing paradigms, that is, changing world-views. Kuhn called a paradigm change in the realm of science a scientific revolution. The move from a geocentric to a heliocentric understanding of the solar system was a paradigm change; the move from a phlogiston theory of combustion to an oxygen theory of combustion was a paradigm change; the move from a creationist to an evolutionary understanding of living nature was a paradigm change. Students, Kuhn claimed, are trained by their professors within the context and confines of a paradigm that is accepted by the scientific community. A paradigm represents a certain general world-view; it partitions the world in certain ways that inform the design of research projects and strategies deemed relevant and promising. As the scientific community continues to explore a paradigm, the continued variation and testing of theories may result in the recognition of an increasing number of exceptions, anomalies, or negatives. The paradigm eventually may slide into a crisis, which triggers a scientific revolution, the relatively sudden and rapid replacement of an old with a different, new paradigm. Note that for Kuhn, science does not slowly and continuously approach a true understanding of the world, but instead jumps from an old paradigm, or world-view, to a new one.

The move from traditional evolutionary systematics to phylogenetic systematics, or cladistics, that played out in the late 1970s and early 1980s was described as a paradigm change as well. Systematics is the science that describes plant and animal species and classifies those in a way that represents their evolutionary relationships. The radical difference between traditional evolutionary systematics and revolutionary phylogenetic systematics is perhaps best illustrated by the case of crocodile classification (see Hennig, 1974; Mayr, 1974). Crocodiles are clearly reptiles, and are thus traditionally classified in a group that includes turtles, the tuatara, lizards, and snakes, that is, animals that share certain anatomical, physiological, and behavioral traits. Birds, by contrast, are traditionally classified as a separate group, a classification that is also reflected in zoos with a separate reptile and bird-house, respectively. Cladists, by contrast, argue that anatomical, physiological, or behavioral traits should not determine classifications, which instead must be based strictly on recency of common ancestry. Everybody involved in this discussion agrees that crocodiles and birds share a more recent common ancestor that is not also ancestral to turtles, the tuatara, lizards, and snakes. This means that crocodiles and birds are more closely related to one another than either of them is to any other reptile lineage. The requirement then arises to include birds in the reptiles in order to render the latter a natural group, that is, a group that includes its ancestor and all, instead of only some, of its descendants. Some have for that reason chosen to drop the name Reptilia, and to use the name Sauropsida instead for the group that includes the traditional reptiles plus birds. We now have feathered dinosaurs from China that indicate that (some) theropod dinosaurs and birds share a common ancestor that is not also ancestral to other reptile lineages. Theropod dinosaurs and birds are thus more closely related to one another than either is to any other reptile lineage, or, in other words, theropod dinosaurs and birds form a group within Sauropsida that does not also include any other sauropsids. In cladistic jargon, it is then possible to claim that dinosaurs are not really extinct, but that they in fact survive in the form of their closest relatives, the birds. Such a conclusion would be anathema to an evolutionary systematist (Dingus and Rowe, 1998).

Kuhn derived his insights with regards to scientific revolutions from a historical analysis of natural sciences, especially physics. Following Kuhn, another philosopher of science, David L. Hull (1935–2010), wanted to study a paradigm change in real time. He chose systematics for his case study and inserted himself into the systematists' peer community becoming a board member of the Society of Systematic Biology (formerly known as Society of Systematic Zoology) and coeditor of its journal. Hull published his observations and conclusions in 1988 in a book titled *Science as a Process: An Evolutionary Account of the Social and Conceptual Development of Science* (Hull, 1988). There is a certain irony here in that as a graduate student, Hull had participated in a seminar taught by Popper. Not surprisingly, he defended a Darwinian view of science, arguing that scientific theories function as entities that—like species—form evolutionary lineages. Yet, his book on the science wars in systematics that eventually resulted in the supremacy of phylogenetic systematics, or cladistics, over competing schools of biological systematics cemented the picture of a scientific revolution, called the *cladistic revolution*. At its center stood Willi Hennig (1913–1976), a German entomologist and—like Darwin—a "cautious revolutionizer" (Schmitt, 2010).

Hennig's contribution to biosystematics has, indeed, been called *paradigm changing*, or *paradigm forming*, such that Hennig's impact as a scientist has been likened to the significance of such historical luminaries as Copernicus and Darwin (Kühne, 1978/1979). The *Dictionary of Scientific Biography* compares the Hennigian with the Darwinian revolution in biology (Dupuis, 1990, p. 408). Others noted that "Hennig undoubtedly influenced systematics like no other biologist since Linnaeus and Darwin" (Richter and Meier, 1994, p. 219). And Hull, the chronicler of the cladistic revolution, writes "Just as Darwin's *Origin of Species* was one long argument against creationism, Hennig's *Grundzüge* was directed primarily against the priority claimed by German idealists for morphology over phylogeny" (Hull, 1988, p. 135). That is an important statement, as it pitches Hennig's *magnum opus* of 1950—*Grundzüge einer Theorie der Phylogenetischen Systematik* (Outlines of a Theory of Phylogenetic Systematics)—not so much against competing schools of biosystematics, but rather against German idealistic morphology. Such, indeed, is the historically correct context, and it also is the topic of this book.

Hennig's paradigm changing ideas about systematics and its importance for the reconstruction of evolutionary (phylogenetic) relationships did not materialize out of nothing. Instead, Hennig looked back on a long-standing controversy between idealistic morphologists and phylogenetic systematists among German biologists, a debate that during the ascent and reign of the Hitler regime acquired distinct ideological and political connotations. So much so that in spite of Hennig coming to this debate as a late outsider, a controversy sprung up as to whether he, too, had been a Nazi or not. There is no record that Hennig ever was a member of the National Socialist German Workers' Party (NSDAP), let alone an outspoken supporter of its ideology, but as a young man he was a member of its paramilitary wing, the SA (Storm troopers or Brown shirts). Such a brown dot on his otherwise white coat was interpreted as "a rather mild form of opportunism" by his biographer Michael Schmitt, motivated as it was by a young researcher's dependence on government grants (Schmitt, 2013, p. 173). This followed, of course, the rapid alignment of German universities and research funding agencies with the ideology of the Third Reich come 1933.[*] The myth of Hennig having been a Nazi, even a member of the NSDAP, was initially spread by the panbiogeographer Leon Croizat (1894–1982) (Platnick and Nelson, 1988, p. 415; see also Morrone, 2000, p. 46; Heads, 2005, p. 106; Williams and Ebach, 2008, p. 95). Croizat was notorious for his ranting about phylogenetic systematics, German Idealism, and Italian Fascism. His allegations, which he never put in print nor ever retracted (Platnick and Nelson, 1988, p. 415), were thoroughly refuted by the research of the Hennig biographer Michael Schmitt (Schmitt, 2001, 2013; see also Schlee, 1978; Wiley, 2008). The silliness of Croizat's accusations is best brought out in his own words. When confronted by the wife of one of the heroes of the cladistic revolution in the late 1970s, Croizat surmised "Hennig was in the German army—he survived the war—he must have been a Nazi."[†] Croizat's claim is as invalid as is his misconstrued syllogism. But again, as a latecomer to the debate between idealistic morphologists and phylogenetic systematists, some of the sources Hennig used in developing his ideas were quite heavily contaminated with Nazi

---

[*] Fact is, however, that the research funding was not significantly tied to party membership during the Third Reich: Deichmann (1996).
[†] E-mail from Gareth Nelson to OR, dated August 8, 2011.

bio-politics—connotations Hennig carefully glossed over in his 1950 *Grundzüge*. But science, at least the kind that seeks to decipher evolutionary relationships, does not happen in the germ-free environment of a clean-lab, pursued by a cold, detached and unbiased intellect. Science is both an eminently collaborative as much as competitive process, involving protagonists who are embedded in their respective social and political environment. Nowhere is this better brought out than in a science that is as easily ideologically instrumentalized as is evolutionary biology.

It is true, though, that Hennig was drafted into the *Wehrmacht*. After he was wounded at the Eastern front in 1942, he served in a pest (insect) control unit first in Germany, later in Greece and, toward the end of the war in Italy, where he was taken prisoner of war by the allied forces (Dupuis, 1990, p. 407). It was during this time in northern Italy that Hennig laid down the first draft of his *Grundzüge einer Theorie der Phylogenetischen Systematik*. The book-length manuscript filled a voluminous Italian notebook, bound in thick cardboard, and "gives the impression of having been written in a continual steady flow" (Schlee, 1978, p. 381). Because of paper shortage in postwar Germany, the publication of the book was delayed until 1950. This motivated Hennig to publish the core essence of his ideas in the form of two essays, which appeared in 1947 and 1949, respectively, in the unlikely periodical *Forschungen und Fortschritte—Nachrichten der deutschen Wissenschaft und Technik* (Research and Progress—Tidings from German Science and Technology). Hennig may well have chosen this outlet to stage systematics as a mainstream science in its own right, not just as an auxiliary science serving other branches of biology. Every science, he claimed, starts with the ordering and classification of the phenomena of interest (Hennig, 1947, p. 276), and in that sense, systematics is an essential part of all sciences, Hennig maintained·(Hennig, 1950, p. 4). For biosystematics, this means more specifically that its aspirations are not exhausted with the mere discovery of order in nature. More important is the rationalization of the world of phenomena, which means to address the manifold that is apparent in nature with the tools of rational thought, that is, the tools of logic. The limited space available forced Hennig to employ in those early essays a focused and concise prose, something that cannot be said of his 1950 book. What becomes apparent in these early two essays, though, is Hennig's penchant for neologisms, some of which would eventually become commonly used household terms of phylogenetic systematics. This urge to introduce new technical terms was rooted in Hennig's quest for a precision language (*Präzisionssprache*) (Hennig, 1949, p. 138) in systematics, where crucial terms were to be fully disambiguated. Such a disambiguation of the language of science was one of the central concerns of logical positivism (logical empiricism) still very much *en vogue* at the time, with formal logic as a special branch of mathematics providing the paradigmatic example. It was then only consequential of Hennig to aspire to a mathematization of systematics "in the sense of a *mathesis universalis*," which he again claimed must be the goal for all true sciences (Hennig, 1949, p. 138). The *mathesis universalis*, the description of the world in purely mathematical (logical) terms is an old Leibnizian ideal that was most famously pursued by Rudolf Carnap among the logical positivists that emerged from the Vienna Circle (Carus, 2007, p. 129)—the latter an illustrious interdisciplinary evening discussion group of the 1920s composed of faculty members from the University of Vienna (Stadler, 1997).

Although Hennig would cite Carnap's doctoral thesis in his 1950 book, he backed up the mandate of a *mathesis universalis* for systematics with a reference to the nature philosopher Bernhard Bavink (Henschel, 1993), an early Nazi sympathizer (later turned critic) and champion of euthanasia (Klee, 2003, p. 33). Bavink hailed "the German nationalist movement, which I support with abandon (and not just since January 30, 1933, but since the ill-fated November 1918)" (Bavink, 1933, p. ix). His support for eugenics Bavink anchored in an organicist understanding of the German National Community as a hierarchically structured complex whole, where the parts willfully submit themselves to the greater good of the whole. This, in turn, set him in opposition to logical positivists such as Carnap and other members of the Vienna Circle, who combatted such—in their view misapplied—organicism as "broadly sympathetic to the authoritarian fascism that was gaining ground throughout Europe" (Carus, 2007, p. 208). Small wonder, then, that the Vienna Circle dispersed with the ascent of the Third Reich, most of its members emigrating overseas.

In his *Results and Problems of Natural Sciences—An Introduction into Modern Nature Philosophy*—a book that went through several editions and that was read and cited by Hennig. Bavink sought to transcend the then-popular materialism—vitalism debate—and to integrate the two mainstream movements pervading German biological sciences at the time, logical positivism, on the one hand, and organicism on the other. Hennig, in his *Grundzüge*, as also in the two antecedent essays, aspired toward the very same goal. Bavink characterized natural science as "the logical connection of facts in thought" (Bavink, 1933, p. 382)—a perspective that not only captured the positivists' philosophical analysis of science but also Hennig's apprehension of phylogeny reconstruction: "the concern for logic was a constant in Hennig's thought" (Dupuis, 1984. p. 19). But when it came to biology, Bavink adopted the holism so characteristic of organicism, and introduced the concept of *Gestalt* to characterize the hierarchically structured complex wholes that are organisms (Bavink, 1933, p. 385). But *Gestalt* is a concept that utterly defies logical analysis, and was most frequently linked to intuitive apprehension instead. Hennig likewise endorsed the concept of enkapsis, describing a nested hierarchy of complex wholes that lay at the core of German organicism, a concept he applied not only to organisms but also to the entire phylogenetic system and its parts. He referred to the latter as phyletic *Gestalten*, which due to their multidimensional multiplicity defy logical or precise mathematical analysis (Hennig, 1950, p. 279). Hennig again cited Bavink in support of his claim that a new calculus, a *mathematics of form*, needed to be developed in order to capture organic or phyletic *Gestalten* (Hennig, 1949, p. 136; 1950, p. 152; see also Bavink, 1933, p. 388; 1941, p. 462), a calculus that was not available to him at the time he wrote his *Grundzüge*.

The cladistic revolution in biological systematics triggered by Hennig is often characterized as a methodological one. But in his *Grundzüge*, Hennig reached far beyond the strictures of a new methodology, seeking to synthesize German comparative biology by building a bridge between the holistic–organicist and logical positivist traditions. In so doing, Hennig dealt with a rich history reaching back to Ernst Haeckel (1834–1919) and beyond. With Ernst Haeckel, we have arrived at a biologist who can rightly be called the first true phylogenetic systematist, one, however, who was notorious for his blending of strict empiricism with an organicist motivated nature mysticism. Through his nature mysticism, many have argued, Haeckel became a trailblazer for the later rise of Nazi ideology, in particular Nazi bio-politics that embraced a crude version of Social Darwinism. While this assessment remains highly controversial, it is nevertheless true that Nazi ideologues reconstructed Haeckel in their own service. It is with Haeckel, then, and his take on idealistic morphology, that the history of phylogenetic systematics must begin, a history that from its beginning evinced a tension between science and metaphysics, between *Logos* and *Gestalt*, with all its social and political implications.

In a popular account of the history of biology, published in 1907, Haeckel was recognized as the first author to have stipulated that animal classification has to be strictly genealogically structured (Krumbach, 1919, p. 970). Yet, the philosopher Theodor Ziehen (1862–1950)—one who massively influenced Hennig—noted that in his teaching, Haeckel would frequently transcend the "naïve materialistic realism" that otherwise characterized his science (Ziehen, 1919, p. 959). Indeed, Haeckel was a fervent devotee of the German culture icon Johann Wolfgang von Goethe (1749–1832), whom he considered to be a predecessor of Darwin (Walther, 1919, p. 948). When Haeckel writes "Whereas substance and force are eternal and infinite, their form is subject to eternal and infinite change and motion" (Haeckel, 1866, vol. 2, p. 442), he positions himself squarely in the tradition of German Idealism (Beiser, 2002). Haeckel has thus been characterized as someone who dressed up the old German *Naturphilosophie* in new empirical clothes (Krumbach, 1919, p. 970; see also Ziehen, 1919, p. 960). The two styles of scientific reasoning that would come to characterize German biology are thus at work in Haeckel's science already. Haeckel ridiculed systematists who were content with the mere description of new species. He wanted to ask big questions, which he sought to answer through a consistent causal-materialistic approach to nature. The *Kausalgesetz* (law of causation)—in the exclusive sense of a *causa efficiens*—he took to be universal, admitting of no exceptions, and hence establishing the unity of God in Nature. Matter and soul for Haeckel were coextensive, nature explicable, and mysterious at the same time. Haeckel's *magnum opus*—his *Generelle Morphologie* of 1866—is consistently monistic-reductionist, materialistic-mechanistic in its approach to nature, yet it is peppered with references to Goethe, which add an idealistic, even mystic undertone.

The tension between reductionism versus holism, between externalism and internalism, was starkly brought out in early studies of *Entwicklungsgeschichte*, of embryonic development. Working with frog eggs subject to mosaic development, the founder of *Entwicklungsmechanik*, Wilhelm Roux (1850–1924) reduced the organism to an aggregate of parts that engage in a Darwinian competition with each other. He saw the developing organism as governed by the laws of physics; blood vessels would branch under the rule of the laws of hydrodynamics, for example. His opponent Hans Driesch (1867–1941), a biologist later turned philosopher, studied the development of sea urchins. Killing off one blastomere at the two-cell stage will result in the formation of a complete organism at half the normal size from the single remaining cell. This to Driesch signaled the presence of a regulative force inherent in the fertilized egg that guides the formation of the complex whole that is the organism, an ultimately (neo-)vitalistic principle he called *entelechy*. Driesch and his friend and collaborator, Jakob von Uexküll (1864–1944)—the latter an animal psychologist—were the main driving forces in getting the organicist–holistic trend in German biology off the ground, a movement which duly attracted the criticism of the (neo-)positivist philosophers of the Vienna circle. The opposition of the positivists was politically motivated, as they sensed that the reawakening of vitalism in biology opened the door to irrational nature mysticism that could be appropriated by the raising *völkisch*-fascist movements of the time. Such indeed happened during the years of the Weimar Republic, a period that most contemporaries experienced as a time of profound crisis—socially, politically, but also scientifically (Forman, 1971).

Inspired as he was by Hans Driesch, it is in the organicist–holistic tradition of German biology that Martin Heidenhain (1864–1949) of distant Jewish ancestry developed his concept of the enkaptic hierarchy. The enkaptic hierarchy is a nested hierarchy of individuals, or complex wholes, subject to the part-whole relation, and built up from divisible components. Cells would divide to form tissues, organs, organ complexes, and the whole organism. Each level of complexity in this enkaptic hierarchy that is the organism is characterized by emergent properties, such that the whole at any level of complexity is never merely the sum of its parts. The enkaptic hierarchy soon became the central metaphor of the organicist–holistic movement in German biology. The concept was seized upon by comparative (idealistic) morphologists, developmental biologists, ecologists, paleobiologists, and phylogeneticists. The developing organism, the biosphere, the Tree of Life, they were all quickly re-conceptualized in terms of an enkaptic hierarchy. As German biologists came under increased ideological pressure, authors in each of these areas of research drew increasingly compromising and utterly misguided causal correspondences between their science and yet another enkaptic hierarchy—that of the German National Community (*Volksgemeinschaft*). Although willfully aligning their rhetoric with *völkisch* Nazi ideology, the organicist–holistic camp of German biologist came under attack from the social-Darwinist camp of phylogenetic systematists, who called for the priority of their science on the basis of its foundational ties to the *Lebensgesetze*, the laws of inheritance and the relevance derived from those for policies in eugenics and racial hygiene. Whereas the organicist–holistic faction declared Goethe as their patron saint, the phylogenetic systematists reconstructed Haeckel in the light of Nazi bio-politics. The conflict most virulently escalated at the University of Tübingen, where the botanist Ernst Lehmann (1880–1957) tried to establish an Aryan Biology, akin to the anti-Semitic (i.e., anti-Einsteinian) Arian Physics inaugurated by Nobel laureates Philipp Lenard (1862–1947) and Johannes Stark (1874–1957). Lehmann eventually succumbed to an intrigue instigated against him by students and colleagues under the influence of Karl Astel (1898–1945), a friend of Himmler and the driving force in establishing Jena University as the prime example of a SS university. Astel was also the leading force in the establishment and enforcement of eugenic policies in Thuringia.

Thanks to the intrigue incited against him, one that even triggered the involvement of the Gestapo, Ernst Lehmann could claim (falsely) that he had been chased all the way to the fence of the concentration camp after the collapse of the Third Reich.* Of his two main adversaries, one went missing in the battle of Stalingrad, and the

---

* *Universitätsarchiv Tübingen*, UTA 126/373, 1. See also Heiber (1991, p. 440) and Bäumer (1990a,b).

other fled to South America after the war had come to an end. Karl Astel committed suicide on April 4, 1945. The leading idealistic biologist in Germany, the botanist Wilhelm Troll (1897–1978), was deported, along with other intellectuals, artists, and business leaders from eastern Germany, to the Western (American) Sector of occupied Germany in June 1945, where he eventually established himself as a professor at the University of Mainz. The collapse of the Third Reich had cleared the ground for a new start. In phylogenetic systematics, the new beginning was ushered in by Willy Hennig, who sought synthesis rather than conflict, but conflict he would provoke.

Author

Olivier Rieppel is Rowe Family Curator of Evolutionary Biology at The Field Museum in Chicago. His main current research interests focus on Triassic marine reptiles from southern China. He also contributed extensively to the comparative anatomy and evolution of modern reptiles, most notably the evolutionary origin of turtles, and snakes. He published widely in the history and philosophy of comparative biology on topics as diverse as species concepts, mid-eighteenth-century French biology, and the history of phylogenetic systematics. Rieppel is on the editorial board of several peer-reviewed scientific journals, and himself published more than 350 scientific papers and 7 books.

1 The Evolutionary Turn in Comparative Anatomy

CARL GEGENBAUR'S IDEALISTIC MORPHOLOGY

After returning from an extended journey to southern Italy and Sicily, where he met up with his teacher and professor Albert von Kölliker (1817–1905), who had in fact suggested this remote *rendezvous* in the interest of collaborative research in marine biology, the young Carl Gegenbaur (1826–1903) obtained the *venia legendi* by means of his *Habilitation* at the Julius Maximilians University of Würzburg in the winter semester of 1853/54. Somewhat ahead of him on his career path was Franz Leydig (1821–1908), already a *Privatdozent* during Gegenbaur's student years, who held the position of *Prosektor* (first assistant) at the Anatomical Institute of Würzburg University—a respected personality and gifted scientist with whom Gegenbaur soon developed a personal friendship. The *venia legendi*, the right—and duty—to teach at the university level and hence to qualify as *Privatdozent* required the *Habilitation*, that is, research culminating in a second thesis following the doctoral degree. To become a *Privatdozent* at a German university of the time did not imply a salaried position, but rather the opportunity to teach courses for which tuition fees were to be paid by the students to the instructor. A *Privatdozent*'s income thus crucially depended on the number of students attending his classes. Since Kölliker, as well as his friend Leydig, along with another young colleague and companion on the Italian voyage, Heinrich Müller (1820–1864), were already covering anatomy and physiology in the courses they taught, Gegenbaur offered a course in zoology in the summer semester 1854, so as to avoid any potential financial conflict and competition for students. Come summer semester 1855, Leydig was promoted to associate professor, thus vacating the position of *Prosektor* for which Gegenbaur eagerly applied. While going through all the qualifying examinations that were required to be hired into this position, Gegenbaur received an offer for employment at the rank of associate professor at the Friedrich Schiller University in Jena. Filled with joy and great expectations, Gegenbaur savored the moment by getting off the train in Apolda in central Thuringia, from where he set out for a three-hour walk down into the Saale valley toward Jena on a beautiful late summer day in 1856. Especially in comparison to Würzburg, located in catholic Bavaria where conservative politics were preached from the pulpit on Sunday mornings, Jena was a liberal university town at the time, with such luminaries as Martin Luther, Johann Wolfgang von Goethe, and Friedrich Schiller among its past residents. Ten years later, when looking back on the time they had by then spent together in Jena, Gegenbaur's friend and colleague Ernst Haeckel characterized the town as the "pulsating heart of the German spirit of freedom," the "nursery of German philosophy and science in a liberal polity," where the two of them had had the privilege to work in rooms that had once been occupied by Goethe.* Gegenbaur was greeted by a particularly welcoming university administration, which after initial resistance granted him his wish to have the teaching duties for physiology and anatomy split and assigned to two separate chairs, so that he himself could concentrate on what he knew he was best at, comparative anatomy.† This administrative move, the first such move to have happened at a German university, cemented an institutional separation of physiology and comparative anatomy, recognizing each as separate and distinct sciences. It was that partitioning of those scientific disciplines that allowed the rise of idealistic morphology to new respectability. Physiology was recognized as an experimental, etiological science; comparative (idealistic) morphology as a primarily descriptive science, with its own, distinct principles and laws of form, rather than of function (Figure 1.1).

FIGURE 1.1   Carl Gegenbaur. (Smithsonian Institution Archives. Image 85-4507.)

---

* Haeckel, 1866, vol. 1, pp. viii–viv.
† Fürbringer, 1903a, pp. 397–404.

In 1859, the same year that Charles Darwin (1809–1882) published *On the Origin of Species*, Gegenbaur published his first textbook,* *Grundriss der vergleichenden Anatomie* (*Fundamentals of Comparative Anatomy*), one that was destined to become a classic. The spirit pervading Gegenbaur's textbook could not have stood in greater contrast to Darwin's *Origin* in terms of its underlying philosophy: German Idealism on the one side, English materialism on the other. Himself a prominent exponent of systematic morphology, Adolf Naef would later characterize the first edition of Gegenbaur's textbook as the culmination and great legacy of classic idealistic morphology,† when in fact it aimed to unclutter classic idealistic morphology from all too fanciful flights of imagination so characteristic of the past. Its opening sentences read: "The science that is concerned with life as it manifests itself in organic nature has for a long time already been split into two branches. One of those investigates the processes that become apparent in life and are brought out through life, while the other is concerned with the material form in which these processes of life become manifest. This is the way in which physiology and morphology, at least as far as the animal kingdom is concerned, have separated."‡ Separated they indeed have—as a consequence of Gegenbaur's negotiations with the administration of Jena University.

Carl Gegenbaur was born on August 21, 1826, in Würzburg. Written years after his conversion to Darwinism, he opened his autobiography§ with a retrospective on his own familial ancestry: "What we are, we have become through inheritance and adaptation, as is also true of everything else that exists organically." His patrilineal bloodline reached back to Fulda through a succession of well-to-do civil servants with higher education. His mother's family had its roots near Aschaffenburg on the Main River, again a well-situated family, his grandmother of noble origins, providing their children with an opportunity to earn a university degree. Gegenbaur thus grew up embedded in the classic *Bildungsbürgertum*—a tradition he himself strove to uphold.¶ Looking back with mixed feelings on his strictly regulated school years at the *Gymnasium* (grammar school), the elder autobiographer sharply criticized modern tendencies to reduce or even abandon the study of classic languages (Ancient Greek and Latin), now pejoratively labeled "dead languages" by some, and emphasized the importance of an intimate knowledge of antiquity as the foundation of all culture. Upon entering the University of Würzburg at the age of 19, Gegenbaur first had to complete a *biennium philosophicum*, then mandatory at Bavarian universities, with classes in philosophy and history, before he could move on to study medicine. The medical faculty there included top-class professors, such as Kölliker and Leydig. Kölliker had come to Würzburg from Zürich in 1847. He had been hired to represent physiology, but he also taught courses in comparative anatomy, histology, and embryology (*Entwicklungsgeschichte*): "I was a most eager

---

* Gegenbaur, 1859.
† Naef, 1926, p. 415.
‡ Gegenbaur, 1859, p. 1.
§ Gegenbaur, 1901. Given his eminent status in German biology, there exist several accounts of Gegenbaur's life, work, and influence, among which a biography by his student Fürbringer (1903a), as also Coleman (1976), Nyhart (1987, 1995, 2003), Hoβfeld and Olsson (2003), Hoβfeld et al. (2003), Laubichler (2003), and Richards (2008).
¶ Nyhart, 1995, p. 147. On the tradition of the *Bildungsbürgertum*, see also Ringer, 1969 (1990).

student,"* Gegenbaur recalled. Unlikely as it was at the time that a "Berlin democrat" should be appointed at a Bavarian University, the leftist activist and pre-eminent pioneer of cancer research Rudolf Virchow (1821–1902) arrived in Würzburg in 1849 to represent pathological anatomy. In his teaching he covered all aspects of anatomy, a subject area "that came to be dominated by the concept of development (*'der Gedanke der Entwicklung'*). This is *Virchow's* great merit."† Gegenbaur repeatedly took classes offered by Virchow, thus witnessing the evolution of Virchow's own thoughts, yet if these courses did indeed sketch morphological transformations or derivations in vague evolutionary terms, such terms did not find their way into Gegenbaur's textbook of 1859.

Gegenbaur's public examination for his doctorate was scheduled for April 15, 1851. He chose to defend a thesis about change in the world of plants, where species often could not unequivocally be delimited from one another due to extensive variation, as for example in the genus *Hieracium* (hawkweed) that was a much discussed example at the time. Such extensive variation, blurring species boundaries, he found to be suggestive of species transformation. Or so Gegenbaur later claimed in his autobiography:

> The idea of development (*Entwicklung*) shows us a way in which not just plants, but also animals could originate, as all that lives has its origins, and the history of development allows us to connect with those beginnings and thus to recognize and comprehend the whole …. These were, more or less, my words, which a few years later, as I remembered them, rendered Darwin's theory quite familiar to me.‡

At the time when he delivered his examination speech, however, it was Schelling's rather than Darwin's voice that resonated in Gegenbaur's words. A luminary of German Idealism, the philosopher Friedrich Wilhelm Schelling (1775–1854) sketched a picture of nature as ever becoming, ever changing, never standing still. Nature, according to Schelling, "is nothing less than living activity, or productivity itself"§: *natura naturans* rather than *natura naturata*. Kölliker, of foreign, that is, Swiss extraction and hence less influenced by German philosophical tradition, was acting as the official commentator of Gegenbaur's presentation. In this capacity he reminded his erstwhile student that he should better stick to empirically well-founded facts, for all that he had proffered about the origins and relationships of various forms of life was highly speculative.

Following his graduation from Würzburg University, Gegenbaur travelled to Berlin to hear the famous physiologist Johannes Müller (1801–1858), whose most brilliant student—in Gegenbaur's assessment—would turn out to be Ernst Haeckel. Gegenbaur found first signs of much of what would later blossom in the work of Haeckel and others—including his own—to have been preconceived in Müller's

---

* Gegenbaur, 1901, p. 46.
† Gegenbaur, 1901, p. 47. Although he later turned against Darwinism for social, political, and scientific reason, "Virchow, as he later liked to observe, had maintained, prior to Darwin, that evolution was a hypothesis suggested by many considerations of modern science" (Richards, 2008, p. 104, n. 76).
‡ Gegenbaur, 1901, p. 53.
§ Beiser, 2002, p. 530.

thoughts. Gegenbaur reminisced: "The development of *Synapta* which he studied in *Trieste* had led this deep thinker to recognize problems that downright tortured him. They had the same source as later the origin of species through the struggle for existence, things that at the time were not yet talked about."[*] Not recognizing the parasitic nature of the minute snails he observed to emerge from sea cucumbers, Müller thought he had witnessed a holothurian species, that is, an echinoderm, give birth to mollusks. Distrusting his senses, he had shown the phenomenon to visiting scientists passing through the Trieste marine biological station, and also to his own son who accompanied him on this trip, but they all corroborated his observations. This was indeed quite incredulous, as mollusks and echinoderms are separated by a vast gap in the natural system, a gap that the great comparative anatomist from the Paris Natural History Museum, Georges Cuvier (1769–1832) had declared unbridgeable. Cuvier had classified echinoderms and mollusks in different branches, or *embranchements*, of the animal kingdom, of which he recognized a total of four. These four *embranchements* Cuvier found to be deeply separated from one another, so much so that no meaningful anatomical comparisons between them were deemed possible. The same conclusion had been reinforced by the eminent embryologist Karl Ernst von Baer (1792–1876), who considered mollusks and echinoderms to represent two profoundly different types of organization that could not be bridged by intermediate conditions of form. Indeed, the embryonic development of mollusks and echinoderms proceeds along radically different trajectories, starting out with fundamentally different cell cleavage patterns. Nonetheless, Müller eventually concluded to the presence of a "molluskigerous organ" in the holothurian *Synapta* that would bring forth snails through a process of heterogony in a manner analogous to the alternation of generations so well known from many marine organisms. What this claim ultimately implied was the saltational origin of a radically different kind of species from another ancestral one. In addition to a monograph presenting these results published in 1852 in Berlin, Müller published an English abstract in the *Annals and Magazine of Natural History* that appeared in London in the same year. His account was followed by an editorial commentary that did not question the eminent physiologist's observations, but urged caution with respect to Müller's conclusions (Figure 1.2).[†]

From Berlin, Gegenbaur continued on via Hamburg to Helgoland, using the opportunity to study marine invertebrates, then back to Würzburg via Düsseldorf and Koblenz. In 1852, he embarked on the trip through Italy down to Messina in Sicily already mentioned, where he met up with Kölliker, and Heinrich Müller. His travelogue is saturated with expressions of enthusiasm and admiration as he visited the famous relics from antiquity and other, more modern monuments of Western European history and culture. Returning from Italy he first taught as *Privatdozent* in Würzburg, then took up his position in Jena in 1856. Gegenbaur sketched in his autobiography how his appointment in Jena almost skidded into jeopardy due to his request to be dispensed from teaching physiology, which he felt unqualified for due to his lack of knowledge of physics and chemistry. Gegenbaur eventually prevailed,

---

[*] Gegenbaur, 1901, p. 57.
[†] Müller, 1852a,b.

FIGURE 1.2   Johannes Müller. (Universitätsbibliothek Tübingen, Bilddatenbank.)

however, with his assessment that anatomy and physiology were profoundly different sciences both in methods and in philosophy, such that Jena University became the first German university to officially recognize the separation of these subject areas through the creation of two separate chairs. Berlin was to follow suite upon the retirement of Johannes Müller: "I leave it to others to defend the importance of physiology, it has been done abundantly ...," Gegenbaur concluded when turning his focus on comparative anatomy.[*]

Gegenbaur's Jena years were marked by his deep friendship and collaboration with Ernst Haeckel:

> ... Haeckel became a close friend. Together we have wandered through the region, many charming locations did I discover with him. As long as the seasons would permit, we would be seen together on a walking tour on Saturdays. It goes without saying that science would not be neglected during the time we thus spent together.[†]

Haeckel in turn acknowledged the fraternal friendship he cherished with Gegenbaur from the day when they first met in the Gutenberg forest near Würzburg in 1853, after Gegenbaur's return from Messina.[‡] Both, in the laboratory as much as on their walks through the gorges, woods, and hills of the Triassic *Muschelkalk* formation

---

[*] Gegenbaur, 1901, p. 95.
[†] Gegenbaur, 1901, p. 96.
[‡] Haeckel, 1866, vol. 1, p. viii.

through which the Saale had cut her way, Haeckel absorbed the knowledge, the wisdom and the enthusiasm for the science of comparative anatomy from his mentor eight years older, while Gegenbaur admired the youthful energy, the boldness, and the fighting spirit that fired up his companion. The influence that Haeckel—often called the German Darwin—exerted on Gegenbaur during the years they spent together in Jena is well documented in subsequent editions of Gegenbaur's textbooks on comparative anatomy, as will be discussed in more detail later. In his autobiography, Gegenbaur acknowledged the publication of Darwin's *Origin* in 1859 to have triggered a monumental change in biology, but reiterated that since he had already toyed with similar ideas years ago, he "found himself unable to greet the arrival of Darwin's doctrine with great enthusiasm and applause."[*] Happy to return to southern Germany, as it was closer to home, Gegenbaur left Jena in 1873 to accept the Chair of Professor of Anatomy at the University of Heidelberg, where he also acted as Director of the Anatomical Institute. Never returning to Jena again, he died in Heidelberg on June 14, 1903.

It is unquestionable that the time he spent in Jena were formative years for Gegenbaur: "everything that I have later achieved has its source [in Jena], which obliges me to lasting thanks."[†] As has been remarked by the historian of biology Lynn Nyhart, Gegenbaur had initiated in Jena during the 1860s, under the influence and with the collaboration of Haeckel, a new research program, "evolutionary morphology that would make comparative anatomy 'scientific' through its infusion with evolutionary meaning,"[‡] but such development and reorientation of Gegenbaur's thoughts demonstrably took place after he published the first edition of his *Grundzüge der vergleichenden Anatomie* in 1859. In his autobiographical recollections, however, penned two years before his death, he portrayed himself as first endorsing evolutionary ideas in his public examination lecture that marked his doctorate in 1851, then as greeting the arrival of Darwin's doctrine in Heinrich Georg Bronn's (1800–1862) German translation of 1860[§] with the coolness of a seasoned *habitué* who could hardly be moved by its purported originality. This apparent paradox is resolved with the understanding that species "origination" and "transformation" meant different things to him in 1851 and 1859 as opposed to 1870, the year when his textbook came out in a second edition—a difference Gegenbaur conveniently glossed over in his autobiography.

Gegenbaur's lifelong interest and expertise concerned the comparative anatomy of animals, both vertebrate and invertebrate. Why would he then put a plant center stage in his doctorate examination presentation? Gegenbaur's interest in botany was awakened at a young age under the gentle tutelage of his mother, who took him out to the fields and forests, taught him the German vernacular names of the native plants and instructed him on how to collect plants and mount them on paper.[¶] In his graduation speech of 1851, Gegenbaur introduced examples from plants rather

---

[*] Gegenbaur, 1901, p. 98.
[†] Gegenbaur, 1901, p. 103.
[‡] Nyhart, 2003, p. 163.
[§] For the history of Bronn's translation of Darwin's *Origin*, see Gliboff, 2008.
[¶] Gegenbaur, 1901, p. 16.

than animals,* capitalizing in particular on the rampant variation manifest within the genus *Hieracium*, which rendered species identification particularly difficult.† *Hieracium* was a notorious case, one on which such luminaries as the botanist Carl Wilhelm Nägeli‡ (1817–1891), and the zoologist Ernst Haeckel§ would later refer to in defense of their contrasting views on species. In 1834, the botanist Anton Friedrich Spring (1814–1872) had won first place in a competition initiated by the philosophical faculty of the Ludwig Maximilians University of Munich for the best prize essay (published in 1838) on the nature of the genus, species, and variety (*Abart*) in biological systematics. In his monograph, steeped as it was in the tradition of German Idealism, Spring noted of *Hieracium murorum* that its outward appearance differed during the spring and autumn bloom respectively, as is the case in some other plant species as well.¶ But the line Spring is most famously remembered for reads: "species do not persist in a state of 'being', but instead are continuously 'becoming'" (*Die Arten sind nicht, sondern sie werden*).** While this was a dynamic conception of species, it was not a Darwinian one,†† but one rooted in German Idealism instead, as articulated in the philosophy of Schelling, for example. And yet, it is a dynamic conception of species that in retrospect could easily be dressed up in Darwinian clothes,‡‡ which seems to be the way the elder Gegenbaur himself wanted to present his own earlier views. The historian of biology Robert J. Richards§§ characterized Schelling as "indeed proposing a real evolution occurring in nature, and [he] seems to have been the first thinker to apply the term [evolution] to species alteration." If Schelling's dynamic conception of nature was an evolutionary one, it was nonetheless not a Darwinian one. Schelling's continuously active and productive *natura naturans* was driven by a universal dynamism inherent in living matter, an internalist perspective, whereas the Darwinian principle of natural selection implies an externalist perspective, as it invokes an evolutionary force that is external to organisms. Similarly, if the 1870 edition of Gegenbaur's textbook was infused with a Darwinian meaning, its 1859 first edition was firmly embedded in the idealistic tradition.

At the heart of Gegenbaur's comparative, or idealistic morphology, lies the desire to search for order in nature, to classify organisms in a system of nature that would reveal the fundamental body plans that characterize more inclusive groups of organisms, and their multivarious manifestations in the species that constitute these more inclusive groups of organisms. It was, thus, a primarily classificatory endeavor, coupled nevertheless with the hope that the resulting classification, and the regularities of structural organization on which it is based, would reveal Laws of Form. Such Laws of Form that would determine anatomically defined body plans, would

---

* "I did not dare to generalize [my views on species origination and transformation] to animals, since I lacked sufficient relevant knowledge" (Gegenbaur, 1901, p. 53).
† "Systematic [species] identification, first exercised on plants, was a perfect means to sharpen not only the visual sense, but also the [taxonomic] judgment" (Gegenbaur, 1901, p. 25).
‡ Nägeli, 1865, p. 32.
§ Haeckel, 1868, p. 223.
¶ Spring, 1838, p. 162.
** Spring, 1838, p. 49.
†† For analysis, see Rieppel, 2011a.
‡‡ For a misguided neo-Darwinian evolutionary interpretation of Spring's writings, see Geus, 2010.
§§ Richards, 2002, p. 145.

be Laws of Coexistence, that is, laws of structural composition, rather than Laws of Succession, where cause must precede effect—as in physiological investigations.[*] In the preface to his *Grundzüge der vergleichenden Anatomie* of 1859, Gegenbaur sketched his understanding of comparative anatomy in some detail: the goal is to reveal the fundamental body plans that underlie the multiplicity of appearances of organisms, to bring to the fore the unity of type and through it the *basic idea* or *Grundidee*[†] that dominates the composition of organisms in their multifaceted apparition. The unity of type is the essential, that is, the universal principle manifest in the varied individual appearances of organisms that nevertheless share the same fundamental body plan. The student is best introduced to such underlying lawfulness of animal organization through dissection, followed by pictorial representation of his observations. His textbook, Gegenbaur explained, would not be structured systematically, as along the lines of Cuvier's classification for example, and unlike Henri Milne-Edwards[‡] (1800–1885), he would emphasize form, not function in his treatise, since the same function can be performed by organs that do not structurally correspond to each other as determined by the unity of type (e.g., animals with a backbone represent a distinct type, the vertebrates, yet the function of breathing in vertebrates can be executed by gills or lungs). The ultimate goal is to illuminate and illustrate the "transformations of organs and organ systems within an animal series in the way these are dominated and constrained by the respective type."[§] Or, in other words, organisms of the same type lend themselves to serial arrangement; tracing the corresponding organs and organ systems through any such series reveals the various modifications under which the universal principle that is the type is expressed in the particular successive links of the series. But when Gegenbaur, in 1859, speaks of the "transformation," or "modification," of certain organs, or organ systems revealed by a series of organisms representing the same type, he does not imply Darwinian transformation as a result of variation and natural selection. He means an ideal transformation, one that does not belong to the space of time and matter, but instead is grasped in timeless conceptual space. The Laws of Form that are revealed by a "transformation" series of organs of organisms representing the same type are not inherent in those organs and organisms, but are primarily descriptive, hopefully eventually explanatory statements about those organs and organisms.

The organizational type is primitively inherent in an organism as it governs its embryonic development, but at the same time the type reveals itself in the shape and arrangement of the organs of the developing organism.[¶] Within the "science of animal form" Gegenbaur recognized two distinct yet complementary fields of investigation, embryonic development (*Entwickelungsgeschichte*), and anatomy of the mature organism. The latter he further divided into zootomy, which dissects organisms into their parts, and comparative anatomy, which seeks the underlying unity of plan of animal organization. Zootomy he characterized as an analytic, comparative anatomy as a synthetic science: zootomy takes things apart, yielding

---

[*] Mainx, 1971.
[†] Gegenbaur, 1859, p. v.
[‡] Milne-Edwards, 1851; see also Bronn, 1858a, p. v.
[§] Gegenbaur, 1859, p. vi.
[¶] Gegenbaur, 1859, p. 34.

singular, individual facts, whereas comparative anatomy puts things together in the search for a lawful regularity among those facts. Such regularity reveals laws of structure that are expressed in corresponding topology and connectivity of the organ and organ systems of organisms that are of the same type. The unity of type that is instantiated in the organism is conditioned by the types of organs that harmoniously combine to form the complex whole. Gegenbaur hence identified unity of type both in component organs or organ systems, and in the organism as a whole. The type of an organ is determined in large parts by its *Entwickelungsgeschichte*, that is, its embryonic development from the same rudiment (*Anlage*) in different organisms that are of the same type. Consequently, the type instantiated by a particular organism is more easily apprehended the more thoroughly the embryogenesis of the organ systems that in their combination constitute the type has been investigated and described.

The type as a universal principle that dominates and regulates the material manifestation of animal form and organization, and that reveals itself through such manifestation in individual organisms and their multifaceted appearances which, when compared across the entire animal kingdom, "reveal a continuous series of transformations from the most simple to the most complex"*: this is classic idealistic morphology that can be traced back straight to Goethe. The type in that sense does not obtain from mathematical, or logical analysis of nature, but from what Schelling had called "intellectual intuition": "the capacity to see the universal in the particular, the infinite in the finite, and indeed to combine both in a living unity."† The universal and the particular interpenetrate each other, and mutually condition each other.‡ There is no real, historical, causally conditioned transformation of animal species here involved. What is implied, instead, is a universal, that is, immaterial principle that shapes organisms according to plan, laws of structure rather than function that determine a nested hierarchy of types within which the more general condition of form subsumes the more specialized condition of form. Based on the concept of the unity of type, that is, on an underlying universal principle at all levels of inclusiveness, the names of such systematic groupings as make up the hierarchy of types are general names, not proper names; the groupings are conceptual constructs, not individuals (complex wholes).

The key concept underlying the unity of type is that of homology. The terms "homology" and "homologue" were introduced by Richard Owen (1804–1892) at the time when he served as Hunterian professor at the Royal College of Surgeons in London. A controversial figure among his peers, Owen went on to become superintendent of the section for natural history at the British Museum, which during his long career he transformed into the British Museum of Natural History in South Kensington (now called the Natural History Museum). Born in Lancaster in 1804, he died after a long and distinguished career as Sir Richard Owen in London (Richmond) in 1892.§ In the glossary appended to his *Lectures on the Comparative*

---

* Gegenbaur, 1859, p. 3.
† Schelling, cited in Beiser, 2002, p. 580.
‡ In Goethe's science, the universal and the particular are "not only intimately connected but ... they interpenetrate one another" (Cassirer, 1950 [1978], p. 145).
§ On Richard Owen, see Desmond, 1982; Rupke, 1994.

Anatomy and Physiology of the Invertebrate Animals that were published in 1843, Owen defined homologues as: "(Gr[eek]. *homos*; *logos*, speech) The same organ in different animals under every variety of form and function."* The "sameness" in that context results not from structural and/or functional similarity, but from a correspondence of parts of organisms in terms of their topological relations, and connectivity relative to other parts. Homology is thus a correspondence relation† between equivalent parts of two or more organisms under every variation of form and function; the parts entering into that relation are homologues. The term "homologue" Owen derived from the Ancient Greek terms *"homos"* for "like," and *"logos"* for speech, meaning that homologous parts in different organisms are to be referred to by the same term, or, in other words, homologues are namesakes. In Owen's words, "[a] 'homologue' is a part or organ in one organism so answering to that in another as to require the same name."‡ Under Owen's definition, the relation of homology thus turns out to be an identity relation. Homology sorts animals (and plants) into groups of like kind: the vertebral column is homologous throughout vertebrates; the jaws are homologous throughout gnathostomes; the tetrapod limbs are homologous throughout tetrapods; the three sound-transmitting ear ossicles are homologous throughout mammals.

In his discussion of homology, Gegenbaur followed Richard Owen's *Principe de l'ostéologie comparée ou recherches sur l'archétype et les homologies du squelette-vertebré* of 1855, a text that had been translated by "Owen's long-term friend and colleague" Henri Milne-Edwards, and that was published in Paris by J.-B. Baillière on the occasion of Owen's induction in the French Legion of Honor.§ For Gegenbaur, *"Homologies will exist with a certain regularity only for organs in animals that belong to the same type."*¶ This is nothing but a reiteration of Cuvier's earlier claim using different terms, that is, that the relation of homology obtains among organisms that belong to the same type, but does not obtain between organisms that belong to different types. Or, in other words, it is the establishment of homology relations that marks out the fundamental types of animal organization. In the body plans that define the basic divisions of the natural system there prevails, according to Gegenbaur, "a general basic form to which the particular organs and organ systems are accommodated, and from which they develop in various directions without ever being able to free themselves from the reign of the type."** The universal principle is what governs the particular while at the same time revealing itself in that particular. Given its status as a universal immaterial principle, the type cannot stand to the organism in the same relation of cause to effect as billiard balls do in classic physics (Newtonian mechanics). Nor does comparative anatomy deal in the causal-analytic laws that are the subject of physiological investigations. Laws of structure in comparative idealistic morphology are laws of coexistence of homologues, not laws of succession that link ancestral with descendant conditions of form instantiated in

---

* Owen, 1843, p. 379; see also the analysis by D.M. Williams (2001, p. 192).
† Brigandt, 2002.
‡ Owen, 1866, p. xii.
§ Rupke, 1994, p. 164.
¶ Gegenbaur, 1859, p. 35; emphasis in the original.
** Gegenbaur, 1859, p. 35.

species transformation. Nonetheless, it is evident how easily such a natural system, based on the principles of idealistic morphology, can be cast in evolutionary terms. Darwin explained:

> ... all true classification is genealogical; that community of descent is the hidden bond which naturalists have been unconsciously seeking, and not some unknown plan of creation, or the enunciation of general propositions .... I believe that the *arrangement* of the groups within each class, in due subordination and relation to the other groups, must be strictly genealogical in order to be natural.... Descent being on my view the hidden bond of connexion which naturalists have been seeking under the term of the natural system.*

The beauty of that conclusion was, for Darwin, that "systematists will be able to pursue their labors as at present."† The systematists that Darwin referred to in this context were the comparative (idealistic) morphologists researching the natural system.

Gegenbaur considered *Analogy* and *Homology* fundamentally different categories, yet both grounded in comparison.‡ Analogy is a category of physiology: the gills in fishes and the lungs in tetrapods are comparable in terms of their function: both serve respiration. But they are not homologous, jointly they don't mark out a type. Homology, in contrast, is a category of comparative morphology: the forelimb of a tetrapod is homologous to the pectoral fin of a fish. Dissimilar in function (swimming vs. walking), they both share the same topological relations, and connectivity with adjacent body parts. Anterior and posterior paired appendages are a key feature of the type of gnathostomes (vertebrates with jaws). Topology and connectivity are tools in the search for homologies in mature organisms. Another tool in homology assessment recognized by Gegenbaur is embryonic development: homologues develop from corresponding embryonic rudiments (*Anlagen*). The early developmental stages of paired appendages in fish and tetrapods are remarkably similar, both starting out with the sprouting of limb buds from the ventrolateral body ridge. But whereas the comparison of mature organisms reveals only laws of coexistence of homologues, *Entwickelungsgeschichte*, the comparative study of embryonic development does have the potential to reveal regularities of succession in the appearance of organs and organ systems in the developing organism. It was Gegenbaur's friend, Ernst Haeckel, who was to immortalize his name with the evolutionary interpretation of such laws of succession as appear to be manifest in animal development.

HAECKEL'S ASSAULT ON IDEALISTIC MORPHOLOGY

Ernst Haeckel was a strong yet colorful personality, of monumental importance in German biology, and held partly responsible for the rise of Social Darwinism in the *völkisch* tradition and particularly under the National Socialist regime. His influence in the field of biological sciences during his lifetime was enormous; his legacy would

---

* Darwin, 1859, p. 420.
† Darwin, 1859, p. 484.
‡ Gegenbaur, 1859, p. 35.

influence German biology for decades to come. As if that were not enough, Haeckel also forcefully engaged in philosophical, social, religious, and political debates, with disastrous consequences in the eyes of some historians and philosophers. There is hardly any episode in his life or any facet of his science that has not been narrated, commented upon, interpreted, and controversially debated.[*] Ernst Heinrich Philipp August Haeckel was born on February 16, 1834, in Potsdam. His father Carl Gottlob was a high-ranking civil servant (*Regierungsrat*), his mother Charlotte *née* Sethe came from a family of lawyers. In 1835 the Haeckel family moved to Merseburg, where Ernst passed through the *Volksschule*, then the *Gymnasium* (grammar school), to qualify for university education. In his parents' house, Haeckel was exposed to political debate, while being encouraged to pursue his interests in travel, literature, and natural history. In the spring of 1852, Haeckel took up the study of medicine in Berlin, but moved on to Würzburg for the winter semester of 1852/53 to study under Franz Leydig, Albert von Kölliker, and Rudolf Virchow. Particularly, Virchow's emphasis on the cellular level of biological organization (Virchow's cellular pathology, and his vision of an organism forming a cell state) would exert a lasting influence on Haeckel, who for some time served as his scientific assistant. On the occasion of his 60th birthday, February 16, 1894, Haeckel reminisced how much Virchow's penchant for an integration of science with philosophy had impressed him, such as Virchow's reduction of spiritual phenomena to material processes.[†] Probably following Gegenbaur's suggestion,[‡] Haeckel returned to Berlin in the spring of 1854, where he studied under Johannes Müller, who taught anatomy and physiology. Haeckel retained a lifelong admiration for this extraordinary scientist, who took him on a trip to the coastal waters of Helgoland where he introduced his student to the riches of marine biology. Johannes Müller's method of pelagic fishing would serve Haeckel well in the years to come, providing him with the material for his epic studies of marine plankton. Also in Berlin, Haeckel heard the botanist Alexander (Carl Heinrich) Braun (1805–1877), who would influence his views on species and biological individuality. Haeckel passed the examination for his doctorate degree in 1857 and, following a trip to Italy, earned the *venia legendi* through his *Habilitation* in 1861 at the University of Jena—at the suggestion and with the support of Carl Gegenbaur. Haeckel would go on to pursue an extraordinary academic career at the University of Jena, a long tenure that lasted until 1909, enriched by frequent travel throughout Europe and the Mediterranean, and punctuated by a couple far-flung expeditions to tropical shores—undertaken in the spirit of the great Alexander von Humboldt, and artistically sketched "through the blue German fairy-tale eye."[§] In the years 1882/83, Haeckel was involved with the creation of an autonomous Zoological

---

[*] The most recent and most comprehensive biography of Haeckel is Richards, 2008. Haeckel's influence and importance in nineteenth-century German biology, especially morphology, was most comprehensively sketched by Nyhart (1995). A recent, short and useful introduction to Haeckel's life and work in its sociopolitical context was provided by Hertler and Weingarten (2001).

[†] W. May, 1904, p. 195. This is the published version of a talk May had delivered at the *Naturwissenschaftlicher Verein* in Karlsruhe on the occasion of Haeckel's 70th birthday.

[‡] Nyhart, 1995, p. 148.

[§] Wilhelm Bölsche (1861–1939), a poet, publicist, and fellow Darwinist, co-founder of the Monist League and member of the *Gesellschaft für Rassenhygiene*; cited in May, 1904, p. 198.

Institute at Jena University; founded by Haeckel, the Phyletic Museum in Jena was built in the years 1907/08. Haeckel died in his house, the "Villa Medusa" (now the *Ernst-Haeckel-Haus*) in Jena on August 9, 1919 (Figure 1.3).

Haeckel first became acquainted with Darwin's work through the translation of Darwin's *Origin* by the German paleontologist Heinrich G. Bronn, published in 1860.[*] In the same year, Bronn published a critical review of Darwin's *Origin* in the *Neues Jahrbuch für Mineralogie, Geognosie, Geologie und Petrefakten-Kunde*, a monthly periodical widely read among German paleontologists, geologists, and mineralogists.[†] Bronn's review of Darwin came two years after the publication of his monumental investigations into the lawful distribution of fossils in the earth's crust, his *Entwicklungs-Gesetze der organischen Welt*, which was translated as *The Laws of Evolution of the Organic World* for an English abstract.[‡] That treatise contained a

FIGURE 1.3   Ernst Haeckel. (Smithsonian Institution Archives. Image 85-4454.)

---

[*] Gliboff, 2008.
[†] Bronn, 1860.
[‡] Compare Bronn, 1858b, 1859. Bronn in these texts does not argue species transformation, but investigated the lawful distribution of species in the Fossil Record. See also Bowler, 1975, p. 101.

graphic representation of the natural system, and its relation to the temporal dimension as indicated by the Fossil Record in the form of a branching tree-like diagram that looks like a phylogenetic tree, but isn't. Still it has been argued that Bronn, and his tree-like representation of the natural system, conceptually prepared Haeckel's mind to adopt an evolutionary perspective in comparative anatomy.[*] In his review of Darwin's theory of evolution, Bronn noted that proof and disproof will be equally hard to come by.[†] The problem that needs to be solved, he insisted, is to explain how anorganic matter can give rise to organic matter of a cellular structure, and how this cellular organic matter can subsequently be transformed into the germs and eggs of species of low-ranking organisms. Perhaps not impossible, he thought, but highly unlikely to be explicable on Darwin's account: "It remains inconceivable how an organism so sagaciously engineered down to its last fiber, such as a butterfly or a horse, etc. should be the product of a blind natural force."[‡] Therewith, Bronn set the tone for the reception of Darwin's theory in German biology for decades to come. Starting with Haeckel, Darwinism was equated with, or reduced to the theory of natural selection, a thoroughly externalist account of evolution that would encounter much skepticism among German biologists.

Haeckel delivered his first lecture on Darwin's theory on the origin of species at the 38th Assembly of German Natural Scientists and Physicians, held in Stettin, Poland, on September 19, 1863.[§] From there on, Haeckel untiringly propagated evolutionary theory in his science and in public, acting as Darwin's Ambassador throughout Germany, indeed throughout continental Europe, with a fervor that earned him the occasional reminder for moderation from Darwin himself. He drew out controversial sociopolitical consequences from the *Entwicklungsgedanke* that would influence teaching schedules in schools and at higher educational levels. Haeckel was an uncompromising *Freidenker* (Freethinker); he emphatically battled Christian religion, based as it is on the premise of a duality of body and soul, and eventually founded the Monist League (*Monistenbund*) at the Zoological Institute of the University of Jena in 1906. With monism came also the claim for a unity of science, the rejection of a hiatus separating anorganic from organic matter, indeed the propagation of a unity of *Weltanschauung* underpinned by a Social Darwinism that propagated eugenic and euthanasia measures, as well as the death penalty as a "truly beneficent"[¶] means of artificial selection. The escalation of these topoi can be followed through the successive editions of Haeckel's highly popular and widely read *Natürliche Schöpfungsgeschichte*, a book that was first published in 1868. The book went through 9 editions, and was translated into 12 languages. In the first edition, the seventh "lecture" (chapter) on natural and artificial selection remained largely free of sociopolitical connotations. The rampage starts in the second edition of 1870, where the health and vigor of the Spartans of Ancient Greece are explained as a consequence of artificial selection, through the killing of unfit offspring. Similar beneficial practices were also reported to have been practiced by native Indian tribes

---

[*] Richards, 2008, p. 478.
[†] Bronn, 1860, p. 113.
[‡] Bronn, 1860, p. 116.
[§] Haeckel, 1864.
[¶] Haeckel, 1875a, p. 155.

of North America, Haeckel's "primitive redskins."* This stands in stark contrast to the "debilitation of modern civilized nations,"† which results from *"military selection"* where the strong and capable die at the front lines for mostly senseless causes such as family feuds, while the weak and incapable stay behind and reproduce. Similarly detrimental is the *"medical selection"* which enables the propagation of "insidious hereditary evil."‡ The same concerns are articulated in the third edition of 1872, where in addition he scolded "human civilizations" that, while condoning warfare, propose the abolition of the death penalty as a "liberal measure": "Not only is the death penalty for incorrigible criminals rightful, it is also a blessing for the superior part of humankind; the same blessing that the eradication of straggling weeds brings to the careful cultivation of a garden."§ The metaphor of pulling out straggling weeds from a carefully cultivated garden would prove to be a forceful and long-lived one in the *völkisch* tradition that came to define the German National Community. In the sixth edition of 1875, Haeckel introduced yet another vehicle of detrimental artificial selection: "Far more dangerous and devastating than the medical is the *clerical* selection process, implemented everywhere by powerful and uniformly organized hierarchies."¶ The influence exerted by the clergy on education, family life, and community life during millennia has resulted "in the decay of the entire culture and convention." This, he claimed, is particularly evident in the case of the Catholic Church, which in the course of its history has pushed science to an all-time low.

Artificial selection can thus have beneficial as well as detrimental effects on human society, and it is the task of politics informed by science to make the right decisions and implement them. The significance of Haeckel's argumentation lies in the fact that Haeckel coupled with the doctrine of natural selection an unbridled progressionism,** which he assumed would carry over into the *Volkspflege* through a pursuit of carefully designed means of artificial selection. Haeckel was member of the right-wing (*völkisch*) *Alldeutscher Verband* and honorary member of the *Gesellschaft für Rassenhygiene*, but also supported the *Verband für internationale Verständigung* in its pacifistic efforts.†† His unwavering confidence in historical progress led him to believe that the continuing evolution of the human brain would one time render armed conflicts a thing of the past. Given Haeckel's extraordinarily rich legacy of publications in science and popular writing, his public involvement in social debate and politics, the intrepid statements he presented with the authority of a scholar of higher standing whose views are firmly rooted in the enlightenment provided by modern empirical science in general and Darwinian biology in particular, and his

---

* Haeckel, 1870a, p. 153.
† Haeckel, 1870a, p. 154.
‡ Haeckel, 1870a, p. 155.
§ Haeckel, 1872a, p. 155.
¶ Haeckel, 1875a, pp. 154–155.
** "Progress is a necessary effect of natural selection" (Haeckel, 1866, vol. 2, p. 257). Haeckel's *Fortschritts-Gesetz* (Law of Progress) was motivated by Heinrich G. Bronn's (1858a,b) studies on the formation of the earth's crust and the organic bodies living on it.
†† Chickering, 1973. Haeckel's pacifistic ambitions are detailed and interpreted by Weikart (2004, pp. 175–176). Weikart's tenor is generally anti-Darwinian, but his account of Haeckel's ambivalent pacifism is consistent with Haeckel's writings.

shifting political and social allegiances, it is not surprising that historical analysis of his work and impact has been highly controversial.* This applies especially to Haeckel's role as a harbinger for National Socialism in Germany.† When Hans Schemm—a Nazi of the first hour and founder of the notorious National Socialist Teachers' League—characterized National Socialism as politically applied biology, he could look back on Haeckel who himself had declared politics to be applied biology. However, to characterize Haeckel as a mastermind of National Socialism—a trailblazer either himself or through his students, such as the school reformer and later Nazi ideologist Paul Brohmer (1885–1965)—is quite another thing than the use, indeed abuse of Haeckel's writings and lasting influence in support of Nazi biologism by Nazi ideologues. As the historian of biology Mario Di Gregorio put it, there have been "periodic attempts to find a historical figure that was responsible for the Nazi and fascist horrors. The same accusation has been made against Richard Wagner, Nietzsche, Hegel and even Martin Luther."‡ On a more general level, however, Haeckel was certainly steeped in the *völkisch* tradition, which was radicalized by National Socialist ideologues without necessarily focusing on Haeckel himself and his writings.§ In his own time, at any rate, Haeckel had to defend himself not as a protagonist of right-wing politics or fascism, but instead of socialism and academic liberalism.¶

The attack on him was mounted by none other than his former professor at Würzburg, Rudolf Virchow, at the 59th Assembly of German Natural Scientists and Physicians, held in Munich in September 1877.** Haeckel had taken the floor on September 18, delivering a talk that offered a comprehensive *tour de force* through his vast theoretical edifice. He opened with the category mistake already chastised in his *magnum opus*, the *Generelle Morphologie* of 1866, forcefully rejecting teleology while declaring adaptation to be the "necessary effect" of a random concatenation of mechanical causes.†† He once again stressed the central importance of the *Entwicklungsgedanke*, as well as the fact that—replacing the mathematical–physical with a historical–genealogical approach—the theory of descent alone had once and for all provided a complete explanation for the origin of mankind, not only with respect to corporeal characteristics, but also with respect to the spiritual endowment of humans. The cell he characterized with Ernst Wilhelm

---

* For a recent detailed and balanced account, see Reynolds, 2008.
† Compare Richards, 2007, 2008, and http://home.uchicago.edu/~rjr6/articles/Myth.pdf (accessed March 10, 2016), with Gasman, 1998, 2002, and http://www.ferris.edu/ISAR/gasman2.pdf (accessed March 10, 2016). For a brief and balanced assessment of Haeckel's influence on, or abuse by, National Socialism and post-War East German (DDR) politics, see Hoβfeld and Breidbach, 2005.
‡ Di Gregorio, 2005, p. 569.
§ Mosse, 1998.
¶ Haeckel's Monist League was a stronghold for freethinkers who besides eugenics propagated ideas quite anathema to National Socialist ideology, for which reason it was banned by the Nazi regime (e.g., Weikart, 2004, p. 70).
** For a detailed account of the exchange and its consequences, see Richards, 2008, pp. 318ff. A detailed and insightful analysis of the social and political context of the debate in relation to the cell state theories held by Virchow and Haeckel was offered by Reynolds (2008, pp. 129ff). The account given here draws from May, 1904, pp. 239ff, with slightly different emphasis.
†† Haeckel, 1877a, p. 15.

von Brücke (1819–1892)[*] as an elementary organism, one that his celebrated teacher Virchow had made the foundation of theoretical medicine. The soul of the cell he defined on a material basis as the tension forces inherent in the protoplasm. On the basis of his monism Haeckel declared the soul to be coextensive with organic matter at all levels of its hierarchical integration, from "plastidules" and cells to the complex whole that is the organism. The *Entwicklungsgedanke*, he trumpeted, would not only unite humanities with natural sciences, but also offer a secure naturalistic basis from which to derive ethical and moral norms of behavior: love, social instincts, and a sense of duty, virtues which are not to be confused with church religion.[†] Given its central and unifying importance, the *Entwicklungslehre* was bound to play a major role in school curricula: "we believe that a far-reaching school reform will be inevitable."[‡] In closing, Haeckel took a stab at Emil Du Bois-Reymond's (1818–1896) analysis of the limits of our empirical knowledge of nature[§]:

> The perspective of unlimited progress that the theory of descent opens up, offers at the same time the best protest against that tiresome *ignorabimus*, which has lately been used against it in many quarters. For nobody can predict which limits of knowledge of nature the human intellect will be able to transcend in the future given its astounding development.[¶]

There was not much in Haeckel's presentation that he had not proclaimed or published on earlier occasions. Virchow took the floor on September 22, at which time Haeckel had already left the meeting. Virchow introduced the celebrated Lorenz Oken, who once founded the Annual Assembly of German Natural Scientists and Physicians, as a victim of repression in academic teaching and research, a circumstance he then contrasted with the laxity that currently prevailed at German universities, an unwarranted latitude in academic freedom that allowed professors such as Haeckel to indulge in the arbitrariness of personal speculations bare of all scientific foundation. Both Haeckel and, before him, the botanist Carl Nägeli had profited from academic freedom when voicing evolutionary thoughts in front of this illustrious convention, but nonetheless, Virchow cautioned, scientists would be well advised "to practice some restraint by renouncing hobbies and personal opinions"—a remark that earned animated acclamation from the audience. Virchow continued by contrasting the "freedom of academic teaching" with the "speculative expansion of research programs," and argued that speculative frontiers in science belong to the realm of research, whereas the lecture hall should serve the dissemination of empirically secured knowledge only. The theory of descent, propagated like a "new religion" by Haeckel, in large parts still charts unsecured territory, and should be disseminated to

---

[*] Brücke, 1861.

[†] Haeckel, 1877a, p. 19.

[‡] Haekel, 1877a, p. 18.

[§] Du Bois-Reymond, 1872. The physician and anthropologist Emil du Bois-Reymond from Berlin had famously issued an *ignorabimus* that referred to the residual reality that will never be accessible to empirical science.

[¶] Haeckel, 1877a, p. 19.

the broader public only once its foundation has been cemented, and a secure building has been erected on it. "Only when it has been made clear to me how the combined properties of hydrogen, oxygen, and nitrogen can endow the protoplasm with a soul would I consider it to be justified to introduce the concept of a plastidule soul in the classroom." Until such time, "we have to tell the teachers: do not teach that!"* For the time being, Haeckel's metaphysical musings remain a mere play on words, Virchow insisted. He concluded with the paradigm that "freedom of research" does not equal "freedom of teaching."

Applied in its current, still highly speculative makeup to our entire worldview, to society and state, the theory of decent posed a great danger, according to Virchow, because of the socialist tendencies inherent in the *Entwicklungsgedanke*. Virchow placated the proletarian revolutionary unrest that caused turmoil in Paris in the spring of 1871 as a possible consequence of socialism radicalized in the wake of the French defeat in the Franco-Prussian War, an interpretation that appeared to be vindicated only too soon by the attempted assassination of Emperor Wilhelm I, in Leipzig on May 11, 1878, by Max Hödel (1857–1878), and in Berlin on June 2, 1878, by Karl Eduard Nobiling (1848–1878).[†] The conservative and religious press immediately sided with Virchow, celebrating his sound conservatism as much as the defeat that the Haeckelians—called "ape fanatics"—had supposedly suffered at the Munich meeting. More progressive media like the *Frankfurter Zeitung* deplored Virchow's point of view, as he had lost an opportunity to support, together with Haeckel, the pending school reform legislation. Haeckel in turn went public with his reply to Virchow, issuing a pamphlet titled *Freie Wissenschaft und freie Lehre* (*Freedom in Science and Teaching*), published in 1878. Haeckel accused Virchow of trying to suspend the freedom of academic teaching, and cited article 20 of the Prussian constitution, as well as section 152 of the constitution of the German *Reich*, which both proclaimed the freedom of science and academic teaching. According to Haeckel, Virchow's rejection of a cell soul as a mere play on words documented nothing but his estrangement from his own mechanistic principles that lay at the heart of his cellular pathology.[‡] Haeckel derided Virchow's appeal to strict empiricism as a philosophical *naïveté*: "what do we know about the true nature of force, or matter?" "The cell," he continued, is an abstract concept, not an observable entity: what is observable is always only this or that cell, differentiated in this or that way. "Should cell theory therefore likewise be banned from the classroom?"[§] Haeckel launched into a broad survey that ranged from physics and chemistry across biology to philosophy, history and linguistics in order to show that in any science, all observation is inextricably intertwined with theory and speculation. Darwinism he

---

* Virchow, 1877, p. 70.
† Forcefully condemning these "insane" attacks by social democrats, Haeckel equally forcefully opposed the Prussian "*Kreuz-Zeitung*" for locating in the theory of descent the motivation for such "treasonable" actions (Haeckel, 1879a, p. xxivf).
‡ Haeckel, 1878a, p. 49.
§ Haeckel, 1878a, p. 61.

found to be—if anything—aristocratic,* not democratic, and certainly not socialist in it underpinnings, emphasizing however that science should be free to progress independent of any political connotations that could be read into, or derived from it.

> Should one want to ascribe to this English theory any political connotations—which is certainly possible—then its inherent tendency can only be aristocratic, certainly not democratic, and even less socialist .... For the theory of natural selection teaches that in human life, as much as in plant and animal life, there is always only a small and privileged minority that can survive and strive, while the vast majority starves and sooner or later perishes miserably.[†]

Haeckel concluded by characterizing Virchow's *Restringamur*-speech in München as the second part (in spirit) to Du Bois-Reymond's *Ignorabimus*-speech, originally held at the 45th Assembly of German Natural Scientists and Physicians in Leipzig in August 1872, and proclaimed: "If Emil Du Bois-Reymond wanted to proclaim his *ignorabimus*, if Virchow wanted to proclaim his even more extreme *restringamur* as the core slogan of science, then let us retort form Jena and from a hundred other educational centers with a resounding: *impavidi progrediamur.*"[‡] Indeed, the reactionary specter Haeckel identified in Virchow's address led him to intensify his "propaganda" in support of the theory of descent, as he explained to his friend, the poet Hermann Allmers (1821–1902).[§]

Some commentators rated the significance of this exchange between Haeckel and Virchow comparable to that of the 1830 debate about the means and goals of comparative anatomy between Georges Cuvier and Geoffroy Saint-Hilaire at the Royal Academy of Sciences in Paris,[¶] which attracted international attention and commentary. Haeckel's pamphlet was accordingly translated into both French and English. Haeckel prefaced the English translation, dismissing Virchow's specific attacks on the shortcomings of the "modern doctrine of evolution" as an old hat, competently refuted many times such that no further discussion would seem to be warranted. On a more personal note, he lamented the discord that had emerged between himself and "a man whose ardent disciple and most enthusiastic follower" he once had been, but then went on to express his bewilderment that a "man who stood so long at the head of a party of progress in science and in politics" had come to publicly defend such a reactionary stance ready to be exploited by politicians and clerics alike "in favor of mental retrogression."[**] In his introduction to the English edition, Darwin's bulldog Thomas Henry Huxley (1825–1895) located in Haeckel

---

* Haeckel's interpretation of the Darwinian theory of variation and natural selection as "aristocratic" had nothing to do with a socially stratified society, governed by old aristocracy. Instead, Haeckel wanted to defend biological inegalitarianism as seemed to result from variation and was thought to be required for natural (or artificial) selection to work in opposition to socialist egalitarianism (see also Weikart, 2004, p. 92; and also the discussion in Reynolds, 2008, p. 135).

† Haeckel, 1878a, p. 73.

‡ Haeckel, 1878a, p. 93: "let us fearlessly go forward" ("courageously let us go forward" in Richards, 2008, p. 327).

§ Weindling, 1989a, p. 321.

¶ Appel, 1987.

** Haeckel, 1879a, pp. xxii–xxiv.

and Virchow the same values that Goethe had identified in Geoffroy Saint Hilaire and Cuvier: "The one intellect is imaginative and synthetic ... the other is positive, critical, analytic." Finding merit and fault on both sides, Huxley felt compelled to apply to "both of the eminent antagonists the famous phrase of a late President of the French Chamber—'*Tape dessus*' (give it to him!)."[*] The French translator was less critical, less balanced in his support for Haeckel, and enthusiastically recapitulated how in the course of embryonic development humans pass through the stages of a worm, a larval sea squirt, a lancelet, a lamprey, a fish, etc.,[†] a process reflected in the Fossil Record and bearing testimony to the grand scheme of life: progression from monad to man through the division of cells and the division of labor. This is the crude sort of recapitulationism that Haeckel's name unjustifiably became associated with, and for which Haeckel earned so much critique. Most notably, however, the issue of socialism underpinning the Darwinian theory of descent with modification was far from being swept off the table through that exchange. Haeckel's critic August Bebel (1840–1913) defended a socialist interpretation of Darwinism that was to remain controversial in German biology well into the National Socialist era.[‡] The reason is that authors who engaged in an opposition to the socialist or social-democratic movement picked up on Haeckel's characterization of selection theory as "aristocratic," a move that resulted in a radical right-wing interpretation of evolutionary theory. The Haeckel–Virchow debate of 1877 has for this reason been identified as the starting point for the development of German Social Darwinism, which would become the backbone for German eugenics and racial hygiene.[§]

Haeckel became engaged to his cousin Anna Sethe in 1858. Having established himself in a secure position at the University of Jena, Haeckel exchanged wedding vows and rings with his beloved Anna on August 18, 1862. Fate struck on February 16, 1864, when less than two years into the happy marriage, the death of Anna separated the spouses. Haeckel was grief stricken, and fought off depression through an enhanced pursuit of research and writing. The efforts of that period brought forth from the depth of Haeckel's pain his *magnum opus*, his *Generelle Morphologie* of 1866. It was certainly not the most widely circulated of his books, outdone by far by such popular writings as his *Natürliche Schöpfungsgeschichte* of 1868, *Die Welträthsel* of 1899, or *Die Lebenswunder* of 1904. But it was the treatise that not only established his reputation as a brilliant synthesizer of contemporary biological sciences within a Darwinian framework, but also one that had enormous influence on German biology in general, followed only by his *Gastraea-Theorie* that appeared in the *Jenaische Zeitschrift für Naturwissenschaft* in 1874, and that provided certain revisions of views and ideas Haeckel published in 1866.

In his widely read history of animal morphology, Edward S. Russell characterized Haeckel's *Generelle Morphologie* as "a belated offshoot of *Naturphilosophie* ... a medley of dogmatic materialism, idealistic morphology, and evolutionary theory."[¶]

---

[*] Huxley, 1879, p. vi.
[†] Soury, 1879, p. xvii.
[‡] On Bebel and like-minded contemporary socialist interpreters of Darwinism like Karl Kautsky (1854–1938) and Eduard Bernstein (1850–1932), see Richards, 2008, pp. 326ff.
[§] Weingart et al., 1992, p. 117.
[¶] Russell, 1916 (1982), p. 248.

The Haeckel biographer Robert J. Richards found Russell's denigrating assessment of Haeckel's work to be tainted by Russell's own adherence to Aristotelian final causes and his consequent abhorrence of brute materialism and mechanism in the explanation of phenomena of life.[*] The *Generelle Morphologie* can indeed be read as written in a renewing spirit to erase any residual traces of German Idealism and Romanticism in biology, in favor of a mechanistic and materialistic conception of organisms and their evolution. At the time of Haeckel's writing, the term "evolution" (*évolution* in French; *Entwicklung* in German) remained ambiguous. Originally, the term referred to embryonic, or individual, development.[†] But in the last sentence of his *Origin* of 1859, Darwin used the term to refer to his theory of descent with modification. To disambiguate the language of biology, Haeckel introduced the terms *ontogeny* to refer to individual development, and *phylogeny* to refer to the process of descent with modification. Haeckel's claim then was that comparative morphology should not be content with the dissection of adult organisms and the search for ideal, that is, conceptual ties between them. Instead, it should seek to understand organismic form in the light of its ontogeny, as only such a synthetic, indeed dynamic conception of organismic form promises to throw light on its phylogenetic origin and relations. Haeckel himself proclaimed the *Generelle Morphologie* to reflect "the duty of our profession ... a resolute intervention in the sacred battle for the freedom of science and for the discovery of truth in nature."[‡] On the other hand, and Haeckel's own claims to the contrary notwithstanding, there are distinct traces of *Naturphilosophie* and idealistic morphology in Haeckel's *Generelle Morphologie*, such as his crystallography of organisms called *Promorphologie*,[§] or the progressionist dynamism that pervades all life, manifest both in ontogeny as well as in phylogeny. On the whole, the book reveals an obsession with conceptual and terminological clarification, coupled with an urge to dichotomously classify all branches of biological science as well as the objects of their investigation in contrasting categories. It reads almost like a housekeeping exercise: to take stock of what there is in contemporary biology, to clarify and classify, to illuminate and to synthesize, in short: to erect a systematically structured science of morphology. The subtitle of the *Generelle Morphologie* reads: *Critical Fundamentals of a Mechanistic Science of the Evolved Forms of Organisms on the Basis of the Theory of Descent*. Haeckel dedicated the first volume of his book to his "esteemed friend and colleague Carl Gegenbaur," and introduced it with a quote from Goethe that placated German Idealism: "Nature forever creates new *Gestalten*; what is present has never been before; what is past will never return: everything is new, and yet forever the same old." Goethe continues to accompany the reader throughout the text.

In the introduction, Haeckel emphasized that he was going to transcend the old dualism that contrasted morphology with physiology on the basis of the principle of causality, and noted how his earlier advisors Johannes Müller in Berlin and Rudolf Virchow in Würzburg, the latter trough his cellular pathology, had inspired in him

---

[*] Richards, 2008, p. 440.

[†] Bowler, 1975.

[‡] Haeckel, 1866, vol. 1, p. xii.

[§] The *Promorphologie* Haeckel's is "a last residue of idealistic-morphological thinking ... [it] is geometry and crystallography applied to unsuitable objects" (Meyer(-Abich), Adolf, 1934, p. 499).

the initial spark of monism. Morphologists should not be content with a description of animal (and plant) form, but should seek to understand the becoming of form through its development. Haeckel rejected the thoughtless fact gathering strategy of strict and pure empiricism, which he found to typically get lost—particularly in botany—in futile debates about species delimitation while completely neglecting the perspective of the complex whole that is tied together through causal and historical relations. Only once the general laws that are manifest in the multiplicity of organic appearances have been recognized and formalized will the art of form description be elevated to a true science of form comprehension. The recognition of lawfulness based on underlying causality is what transforms artful *Morphographie* into scientific *Morphologie*. The universe is built on matter and force: there is no matter without inherent force, nor is there any force separated from matter and thus devoid of function. This is true for the anorganic as much as for the organic realm. Morphology had traditionally been conceived as the science of static form, physiology as the science of dynamic function. However, the study of *Entwickelungsgeschichte*, or ontogeny, will always remain an essential part of the science of morphology, and for Haeckel it is through the study of ontogeny that comparative morphology is most intimately connected with physiology. The separation Gegenbaur had so forcefully pursued in his negotiations for employment in Jena was sacrificed by Haeckel on the altar of monism, in the name of a unifying principle of causality. Other distinctions Gegenbaur had introduced Haeckel adopted and refined. The science of morphology he divided into two branches, the static science of anatomy, and the dynamic science of morphogeny. Within anatomy, dissecting zootomy is analytic; it catalogs facts. Comparative anatomy is synthetic; it explains form. Form can be analyzed, and explained, along two different lines: one is to look at the way organisms are constructed out of individual structural components, the science Haeckel called *Tectologie*. The other line is *Promorphologie*, the science that investigates the invariable stereometrical properties of the basic, elementary forms that combine to form the form of the organism as a whole. But since all being is only fully comprehended in its becoming, a dynamic conception of morphology must complement the comparative anatomy of adult organisms. This dynamic *Morphogenie* further divides into two branches, the individual *Entwickelungsgeschichte* or ontogeny, and the collective *Entwickelungsgeschichte* or phylogeny.

Turning to the relation of morphology to systematics, Haeckel ridiculed the sorry bunch of museum systematists who painstakingly label and catalog the specimens represented in their dusty collections, the needled bugs, the pickled frogs, or the dry and faded plants stuck on herbarium sheets, each identified to its species. During Haeckel's youth, "mindless systematic work dominated zoology at many German universities," until around the middle of the nineteenth-century new developments in comparative anatomical and physiological work became prevalent, and "found their outward manifestation in the initiation of a new journal, the *Zeitschrift für wissenschaftliche Zoologie*."[*] Through Haeckel's work, the pedantic and uninspiring "species-zoology" was suddenly confronted with a much more aspiring program

---

[*] R. Hertwig, 1919, p. 952.

sketched in short yet powerful strokes.* Haeckel complained that the systematics community of biologists tended to concentrate on the outer appearance of organisms only, without the slightest interest concerning the detailed anatomy, the development, physiology, life history, and phylogeny of their organisms. In nature, the outer appearance and inner structure of an organism are intimately interwoven, he claimed. To concentrate on one aspect only and not also on the others will result in a mistaken comprehension of the whole. This, according to Haeckel, had led to a schism between anatomy and systematics to the detriment of the latter.† The goal of comparative morphology and systematics cannot be the analytic ordering of species diversity in an artificial system that resembles an identification key rather than conveying knowledge about real, that is, historical relationships. Instead, the science of morphology must pursue the reconstruction of a natural system on the basis of a synthesis of comparative anatomy and embryology—a natural system that would seek to represent the phylogeny, the *Stammbaum*, of the species recognized by the systematist. This is how, according to Haeckel, comparative morphology can fruitfully enter into an intimate relationship with systematics. Comparative morphology in its static and dynamic approach offers the tools to decipher the blood relations (*Blutsverwandtschaft*) that prevail among species. It is these blood relations that will have to determine the structure of the natural system, one that can be depicted as a phylogenetic tree as well as translated into an hierarchically structured index of terms and names (*Sach- und Namensregister*) which convey information not only about historical relationships among species and groups of species, but also about the morphological characteristics of these species and groups of species that reveal those phylogenetic relationships.

Having thus laid out his agenda over the first 60 or so pages of his *Generelle Morphologie*, Haeckel had brought into play a number of core concepts of his science of morphology that require more detailed comment and analysis, as they remained highly influential throughout the succeeding decades of systematic biology. Central to his argument is, of course, Haeckel's monism, from which derives the claim for a unity of science that reflects the unity of nature governed by natural laws. The task of science is to discover the laws of nature that are grounded in the principle of causality, a materialistic and mechanistic perspective averse to all auxiliary vitalistic constructions in the explanation of living phenomena. Next is the concept of biological, that is, historical—as opposed to logical—individuals. Haeckel cited Thomas H. Huxley's definition of the *organic* individual‡ as "the organism in all its transient developmental stages from conception to reproduction." Such an organic individual could be represented by a single, or—in the case of organisms subject to alteration of generations—by a string of physiological individuals. The particular organism from conception to death thus is, for Haeckel, the paradigmatic example of a physiological individual, but what about the constituent parts of an organism, its organ systems, organs, or cells? Conversely, trading upward in the scale of complexity, what about

---

* R. Hertwig, 1919, p. 955.
† Nyhart, 2009a, described the decline of systematics and the rise of ecology in the second half of the nineteenth century. Systematics in this context means alpha taxonomy, that is, systematics at the species level.
‡ Haeckel, 1866, vol. 1, p. 56.

the individuality of animal colonies, of species, and of *Thierstämme* (phyla)? The species itself was a highly problematic category for Haeckel. Ever since Linnaeus, the species was considered the basic currency of comparative biology. But just as Gegenbaur had experienced as a child at the side of his mother, so Haeckel recounts how as a 12-year-old boy he tried in vain to distinguish "good" from "bad" species of blackberries and willows, of roses and thistles.* What are species, and how should they be understood? Whatever museum systematists might be telling us: if Darwin is right, can species in fact exist as discrete and independent entities in nature? And finally there is the natural system, the ultimate goal of morphological research. How can it be built up from species through groups of species of ever increasing inclusiveness all the way to the animal phyla recognized by Georges Cuvier and Karl Ernst von Baer—and beyond? Is *all* of life one complex whole? What are the clues to the blood relationships (*Blutsverwandtschaft*) that structure the phylogenetic tree and, through causal and historical relations, endow it with reality? And worst of all: if these phylogenetic or blood relationships are historical in nature, and hence no longer directly observable, how can they become the object of empirical science?

THE GEGENBAUR TRANSFORMATION

The notion of the "Gegenbaur transformation"† was first introduced by the holistic–organicist bio-philosopher Adolf Meyer(-Abich) (1893–1917) in his widely read axiomatization of biology of 1934.‡ The term was quickly adopted by the *völkisch*–organicist botanist, idealistic morphologist, and editor of Goethe's scientific writings Wilhelm Troll,§ who in his 1937 textbook chastised phylogenetically motivated morphology as a category mistake. The Swiss zoologist and idealistic morphologist Adolf Naef noted in his critical review of the concept of homology published in 1926: "In the year 1870 (the year of the second edition of his *Grundzüge*) Gegenbaur completed his transformation to a phylogeneticist under the influence of Haeckel."¶ The holistic/organicist and neo-Lamarckian functional anatomist Hans Böker** (1886–1939), a former associate of Robert Wiedersheim's†† (1848–1923) institute at the Albert Ludwigs University in Freiburg who became full professor and director of the Anatomical Institute at the University of Jena in 1932, praised the first edition of Gegenbaur's *Grundzüge* (1859) for having brought much more precision and conciseness to comparative anatomy and its concept of body plan (*Bauplan*) than had been customary before—"but nobody was bold enough to conclude that there could be a connection between body-plans and blood-relationships."‡‡ This would change with the second edition of Gegenbaur's *Grundzüge* published in 1870: here, the *Bauplan* became evidence of blood relationship, comparative anatomy became evolutionary

---

* Haeckel, 1866, vol. 1, p. xvi.
† For an earlier historical analysis, see Coleman, 1976; Rieppel, 2011b.
‡ Meyer(-Abich), 1934, p. 516.
§ Troll, 1937, p. 45.
¶ Naef, 1926, p. 416.
** R. Mertens, 1953, p. 397.
†† Nyhart, 1995, pp. 216ff.
‡‡ Böker, 1924, p. 3.

morphology, and the goal of the whole enterprise became the reconstruction of the natural system in a way that reflects phylogenetic relationships.

Max Fürbringer[*] (1846–1920), Gegenbaur's former student and assistant, then professor in Jena and later Gegenbaur's successor at Heidelberg University, characterized his deceased professor's research program as one where "nothing is torn out of its natural context. Everything is alive, permeated and united through the vital principles of causation and correlation."[†] Causation here refers to the phylogenetic framework, correlation to comparative anatomy or, more specifically, to the relation of homology. The "theoretical replacement"[‡] of idealistic morphology or typology with phylogenetic morphology needed not only to explain how abstract (i.e., spatiotemporally unrestricted) concepts such as the type could be transformed into spatiotemporally located objects such as an ancestral species, but it also had to explain how structural correspondence could support an inference to causation. Homology is a relation of coexistence, causation—at least in phylogeny—is a relation of succession.[§] By his own admission, Gegenbaur attributed[¶] his conversion to Darwinism, and through it, to evolutionary morphology not only to Bronn's translation of Darwin's "famous book,"[**] but also—and probably even more so—to chapters 19 and 20 in the second volume of Haeckel's *Generelle Morphologie*. Haeckel's monism certainly caught up with him, as did Haeckel's consequent insistence that biological processes—both genealogical as well as physiological—are to be explained exclusively on the basis of mechanistic causation, without any recourse to vitalistic and teleological principles. And yet, Gegenbaur's mind was imbued with Haeckelian progressionism, all development—embryonic or evolutionary— lawfully ascending to higher levels of complexity and perfection. Gegenbaur was further attracted by Haeckel's insistence of the importance of empirical research as the basis for powerful induction hat would lead to the formulation of natural laws, although he was more hesitant than Haeckel to attribute to such biological laws, and especially to laws of evolutionary (i.e., historical) transformation, a nomological status that is on par with the laws of physics (mechanics). Gegenbaur offered little comment on issues of variation and inheritance, but he did concur with Haeckel that natural selection is what drives species transformation, as well as a lawful progression toward perfection. In his estimation

> The theory of descent will usher in a new era in the history of comparative anatomy. That theory will mark an even more significant turning point in this science than any other previous theory had ever brought about before. The theory of descent has deeper significance than any other theory of comparative anatomy, such that there is no aspect of morphology that would not be most intimately touched by it.[††]

---

[*] On Fürbringer, see Nyhart, 1995.
[†] Fürbringer, 1903b, pp. 604–605.
[‡] Meyer(-Abich), 1929, p. 156.
[§] Mainx, 1971; on Mainx, see Sloan, 2002, p. 226.
[¶] Gegenbaur, 1870a, p. 76.
[**] Gegenbaur, 1870a, p. 76.
[††] Gegenbaur, 1870a, p. 19.

Although touching on underlying philosophical issues in the various editions of his *Grundzüge der vergleichenden Anatomie* (*Grundriss*as of 1874), Gegenbaur dealt with those most succinctly in 1876, in the introduction he wrote for the first volume of the new journal he had founded, the *Morphologisches Jahrbuch. EineZeitschrift für Anatomie und Entwickelungsgeschichte*[*] published by Wilhelm Engelmann in Leipzig. In the science of morphology, Gegenbaur stressed, the "consideration of the particular enhances the understanding of the whole, just as the whole deepens the understanding of the particular"[†]: a hermeneutic circle that was intended to picture induction and deduction as intimately engaging with each other just as analysis and synthesis do—or should do. As is also the case for all other sciences, the initial step of morphological analysis is descriptive, and as such analytic. This, however, cannot be the final step, but must instead lead to a comparative approach, which constitutes the synthetic component of morphological research. This synthetic step is critical, since the goal is to unveil a *causal* relation in morphological correspondence, that is, in homology. Homology is thus no longer understood as a mere correspondence relation, but is now interpreted as a causal relation indicative of common ancestry. Darwinism in Gegenbaur's hands did not motivate studies of variation or adaptation, but simply provided a new theoretical, that is, explanatory foundation for the old research program, that is, classic (idealistic) comparative morphology—just as Darwin had predicted would be the case. Following the theory of descent, one can no longer accept that each species, or each body plan, was specially created, each an instantiation of a Divine idea. Instead, species as much as the body plans they instantiate, must be assumed to be related to each other through bloodlines. Evolutionary causation of morphological transformation is best revealed by casting the net of morphological comparison as wide as possible when seeking for congruence in the serial arrangement of organisms, their organ complexes or their organs according to the conditions of form they instantiate.[‡]

> Wherever we can on the basis of precise comparison recognize a correspondence of organization, the latter, based as it is in inheritance, indicates common ancestry. Such comparison must not be restricted to singular, arbitrarily chosen organs, however. Instead, the organisms must allow the possibility of genealogical relationship in all of their parts.[§]

The evolutionary morphologist was thus called upon to avoid concentrating on a single organ or organ system in his comparative analysis. Inference to common descent is strengthened through the consilience of evidence, that is, evidence that derives from the comparative analysis of as many organs or organ systems as can be accommodated in a given study. The more organs, or organ systems of organisms representing the same type can be serially arranged along transformation series that run parallel relative to one another, the stronger is the inference that the morphological

---

[*] After Gegenbaur's death in 1903, the journal was renamed *Gegenbaur's Morphologisches Jahrbuch.* The journal ceased publication in 1990.
[†] Gegenbaur, 1876, p. 2.
[‡] Gegenbaur, 1876, p. 12.
[§] Gegenbaur, 1870a, p. 19.

evidence can be causally, that is, historically explained. To explain morphological transformation as the result of phylogenetic processes presupposes mechanisms of inheritance, yet these mechanisms remained unknown, which rendered all explanations in terms of inheritance highly conjectural. But, argued Gegenbaur, there could be no science whatsoever if scientific reasoning would be tied to the requirement to abstain from all hypothetical inference.* The theory of descent underpins comparative morphology with the universal principle of causality, while the success of comparative morphology provides one of the most important pillars in support of the theory of descent.†

Gegenbaur considered anatomy to be analytic and descriptive; comparative anatomy in contrast is synthetic, its goal is to arrange organisms and their parts into continuous series of form, which allow the inference of an underlying process of transformation. The crucial step in the Gegenbaur transformation, according to Meyer(-Abich), is the replacement of an idealistic arrangement of organs, organ systems, or organisms (at any stage of their development) into a continuous, uninterrupted series of conditions of form, by the inference from such a series to an underlying historical process of transformation. A purely idealistic, typologically conceived series of form conditions was for Meyer(-Abich) a logical construct devoid of time and hence of directionality: it could be read in either direction. Not so if that series of form conditions was interpreted as a result of historical transformation. In this case, the arrow of time allows only one direction in which the series must be read.‡ For Gegenbaur, the transformation series lawfully progressed from the simple to the more complex, from the primitive to the derived condition of form, from the "lower" to the "higher" level of organization.§ The assumption of an underlying phylogenetic process provided the platform from which to ascend through induction to the inference of laws that range over these transformation series. Once those laws are inductively secured, comparative anatomy has reached maturity through the acquisition of a basis for deduction:

> Comparative anatomy seeks to derive conditions of form from events, and aspires to discover the laws that govern those events. Even if those laws cannot, perhaps only not quite yet, be formulated in the language of mathematics, the significance of their scientific content may only be modified with respect to other sciences [i.e., physics and physiology], but not with respect to its own proper science. Any dissenting opinion would also have to dispute the scientific status of other historical sciences, or even of natural sciences such as [historical] geology.¶

If the science of morphology had all the forms that live and ever lived on earth at its disposal for comparison, it should be able to connect those in a way that corresponds to their history, that is, in a way that reflects the Tree of Life. Alas, these forms are not available, as the Fossil Record in particular remains woefully incomplete.

---

* Gegenbaur, 1876, p. 14.
† Gegenbaur, 1876, p. 19.
‡ Meyer(-Abich), 1934, pp. 514–515.
§ Gegenbaur, 1870a, p. 75.
¶ Gegenbaur, 1870a, p. 7.

If all the forms that ever existed were available for comparison, "many riddles would find their solution, but at the loss of the science of morphology," because the latter would lose its theoretical significance, Gegenbaur maintained.[*] Empirical gaps in the theoretically continuous and uninterrupted series of forms will therefore have to be bridged conceptually. In comparative morphology, a conceptual comprehension of the multiplicity of form must be established before phylogenetic, that is, causally grounded transformation may be invoked.[†] "Where sensory experience fails, combinatorial intellectual activity must begin"[‡]: the tool to bridge the morphological gaps that exist between organisms, their organs, and organ complexes is the relation of homology. And it is in the definition of homology that Gegenbaur's transformationist rhetoric becomes most apparent, and most controversial. He continued to refer to Richard Owen when defining *"special homology"* or *"homology in the strict sense,"* but the definition now reads quite differently than in the first edition of his *Grundzüge*:

> We call this the relation between two organs that are of common origin, which therefore have developed from the same embryonic rudiment (*Anlage*).[§] Special homology differs from all other previously mentioned kinds of homology ... in that the comparison here requires the detailed demonstration of phylogenetic relationships (*da die Vergleichung hier genaue Nachweise der verwandtschaftlichen Beziehungen erfordert*) ... The demonstration of special homology is one of the main tasks of comparative anatomy.[¶]

Gegenbaur clarified the salient difference between his new definition of homology and his previous uses of the concept in his next treatment of comparative anatomy, an abbreviated version of his *Grundlagen* now called *Grundriss der vergleichenden Anatomie* published in 1874. When pre-Darwinian, that is, Goethean idealistic, morphology had talked about relationship, all that had been implied was an ideal-form relationship. In contrast, his new definition of homology implied blood relationship, he argued: "We thus substitute the concept of relationship for the concept of correspondence or identity of organization ... the science of the relationships of organisms, i.e., the theory of descent or phylogeny, is based on the law of inheritance."[**] Had pre-Darwinian theoreticians read teleology into embryonic development, this would now be substituted with phylogenetic explanations. Derivation from the same embryonic rudiment continued to provide the key to special homology, yet again Gegenbaur stated confusingly: "the search for special homologies requires the precise validation of phylogenetic relationships."[††] For his readers and critics, the issue, which reappeared in unaltered words in the second edition of his *Grundriss*[‡‡] of 1878, lay in the fact that Gegenbaur's

---

[*] Gegenbaur, 1870a, p. 76.
[†] Gegenbaur, 1876, p. 10.
[‡] Gegenbaur, 1870a, p. 76.
[§] The link of homology to the derivation of homologous organs from equivalent embryonic rudiments (*Anlagen*) was central in the controversial discussion of Haeckel's *Gastraea* theory (Nyhart, 1995).
[¶] Gegenbaur, 1870a, p. 80.
[**] Gegenbaur, 1874, p. 5.
[††] Gegenbaur, 1874, p. 64.
[‡‡] Gegenbaur, 1878, p. 68.

definition seemed to imply that prior knowledge of phylogenetic relationships was required in order to detect special homology. This would put the cart before the horse, as homology is supposed to indicate phylogenetic relationship and not the other way around. What Gegenbaur really wanted to say is that homology and ancestry stand in a relation of reciprocal illumination, subject again to the consilience of evidence. This becomes apparent in his most mature work, the *Vergleichende Anatomie der Wirbelthiere* of 1898, published three years before his death.

"The task and the goal of comparative anatomy is phylogeny, and therewith the discovery of the lawful relations that prevail amongst organisms"; comparative anatomy investigates the history of organs, and, through the sum of all organs, the history of organisms.[*] In that sense, comparative anatomy is a historical science, on par with geology.[†] Descent with modification, it is true, is not directly observable, but to conclude from theoretical premises is characteristic of all sciences, and in the case of comparative anatomy, these premises are built on three classes of observed facts: the structural correspondence of organs and organ systems among organisms; the derivation of organs or organ systems from corresponding embryonic rudiments[‡]; and the Fossil Record which, in spite of its incompleteness, delivers a "quite forceful argument."[§]

Structural correspondence and embryonic derivation remain the key to homology, but:

> for the demonstration of the homology of an organ, the consideration of the remaining phylogenetic relationships of the organisms under consideration is of greatest importance; because the relation of homology is governed by common ancestry, homologous organs are derivatives from the same origin that have either departed from the common original condition of form to the same degree, or have done so one to a greater, the other to a lesser degree ... the task is further complicated by the fact that if an organism progresses to a higher level of complexity, it does not necessarily enter this level of differentiation with all its organs [simultaneously]. At higher levels of complexity one or the other organ may persist in a lower condition of from, as also at lower levels of complexity particular organs may exhibit a higher level of differentiation. Greatest circumspection is therefore required to avoid erroneous conclusions.[¶]

What Gegenbaur is saying here would prove to express a most fundamental insight that would acquire central importance in the future development of phylogenetic systematics. According to Gegenbaur, homologous organs, which allow the inference of common origin, are derived from the same ancestral condition of form that is either revealed through their developmental derivation from corresponding embryonic rudiments, or is reconstructed from their correspondence in terms of topology and connectivity in adult organisms, or both. As homology is governed by phylogeny, the prediction derived from the Law of Progression would be that all the organs under consideration of the organisms under comparison would congruently

---

[*] Gegenbaur, 1898, pp. 2–3.
[†] Gegenbaur, 1898, p. 27.
[‡] Gegenbaur, 1898, p. 17.
[§] Gegenbaur, 1898, p. 21.
[¶] Gegenbaur, 1898, pp. 22–23.

reveal the same progress toward increasing complexity. In other words, the transformation series of the various organs under consideration of the organisms under comparison would run parallel relative to one another. This is not necessarily the case, however, as Gegenbaur recognized that some organs may either progress beyond, or lag behind relative to others in their phyletic development. The corollary of that insight is that whereas the comparative study of embryonic development can, indeed, illuminate morphological comparison, "all the praise for embryology as the lantern that illuminates anatomy must be received with great caution."[*] Phylogeny cannot simply be gleaned from embryonic development, a warning Gegenbaur issued with increasing emphasis during the long course of his career.[†]

Such unequal development of organs or organ systems is exemplified by Gegenbaur's defense of his "gill arch theory" of the origin of paired appendages in vertebrates, against the rival "lateral fin-fold theory."[‡] The "lateral fin-fold theory" derives the paired appendages of vertebrates through compartmentalization and differentiation from the anterior and posterior end respectively of a primitively continuous fin-fold that is hypothesized to have run along the ventrolateral body edge of ancestral forms. According to Gegenbaur, the paired appendages of vertebrates had evolved from gill arches that had moved backward along the trunk. The gill arches themselves would have transformed into the endoskeletal pectoral and pelvic girdle elements, the gill rakers would have evolved into the endoskeletal support structures (radials) of the fins. Gegenbaur's "gill arch theory" was based on the presumption that, among the types of fin structures seen in fishes, the archipterygium[§] is the primitive one. In contrast to the metapterygium—the broad-based "fan-shaped" fin structure seen in actinopterygians such as sturgeons (*Acipenser*), bowfins (*Amia*), gars (*Lepisosteus*), and teleosts (e.g., salmon or trout)—the archipterygium is characterized by a narrow base supporting an outward pointing axis that carries pre- and/or postaxial radials. Among extant vertebrates the archipterygium is exemplified by the coelacanth, lungfish, and tetrapods, but it also occurs in the fossil relatives of the coelacanth, in fossil lungfish, and in some Paleozoic sharks.[¶] In 1879, Gegenbaur published in his *Morphologisches Jahrbuch* a monograph by Michael von Davidoff, his student and assistant, on the comparative anatomy of the pelvic fin in fishes. Davidoff sought to document that the pelvic fin is less prone to phylogenetic transformation, that is, is more conservative in its evolution than the pectoral fin. This, of course, implies that in the course of the progressive evolution of fishes, the pelvic fins would lag behind the pectoral fins in terms of evolutionary transformation and progression. As a consequence, the pelvic fin would approach the ancestral morphology more closely than the pectoral fin, and should therefore play a

---

[*] Gegenbaur, 1876, p. 14.

[†] See the analysis by Nyhart (1995, 2003).

[‡] A detailed historical analysis of that debate was given by Nyhart (1995, pp. 251ff).

[§] The term was introduced by Gegenbaur (1870b (1912), p. 370) for a hypothetical ancestral fin structure derived from a comparative analysis of sharks and their relatives (skates, rays, and chimaeras).

[¶] The debate still continues. The "lateral fin-fold theory" enjoyed greater acceptance for most of the twentieth century, especially after Goodrich's (1930) influential review. Gegenbaur's "gill arch theory" was revitalized on paleontological grounds by Zangerl and Case, 1976, on genetic grounds by Gillis et al. (2009).

more important role than the pectoral fins in discussions of the origin of the paired appendages in vertebrates. In the opening section to his paper, however, Davidoff professed allegiance with Gegenbaur's conception of the origin of vertebrate paired appendages, on which basis he declined to further comment on the rival views, referring to Gegenbaur's relevant papers instead.[*]

Commenting on his student's work, Gegenbaur introduced additional support for the latter's conclusions by mentioning salamanders in which the foot retains the primitive five digits, as opposed to the four digits that remain in the hand. He also pointed to the fact that it is the forelimb that transformed into wings in pterodactyls, birds, and bats, all of which retain a comparatively primitive hind limb structure.[†] But that was not the only motivation for his postscript. Instead, he wanted to comment on the lateral fin-fold theory that had so far not played any role in his deliberations on the origin of vertebrate paired appendages, as he had only recently learnt of it through "BALFOUR's excellent investigations."[‡] The "lateral fin-fold theory" is generally associated with the names of the British morphologist Francis Maitland Balfour (1851–1882)[§] and the American James K. Thacher (1847–1981),[¶] who were the first to challenge Gegenbaur's "gill arch theory."Gegenbaur emphasized that he had based his investigations on sharks, which he considered the most primitive living gnathostomes. Thacher, in contrast, had made the sturgeon the object of his studies, which was considered a much more advanced fish by Gegenbaur. Gegenbaur had acted in the belief that the shark fins would most closely approach the hypothetical ancestral condition, but he admitted with respect to sturgeons that "it would be erroneous to conclude that all of their organ systems express greater progression in their differentiation." Therefore one had to at least countenance the possibility that whereas the sturgeons do exhibit greater progression relative to sharks in many of their organ systems, they do not do so in the structure of the vertebral column, and similarly might not do so in the structure of their fins.[**] By comparison to sharks, sturgeons would thus show a mosaic pattern of organ differentiation, some progressive, others lagging behind. Indeed, Gegenbaur would admit a similar incongruence of organ evolution in the objects of his own investigations. No living shark, and it is on those that he initially based his "gill arch theory," shows an archipterygium. Instead, living sharks show broad-based fins which, at first sight, would seem to lend much greater support to the "fin-fold theory." The implications of the latter theory that Gegenbaur found most objectionable were the *de novo* formation of (endo-)skeletal elements (limb girdle elements and fin support structures) and their coming together to form a structural and functional unit, "two hypothetical steps that have never gained any observational basis."[††] However, his own original sketch of the archipterygium had no

---

[*] Davidoff, 1879, p. 451.

[†] Gegenbaur, 1879, p. 522.

[‡] Gegenbaur, 1879, p. 524, took the lateral fin-fold to mark the way, "segment by segment" that the gill arch migrated backward to form pelvic and pectoral girdles and fins—an interpretation he later abandoned.

[§] Gegenbaur, 1879, p. 521.

[¶] Gegenbaur, 1879, p. 521; see also Thacher, 1877a,b.

[**] Gegenbaur, 1879, p. 522.

[††] Gegenbaur, 1879, p. 523.

observational basis either. The archipterygium initially was a theoretical construct that Gegenbaur derived from the morphology of the pectoral and pelvic fins in sharks through a stipulated series of rather convoluted transformations.* Motivated by the desire to establish a unity of plan that encompasses fins as well as tetrapod limbs, Gegenbaur turned to other fishes that would represent the archipterygium in a more naturalistic fashion than extant sharks, and found this to be the case in lungfishes,† particularly in the Australian lungfish *Neoceratodus*.‡ This implied that a group of fishes he considered far advanced over the shark's level of organization nevertheless most closely approached the ancestral fin structure.§ Again, an otherwise much-advanced fish he thought to retain a primitive structure of its paired appendages.

Given Gegenbaur's progressionism, this discussion implies that any particular organism, instantiating as it does a certain type of organization, could be composed of organs or organ systems that could be more, or less, advanced on the ascending scale of complexity and perfection. A hypothetical ancestral condition of form could thus not simply be read off of embryos, but ultimately had to be inferred on the basis of comparative anatomy. This is the reason why, according to Gegenbaur,¶ comparative embryology will never be able to render comparative anatomy redundant. This conclusion of Gegenbaur's reflects a general debate, or conflict, pervading evolutionary morphology in German biology of the late nineteenth century.** Gegenbaur recognized two clues that could lead to the recognition of homology—comparative anatomy and ontogeny, the derivation of organs from corresponding embryonic rudiments. The conflict, or debate, turned on the issue which one of those two clues should be accorded epistemic priority over the other. Gegenbaur's turn from an idealistic to an evolutionarily motivated morphology was unquestionably heavily influenced by Haeckel, who in turn earned a place in the history of biology through his expositions on the relations between ontogeny, phylogeny, and the Fossil Record. And so it is to Haeckel we must now turn.

---

* Gegenbaur, 1870b.
† Gegenbaur, 1870b (1912), p. 370. Gegenbaur used the genus name *Ceratodus* for the extant Australian lungfish; in modern taxonomy, *Ceratodus* stands for a Paleozoic lungfish genus.
‡ Gegenbaur, 1872 (1912), p. 529. The pectoral and pelvic fins of the African (*Protopterus*) and South American (*Lepidosiren*) lungfish are paedomorphic, that is, retarded in their development. Gegenbaur used the old genus name *Ceratodus*, which today is replaced by *Neoceratodus*.
§ For further historical analysis, see Nyhart, 1995; Bowler, 2007.
¶ Gegenbaur, 1889, p. 4. The essay is dated September 1888 (Nyhart, 1995, p. 250, n. 2).
** Concerning the *Competenzconflict*, see Nyhart, 1995, pp. 262ff; see also Nyhart, 2002.

2 Of Parts and Wholes

BEOBACHTUNG UND REFLEXION

A science that is content with description, enumeration, and a catalog of what is immediately given here and now in perception amounts to bean-counting empiricism, said Haeckel; the dusty drawers of museum systematists full of specimens, all labeled but none understood, was one of his prime examples of such meaningless pursuit of knowledge. Haeckel, who wanted to ask big questions, brought into play the admired embryologist Karl Ernst von Baer, who had subtitled his lasting 1828 monograph on animal development with *Beobachtung und Reflexion*, observation and reflection, which necessarily must go hand in hand if science is to prove fruitful and to generate interesting insights and results. This is a classic Kantian theme, which, while paying homage to the immortal grandmaster of German philosophy, Haeckel deployed in what he liked to portray as a forward-looking positivist rather than a retrospective Kantian perspective. For Kant, what is immediately apprehended by sensory experience is a mere intuition (*Anschauung*), not knowledge, for "thoughts without content are empty, intuitions without concepts are blind."[*] Raw observation—if at all possible—would yield nothing but a chaotic multiplicity of appearances without sense or context. That is obviously not good enough for science. The task of science, according to Kant, is to search for regularities in nature that would reveal the laws reigning over such regularities. True science, interesting science, is ultimately concerned with the discovery of universal laws of nature, and those are preferably to be formulated in the language of mathematics. For Kant,[†] the arm of natural science reached only as far as it is capable to capture nature in mathematical terms. For Haeckel, these laws of nature had to be grounded in causality, not in idle metaphysical speculation.

There was, indeed, a rising tide building against idealistic *Naturphilosophie* in mid-nineteenth-century German biology. The influential Johannes Müller, professor at the University of Berlin and a self-proclaimed disciple of Goethe, thus conscious of the pitfalls and limitations of scientific experimentation, is nonetheless celebrated as the godfather of comparative physiology in German biology.[‡] Other than Gegenbaur and Haeckel, his students included such celebrities as Matthias Jacob Schleiden (1804–1881), Rudolf Virchow, Emil du Bois-Raymond (1818–1896), and Hermann von Helmholtz (1821–1894). After Johann F. Meckel's death in 1833, Johannes Müller took over the editorship of the *Archiv für Anatomie und Physiologie*, which he renamed as *Archiv für Anatomie, Physiologie, und wissenschaftliche Medicin* for its 1834 volume. He arranged for the publication in 1838 of Schleiden's discovery that all plant development starts from a cell in his journal. This was the first step toward a cell theory

---

[*] Kant, cited from Gardner, 1999 (2008), p. 68.
[†] Kant, 1786, p. viii.
[‡] Cassirer, 1950 (1978), p. 186.

35

that would forever change the face of biology. Schleiden opened his seminal paper with a sentence that again brought up a Kantian concern: "The universal law of human reasoning that reigns over all the sciences, which is the persistent pursuit for unity in its findings, has also manifested itself in the realm of organisms"[*]: the unity of cognition is reflected in the unity of the plant that ultimately originates from a single cell. The cell was for Schleiden the paradigmatic biological individual; the whole plant that develops from it is an aggregate of individuals. Plant growth is lawfully, that is, necessarily the result of a multiplication of cells. Schleiden admitted to cases in the plant world where the formation of cells eluded direct observation through the microscope, a fact that "has no impact, however, on the general validity of the law, because this is required by analogy, and also because we can fully specify the reasons for such impossibility of direct observation."[†] When Schleiden communicated his findings to his fellow student Theodor Schawnn (1810–1882), the latter conveyed similar observations he had made in animal tissues. The result of that exchange was the "Schleiden-Schwann" cell theory, which Johannes Müller quickly incorporated in his teaching. Virchow would later, in 1858,[‡] generalize cell theory with his slogan *omnis cellula e cellula* (all living cells originate from living cells), and recognize pathologies as cellular malfunctions.

Schleiden himself expounded his vision of modern botany in his major work of 1842/43, *Grundzüge der wissenschaftlichen Botanik*, two volumes, in which he took a stand against two trends prevailing in botany at the time, where botanists either perceived value in species descriptions in and of themselves, cataloguing plant diversity at home and abroad without further synthesis, or else indulged in empirically unfounded *naturphilosophisch* speculation.[§] His treatise opened with an exposition of the proper philosophical foundation for scientific botany, which Haeckel found as "pathbreaking"[¶] as Johannes Müller's reformation of zoology. Schleiden wanted to frame botany as a thoroughly inductive science, in which context he took up the von Baerian theme of *Beobachtung* and *Reflexion* again, something Haeckel would later likewise do with great fervor[**]:

> Without empirical investigation, without observation there cannot be any empirical science, but a collection of naked facts is still far from being science, just as mere building material is not yet a temple. What is required beyond observation is a separation of the questionable from the secure, of the accidental from the essential, is the derivation of rules and laws—all of which are intellectual operations of which pure observation knows nothing and cannot know anything.[††]

Schleiden wanted to make causal analysis the centerpiece of scientific botany,[‡‡] an agenda applauded by Haeckel who cited[§§] Schleiden's *Grundzüge der*

---

[*] Schleiden, 1838, p. 137.
[†] Schleiden, 1838, p. 156.
[‡] Virchow, 1858.
[§] Jahn, 2001, pp. 313, 324.
[¶] Haeckel, 1866, vol. 1, p. 64.
[**] Haeckel, 1866, vol. 1, p. 65; see also Haeckel, 1875a, pp. 1ff.
[††] Schleiden, 1842, p. 11.
[‡‡] Cassirer, 1950 (1978), p. 155.
[§§] Haeckel, 1866, vol. 1, p. 63.

wissenschaftlichen Botanik in support of his own views on the demarcation of science from metaphysics. Schleiden had emphasized the necessity of philosophical training, especially in logic, for every scientist, and in a Baconian tradition proclaimed that scientific inquiry may inductively ascend from the particular to the universal, or deduce the particular from the universal. It is this interplay of an ascending and descending search for knowledge that has led to the great discoveries in physics. "But if we compare morphological sciences with the theories of physics," Schleiden continued, "we have to admit that the first lags infinitely behind." This is not only the result of the fact that the complexities of life dealt with by comparative anatomy elude mathematical analysis, but even more so the result of a general neglect "of methodological, conceptual and general philosophical"* analysis. Haeckel could not have agreed more, and further cementing his views of what proper scientific inquiry is or ought to be, he turned to his celebrated teacher again, Johannes Müller, the "greatest physiologist and morphologist of the first half of the 19th century."† Müller, in his most important and highly influential *Handbuch der Physiologie des Menschen* of 1840, had listed gravity and inertia as paradigmatic laws of physics (mechanics). But what was the *nature*, that is, the substantial, or mechanical basis of gravity—a force acting at a distance? Müller referred to "André-Marie Ampère (1775–1836) [who] laid the foundations for the discovery of natural laws from which the phenomena of electromagnetism can be deduced with the same certainty as geometrical truths can be deduced from their underlying axioms ... And yet, the nature of electricity remains elusive." The same, Müller asserted, is true of the mechanics of innervation, of the life of a cell, of the mechanics of plant and animal development, and of the mechanics of psychological processes. However, not knowing the substantial, that is, material basis of natural causation does not prevent science to generalize from empirical observations to laws with predictive power. Science, Müller continued, cannot progress through philosophical analysis of concepts alone, nor is raw observation sufficient by itself. What is required instead is a conceptual frame that structures sensual experience, allowing the separation of the accidental from the essential, that is, a *philosophication* of empirical experience.‡ Haeckel drew several major conclusions from these authoritative and foundational textbooks by Mathias Schleiden and Johannes Müller: science is about the discovery of natural laws; induction and deduction are complementary ways to gain scientific knowledge; nature can reveal lawfulness even if the complexity of the phenomena does not (yet) allow to express these laws in mathematical terms; and laws can have predictive power even if the mechanics of the processes over which they range are not (yet fully) understood. Haeckel characterized science thus conceived as "philosophical empiricism" (*philosophische Empirie*) or "empirical philosophy" (*empirische Philosophie*), which is the same as both result from an intimate reciprocal illumination of observation and reflection, of the "description of nature" (*Naturbeschreibung*) and the "philosophy of nature" (*Naturphilosophie*).§ Such philosophical foundation appeared to Haeckel to be the

---

* Schleiden, 1842, p. 40.
† Haeckel, 1866, vol. 1, p. 94.
‡ J. Müller, 1840, p. 522.
§ Haeckel, 1866, vol. 1, p. 64.

only way in which to bring the *Entwicklungsgedanke* that applies to living entities within the reach of the arm of science. *Naturphilosophie?* Yes, said Haeckel, but *not* in the sense of a *Naturphantasterei* (nature phantasy) as articulated by Schelling or Lorenz Oken (1779–1851). Haeckel countered German Idealism with a conceptual analysis and classification of his *Naturphilosophie*. Analysis and synthesis are two complementary roads to scientific discovery, where analysis concerns the particular, the individual, synthesis the universal, the law that ranges over individuals of the same kind: "It is only through synthesis that the most important universal laws will be revealed, insights that analysis will never lead us to."[*] Induction and deduction are likewise reciprocally illuminative methods of inference as Schleiden had argued. Induction ascends from the analysis of particulars through comparison and generalization to universal statements, which are subsequently tested by deduction. In that sense, induction stands at the beginning of scientific inquiry, deduction at its conclusion. Analysis and synthesis, induction and deduction: they relate to scientific theory construction as essentially as inhalation and exhalation relate to life—a Goethean metaphor, one among many adopted by Haeckel.[†]

SINGLE CAUSE, COMPLEX EFFECT

Motivated by his monism Haeckel eventually slipped back into muddled waters, proclaiming a unity of matter and soul that pervades all of organic nature and even beyond as it extends to crystals (a "naive hylozoism"[‡]). Given his burning enthusiasm for Darwin, the young Haeckel denied in his *Generelle Morphologie* of 1866 a clear-cut separation of the anorganic from the organic realm: the latter must, after all, have emerged from the former at some point in time. There was, in his opinion, absolutely no need to invoke any vitalistic forces to explain the intricate phenomena manifest in living substance. In his time (i.e., pre-Darwin), the unmatched Johannes Müller may at heart still have been a vitalist, thought Haeckel, but given some of the positions he forcefully articulated in his *Handbuch der menschlichen Physiologie*, he at least brought out the glaring contrast between vitalism and mechanism like nobody else had done before, thus forcing a decision which one of the two alternatives to embrace.[§] Had there ever been a satisfactory explanation for the existence of functionless rudimentary organs? The *nisus formativus* of Blumenbach, the *loi du balancement des organs* of Étienne Geoffroy Saint-Hilaire, the *force formatrice* of Étienne Serres, or worse, the entirely metaphysical principle of plenitude invoked by Agassiz in his *Essay on Classification* of 1857—in Haeckel's eyes, none of those proposals could possibly rival the explanatory force grounded in mechanistic causality that characterizes Darwin's theory of variation and natural selection. On Haeckel's analysis, Darwin's principles exclusively imply one cause and one cause only, the *causa efficiens* that is grounded in the material fabric of nature. Darwinian theory leaves no room for goal directedness, for teleological processes driven by a *causa finalis* as is implied by all the other competing explanations of rudimentation.

---

[*] Haeckel, 1866, vol. 1, p. 79.
[†] Haeckel, 1866, vol. 1, p. 83.
[‡] Cassirer, 1950 (1978), p. 163; see also extended discussion in Richards, 2008.
[§] Haeckel, 1866, vol. 1, p. 95.

Here again, Haeckel would violate his own first principles with his blind faith in the progressive nature of the evolutionary process, driven by a struggle for existence that only the best could and would survive. But in his rejection of vitalism, he had another ax to grind: "We recognize in Darwin's discovery of natural selection in the struggle for existence the ultimate proof for the exclusive validity of mechanically efficient causes in the entire field of biology; we recognize therein the demise of all teleological or vitalistic interpretations of organisms." Monism does not allow for a plurality of causes, but for one type of causality only, that is, the mechanistic one. A monistic science then is implicitly a mechanistic science.* To reduce the phenomena of life to mechanistic causality not only made Haeckel's *Promorphologie*, a crystallography of organisms,† possible in the first place, but also allowed biological laws to aspire to the nomological status that is enjoyed by the laws of physics: "what the theory of gravity achieves with respect to anorganic components of the world (*Weltkörper*), the theory of descent accomplishes for organic nature."‡

The theory of descent was at the time, and would continue to be, highly critically received in German biology, because it issues statements about ancestry and descent that reach into the past, into "deep time" that is no longer accessible to direct observation. Bronn, for example, in his 1860 review of Darwin's *Origin*, characterized the book's contents as the development and exposition of a basic thought that most likely will bring similar emotions to scientific discourse as had once been triggered by Lyell's *Principles of Geology* (1830–1833)—"whether with the same success may be doubted, for there is no way to adduce irrefutable evidence in its support ... nor is it of course possible to deliver decisive counterevidence."§ In defense of the unflinching certitude in his positive evaluation of Darwin's theory, Haeckel appealed to one of the first principles of inductive inference, the consilience of logically independent lines of evidence.¶ "The theory of descent," according to Haeckel, "postulates no processes that are not empirically accessible; instead, it only generalizes over the results of innumerable congruent empirical observations and draws from those a powerful inductive inference, which is as certain as any other well-founded induction. It subsumes the multitude of known phenomena in the diverse organic world (*Formen-Welt*) under one single explanatory thought that is not contradicted by any known fact."** *"The validity of Darwin's theory of selection requires no further proof."*††

---

* Haeckel, 1866, vol. 2, p. 150.
† The analogy of organismal development and growth with crystals or minerals, still much discussed in Haeckel's time, for example by Johannes Müller (1837, p. 22) in his *Prolegomena* to his *Physiologie des Menschen* (see also Bronn, 1853, vol. 1, p. 26, and especially Bronn, 1858b, as well as J. Müller's student H. Jordan, 1842), had illustrious forerunners in the eighteenth century such as the theories of generation proposed by Pierre-Louis Moreau de Maupertuis (1698–1759), and Georges Buffon (1707–1788): Roger, 1971; Rieppel, 2011c. It ultimately reaches back to Ancient Greek and Roman atomistic philosophy (e.g., Lucretius's *De rerum natura*).
‡ Haeckel, 1866, vol. 2, p. 150.
§ Bronn, 1860, p. 112.
¶ The principle is usually attributed to the English philosopher William Whewell (1794–1866), not named by Haeckel, 1866.
** Haeckel, 1866, vol. 2, p. 292.
†† Haeckel, 1866, vol. 2, p. 293.

Darwinian natural selection plays out in the struggle for existence, which in turn results from the competition of individual organisms for resources. In Haeckel's view, it is *physiological individuals** that fight for survival, that is, individuals that come into existence through their conception, and that come to an end when they die. But what exactly is an individual in asexually reproducing organisms, in plants, or in colonial animals, asked Haeckel? Schleiden in his early groundbreaking paper found the application of the concept of "individual" to plants highly problematic,[†] and further noted an indeterminacy that had become apparent in the vocabulary of plant physiologists, as such intuitively obvious morphological units as root, stem, or bud have become increasingly difficult to precisely delineate and characterize. Schleiden touted the advantages of his cell theory as it provided the unit, the cell, which alone can be called an individual in the strict sense of the term. Alexander Braun, whom Haeckel had heard in Berlin, presented a talk on the subject at the Royal Academy of Sciences on November 25, 1852, which consisted of a lengthy historical review that concluded with reference to Schleiden's distinction of a hierarchy of individuals in plants: the cell as a plant of first order, the sprout as a simple plant of second order, and the whole plant (*Stock*) as a composite plant of third order. With this Braun announced his intention to investigate as to whether the distinction of such "relative individualities"[‡] can, indeed, be justified. The results of his analysis he discussed in a talk read on February 3, 1853, with the general conclusion that a plant is a hierarchically structured aggregate of several generations of polymorphically differentiated individuals: plants form "organically interconnected family trees (*Familienstöcke*), variably differentiated in their branching patterns, which comprise countless generations of individuals that complement each other with their variant vocations."[§] The proof for Braun lay in the close similarities that plants exhibited with polymorphically differentiated colonial invertebrates, organisms that had been so superbly described and analyzed by Rudolf Leuckart (1822–1898).[¶] Leuckart warmly acknowledged the friendship, help, support, and hospitality Braun had extended to him during his stay at the University of Giessen in 1850/51, and credited him with the insight that plants are aggregates of individuals.[**] Of the invertebrates he studied, Leuckart "recognized in them a connected union of individuals ... which originate from the same morphogenetic laws, but which do not agree in form and function, but which adapt themselves to the physiological requirements of the entire union in many variable ways."[††]

---

* The term "physiological individual" was introduced by Spring (1838, p. 35) in contradistinction to "systematic individual." The physiological individual comprises the organism in all its developmental stages from conception to death; the systematic individual represents an organism only at some one stage of its development. For critical comments, see Braun, 1853, p. 27, n. 3.
† Schleiden, 1838, pp. 168, 170.
‡ Braun, 1853, p. 34.
§ Braun, 1853, p. 89: "*Was die Pflanze auszeichnet, das ist die ... auftretende Bildung der Familienstöcke, als organisch gebundener, in ihren Verzweigungen mannigfach geordneter Stammbäume, die zahlreiche Generationen durch verschiedene Begabung sich ergänzender Individuen umfassen.*"
¶ Braun, 1853, pp. 69ff.
** Leuckart, 1851, p. 35, n. 40. For more detail, see Nyhart and Lidgard, 2011.
†† Leuckart, 1851, p. 330; the translation is from Nyhart and Lidgard, 2011, pp. 436–437.

LEVELS AND MODES OF INDIVIDUALITY

Haeckel gratefully acknowledged[*] the instructions and advice he had received from Braun, who had moved to Berlin in 1851, and placed his former teacher at the top of a list of names that included Aristotle, Hippocrates, Goethe, Linné, and Erasmus Darwin, all of whom according to Haeckel had conceived of the plant as an "aggregate or colony of individuals."[†] In Leuckart's "superb monograph" of 1851 on colonial invertebrates Haeckel found confirmation that these creatures, too, "form colonies of polymorphic individuals," and further the insight that it was necessary in the realm of biology not only to sharply distinguish between *morphological* and *physiological* individuals, but also to recognize *"individuals of different rank."*[‡] Braun and Leuckart thus provided the starting point for Haeckel's analysis of the intricacies of biological individuality, which has been characterized as betraying "a kind of mania for puzzle solving."[§] For a philosophically minded intellect such as Haeckel's, biological individuality was, indeed, a puzzle, as much as it remains one today.

In classical philosophy, the logical individual is a particular, denoted by a proper name, and nothing more—the antithesis to a universal. In that sense, individuality is a classic category of logical analysis. The (self-) identity of a logical individual is a numerical one, not one anchored in any intrinsic properties. While the classical concept of a logical individual ("a particular, a thing, denoted by its name and nothing else"[¶]) can be—and has been[**]—applied to biological entities, it needs to be enriched with *natuurhistorisch* content in order to make it interesting. But when it is rendered in flesh and blood, it is no longer a logical individual, but rather becomes a biological individual, a concept first introduced by Anton Spring in his prize-winning essay on the nature of species that was published in 1838 (see the discussion in Chapter 1). Spring's essay was the starting point for Alexander Braun's investigations of the nature of biological individuality, who in turn influenced Rudolf Leuckart, and Ernst Haeckel.[††] Schleiden, Braun, and Leuckart all emphasized generative mechanisms, such as relations of division, growth, and differentiation, when discussing colonies, or hierarchies of individuals. Haeckel agreed: a comparison of phanerogams and higher cryptogams with coelenterates, in particular hydromedusae, immediately reveals the identity of the relations in which the sprouts and buds stand to one another on the one hand, and to the whole on the other. Irrespective of the polymorphic differentiation of the parts, the generative mechanisms of the formation of the whole are the same.[‡‡] For Haeckel, both plants and colonial invertebrates form, not aggregates nor colonies, but hierarchies of individuals, complex wholes each composed of parts that are tied together by the same shared generative mechanisms. And the same is true of whole organisms:

---

[*] Haeckel, 1866, vol. 1, p. xxii.
[†] Haeckel, 1866, vol. 1, p. 243.
[‡] Haeckel, 1866, vol. 1, p. 257; also p. 264.
[§] Richards, 2008, p. 134.
[¶] Hull, 1989, p. 183.
[**] See Ghiselin, 1974, 1997; Hull, 1976, 1989, 1999.
[††] Haeckel's analysis of biological individuality "was closely modeled on existing schemes proposed by botanists" (Nyhart, 1995, p. 136).
[‡‡] Haeckel, 1866, vol. 1, p. 243.

The body of the great majority of organisms living today represents an entangled construction composed in a most complicated way of parts or organs of the same and different kinds. Quite generally we can classify these *"Partes similares et dissimilares"* in subordinated categories, such that every higher category forms a discrete and self-sufficient unit, while at the same time being composed of a multitude of units of the next lower category. These categories we consider to represent different levels or orders of organic individuals. We can thus define *Tectology* or structure theory as the science of the composition of organisms from organic individuals of different order.[*]

Haeckel's classification of biological individuals—a "theory of relative individuality"[†] that would hold across the plant and animal kingdom—betrays the same obsession with conceptual clarification and classification that runs through his analysis of the science of morphology. Just as he recognized a static (anatomy) and dynamic (morphogeny) branch of morphology, so did he recognize static (morphological individuals) and dynamic (physiological individuals) manifestations of biological individuality. A morphological individual is a *form*-individual, a unified complex whole composed of parts none of which can be removed without compromising the integrity of the whole. As such, the form-individual is to be apprehended as a static unit.[‡] The physiological individual is a performance-individual or a life-unit (*Leistungsindividuum*), and as such a dynamic individual. Whereas the form-individual is apprehended in terms of a static time-slice, the physiological individual is apprehended throughout its (transient) duration.[§] Of morphological individuals there are six orders, which each represent at some level of organismic complexity a mature physiological individual as well. At higher levels of complexity, which are species specific, the orders of morphological individuals collectively form a nested hierarchy, which, in its totality, represents the mature physiological individual. Thus, the first order of *organic* individuality is represented by the cell (*Plastiden*), which is a morphological individual nested within multicellular organisms, while also representing the mature physiological individual in unicellular organisms. The second order of organic individuals is organs or organ systems (*Werkstücke*) composed of cells (*Zellstöcke*, in analogy to *Tierstöcke*); they form mature physiological individuals at the level of protists, algae, and coelenterates. The third order individuals are antimeres (*Antimeren*), that is, corresponding (*homotypical*) morphological structures across a symmetry plane, such as the left and the right parts of organs and organ systems in bilaterally symmetrical organisms. Antimeres form, according to Haeckel, mature physiological individuals at the level of protists, as well as lower plants and animals. Fourth order individuals are metameres (*Metameren*), that is, corresponding (*homodynamic*) parts along the axis of an organism such as the internodes of the vegetative stem in *Equisetum*, the segments in annelid worms, or the somites in vertebrates. Metameres form mature physiological individuals at the level of mollusks, lower worms, and algae. The fifth order of organic individuals is represented by persons (*Personen*), that is, sprouts in plants and coelenterates, or

---

[*] Haeckel, 1866, vol. 1, p. 242.
[†] Haeckel, 1866, vol. 1, p. 264.
[‡] Haeckel, 1866, vol. 1, p. 265.
[§] Haeckel, 1866, vol. 1, p. 265.

"individuals" in the sense of bodies in higher organisms, at which level they also represent a mature physiological individual; the fifth level never forms a mature physiological individual in plants. The sixth order of organic individuality is colonies or corms (*Cormen* or *Tierstöcke*) such as trees and bushes, polyp colonies (*Polypenstöcke*, i.e., polyps connected by a hydrocaulus), or a chain of salps. This is also the level of organic individuality that forms the physiological individual in most plants and coelenterates. Haeckel's classification of organic individuals may seem puzzling and eclectic, but it has an ingenious core: higher animals without complex life cycle successively realize, through a process of multiplication and differentiation, the lower levels of form individuality during their development, while each of these lower levels of form individuality represent a mature physiological individual at successive levels of plant and animal organization. This is, in essence, Haeckel's biogenetic law in its early incarnation.

SPECIES, INDIVIDUALS, HISTORY, AND REALITY

Going from the single cell to animal colonies, Haeckel recognized six orders of individuality, emphasizing all the way that biological individuality is hierarchically organized, subject to the part—whole relation, and anchored in the parenthood relation: biological individuals descend from one another through such processes as growth, division, and differentiation. The question naturally arises whether such a concept of biological individuality could not also be applied to yet higher levels of complexity and inclusiveness, such as species, phyla, indeed the whole of life? It was the renowned botanist Carl Wilhelm von Nägeli, who played a crucial role in the transformation of the concept of a plant or animal species from an abstraction to "a thesis about concrete and ontologically real species, conceived as dynamic holistic entities analogous to organic individuals."[*] Nägeli was born on March 27, 1817, in Kilchberg near Zurich, where he passed through the school and university system. While a student, Nägeli took classes from Alphonse de Candolle (1806–1893) in Geneva. In 1842, Nägeli went to Jena to collaborate with Schleiden, who had just published his foundational *Grundzüge der wissenschaftlichen Botanik*. From 1845 through 1852, Nägeli taught at the University of Zurich. In 1852, he became full professor at the University of Freiburg, Freiburg im Breisgau, Germany, and in 1857 was offered the Chair of Botany at the University in Munich. He died in Munich on May 10, 1891.[†] His major work, the *Mechanisch-physiologische Theorie der Abstammungslehre* of 1884, was widely read and highly influential, famous for its purely mechanistic Principle of Perfection (*Vervollkommnungstrieb*) that Nägeli believed to be inherent in the *idioplasm* of the cell.[‡]

In his pre-Darwinian study of individuality in nature, Nägeli[§] found the "question whether species originated from one another or not of greatest significance, both for theoretical as well as practical reasons ..."[¶] Going beyond Schelling and Spring,

---

[*] Sloan, 2009, p. 84.
[†] Olby, 1974.
[‡] Nägeli, 1884, p. 183.
[§] Nägeli, 1856, p. 203.
[¶] Nägeli, 1856, p. 203.

Nägeli maintained that "the species, just as any other natural entity (*Erscheinung*), can never persist in a state of complete stasis... Therefore, the species is itself an individual, that 'develops' (*sich entwickelt*) through continuous change, that finds a limitation through this change, and in this limitation gives rise to other species; [the species is] an individual, which consists like a tree of numerous past and future generations of partial individuals (*Theilindividuen*)."* The most famous and widely cited insight from Nägeli's post-Darwinian writing on species unquestionably was to anchor the reality and consequent individuality of species in the genealogical descent of species from one another, that is, in their history. Descent implies origin, and origin implies a place and a point in time; thus located in space and time, the species, just as "the genus and higher categories are not abstractions but concrete things. They are complexes of related species (*Formen*) that have a common origin."†

Haeckel‡ cited Nägeli in his discussion of biological individuality, yet the *Jenenser Darwinist* did not think that species exist at all. Looking back on his youth, when at the age of 12 he attempted to create a herbarium, Haeckel proclaimed that at that time already he was fascinated by the problem of the constancy and change of species. Confronted with the variability of the plants he had collected, he asked himself whether he should only keep those specimens which best represent the species, just as the "good" professional systematists tended to do, and throw out the remaining ones, or whether he should admit the whole variety of specimens he had collected into his herbarium, which would have allowed him to reconstruct complete and continuous series of intermediates that bridge the gaps between the *bona species* of systematists. His solution was to create two herbaria, an "official" one that contained only the "typical" specimens, and a secret one, accessible only to his best friends, that contained those specimens which Goethe had aptly epitomized as "characterless and dissolute."§

Turned a Darwinian, Haeckel called the belief in the constancy of species an "absurd dogma,"¶ popular with those "thoughtless systematists"** who blindly follow Linné in their zest to describe and catalog as many species, subspecies, varieties, and so on as can be distinguished. "Although these species manufacturers spend their entire life distinguishing and naming species, the majority of them is completely incapable of telling us what they mean by the term 'species'."†† Having dealt with morphological and physiological individuals in the first volume of his *Generelle Morphologie*, Haeckel turned his attention to genealogical individuals in the second volume, where he again characterized them in terms of a nested hierarchy of increasing complexity. Borrowing from Thomas H. Huxley, Haeckel defined the individual reproductive cycle as a genealogical individual of first order: "all the various differing forms that the egg passes through in its development and differentiation together form a continuous chain and thus belong to a single, temporally variable but materially

---

* Nägeli, 1856, pp. 210–211; see also Temkin, 1959, p. 343.
† Nägeli, 1865, p. 32.
‡ Haeckel, 1866, vol. 1, p. 264; Nägeli is cited alongside Schleiden and de Candolle in this context.
§ Haeckel, 1866, vol. 1, p. xvi.
¶ Haeckel, 1866, vol. 1, p. 170.
** Haeckel, 1866, vol. 1, p. 193.
†† Haeckel, 1866, vol. 2, p. 329.

contiguous unity of form (*Formeinheit*), which comes to an end with the sexual maturity of the developing organism."* Although a genealogical individual of first order, the reproductive cycle is not a closed, but an open system, as it gives rise to descendant reproductive cycles. The same is true of Haeckel's genealogical individual of second order, the species, which is a historically interconnected multitude of reproductive cycles. The species, according to Haeckel, is "the totality of all reproductive cycles, which under the same conditions of life preserve the same form, or forms which at most differ from one another through polymorphism of their life stages (*adelphischer Bionten*)."† Again a historically conditioned material system tied together through the relation of reproduction, the species forms an open system as, on Darwin's theory, it gives rise to daughter species. In his *Generelle Morphologie* of 1866, Haeckel still accepted the four divisions (*embranchements*) of the animal kingdom that Cuvier had recognized, and that von Baer had likewise identified as fundamentally different types of animal organization. The four *embranchements* recognized by Cuvier were the vertebrates, the mollusks, the articulates (insects and crustaceans), and the radiates (echinoderms). Haeckel at that time thought each of these four divisions to have had a separate (spontaneous) origin from a macroscopically undifferentiated clump of protoplasm (a *Monere*), and he introduced the term *Phylon* to refer to these divisions which he defined as genealogical individuals of third order: The phylon "is the sum of all organic species, which have descended from one single autonomous form of *Monere*."‡ Implicit in that definition is Haeckel's concept of monophyly: a group that comprises all and only those species that have descended from a single common ancestor. This concept, which evolved into a core concept of phylogenetic systematics, will require more discussion. At this juncture it will be noted that the concept of monophyly introduces an asymmetry between genealogical individuals of first and second as opposed to those of third order. Again a historically conditioned material system tied together by the relation of descent, the monophyletic phylum—in contrast to the individual reproductive cycle and the species—is a closed system. Organisms give rise to organisms, species give rise to species, but on Haeckel's account, a phylum does not give rise to other phyla: "the species itself is not a closed unit, but a real and completely closed unit is, indeed, *the sum of all species that have gradually evolved from a common ancestral form (Stammform)* as for example the vertebrates. This sum we call a stem (*Stamm*) or *Phylon*,"§ the Ancient Greek term for tribe or nation. A clear tension was starting to build at this point in Haeckel's argumentation: although he recognized species as genealogical individuals of second order, he considered only phyla, that is, genealogical individuals of third order, as *real*, because phyla, in contrast to species, form closed rather than open systems.

This asymmetry of species versus phylum must have left Haeckel dissatisfied. The individual reproductive cycle may be an open system, but at least it is precisely defined: from the first cleavage of the egg cell to the first act of reproduction (in whichever way). Species could only be considered closed systems if their variability

---

* Haeckel, 1866, vol. 2, p. 27. Haeckel discussed the various modes of reproduction in the plant and animal kingdoms to generalize his concept of a genealogical individual of first order.
† Haeckel, 1866, vol. 2, p. 359.
‡ Haeckel, 1866, vol. 2, p. 30.
§ Haeckel, 1866, vol. 1, pp. 28–29.

were ignored.* During his earlier studies of radiolarians, Haeckel had realized the extensive variation that can exist within species, which led him to the conclusion: "the vacillating concept of the species is at bottom nothing but an arbitrary abstraction, and is just as much left to subjective discretion, as are the concepts of the individual, the genus, family, order, etc."† After his adoption of Darwinism, Haeckel saw variation as a signal of species formation: *"Varieties therefore are incipient species."*‡ This conclusion echoed Darwin's own, who had proclaimed: "a well-marked variety may be justly called an incipient species."§ On Darwin's theory, no sharp demarcation of species was possible. Unlike phyla, each representing a distinct type marked out by their special body plan (*Special Bauplan*¶) that had allowed their recognition and definition by Cuvier and von Baer, there are no essential (*wesentliche***), that is, invariable and immutable traits that characterize any one species and distinguish it from all the others. The continuity that prevails in the descent of a new species from an ancestral one, a transformation that takes place in a succession of small steps subject to natural selection, necessarily renders species demarcation arbitrary: "I have myself shown that amongst calcareous sponges one can arbitrarily distinguish three, or 21, or 111, or 289, or 591 species."†† Indeed, in his monograph on *Die Kalkschwämme* of 1872, Haeckel retracted his own species definition he had given in his *Generelle Morphologie* of 1866: "All previous attempts to specify the meaning of the concept of 'species' (*Species-Begriff*) have failed ... my own attempt to capture the meaning [of 'species'] genealogically is as insufficient and untenable as are all other attempts. The reason lies in the very nature of the species. The species is as much an arbitrary *abstraction*, depending on the subjective views of the author, as much a category with a *relative* meaning only as are the varieties, the genus, the family, etc."‡‡ This again echoes Darwin's view of "the term species, as one arbitrarily given for the sake of convenience to a set of individuals closely resembling each other."§§

Such is not the case for the genealogical individuals of third order, "divisions which we called *stems or phyla*, [as] the *only real categories* that we can recognize."¶¶ Each phylum has a discrete origin through spontaneous generation (*Generatio spontanea*), from which point on each phylum remains fully separate from any other throughout its history. Each phylum has a distinct body plan marked out by essential (*wesentliche*) characters that allow to sharply and unequivocally distinguish it from any of the three other phyla. From its point of origin, each phylum diversifies separately, this diversification resulting from the gradual and successive formation and diversification of new species that would group into genera, families, orders, and classes. But within each phylum, these systematic categories, from species on up, are

---

* Haeckel, 1866, vol. 2, p. 392.
† Haeckel, 1862, vol. 1, pp. 232–233; see also Nyhart, 1995, p. 136, n. 811.
‡ Haeckel, 1866, vol. 2, p. 392; see also Haeckel, 1868, p. 223.
§ Darwin, 1859, p. 52.
¶ Haeckel, 1866, vol. 2, p. 288.
** Haeckel, 1868, p. 221.
†† Haeckel, 1873, p. 246.
‡‡ Haeckel, 1872b, p. 477.
§§ Darwin, 1859, p. 52.
¶¶ Haeckel, 1866, vol. 2, p. 393.

all arbitrary and hence artificial constructs, except for the phylum itself. The reason is the unbroken continuity of species diversification, which blurs all sharp boundaries drawn by systematists. What is real is not the systematic categories recognized by systematists, but the phylogenetic tree (*Stammbaum*) that results from species multiplication and diversification. Each phylum has its own and independent phylogenetic tree that traces the continuous blood relationships that connect all the species in the phylum back to their common point of origin.[*] These bloodlines are not accessible to direct observation, but they are revealed by the science of morphology, that is, comparative anatomy and morphogeny, comparative embryology.[†] To doubt the validity of such inference of blood relationship from relative degrees of morphological similarity reveals nothing but a deficiency in philosophical sophistication. The evidence in favor of such inductive inference is overwhelming, as it is supported by an imposing consilience of independent lines of inquiry that come together in a threefold genealogical parallelism[‡]: a parallelism between *phyletic* (paleontological), *ontogenetic* (individual), and *systematic* development, the latter manifest in ascending plant and animal classification. What Haeckel here referred to was the threefold parallelism invoked by the Swiss-born paleontologist Louis Agassiz (1807–1873), founder of the Museum of Comparative Zoology at Harvard University and staunch critic of Darwin.[§] In his widely read *Essay on Classification* of 1857, Agassiz had discussed in creationist terms the phenomenon that the successive attainment of increasing levels of complexity in the course of ontogeny is mirrored both in the Fossil Record and in organismic classification. Haeckel, in contrast, found this threefold parallelism to result from the Darwinian mechanisms of evolution, that is, variation and natural selection. Since general theories of evolution, or species transmutation, had variously been proposed before, Haeckel identified Darwinism not with theories of species origination and transformation, but specifically with the doctrine of natural selection instead: "a mechanistic explanation of species transformation based on two physiological factors: inheritance and adaptation."[¶] Natural selection in turn Haeckel tied to two subordinate laws of nature: the Law of Divergence, which results from the struggle of existence; and the Law of Progression, which the "unsurpassable"[**] Bronn[††] had so thoroughly investigated in his prize-winning monograph.

Haeckel would later, in his paper on the *Gastraea* theory of 1874, reject the "type-theory of CUVIER and BAER, which for half a century to the present day formed the basis of the zoological system,"[‡‡] and introduce "a new system on the basis of phylogeny"[§§] that recognized a common origin of all metazoans. Haeckel sketched the first step in that direction in his monograph on calcareous sponges of 1872,[¶¶]

---

[*] Haeckel, 1866, vol. 2, p. 388.
[†] Haeckel, 1866, vol. 2, p. 374.
[‡] Haeckel, 1866, vol. 2, p. 371.
[§] Hull, 1973, pp. 428ff.
[¶] Haeckel, 1866, vol. 2, pp. 166ff.
[**] Haeckel, 1866, vol. 2, p. 257.
[††] Bronn, 1858b.
[‡‡] Haeckel, 1874a, p. 14.
[§§] Haeckel, 1874a, p. 15.
[¶¶] Haeckel, 1872b, p. 464.

but neither in that publication, nor in his 1874 paper did he elaborate on the consequences of this new arrangement for genealogical individuals of third order.

CHANGING METAPHORS OF ORDER IN NATURE: THE LADDER, THE TREE, AND THE WEB

In his treatment of biological individuals, Haeckel recognized a nested hierarchy
of (morphological) "form individuals," in which each level of increasing inclusiveness simultaneously represents a mature (functional) "physiological individual"
at ascending levels of complexity in the ladder of life. The metaphor behind this
analysis is that of the Great Scale of Being,* one that is still very much alive in
Haeckel's phylogenetic morphology. Yet as early as 1828, in his landmark treatise
Ueber Entwickelungsgeschichte der Thiere (On the Development of Animals), Karl
Ernst von Baer[†] narrated the episode of how he had stored early embryos of a lizard
and a bird species respectively in unlabeled flasks. Going back to his embryo collection at a later point in time, he found himself unable to identify these embryos even
to the class to which they belong.[‡] Von Baer did not take this as evidence for the fact
that birds pass through a lizard stage during their development, as was claimed by
the older doctrine of recapitulation,[§] one that von Baer was in fact attacking. Instead,
he claimed that lizards and birds share a similar early developmental stage, evidence of a rather close affinity of reptiles and birds, but that during their subsequent
development, the lizard and bird embryos would deviate from one another in a process von Baer identified as one of individuation. Historians of biology have dubbed
this von Baerian model of development, one of individuating deviation from common embryonic origins, a process of *differentiation* rather than of *recapitulation*.[¶]
Von Baer found that the dichotomously structured hierarchy of the animal kingdom
that had been worked out by Georges Cuvier and his collaborators was in its most
general structure reflected in the path of embryonic development of animals: "The
development of the embryo relates to the type of organization as if it [the embryo]
passed through the animal kingdom according to the *méthode analytique* of the
French systematists."[**] The basic idea was captured in the first two of von Baer's four
laws of individual development: (1) "the more general condition of form characteristic of a larger animal group [i.e., an *embranchement* in Cuvier's system; a *type* in
von Baer's language] develops prior to the more special condition of form"; and (2)
"from the most general condition of form develops the less general condition of form
and so on, until at last the most special condition of form makes its appearance."[††]

---

* Lovejoy, 1936.
[†] For more on von Baer, see Lenoir, 1982; Gliboff, 2008.
[‡] von Baer, 1828, p. 221.
[§] Known as the Meckel-Serres-Law, named after Johann Friedrich Meckel (1781–1833) and Étienne Serres (1786–1868); see Gould (1977) for more details.
[¶] Richards, 1992; Nyhart, 1995, 2009b. Biologists have instead adopted Løvtrup's (1978) distinction of Haeckelian versus Baerian recapitulation.
[**] von Baer, 1828, p. 225. *Zoologie analytique, ou méthode naturelle de classification des animaux* was the title of a book published in 1805 by Constant Duméril (1774–1860), a close collaborator of Cuvier.
[††] von Baer, 1828, p. 224.

The insight, then, is that organisms of the same type (those being the vertebrates, the mollusks, the articulates [insects and crustaceans], and the radiates [echinoderms]) share closely similar early ontogenetic stages. But the embryos of organisms of different subgroups within a type (e.g., reptiles and birds within vertebrates) successively deviate, or differentiate, from one another during development according to the degree of their affinity. That is to say: while embryos of organisms from the same type all have the earliest developmental stages in common, more closely related organisms within the type will share a common developmental trajectory over a greater stretch of embryonic differentiation than less closely related organisms (on von Baer's reading, "related" means in terms of their form, not in terms of their evolutionary history). For example: bird and lizard will start deviating from one another during their embryonic development at a later stage, one that is arrived at after an earlier stage at which the more distantly related fish has already started to deviate. Embryonic development thus reflects the same dichotomous–hierarchical pattern as do Cuvierian classifications: a successive divergence of embryonic differentiation according to the degree of form relationships of the organisms. The all-important difference between recapitulationist models of development versus the von Baerian model of differentiation is best brought out through their pictorial representation. The graphic rendition of the recapitulationist model will result in a serial arrangement of animal forms, picturing a ladder of life. Von Baer's model of differentiation, in contrast, will be pictured as a dichotomously branching diagram, as was first done by the Scottish physician Martin Barry (1802–1855) in a 1837 monograph on animal development.* Robert Chambers (1802–1871), who had learnt about von Baer from the physiologist and zoologist William Carpenter (1813–1885),† combined in his notorious and widely read *Vestiges of the Natural History of Creation*‡ a von Baerian perspective with a recapitulationist developmental view.§ The von Baerian component he again famously illustrated with a branching diagram.¶ The influence of the von Baerian model of embryogenesis on Darwin's adoption of the Tree of Life metaphor can hardly be overestimated.** But this is also the juncture at which Ernst Haeckel, his phylogenetic trees, and his controversial Biogenetic Law enter the story. Haeckel's choice of the title *Natürliche Schöpfungsgeschichte* for his second book, the one in which he introduced the Biogenetic Law, may indeed have been motivated by Haeckel's friend Carl Vogt's (1817–1895) translation of Chambers *Vestiges* into German under the title *Natürliche Geschichte der Schöpfung.*††

Eager to spread the Darwinian gospel in Germany, Haeckel wanted the science of morphology to become a dynamic one, called *Morphogenie* rather that *Morphologie*, a terminological change that should lead away from the philosophical–transcendental component of idealistic morphology, and emphasize the historical aspect of form generation instead. In that dynamic rendition, *Morphogenie* becomes synonymous

---

* Richards, 1992, p. 109.
† Secord, 1994, p. xvii; see also Secord, 2000, p. 44.
‡ Chambers, 1844. On the genesis and impact of the book, see Secord, 2000.
§ Nyhart, 2009b.
¶ Chambers, 1844, p. 212.
** Ospovat, 1981; Richards, 1992.
†† Richards, 2008, pp. 225–226.

with *Entwickelungsgeschichte*, of which Haeckel recognized two interdependent branches. The *Entwickelungsgeschichte* of individuals, from conception to death, Haeckel termed *Ontogenie* (ontogeny); the *Entwickelungsgeschichte* of animal phyla (the *embranchements* of Cuvier or the *types* of von Baer, but now dynamically–historically conceived) he called *Phylogenie* (phylogeny).[*] His Biogenetic Law was accordingly expressed in the famous slogan *ontogeny recapitulates phylogeny*.[†] This was a crude yet popular rendition of Haeckel's vision, and his drive to disseminate his insights among the populace of German cities earned him the accusation of scientific fraud with respect to the "idealized" or "semi-diagrammatic" illustrations of embryos he used in support of his Biogenetic Law.[‡] Richard Hertwig, for example, reminisced how Haeckel had used the embryo of a tailed monkey to illustrate the embryo of a Gibbon, and did so by simply shortening the tail in its pictorial rendition: "Haeckel saw in the particular object always only the instantiation of the typical ... his illustrations were too stylized ... and yet, Haeckel in no way intended to deceive his readers. He only wanted to express what for him was clear beyond doubt."[§]

Haeckel introduced the term *Biologisches Grundgesetz* (biogenetic law) in the second edition of his *Natürliche Schöpfungsgeschichte* of 1870, a sequel to his *Generelle Morphologie* written for a broader, more general readership, the first edition of which appeared in 1868.[¶] He did, however, discuss the relation of ontogeny to phylogeny in his *magnum opus* of 1866, opening the relevant chapter with a quote from von Baer's 1828 monograph: "The *Entwickelungsgeschichte* is the true light beam that illuminates the investigation of organic bodies."[**] In that same context, von Baer had also emphasized that "the current understanding of the law of development is necessarily linked with the concept of an ascending scale of animal organization. However, the fish is not just an imperfect vertebrate, it also has the specific character of a fish, as its *Entwickelungsgeschichte* unquestionably demonstrates." This is the concept of von Baerian differentiation, where the fish develops its distinctive character starting out from an early ontogenetic stage that is common to all vertebrates. To this, Haeckel hastened to add: "Obviously one can reach a complete understanding of these important relations, and a correct evaluation of their supreme significance only from the perspective of the [Darwinian] theory of descent." The reason is that "*[t]he ontogeny or Entwickelungsgeschichte of physiological individuals* is most

---

[*] Haeckel, 1866, vol. 1, p. 30.

[†] The Haeckelian slogan was famously reversed by the British marine biologist Walter Garstang (1868–1949): "ontogeny does not recapitulate phylogeny: it creates it" (Garstang, 1922, p. 81); see also Hall, 2000.

[‡] The debate, which started with Ludwig Rütimeyer (1825–1895) from Basel, is narrated and analyzed in detail in Richards (2008, pp. 278ff) and Hopwood (2015). Rütimeyer was professor of anatomy first in Berne, then at the University of Basel. He also served as director of the Natural History Museum in Basel, where he pursued his research on fossil turtles and mammals of Switzerland. Haeckel (1877b, p. xxii) singled out Rütimeyer as his most formidable opponent, and retorted that on his and others' accounts, all semi-diagrammatic illustrations as they are widely used in textbooks and teaching would be fraudulent: "All schematic representations are, as such, 'invented' " (Haeckel, 1877b, p. xxv; see also Haeckel, 1875b, p. 36).

[§] R. Hertwig, 1919, pp. 953, 954.

[¶] Haeckel, 1870a, pp. 361–362; see also Richards, 2008, p. 148, n. 98.

[**] von Baer, 1828, vol. 1, p. 231.

intimately tied to the *phylogeny or Entwickelungsgeschichte of the genealogical tribes (Phyla)."* *"Only the theory of descent can explain the ontogeny of organisms"*: importantly, the explanation called for in that context must be a causal one, rooted in the laws of nature rather than in a *philosophication* of animal form. Haeckel's well-known general conclusion was: *"Ontogeny is nothing but an abbreviated recapitulation of phylogeny."** It is evident that Haeckel's Biogenetic Law was derived from Agassiz's threefold parallelism. But there was a glitch that Haeckel was only too painfully aware of. It was Fritz Müller's "test" of Darwin's evolutionary theory, or rather his application of Darwinian principles to crustaceans observed and collected on the south coast of Brazil.

Fritz (Johann Friedrich Theodor) Müller[†] was born on March 31, 1822, in a small village near Erfurt in Thuringia, Germany. Having passed through the *Gymnasium* in Erfurt, Müller first became an apprentice in a pharmacy in Naumburg, but then turned to the study of mathematics and natural sciences at the Universities of Berlin and Greifswald. While in Berlin, he took classes in anatomy from the famous Johannes Müller. He then turned to the study of medicine, but was refused admittance to the Board Examinations (*Staats Examen*) for political and religious reasons, or more precisely, the *Freidenker* (freethinker) Müller refused to submit to a religious oath required for admission to the medical profession. As political and religious intolerance hardened in Prussia, Müller decided to leave Germany for Blumenau, a German settlement on the Itajai River in southeastern Brazil, where he supported his family as a farmer. He went on to pursue a checkered career in Brazil, with—among other activities—transitory employment as a mathematics teacher at the Lyceum in Desterro (now Florianópolis) on Santa Catarina Island, and as a field naturalist for the National Museum in Rio de Janeiro. Müller died impoverished in Blumenau on May 21, 1879. His legacy as a scientist comprises publications in botany, invertebrate zoology, and entomology. He earned a place in the history of biology through his discovery of Müllerian mimicry in butterflies. Another milestone in his career was the publication of his classic monograph *Für Darwin* in 1864,[‡] a book that Haeckel warmly recommended to Darwin,[§] an episode that triggered a lively correspondence between Darwin and Müller. Darwin would mediate the publication of communications concerning botanical or entomological observations he received from Müller in the science magazine *Nature* in London. He was impressed with Müller's 1864 monograph to the degree that he arranged for, and financed, its publication in English; it appeared in 1869 under the title *Facts and Arguments for Darwin*. Müller's key publication of 1864 also greatly helped to promote Darwinism in Germany, and brought Müller into contact with Ernst Haeckel in Jena, with whom he shared important ideological commitments (Figure 2.1).

It was during his years in Desterro that Müller began to link his studies of crustaceans to Darwin's theory of evolution. On the opening page of his *Origin*,

---

[*] Haeckel, 1866, vol. 2, p. 7.
[†] On Fritz Müller, see Blandford, 1897; Heβ, 1906; Möller, 1915, 1920, 1921; McKinney, 1974; Richards, 2008.
[‡] F. Müller, 1864.
[§] Richards, 2008, p. 100, n. 67.

FIGURE 2.1   Fritz Müller. (Smithsonian Institution Archives. Image 85-4522.)

Darwin had promised "to throw some light on the origin of the species—that mystery of mysteries, as it has been called by one of our greatest philosophers."[*] This philosopher was the astronomer Sir John Herschel (1792—1871), who in a letter to Charles Lyell of 1836 applied the phrase "mystery of mysteries" to the regular disappearance of species from the Fossil Record, followed by the reappearance of different species, a pattern of faunal succession that had been ascertained by Lyell in his classic three volume textbook on the *Principles of Geology*.[†] The proclaimed goal of Müller's 1864 study was the same, namely, to investigate the origin of species through a detailed application of Darwin's theory to a particular group of organisms, that is, the crustaceans easily accessible from his base on Santa Catarina Island. What Müller came up with, however, was less a study in variation and natural selection, but rather a study

---

[*] Darwin, 1859, p. 1.
[†] Cannon, 1961.

in classification and recapitulation. As recognized by Blandford* in his obituary for Müller, the project was to build up a *Stammbaum* (a phylogenetic tree in Haeckel's later vocabulary) for a group of animals even in the absence of a Fossil Record, and then to see how well the pattern of character transformation revealed by that tree would match the observed facts of development.

As far as the first step was concerned, Müller emphasized that any classification built on the theory of descent had to be a coherent one, free of internal contradictions. Since there could have been only one unique historical process of evolution, which at the same time was not amenable to direct observation, the *Stammbaum*[†] would have to carry the warrant of truth within itself in terms of a coherence of the established facts; the degree of coherence would furthermore have to increase with the degree of completeness to which all the described species of the group under study have been included. Müller illustrated the method he had in mind with reference to three clusters of species of amphipods in the genus *Melita*, which he assumed to share a common ancestry: *Melita palmata* and its closest relatives, *Melita exilii* and its closest relatives, and *Melita fresnelii*. Among those, *Fresnelii* and the *exilii* group he found to be united by a rather striking characteristic, the asymmetrical unilateral differentiation of a massive claw on only one of the two in the second pair of legs—a specialization that appeared to be unique among amphipods. Conversely, however, the *palmata* and *exilii* groups appeared to be united by a reduction in the differentiation of the anterior antennae. There existed, therefore, among these three species groups, a conflicting character distribution. Confronted with such incongruence, Müller called for a carful character analysis across a wider spectrum of amphipods, which led him to the conclusion that the differentiation of the anterior antennae is rather variable throughout the group, in contrast to the unique differentiation of the claw in *Fresnelii* and the *exilii* group. There could in his view be no doubt that *Fresnelii* and the *exilii* group therefore share a common ancestor that they not also share with the *palmata* group. Other systematists at the time, when confronted with a conflicting character distribution within a group of organisms under study, would resort to the reconstruction of reticulated affinities of form which, as Darwin had clearly recognized, would not lend themselves easily to an explanation in terms of common ancestry, although hybridization would always remain a possibility of course.[‡] Müller instead advocated a careful assessment of the evolutionary significance of characters for classification not on the basis of their preconceived physiological importance, as Cuvier had done, but in terms of the regularity of their distribution across the group under examination. The apparent occurrence of character conflict could thus no longer be used to reject Darwin's claim to use comparative anatomy and embryology in search for common ancestry, which, once established, could guide the investigator in the reconstruction of the ancestral form. Only when such an ancestral form has been reconstructed or recognized, and

---

* Blandford, 1897, p. 546.
† Writing before Haeckel (1866), Müller (1864, p. 2) did not himself use the terms "phylogeny" and "phylogenetic tree."
‡ Stevens, 1994.

compared to its descendants, has a basis been established for a comparison of the *Stammbaum* to the embryonic development of the species involved.[*]

Turning to this second step in his project, Müller opened his discussion with reference to his former teacher, Johannes Müller in Berlin, who had rejected recapitulationist models of development and defended the von Baerian model of differentiation instead: similarity of early embryonic stages that would be masked as a consequence of divergent differentiation during subsequent development. Darwin had expressed similar thoughts in his *Origin*[†]: "The embryo is the animal in its less modified state ... community of embryonic structure reveals community of descent[‡] ... Even the illustrious Cuvier did not perceive that a barnacle was, as it most certainly is, a crustacean; but a glance at the larva shows this to be the case in an unmistakable manner."[§] Müller appears to have initially been skeptical toward Darwin's evolutionary theory,[¶] but was soon turned around by an observation of his own: if the great variety of crustaceans should, indeed, share a common ancestry, they should all share an early larval (developmental) stage known as *nauplius*. But such a larval stage had not previously been described for the largest section of crustaceans, the malacostracans. On January 24, 1862, a swarm of larval organisms was captured in the ocean near Desterro, among which Müller recognized a single *nauplius* larva.[**] Using additional specimens from the same sample, Müller reconstructed a series of developmental stages that linked the *nauplius* to the prawn genus *Panaeus*, a malacostracan. His study was received as carefully conducted and convincing, yet not beyond doubt since he did not actually rear the adult from the *nauplius* larva. Although he had been able to nurse his *nauplius* along to a somewhat later stage, the remaining gaps of the life cycle of *Paenaeus* were closed by rearing experiments only some 20 years later.[††] But it was good enough for Müller to embark on a comparative developmental study of crustaceans in support of Darwin's theory, an effort that culminated in his 1864 monograph. In summarizing his bewildering findings of his broad survey, he found profound differences in developmental stages of closely related forms as much as closely corresponding stages in forms that differ widely as adults. This may not have been what he had bargained for, but it again chimed with what Darwin had cautioned with respect to comparative embryology:

---

[*] F. Müller's call—expressed in modern terms—for congruence of character distribution in the context of dense taxon sampling, the exemplification of his method in terms of a three taxon statement, and his method of character analysis in view of a conflicting character distribution, can in retrospect be interpreted as an early rendition of the theory and methods of modern phylogenetic systematics (cladistics). Craw (1992) did, indeed, call Fritz Müller a pioneer of cladistics, a claim that not only neglects 100 years of history, but also the fact that as important as Müller's essay was in the 1860s, it had no historical connection to the rise of cladistics (see Hull, 1988, p. 376). Important in its proper historical context, however, is Müller's (1864) insistence on the analysis of multiple characters and the search for congruence among them in order to obtain a phylogeny free of internal contradictions.

[†] For the significance of the influence of von Baer's model of development on Darwin, see Ospovat, 1981; Richards, 1992.

[‡] Darwin, 1859, p. 449.

[§] Darwin, 1859, p. 440.

[¶] Blandford, 1897, p. 546; Winsor, 1978, p. 149.

[**] F. Müller, 1863. For a detailed account of the wider implications of this discovery, see Winsor, 1976, pp. 149ff.

[††] Brooks, 1883, 1887.

"The case is different when an animal during any part of its embryonic career is active, and has to provide for itself. The period of activity may come on earlier or later in life, but whenever it comes on, the adaptation of the larva to its condition of life"* breaks the parallelism that might otherwise prevail between classification and embryonic development. Was crustacean diversity a case where "God writes straight in crooked lines," as a Portuguese proverb said? "Perhaps a keener eye than mine," said Müller, "can recognize, with Agassiz, in this diversity and variation of animal form 'a plan [of Creation] fully matured in the beginning and undeviatingly pursued' "—a quote he took from a textbook on the *Principles of Zoology* by Louis Agassiz and Augustus A. Gould (1805–1866). In accordance with his youthful rejection of religion, Müller concluded "spectacles of faith rarely fit eyes that are trained to look through the microscope."† Small wonder Müller eventually got into trouble with Jesuit leadership in Desterro, which resulted in the termination of his employment at the *Lyceum* there in 1876.

Müller wanted to use his study of crustacean development as a critique of Agassiz's threefold parallelism, which in the latter's hand was steeped in what Haeckel called "bizarre theological-theosophical speculations"‡ and enriched with Cuvierian values. Agassiz adopted von Baer's model of a parallelism between a hierarchically structured classification and successive ontogenetic differentiation, to which he added the parallelism of a third succession, that of the Fossil Record. This threefold parallelism he further enriched with Cuvier's functionalism, which resulted in the claim that the more important an organ or organ system is in terms of its physiological function, the more inclusive would be the systematic group which it demarcates, the earlier would it appear during embryonic development, and the earlier would it appear in the stratified Fossil Record.

Müller raised three principal objections against Agassiz: the provability of those assertions on an empirical basis (e.g., that the functionally most important organs appear first both in the embryo and in the Fossil Record); the many exceptions that are manifest in nature and exemplified by his crustacean studies; and the fact that a natural system must be based not on a few characters selected *a priori* according to their putative physiological importance, but on as many characters as possible. While conflicting character distribution may mask the Plan of Creation for those searching for it, Müller took character incongruence as a fact of nature that should not be ignored, but instead should be properly analyzed within the framework of a dichotomously structured argumentation scheme. For his key publication Müller had chosen crustaceans that are characterized by a complex life cycle: different larval stages display transient adaptations to their environment and thus deviate from a recapitulationist model of development. Müller's major conclusion was that the signal of evolutionary history not only becomes progressively erased in embryonic development, but also distorted through the special adaptations of larval stages.§ While Darwin expressed much enthusiasm about Müller's masterpiece, Haeckel greeted

---

* Darwin, 1859, p. 440.
† F. Müller, 1864, p. 65.
‡ Haeckel, 1866, vol. 2, p. 371.
§ F. Müller, 1864, p. 77.

the monograph as an "unsurpassed," "exemplary," and "masterful" study,[*] the conclusions of which he was quick to assimilate. Haeckel consequently introduced the distinction of *Palingenie* (*Auszugsgeschichte*) from *Cenogenie* (*Störungsgeschichte*) in his discussion of ontogenetic recapitulation.[†] Palingenesis is the abbreviated, "short recapitulation of phylogeny" in ontogeny,[‡] which alone can offer clues for the reconstruction of the ancestral condition of form; caenogenesis refers to the intercalation, in ontogeny, of specially adapted stages, especially larval stages in organisms with a complex life cycle. Haeckel first introduced the core concept of his *Palingenie* in terms of a Law of the Abbreviated or Simplified Development: "*The chain of inherited characters, which develop during ontogeny successively and in a particular order, becomes shorter over time, as certain links of that chain are deleted.*"[§] The concept of palingenesis nicely documents how the image of the *scala naturae*, the Great Chain of Being, continued to underlie Haeckel's phylogenetic trees—a consequence of his progressionism.

MONOPHYLY: THE EVOLUTION OF SPECIES AND LANGUAGES

Haeckel's goal was to transform idealistic morphology into a phylogenetic morphology based on a materialistic–mechanistic worldview: the type is not a Divine idea instantiated in nature, but a historically conditioned complex whole. With his *Generelle Morphologie*, Haeckel defined the agenda for phylogenetic research in German biology for decades to come. And yet, as is indeed well known, Haeckel was quite unable to obliterate all traces of German Idealism in his work. His progressionism, his monism that bordered on pantheism *cum* panpsychism, his monadology underpinning his discussion of spontaneous generation, his crystallography of organisms (*Promorphologie*) that looked back on Heinrich G. Bronn[¶] and Carl G. Carus (1789–1869),[**] his relentless appeal to laws of nature that unify the phenomena in an evolving universe—even the roots of his threefold parallelism, can be traced back to Schelling, Goethe, and others.[††] But in his youthful enthusiasm, and in spite of his plea for a philosophical underpinning of natural science, Haeckel rode roughshod over issues entirely internal to his phylogenetic morphology. The asymmetry between species and phyla was left unresolved. The phylum, the only category that can be accepted as really existing in nature, is a genealogically conditioned complex whole (genealogical individual of third order), and yet is marked out by essential characteristics universally and invariably instantiated in all its parts. This would later be recognized as a category mistake in the ensuing discussion of Haeckel's legacy among (neo-)idealistic, systematic, and phylogenetic morphologists. But no other concept introduced by Haeckel with an iridescent multiplicity of meaning might have created as much confusion as monophyly.

---

[*] Haeckel, 1866, p. 185, n. 1.
[†] Haeckel, 1875c, p. 409.
[‡] Haeckel, 1866, p. 7.
[§] Haeckel, 1866, p. 184; emphasis in the original.
[¶] Bronn, 1858a.
[**] Richards, 2008, pp. 470ff.
[††] Richards, 2002, 2008.

The term first appears in the figure caption that refers to Plate 1 in the second volume of his *Generelle Morphologie*: "Of the many possible hypotheses that can be construed about the number and connections of the organic phyla (*Stämme*) [the third one here proposed] is the monophyletic hypothesis of a unified origin of all organisms."[*] Haeckel was among the first authors to draw up phylogenetic trees. Heinrich G. Bronn had earlier presented a tree-like diagram in his monograph on the *Laws of Development* (*Entwicklungsgesetze*) *of the Organic World* of 1858,[†] but this tree was not based on species transmutation or origination. Instead, it was meant to depict the hierarchical structure of the Natural System and the corresponding parallelism of the appearance in the Fossil Record of progressively more complex organisms and the systematic categories they represent.[‡] In contrast, Haeckel's trees were explicitly built on species origination, transformation, and diversification: "a mixture of fancy, sentimental romanticism, unintentional inexactitude and sincere trials to interpret certain facts."[§] Much has been written on the origin of Haeckel's phylogenetic trees,[¶] but whether or not he was inspired by Darwin's highly stylized branching diagram in chapter four of the *Origin*, derived from his famous sketch titled "I think" in his "Transmutation Notebook B" of July 1837,[**] Haeckel's trees are rendered much more succinctly in a naturalistic representation. It is well known that Haeckel recommended to his friend, the *Jenenser* philologist August Schleicher (1821–1868),[††] Bronn's German translation of Darwin's *Origin* published in 1860, as Haeckel thought that the passionate gardener that his friend was might find it interesting.[‡‡] Schleicher surprised his friend with an open letter he published in 1863,[§§] where he acknowledged his experience of the benefits of artificial selection when weeding in his garden. He also respected the power of the struggle for existence, as when one plant would expand its territorial holdings at the expense of others—sometimes much to the chagrin of the gardener. But that, he said, is not what surprised him most in Darwin's book. Instead, it was the close parallels Schleicher perceived between the thoughts Darwin expressed on species transmutation, and his own research on language evolution. Schleicher welcomed the stringency of Darwin's argumentation, and emphasized that following the example set by Schleiden in his *Grundzüge der Botanik* he had aspired to the same stringent scientific standards in linguistics. Already in his 1860 book on the German language Schleicher professed to have used a vocabulary very similar

---

[*] Haeckel, 1866b, vol. 2, p. 417, n. 1.
[†] Bronn, 1858b, p. 481; see also Richards, 2008, pp. 159, 478.
[‡] Gliboff, 2008, pp. 80–83.
[§] Lam, 1936, p. 155.
[¶] Gould, 1989; Richards, 1992.
[**] Darwin, Notebook B, p. 36, Cambridge University Library, #DAR121.
[††] Schleicher was born on February 19, 1821 in Meiningen, Saxony. He studied theology and oriental languages at the University of Leipzig, from which he was expelled in 1841 following his participation in student revolts. He moved on to the Universities of Tübingen (study of Sanskrit, Hebrew, and Persian) and Bonn, where he obtained his doctorate in 1846. After a detour through Prague he arrived in Jena in 1857, where by the time Haeckel arrived he had established himself as a distinguished linguist and respected gardener (Di Gregorio, 2005, p. 99).
[‡‡] Bouquet, 1996, p. 58; Richards, 2008, p. 125, n. 34.
[§§] Schleicher, 1863.

to Darwin's. Schleicher talked about "language organisms"* (*Sprachorganismen*) and their origins, their transformation and differentiation, their expansion, and their extinction. He talked about "proto-languages" (source languages, *Ursprachen*) and used branching tree diagrams to depict the derivation of families of related languages therefrom.[†] Languages have an origin, diversify, and go extinct—a cycle that not only parallels the life of an organism, but according to Haeckel also characterizes the life of a species[‡] as well as the stages of phyletic development.[§] This parallelism, again a rudiment of German Idealism, is also apparent in linguistics according to Schleicher. In his open letter to Haeckel, Schleicher emphasized the close analogies between his linguistic concepts and the evolutionary concepts developed by Darwin, yet cautioned that in spite of all these striking parallels, the kingdom of languages is too different from the plant and animal kingdoms for it to be possible to apply all of Darwin's principles to linguistics. There is, nevertheless, a striking similarity of the branching tree diagrams published by Bronn in 1858, and Schleicher in 1860, both sources that have influenced Haeckel.

> The real relations which connect all living and extinct organisms with the main categories of the [natural] system are genealogical in nature; their form-relationship is blood-relationship; the natural system is consequently the phylogenetic tree (*Stammbaum*) of the organisms, or their *Genealogema*.[¶]

In this passage, Haeckel boldly fills epistemological concepts with ontological content, without any further analysis. Degrees of form relationship as revealed by comparative morphology are equated with degrees of blood relationships; a nested system of classes marked out by shared properties is equated with chunks of a genealogical continuum represented by the phylogenetic tree. The motivation for Haeckel to do so derived from the principle of continuity (his *Continuitäts Theorie*) that governs Darwinian evolution. As many historians would do after him, Haeckel recognized the major relevance this Aristotelian principle acquired in Charles Lyell's uniformitarian geology, "from which followed with necessity the theory of descent as it was eventually fully fleshed out by Darwin."[**] If continuity prevails in nature, there cannot be sharp boundaries between species, genera, families, and classes. There can only be sharp boundaries between the four phyla, each with its separate origin. A system that breaks the genealogical continuum into a hierarchy of sharply demarcated groups of greater or lesser inclusiveness such as species, genera, families, and so on must necessarily be an artificial system. The natural system instead must correspond to the phylogenetic tree, where the passage from a more internal, thicker branch (genus) to a more peripheral, thinner twig (species) is as gradual and continuous as was the growth of the phylogenetic tree itself through the process of descent with modification.[††] The species are the distal twigs or leaves, connected at

---

* Schleicher, 1860, p. 44.
† Schleicher, 1860, pp. 81, 94.
‡ Haeckel, 1866, vol. 2, p. 361.
§ Haeckel, 1866, vol. 2, p. 366.
¶ Haeckel, 1866, vol. 2, p. 419.
** Haeckel, 1866, vol. 2, p. 314.
†† Haeckel, 1866, vol. 2, p. 397.

the bottom of the trunk to their common ancestor, the *Urform*. Monophyletic is a system, a phylogenetic tree, if it comprises all the species and only those that derive from a common *Urform*.

In spite of his near excessive penchant for conceptual analysis, definition, and classification, which resulted in a plethora of neologisms such as "phylum" and "monophyletic," Haeckel himself never properly defined his concept of monophyly. Using a contextual analysis throughout his *oeuvre* to illuminate its meaning reveals the untamed proliferation of phylogenetic trees to be matched by an equally vacillating and diverse use of the concept of monophyly.[*] Although Haeckel first used the term "monophyletic" in the second volume of his *Generelle Morphologie*, the concept was already implied in his 1863 presentation on Darwin's theory. Darwin's conclusion that "life, with its several powers, having been originally breathed into a few forms or into one,"[†] Haeckel rendered in his *résumé* in a picture that shows all extant and extinct species of plants and animals to have gradually evolved "from a few, perhaps even from *one single stem-form*, a most simple anorganism [proto-organism]..."[‡] Recognizing the extensive variation discovered in his early systematic studies of radiolarians that precluded any sharp species delimitation, Haeckel confronted the possibility that "the great variety [of radiolarians] *could be* derived from such a common primitive form,"[§] but hastened to add a cautionary note that echoed Bronn's review[¶] of Darwin's *Origin*: "The biggest omission in *Darwin's* theory is that it does not provide any clues as to how a primitive organism [*Urorganismus*]—most probably a simple cell—could have originated from which all other forms of life could have gradually emerged."[**] Again, monophyly is not spelled out, but nevertheless implied. From these early concerns onward, Haeckel used monophyly to denote the origin of species, organisms, or organ systems from a unique ancestral source, where this *Urform* could be a macroscopically undifferentiated lump of protoplasm, another organism, another species or a higher taxonomic grouping, a phylotypic stage of development, or another organ system. Haeckel even moved on to apply the concept of monophyly to characterize the views of his linguist friend August Schleicher: "the common descent of all these languages from a common primitive language (*Ursprache*) ... this is the consonant monophyletic view of all eminent linguists."[††]

In his *Generelle Morphologie* Haeckel stipulated the separate origin of the phyla from *Moneren*, morphologically but perhaps not chemically undifferentiated lumps of protoplasm which themselves originated spontaneously.[‡‡] Moving away from the Cuvierian—von Baerian types and toward his *Gastraea* theory, Haeckel specified in the preface to the fourth edition of his *Natürliche Schöpfungsgeschichte* that

---

[*] For a more detailed account, see Rieppel, 2011d.

[†] Darwin, 1859, p. 490.

[‡] Haeckel, 1864, p. 17.

[§] Haeckel, 1862, p. 233.

[¶] "Make organic matter with cellular structure from inorganic matter ... Darwin's theory remains all the more implausible as it does not come closer to a solution of the major problem of Creation" (Bronn, 1860, pp. 115–116).

[**] Haeckel, 1862, p. 231.

[††] Haeckel, 1877b, pp. 349–350.

[‡‡] Haeckel, 1866, vol. 1, p. 202.

"the *Gastrula* has been recognized as the common stem-form, from which all animal stems [phyla] can be derived without any difficulties (except for the lowermost group of primitive animals [i.e., the protists])."* Here, the common root of ancestry that ties metazoans together through their bloodlines is a particular phylotypic stage of embryonic development, rendered as a mature physiological individual to serve its ancestral role.† As Haeckel explained in the first volume of his 1872 monograph on calcareous sponges, it was the fact that all the metazoans investigated pass through a gastrula stage that led him to abandon the Cuvierian—von Baerian types: "I conclude on grounds of the biogenetic law to the common descent of all animal phyla from a single unknown ancestral form, which essentially resembled the gastrula: Gastraea."‡ "*Gastrula*" and "*Gastraea*" are both neologisms, introduced by Haeckel in his 1872 monograph, where, with respect to his earlier classification, the gastrula would represent a morphological individual of second order, the *Gastraea* the corresponding mature physiological individual which functions as ancestor of all metazoans. As he fleshed out his *Gastraea* theory in greater detail, Haeckel specifically spelled out the monophyly of metazoans, while he still allowed protozoans to have had multiple origins through spontaneous generation.§

In the third edition of his *Anthropogenie* of 1877, Haeckel tied the monophyly of vertebrates to an ancestral species instead: "If the theory of descent is at all true, then all vertebrates including humans can only be descended from a common stem-form (*Stammform*), from a single primitive species of vertebrate [*Urwirbelthier-Art*]."¶ But a hundred pages on, he derived the monophyletic mammals from a more inclusive taxonomic group, rather than from a species: "... all mammals including humans are to be derived from a single primitive mammalian stem-form [*Säugethier-Stammform*] ... the *Promammalia*."** As if that was not enough, Haeckel added to the confusion by anchoring the monophyly of mammals, a group that is to include humans, in the "monophyletic origin of their skeleton": "all the elements" that characterize the human skeleton "are also found in the skeleton of other mammals, subject to different shape, but in corresponding position and relation."†† The same equivocality in the use of the concept of monophyly is apparent throughout Haeckel's later work on radiolarians, where the stem-form was identified with a species, then with a more inclusive taxonomic group, further with a structural body plan, and where monophyly of a systematic group could be decoupled from the monophyly or polyphyly of structural systems (skeletal and plasmatic features) characterizing the group.‡‡

In his *Generelle Morphologie* Haeckel accorded reality and unity only to the phylum in virtue of its discrete monophyletic origin: only the phylum is a closed system,

---

* Haeckel, 1873, p. xlii.
† For more analysis, see Dayrat, 2003; Breidbach, 2003.
‡ Haeckel, 1872b, p. 467. For more analysis of Haeckel's *Gastraea* theory, and its controversial reception, see Nyhart, 1995.
§ Haeckel, 1874a, p. 11.
¶ Haeckel, 1877b, pp. 398–399; rendered with emphasis in the original.
** Haeckel, 1877b, pp. 489–490; rendered with emphasis in the original.
†† Haeckel, 1877b, p. 597.
‡‡ Haeckel, 1887, 1888; for more detail, see Rieppel, 2011d.

a genealogical individual of third order."It is the material bond of blood relationship, which ties together and wraps around all members [*Glieder*] of a phylum."* This raises the question whether Haeckel would accord unity and reality to all monophyletic groups, once he had abandoned the Cuvierian—von Baerian types? In that respect again his writings over the years reveal ambiguity. In his early review of Darwinian theory, Haeckel had characterized it as one according to which "all organisms which live today and which have ever lived in the past together form *a unified large whole*, a single, very old and much ramified tree of life ... the entire natural system is a single, large, organically arranged body."† Thirty years later he returned to the question of unity of the organic world: "Comparative anatomy and ontogeny of organisms, in coherence with comparative physiology and psychology, lead us to the monistic conviction of a complete unity of the organic world."‡ But was such anatomical and physiological unity also genealogically conditioned? Haeckel specified with respect to "protophytes" or *Urpflanzen* that the assumption of their polyphyletic origin need not contradict their morphological unity.§ With respect to humans, Haeckel drew similar distinctions: "the assumption of a *monophyletic* origin of all humans proves nothing with respect to the unity of the human species ... unquestionably we must distinguish *several* species of humans as we examine the morphological differences of the so-called 'human races' and weigh them impartially."¶ As with monophyly, so with the concept of unity, Haeckel continuously moved between categories without concern for conceptual ambiguity, his use of concepts freely crosscutting morphological and physiological, ontogenetic, and phylogenetic categories. A genealogically monophyletic group could have organ systems that are of polyphyletic origin, just as a group of polyphyletic origin could be subject to morphological unity.

Starting with his *Generelle Morphologie*, Haeckel opened the floodgates to phylogenetic speculation in German comparative biology. He derided systematic research that was restricted to species level taxonomy, and let his phantasy free reign when drawing up alternative and ever-changing phylogenies that ranged from protists to humans; he drew freely on morphological, embryological, and physiological evidence in their support, whatever served his purposes best. In spite of his zeal for conceptual as well as terminological clarification, his later critics located in his work a lack of methodological discipline, a carefree use of evidence, and conceptual confusion all in defense of his increasingly ideologically tainted Darwinism.

REACHING OUT BEYOND JENA

Among the many international students Haeckel attracted to Jena was Eduard Strasburger, born in Warsaw on February 1, 1844. After two years of studies at the Sorbonne in Paris, he turned to Bonn from where he moved on to Jena to complete his graduate studies under the botanist Nathanael Pringsheim (1824–1894), earning his PhD in 1866. The following year he earned the *venia legendi* at the University of Warsaw. In 1869, age 25,

---

* Haeckel, 1866, vol. 2, p. 393.
† Haeckel, 1864, p. 28.
‡ Haeckel, 1894, p. 88.
§ Haeckel, 1894, p. 94.
¶ Haeckel, 1895, p. 653.

he was appointed successor to Pringsheim at the University of Jena, but at the rank of associate professor only. When he received the offer of a full professorship in botany at the University of Lemberg (now Lviv in the Ukraine) three years later, Haeckel successfully pressured the administration to promote Strasburger to full professor to keep him in Jena. In 1881 he moved from Jena to Bonn, where he had accepted the offer of a full professorship at the Botanical Institute located in the Poppelsdorfer Schloß. Strasburger passed on recruiting efforts by the Universities of Tübingen and Munich, and died in Bonn on May 18, 1912. During his extremely productive years in Jena, Strasburger taught in the same building as Ernst Haeckel, with whom he maintained a lifelong friendship. Much later in life he once thanked Haeckel in a letter for the inspiration and advice he had received from him as a student, and for Haeckel's support and friendship when he had become his colleague.[*]

On the occasion of his promotion to full professor in 1873, Strasburger gave a public inaugural lecture that explored the phylogenetic method for the investigation of living beings. Originally skeptical of phylogenetic morphology, Strasburger by that time had "become a complete convert": his speech "could have been dictated by Haeckel— only the examples came from botany rather than zoology."[†] Or perhaps not quite so. Strasburger characterized the *Jenenser naturwissenschaftliche Schule* as firmly rooted in the phylogenetic approach. Since phylogeny was not directly observable, indirect methods had to be used, as is also the case in other historical sciences such as linguistics.[‡] Not only Haeckel, but August Schleicher might have loomed in the back of the auditorium. And of course: if phylogeny stretched through thousands of years, its course is nevertheless revealed in ontogeny, which unfolds "under our eyes." He went on to urge a phylogenetic perspective in physiological research as well, but then, with respect to morphology, he specified: "The phylogenetic methods we use remain the same, as far as the *modus procendi* is concerned, as our earlier ones: we still operate in the same ways, which become new only through the [phylogenetic] background that we ascribe to them."[§] In Strasburger's speech, the Gegenbaur transformation had thus come full circle. A few years later, Oscar Hertwig—another former Haeckel *cum* Gegenbaur student—put his finger on this sore point: hypothesis rather than observation takes center stage in such a conception of morphology, as all genealogical inferences must necessarily remain hypothetical. Homology, however, is first and foremost a morphological relation, rooted in part at least in observation, and only secondarily underpinned with phylogenetic meaning. Hertwig characterized Strasburger's expositions in terms of a quote he took from the botanist Alexander Braun. Commenting on Strasburger's phylogenetic morphology, Braun found it understandable that, once the natural system had been recognized as expressing blood relationships,

> the investigation of phylogenetic relationships became of greatest importance as one
> sought a new and firmer basis for an understanding of organic form, just as it had earlier
> been thought, and some still think so, that the same goal could be achieved through

---

[*] Mägdefrau, 1973, p. 155; see also http://www.botanik.uni-bonn.de/strasburger.html, accessed July 16, 2012.

[†] Nyhart, 1995, p. 163.

[‡] Strasburger, 1874, p. 57.

[§] Strasburger, 1874, p. 60.

the investigation of the individual development (the ontological [*sic*] development) of single organisms. But the significance of this point of view has been exaggerated as one expected from the theory of descent more than it can deliver; one thought to have found a new method, when indeed there is only an interpretation of results obtained with the previous method from an enhanced point of view.* ... Therefore, if STRASBURGER claims that morphological questions can *exclusively* be decided from a phylogenetic perspective, he seeks the decision in the realm of the unknown that is rendered accessible only through the clarification of morphological issues. It is not descent which decides morphological issues, it is morphology which has to decide about the possibility of descent.[†]

This, Hertwig found, was essentially Strasburger's conclusion.[‡] But Strasburger, it seemed to Braun, wanted even more: the very act of comparison, he had proclaimed, is already a component of phylogenetic research, as any comparison of morphological structures in search for homology makes sense only under the presupposition that those elements could have had a common phylogenetic origin. Salt cubes, Braun continued, that crystallize in a container filled with a saturated sodium chloride solution have a common origin as well, but are not historically related. It would consequently seem, Hertwig agreed with Braun, that morphology could just as well investigate organic form in search of typicity, striving to discover the lawfully constrained unity of plan, without any underlying phylogenetic assumptions[§]—a pure science of form, one that was destined to evolve into a renaissance of idealistic morphology. Strasburger's claim to fame was not based on phylogenetic morphology, however. Instead, he turned the Poppelsdorfer Schloβ in Bonn into an international center for (botanical) cytology.[¶]

Another foreign admirer of Haeckel was Adolf Lang, born on June 18, 1855, in Oftringen, Canton Aargau, Switzerland. In 1873, one year before the federal qualifying examination (*Matura*), he enrolled at the University of Geneva to study zoology and botany. His premature admission to the university had been mediated by Carl Vogt (1817–1893), a marine biologist renowned for his German translation of Robert Chamber's (1802–1871) notorious *Vestiges of the Natural History of Creation*, published anonymously in 1844.[**] Reading Haeckel's *Generelle Morphologie*, Lang was highly impressed, enough so to transfer to Jena University in 1874 to continue his studies under Haeckel and Strasburger. Haeckel reminisced how Lang had arrived in Jena in the spring of 1874, armed with a "warm letter of recommendation from this friend of mine," the "famous naturalist Carl Vogt."[††] He characterized Lang as the model of a hard working student, who earned his degree in March of 1876 with the

---

* Braun, 1876, p. 245.
† Braun, 1876, pp. 246–247. The same view was expressed by Karl von Goebel (1855–1932), at that time professor of botany in Rostock (later Marburg and Munich); see Goebel, 1884, p. 133.
‡ O. Hertwig, 1906a, p. 54.
§ O. Hertwig, 1906a, p. 58.
¶ Mägdefrau, 1973, pp. 155ff.
** On Chambers and his *Vestiges*, see Secord, 2000. According to Richards (2008, pp. 223–224), Vogt's German title for Chamber's book, *Natürliche Geschichte der Schöpfung*, may have inspired Haeckel's *Natürliche Schöpfungsgeschichte*.
†† Haeckel, 1916, p. 1.

highest accolade possible, *Magna cum Laude*. Two months later, in May, Lang was awarded the *venia legendi* at the University of Berne. From 1878 through 1885 he worked as a scientific officer (*wissenschaftlicher Beamter*) at Anton Dohrn's marine biological station in Naples, before returning to Jena in 1885 to collaborate with Haeckel as his assistant, while teaching in the capacity of *Privatdozent*.

> But then a happy fate arranged that in the following year already, a generous benefactor, Dr. Paul von Ritter (originating from Lübeck but at that time in Basel), donated 300,000 Marks to the University of Jena, with the restriction that the revenue (10,000 Marks annually) would be used exclusively to support phylogenetic research in zoology, in any way that I [i.e., Haeckel] deem appropriate. The first thing I did with these funds was to establish an associate professorship for phylogeny, and to appoint in that position my excellent assistant, the *Privatdozent* Arnold Lang.[*]

In 1889 Lang accepted an offer from the University of Zurich, where he was named full professor of zoology and anatomy, and director of the zoological collections. At the University of Zurich, Lang pursued research and advised graduate students in comparative anatomy and experimental genetics. In 1894, he became dean of the faculty for natural sciences, and from 1989 through 1900 he served as president of the University of Zurich. In 1908, Lang decided against the offer to become Haeckel's successor in Jena. He died in Zurich on November 30, 1914.[†]

With respect to the *Ritter-Professur für phylogenetische Zoologie*, Haeckel had made it a requirement that an annual public lecture be delivered in May or June in honor of the donor. In his first *Ritter-Vorlesung*, delivered on May 27, 1887, Lang dealt with "Means and Ways of Phylogenetic Discovery," a "superb" lecture in Haeckel's judgment.[‡] Lang opened his lecture with a reference to Haeckel's *Generelle Morphologie*, calling its author the first to have drawn the full consequences that follow from Darwin's theory for the entire field of biology. Embryology, that is, ontogeny, Lang clarified, is our proud leading light through the dark thickness of phylogeny, a role it first took on in the hands of the immortal Karl Ernst von Baer. He moved on to Fritz Müller's cautionary remarks in that respect, and Haeckel's answer, which was to recognize caenogenesis as a factor erasing the historical signal inherent in ontogeny. From there, he launched into an account of diverse caenogenetic phenomena and their significance as adaptations to the organisms' conditions of existence. From the link he established between caenogenesis and adaptation, Lang concluded to a struggle of existence that may play out between different life stages of organisms that undergo metamorphosis or are parasitic at some stage of their individual development, but may also pitch the adult against juvenile stages within the

---

[*] Haeckel, 1916, pp. 9–10. For reminiscences of Paul von Ritter, who first endowed a Ritter-Professorship for zoology, later a Haeckel-Professorship for geology, see also Walther, 1919, p. 949. Ritter is characterized as an eccentric benefactor who dabbled in science. When confronted with his inadequacies in scientific publishing, he withdrew from Haeckel's sphere of influence, and the city of Jena cancelled the project to erect a monument in his honor. For more on the social hygienist Paul von Ritter and his endowment, see Hoβfeld, 2005a, p. 239, n. 103.

[†] Hescheler, 1916; Kuhn-Schnyder, 1982; see also Rübel, 1947, and http://www.library.ethz.ch/de/ Ressourcen/Digitale-Kollektionen/Kurzportraets/Arnold-Lang-1855-1914 (accessed July 17, 2012).

[‡] Haeckel, 1816, p. 11.

same species. But then, leaving caenogenesis and its biological significance behind, Lang turned from ontogeny to comparative anatomy as the second way in which to acquire phylogenetic knowledge: "The comparative anatomy of adult animals, of the terminal stages of ontogeny, seeks and discovers in those the typical, the ancient and deeply entrenched, i.e., that which has been inherited from common ancestors, just as well as embryology does in its comparison of developmental stages." The reason: "The adult animals, the terminal stages that live today are nothing but the descendants of forms that have lived in earlier geological epochs."[*] Having traveled the distance between Jena and Naples in his career, Lang, it seems, sought to bridge the bitter rift that had opened between Haeckel and Gegenbaur on the one side, Dohrn on the other, with respect to the priority of comparative anatomy versus embryology. Lang did not mention these most prominent protagonists in the *Comptenzconflikt*, however, but instead rejected Alexander Goette's (1840–1922) recent claim that comparative anatomy had absolutely no phylogenetic significance.[†] Lang illustrated the value of comparative anatomy with a review of current debates concerning the origin of annelids and crustaceans, respectively, before switching to paleontology as the third way in which to pursue phylogenetic research. Its superior advantage lies in the fact that it deals with fossil remains of organisms that actually lived in the distant past; its obvious disadvantage is the incompleteness of the Fossil Record. Additional supplementary ways to seek phylogenetic knowledge are provided by ecological and biogeographical research (*Oecologie* and *Chorologie*, both terms that were first introduced by Haeckel in the second volume of his *Generelle Morphologie*).

> As we look back on the ways and means of phylogenetic research, it becomes clear that there is no single key, no master key, that opens all the doors on all the floors in this enormous building that is phylogeny. The keys to many rooms we will have to get from ontogeny, other rooms we will open by means of comparative anatomy, others through paleontology, ecology, biogeography and physiology etc. ... Many doors have a most complicated lock. Several keys will have to be used in the right place and in the right succession so that the gate may be opened to phylogenetic insights.[‡]

Many years later a young man, a grade-school teacher who had enrolled in the University of Zürich in the spring of 1904 to study philosophy and literature, but who soon switched to biology, knocked on Arnold Lang's door, inquiring whether he could pursue graduate studies under his guidance. With Lang's agreement started the career of Adolf Naef, who would make it his life's work to fit those keys together that

---

[*] Lang, 1887, p. 46.

[†] As a *Privatdozent* in Strasbourg, Goette had published in 1875 a monumental study on the ontogeny of *Bombinator igneus* (the European fire-bellied toad now called *Bombina bombina*), which earned him the promotion to associate professor (Richards, 2008, p. 291). In a concluding section to his monograph, Goette mounted a sharp attack on Haeckel, launching into an in-depth analysis and critique of Haeckel's writings. Believing in a lawful causal determination of ontogeny, Goette proclaimed that "the entire morphology of animals is based on necessary natural processes. With this insight certainly nothing has yet been explained, but the perspective is specified from which the investigation into causal determination must proceed" (Goette, 1875, p. 857). From there, Goette (1875, p. 904) concluded: "The individual development of organisms alone accounts for and explains the entire morphology of those."

[‡] Lang, 1887, p. 63.

Lang had talked about in his *Ritter-Vorlesung*, using cephalopods as the empirical basis for his research, creatures that he again studied at Dohrn's station in Naples where Lang had sent him. In the process, Naef's ambitions were heightened by the desire not only to resuscitate systematics from the growing disdain that this science engendered among contemporary biologists, but also to cleanse Haeckel's "naïve phylogenetics"* from the fundamental category mistake that he found Haeckel to be guilty of having committed.

* Naef, 1919, p. 3.

3 The Turn against Haeckel

THE POVERTY OF SYSTEMATICS

On the evening of November 18, 1908, a group of biology professors from the University of Vienna and researchers from the Vienna Natural History Museum met for an informal discussion of a nagging problem: what is the meaning of monophyly? Haeckel had introduced the term, but had used it in such multifarious ways that its meaning had remained quite indeterminate. The evening session was chaired by the paleobiologist Othenio Abel (1875–1946). Other than Abel, the more important voices among those participating in the discussion[*] included the botanists Richard von Wettstein (1863–1931), August von Hayek (1871–1928), and Rudolf Schrödinger[†] (1857–1919), the latter an independent scholar and father of the famous physicist and Nobel Laureate Erwin Schrödinger. Also present were Schrödinger's friend, the entomologist and paleoentomologist Anton Handlirsch (1865–1935), and the invertebrate zoologist Berthold Hatschek (1854–1941), the latter carrying the Haeckel virus in his system, the same that had caused the "Jena epidemic"[‡] some fifty years ago. The men who gathered were members of the "Section for Paleozoology" that was part of the *K. K. Zoologisch-Botanische Gesellschaft in Wien*; in 1910, the section would be renamed *Sektion für Paläontologie und Abstammungslehre* (Section for Paleontology and the Theory of Descent). It had been founded on Othenio Abel's initiative on February 27, 1907, under the patronage of botanist Josef Brunnthaler (1871–1914), who at the time served as Secretary General of the *Zoologisch-Botanische Gesellschaft*. The latter, today called *Zoologisch-Botanische Gesellschaft in Österreich* (ÖZBG), was founded on April 9, 1851, on the initiative of Georg Frauenfeld (1807–1873), a self-taught biologist and explorer with botanical and entomological interests.[§] In his review of the fate of zoology at Austrian universities during the first 50 years of the society's existence (1851–1901), Handlirsch deplored the increasing specialization, fragmentation, and the consequent increasing competition within the field[¶]: "Systematics, for example, was pushed to the background under the influence of the prolific school of Johannes Müller, whereby it was overlooked that this field of research had not been left untouched by the general progress in biological sciences. Indeed, systematics seeks the same that comparative anatomy and embryology are

---

striving for: to discover the true relationships of animals."* Handlirsch went on to praise the tradition of zoological systematics that had flourished at the Viennese *Naturaliencabinete*, the precursor of the Natural History Museum, founded and directed by the entomologist Vinzenz Kollar (1797–1860), and culminating in the work of such people as the entomologist Friedrich Brauer† (1832–1904). But then, Handlirsch continued, the arrival of Carl F. Claus‡ (1835–1899) in 1873 in Vienna, and of Franz Eilhard Schulze§ (1840–1921) in the same year in Graz, "brought the program of Johannes Müller's school [physiology in the broadest sense, including experimental developmental biology] to Austrian universities, where it soon rose to dominance. It was therefore of special importance that the alternative research program (the morphological-systematic and phylogenetic program) found a permanent and fruitful future in the newly expanded *Hofmuseum* under the direction of Friedrich Brauer and the gracious patronage of His Excellency."¶ It is clear from the tone of Handlirsch' reflections that he resented the spread and rise to dominance of Johannes Müller's legacy in zoology at Austrian universities, and he wanted to use the *Zoologisch-Botanische Gesellschaft* to promote the morphological-systematic research program. Handlirsch had himself played an important role in the *Zoologisch-Botanische Gesellschaft*, which he had reorganized through drafting of new bylaws, and restructuring of the board of directors, in 1896. His new bylaws provided for the option of naming honorary members of the society. Among those elected for this honor at the first plenary meeting under the new rules, on April 1, 1896, was Friedrich Brauer, who had served as president of the society in the years 1867 and 1868. Brauer had earlier already put his pen to paper in defense of the importance of systematics: "systematics is not an outdated science," Brauer insisted against Haeckel, who accused it of dealing with concepts—like that of the species—that Darwin supposedly had rendered obsolete. Instead, systematics "has been rejuvenated and revived by the theory of descent."** Still, Brauer expressed his satisfaction that both botany and zoology had left behind their *Sturm und Drang* period of hastily sketched phylogenetic trees, a period that on his account had been ushered in by Haeckel.††

The discussion evening on monophyly of November 1908 had been the first of its kind in Abel's section. It was followed by discussions on monstrosities (1909), adaptation (1909), gigantism (1910), atavism (1913), and aspects of orthogenesis (1915). The topics chosen for discussion, as also the list of lectures given in the section during

---

* Handlirsch, 1901, p. 249. For an account of a similar institutional decline of systematics in Germany during the second half of the nineteenth century, see Nyhart, 2009a.
† Handlirsch, 1905.
‡ Carl Claus was a former student of Leuckart who was offered the Chair in Zoology at the Vienna University in 1873. He also served as director of the newly founded Austrian marine-biological station in Trieste; he accepted much of Haeckel's program, but was critical of Haeckel's *Gastraea* theory (Hatschek, 1888, p. 39; Gicklhorn, 1957, pp. 268–269; Salvini-Plawen and Mizzaro, 1999, p. 26).
§ After studies in Rostock, Schulze earned the *venia legendi* at the University of Bonn in 1864. He became professor in Graz in 1873, and together with Carl Claus founded in 1875 the Austrian marine-biological station in Trieste, serving as its co-director with Claus until 1885, after which Claus became the sole director. In 1884, Schulze became *Ordinarius* for zoology in Berlin (Höflechner, 2007, pp. 723–724).
¶ Handlirsch, 1901, p. 250.
** Brauer, 1885, p. 272.
†† Brauer, 1887, p. 581.

the past twenty years,[*] strongly reflect Abel's own emphasis on paleobiology rather than phylogeny, although phylogeny reconstruction he considered an integral part of his paleobiology.[†] Abel had earned the *venia legendi* at the University of Vienna in 1901, and was promoted to tenured associate professor of paleontology in 1907.[‡] By that time, he had alienated his former supervisor, Eduard Sueβ (1831–1914) and other staff in the Geological-Paleontological Institute, who saw in Abel's paleobiology a competing program to their own emphasis on stratigraphic research. Abel wanted autonomy for paleozoology, and derided stratigraphy as a science subservient to geology. In so doing he took his inspiration from his close friend and colleague, the Belgian paleontologist Louis Dollo (1857–1931), who would proclaim in his inaugural speech held at the University of Brussels on October 20, 1909: "Paleontology is not a branch of geology, but instead a branch of biology … it is a purely biological science … in fact, animal paleontology is nothing but the zoology of fossils … in one word: chronology [stratigraphy] to the geologists, real paleontology to the biologist!"[§] Another senior colleague and later friend of Abel, the North American paleontologist Henry Fairfield Osborn (1887–1969), who rose to the rank of president of the American Museum of Natural History in New York, held similar views: "Paleontology is not geology. It is zoology; it succeeds only in so far as it is pursued in the zoological and biological spirit."[¶] Turning away from his local fellow paleontologists, Abel enjoyed the support he found among the zoologists and botanists he interacted with in the *Zoologisch-Botanische Gesellschaft*, and in particular relished the stimulating atmosphere of the special discussion evenings.[**] On May 3, 1927, the twentieth anniversary year of his section in the *Zoologisch-Botanische Gesellschaft*, Abel signed a contract with Emil Heim & Co., publisher in Vienna and Leipzig, which kicked off a new journal, Abel's *Palaeobiologica*.[††]

Like Abel, Richard von Wettstein had risen through the ranks at the University of Vienna, where he enrolled in 1881 to study medicine and natural sciences. In the course of his studies, he took a class from Abel's later supervisor Eduard Sueβ, with whom he retained a lifelong friendship.[‡‡] He earned his PhD in 1884, the *venia legendi* in 1886. He was appointed full professor of Botany and director of the Botanical Garden at the German Karl Ferdinands University in Prague in 1892, but returned in the same capacity to the University of Vienna in 1899 to succeed his former advisor, Anton Kerner von Marilaun (1831–1898).[§§] In 1901, Wettstein became president of the *Zoologisch-Botanische Gesellschaft*. When tensions between Abel and his

---

[*] Abel, 1927.

[†] "The paleontologist is a historian; this is why the unraveling of the history of animal phyla is one of his most important tasks" (Abel, 1907, p. 74).

[‡] Abel was promoted to full professor in 1912, and was named *ad personam* professor of paleobiology in 1917 (Ehrenberg, 1975).

[§] Quoted by Abel (1928a, p. 10).

[¶] Osborn, 1905, p. 227.

[**] Ehrenberg, 1975, p. 65.

[††] Ehrenberg, 1975, p. 89.

[‡‡] On June 17, 1914, Wettstein presided over a commemoration for the deceased Grandmaster of Austrian geology, Eduard Sueβ, in the large ballroom of the university. The celebration of Sueβ's life was attended by many local dignitaries, including the secretary of education (Cernajsek and Seidl, 2007, p. 257).

[§§] Porsch, 1931.

colleagues rose in the Geological-Paleontological Department, Wettstein became one of the loyal supporters of Abel,* who himself had strong botanical interests, which he claimed to have inherited from his forefathers, who over several generations had been gardeners and landscapers—such inheritance would of course have had to rely on some Lamarckian mechanism.[†] Historians have identified Wettstein as the leading "phylogenetic systematists" among the botanists of his time.[‡] Phylogeny was his preferred topic: it was said that "he lived for evolution."[§] He published his monumental *Handbook of Systematic Botany* in 1901, a text he had written in Prague that would go through several editions in the years to come.[¶] In the preface, he emphasized that botanical systematics should be guided by phylogenetic insights,[**] yet allowed for ambiguity in classification. Plants should be classified in a system that should on the one hand comply with scientific requirements, that is, reflect phylogeny, while on the other hand it should satisfy practical purposes.[††] From the point of view of a strictly "phylogenetic systematics," he argued, polyphyletic groups "must be dissolved once they have been recognized as such."[‡‡] Should the multitude of monophyletic lineages threaten to obscure the clarity of the system, though, practical concerns may require the retention of polyphyletic entities.

Anton Handlirsch, another important voice in Abel's discussion group, initially pursued a career in pharmacy, before turning to entomology under the influence of Friedrich Brauer, whom he later remembered as "my unforgettable teacher."[§§] In 1892, he became an assistant at the Natural History Museum in Vienna, where in the course of his successful career he rose through the ranks all the way to being appointed its director in 1922. He earned the *venia legendi* at the University of Vienna in 1924, and was named an associate professor in 1931.[¶¶] In a preliminary note on fossil and recent hexapods (a group of apterygota insects), read on October 22, 1903, at the Academy of Sciences in Vienna, Handlirsch expressed his surprise that fundamentally different opinions had been voiced regarding the phylogeny of hexapods by authors who all embraced the same basic Darwinian principles. He attributed this disparate state of affairs to the increasing specialization and fragmentation of biological sciences, such that independent phylogenetic systems had been based on morphology, embryology, or biology (i.e., ecology) respectively. In contrast, Handlirsch urged that the whole, fully developed organism and its entire ontogeny should be taken into consideration in phylogeny reconstruction, and that the phylogenetic information content of the characters must carefully be weighted, in particular with respect to the question as to whether the characters under consideration are inherited (primary) or acquired

---

* Ehrenberg, 1975, p. 61.

† Ehrenberg, 1975, pp. 12, 22.

‡ Mägdefrau, 1973, p. 193.

§ Klein, 1932, p. 2.

¶ The fourth edition, published in 1933 (reprinted in 1962), was edited by his son Fritz von Wettstein (1895–1945).

** Wettstein, 1901, p. iv.

†† Wettstein, 1901, p. 1. The neo-idealistic plant morphologist Wilhelm Troll would later declare these two goals as incompatible (Troll and Meister, 1951, p. 109).

‡‡ Wettstein, 1901, p. 15.

§§ Handlirsch, 1925a, p. 14.

¶¶ Jahn, 1966, p. 608.

(secondary). The larger opus he had announced in 1903,[*] Handlirsch's landmark book on *Fossil Insects and the Phylogeny of Extant Forms*, was published in 1908. There, he introduced paleontology as the ultimate arbiter in all controversies that arise from morphological-phylogenetic analysis of exclusively extant forms.[†] He called Haeckel's first attempt to sketch a phylogenetic tree of insects "ingenious," opening the door to a "scientific revolution," one that in his opinion was carried forward by such people as Fritz Müller, Anton Dohrn, and Friedrich Brauer.[‡]

At the time the evening discussion of monophyly took place, Wettstein was at the zenith of his career. While remaining highly productive in botany, he would not change his views on systematics anymore, as is revealed by a comparison of the successive editions of his successful *Handbuch* of 1901. Abel on the other hand was just at the beginning of a stellar career that was to take him to the very top of pre-World War II German paleontology. He had outlined his research program two years ago, on the occasion of the founding of the Section for Paleozoology in the *Zoologisch-Botanische Gesellschaft*.[§] This sketch he would flesh out in the course of his highly productive, yet politically tempestuous life.[¶] His life, work, and influence will have to be taken up again in its proper context. In contrast, Hatschek's productivity declined as he slipped into an ever-deepening depression. After the *Anschluss* of Austria to Hitler Germany in the spring of 1938, Hatschek fell victim to the Nazi purge of Jewish staff at Greater German universities. On account of his Jewish wife *née* Rosenthal, Hatschek was expelled from the University of Vienna and soon thereafter expropriated.[**] Handlirsch, on the other hand, continued his work in insect systematics, and once again gained attention and was applauded well beyond the boundaries of his own discipline with his highly influential entries in Christoph Schröder's widely distributed *Handbuch der Entomologie*, published in three volumes.[††] The third volume that appeared in 1925 opened with Handlirsch' review of the history of entomological systematics. Matthias Jacob Schleiden, Karl Ernst von Baer, Martin Rathke (1793–1860), and Johann Müller were singled out among those who gave comparative anatomy and physiology a new face through the "intellectual penetration of facts," with the consequence that systematics had been pushed into the background at German universities.[‡‡] This, Handlirsch claimed, happened in spite of the fact that Darwin had delivered the foundations for a "precise *phylogenetic systematics*"[§§]:

> I have never really understood why systematics was often considered an inferior science. Systematics concerns the inference of complex and long past processes ... In that sense, the method of systematics resembles the resolution of a highly complicated case on the basis of circumstantial evidence.[¶¶]

---

[*] Handlirsch, 1903, p. 718.
[†] Handlirsch, 1908, p. i.
[‡] Handlirsch, 1908, pp. 1206–1207.
[§] Abel, 1907.
[¶] For an account of the rise and decline of Abel's paleobiology, see Rieppel, 2012a.
[**] Salvini-Plawen and Mizzaro, 1999, p. 32.
[††] Handlirsch, 1925a,b.
[‡‡] Handlirsch, 1925a, p. 11.
[§§] Handlirsch, 1925a, p. 12.
[¶¶] Handlirsch, 1925a, pp. 14–15.

In a following entry, Handlirsch went about to sketch the principles and methods of a "rational phylogenetic systematics"* that would reveal their influence and importance most strikingly a quarter century later through the pen of another young entomologist, Willi Hennig.

On the occasion of the evening discussion of November 1908 in the Section for Paleozoology of the *Zoologisch-Botanische Gesellschaft*, the issue of monophyly still remained a rather muddled one, however. Abel—using examples at a lower taxonomic level—defined monophyly as a group of species that can be united in a genus on the basis that they all derive from a "single stem-species," even if those species descended from different parts of the stem-species, that is, from a different pair of parent organisms in different parts of the distributional area of the stem-species.† Hatschek, a former Haeckel student, clarified what he considered to be a widespread misunderstanding, which is that monophyly requires an origin from a single pair of parents. This misunderstanding he claimed to be rooted in a conception of species propagated by Haeckel and others, one according to which the species is an abstract concept, a mere aggregate of individual organisms. Instead, Hatschek argued, the species is a physiological concept, that is, a physiological individual *sensu* Haeckel, comparable to the individual organism.‡ In that sense, a species forms a reproductive community (*Zeugungskreis*) that can subdivide to form different descendant reproductive communities. Such a conception of species Hatschek found more fruitful in the discussion of monophyly than the highly unlikely origination of a species from a single pair of parents.§ Monophyly then means an origin from a single reproductive community, polyphyly an origin from more than one reproductive community.¶ Schrödinger supported Wettstein's position, who wanted the system of plants not only to represent phylogenetic history, but also to serve practical purposes by being transparent. He consequently distinguished the "practical systematist" from the "phylogeneticist" (*Deszendenztheoretiker*).** Handlirsch, in contrast, was uncompromising: "from the point of view of phylogenetic systematics, 'polyphyletic' groups have unhesitatingly to be designated as unnatural, and have to be rejected or rather dissolved."†† Hayek was a medical doctor who pursued a PhD, later the *Habilitation* in botany, and was appointed first as honorary professor, then as associate professor (in 1926) at the University of Vienna.‡‡ He stressed the importance to distinguish a species concept based on similarity, that is, on morphology, as opposed to a species concept based on phylogeny: morphology is often decisive in the taxonomy of fossils, in contrast to Abel's portrayal of extant species and genera as terminal stages of phylogenetic lineages: two entirely different things according to Hayek.§§ Wettstein retorted that "phylogenetic (*deszendenztheoretische*) investigations are not pure speculations, but as much inductively secured as the purely descriptive method."¶¶

---

* Handlirsch, 1925b, p. 61.
† Abel, 1909, pp. 245–246.
‡ For an account of the German tradition to consider species as individuals, see Rieppel, 2011a.
§ Hatschek, in Abel, 1909, p. 250.
¶ Hatschek, in Abel, 1909, p. 254.
** Wettstein, in Abel, 1909, p. 251.
†† Handlirsch, in Abel, 1909, p. 252.
‡‡ Dolezal, 1969, pp. 151–152.
§§ Hayek, in Abel, 1909, p. 253.
¶¶ Wettstein, in Abel, 1909, p. 255.

But monophyly was not the only issue that was left unresolved in these discussions. Another confusing issue, reflected in Hayek's distinction of morphological versus phylogenetic species concepts, was the relation of the phylogenetic tree to the natural system, the latter expressed in the classifications of plants and animals. The issue at stake was the desired reconciliation of horizontal and vertical perspectives in systematic research—should such at all be possible.

Othenio Lothar Franz Anton Louis Abel was born in Vienna on June 20, 1875, the only child of Lothar Abel and his wife Franziska Antonia *née* Schneider. His father's wish that he attend Law School clashed with the interest he had developed, as a child already, for nature studies and archeology. As he recounts in his autobiography,[*] Abel started to collect butterflies and beetles when eight years old, under the solicitous guidance of his father. Collecting minerals was another one of his passions. In the summer of 1889, Abel met an old acquaintance he had come to know in Budapest in 1880, called Alfred Simmer, who agreed to have Abel take possession of some of his (invertebrate) fossils in exchange for some of Abel's artifacts. With this exchange started Abel's lifelong fascination with fossils, further encouraged by a book he received from his father on Christmas 1889, Oscar Fraas' (1824–1897) popular *Before the Flood* (*Vor der Sündfluth*, published in 1866).[†] In compliance with his father's wishes, Abel enrolled for studies in Law, but in parallel took classes first in botany, later and with greater focus in geology and paleontology. Still pursuing a degree in Law School, Abel was able to secure an appointment as scientific assistant in Eduard Sueβ's laboratory at the Geological Institute of the University. Abel proudly informed his mother[‡] that as of February 1, 1898, he would be drawing a monthly salary of 57 *Gulden* and 63 *Kreutzer*—the news did not go over well: "What? You rogue! You wretched individual! In the past you have caused me nothing but grief and anger! And now this disgrace! Instead of preparing for a Diplomatic career!,"[§] she saw her son drifting into breadless science. In 1899, Abel was awarded his PhD after having successfully passed examinations in paleontology, geology, and botany (Figure 3.1).

One day soon after his appointment as research assistant, Sueβ entered the lab and announced that a crate had arrived with the remains of a fossil dolphin skull from the Miocene of Eggenburg, some distance north of Vienna: "This one you should describe," Sueβ told Abel.[¶] Without experience or training in vertebrate osteology, Abel educated himself about the cranial osteology of cetaceans through comparisons with skulls of living species, which he obtained from the zoological museum of the university. This crucial experience laid the foundation for Abel's pursuit of paleobiology, which seeks to investigate fossils in the light of findings obtained from extant biota. A biological, or rather zoological approach to the study of fossils became not only Abel's passion, but was also the approach to vertebrate paleontology championed by the famous Belgian paleontologist Louis Dollo. Dollo called his research program *"La paléontologie éthologique,"*[**] after the ethological method he

---

[*] Published in Ehrenberg, 1975.
[†] On the broader effects of the book, see Nyhart, 2009a.
[‡] Othenio's father Lothar Abel died on June 24, 1896 (Ehrenberg, 1975, p. 23).
[§] Abel, in Ehrenberg, 1975, p. 44.
[¶] Abel, in Ehrenberg, 1975, p. 46. The results of this early work were published in Abel, 1900.
[**] See Dollo, 1909.

FIGURE 3.1   Othenio Abel, in 1927 or 1928. (Courtesy of Gernot J. Abel, Copenhagen; photo by Albin Kobe, Vienna.)

had learnt as a student of Alfred Mathieu Giard (1848–1908) at the University of Lille.[*] Abel's work on the *Eggenburger* dolphin remains that led to a monograph Abel published in 1900 in the prestigious *Denkschriften der Kaiserlichen Akademie der Wissenschaften* in Vienna earned him in the same year an invitation to describe the fossil whales kept at the *Musée Royale d'Histoire Naturelle de Belgique*, the workplace of Dollo. Abel immediately recognized the like-mindedness of his Belgian senior colleague; this invitation marked the beginning of a lifelong friendship between the two men, Dollo eventually arranging for Abel's appointment at the Brussels Museum of Natural History as a foreign collaborator.[†] Abel submitted the first monograph that resulted from his work on the cetaceans kept at the Brussels Museum as his *Habilitation* thesis to the University of Vienna in 1901.

Abel greatly admired Dollo who exerted a profound influence on him, so much so that Abel dedicated, somewhat belatedly, the first volume of his newly founded journal *Palaeobiologica* that appeared in 1928 to Dollo's 70th birthday.[‡] In his expositions on the founding evening for his Paleozoological Section in the *Zoologisch-Botanische*

---

[*] Giard advised and supported Dollo in ethological matters throughout his career (De Bont, 2010, p. 23).
[†] *Collaborateur étranger*. Ehrenberg, 1975, p. 58.
[‡] Abel, 1928a,b.

Gesellschaft, Abel praised Dollo's Law of the Irreversibility of phylogeny as a most important insight.[*] This law, one of Dollo's lasting claims to fame, was rooted in his ethological method. Dollo was a neo-Lamarckian, a philosophy Abel would likewise adopt. Lamarck himself had propagated the inheritance of acquired characteristics, something deemed impossible by the Darwinians after August Weismann had postulated an impenetrable barrier between body (*Soma*) and the germ line (*Keimplasma*), that is, the phenotype (*Erscheinungsbild*) and genotype (*Erbbild*) in Wilhelm Johannsen's[†] (1857–1927) terms. According to the neo-Lamarckian doctrine, function determines form, and not the other way around. In the context of Dollo's ethological method, behavioral changes trigger morphological change. To reverse such morphological change would require the organism to pass through the sequence of behavioral alterations in reverse as well, which to him seemed a sheer impossibility.[‡]

Yet, there was another law formulated by Dollo that caught Abel's intense interest. Dollo called it the law of the *"chevauchement des spécialisations,"*[§] a term which Abel in his opening remarks to the 1908 evening discussion of monophyly translated as *Spezialisationskreuzungen*. The issue had already resonated in Gegenbaur's discussion of heterochrony, where homologous organs and organ systems undergo evolutionary transformation at different rates in different organisms (see the discussion at the end of the first chapter). The result is a mixture of more, and less advanced characteristics, which is exactly what Dollo had observed in his study of the phylogeny of lungfish, where tooth plates and fins for example did not exhibit synchronous, correlated evolutionary specializations. Abel presented a similar case to his discussion partners derived from his study of the sirenian genus *Metaxytherium*, and spelled out the consequences very clearly:

> Let it be emphasized that the several species cannot be *directly* related because they show *Spezialisationskreuzungen* (*A* is more specialized in its dentition but less specialized in the structure of its pelvis, *B* is more specialized in the structure of the pelvis but less in its dentition).[¶]

What Abel was saying here is that the two species, *A* and *B*, cannot stand in an ancestor–descendant relationship to one another, but can at best be derived from a hypothetical common ancestor. This was an enormous conceptual step away from the Haeckelian search for ancestors and descendants, and toward the search for common ancestry instead.[**] Abel clarified his thoughts further in a talk he contributed to a series of lectures on the theory of descent that the former Haeckel student Richard Hertwig (1850–1937; the brother of Oscar Hertwig), professor of zoology at the University of Munich, had organized for the winter semester 1919/11 in the Society for Natural History in Munich. Gegenbaur had transformed idealistic morphology into phylogenetic morphology by reading evolutionary transformation into the seriability of organisms, organ systems,

---

[*] Abel, 1907, p. 76.
[†] Johannsen, 1909, pp. 123, 127.
[‡] De Bont, 2010, p. 24.
[§] Dollo, 1895.
[¶] Abel, 1909, p. 245.
[**] See Nelson, 2004, pp. 131ff, for more analysis.

or organs. But, explained Abel, if in the case of four genera A, B, C, D, the transformation series for one organ reads A, C, D, B, and the transformation series for a second organ reads A, D, B, C, then the inferred phylogeny of the two organs cannot correspond directly to the phylogeny of the genera. The four genera cannot form a lineal sequence of ancestors and descendants.[*] In his most mature work, *Paläobiologie und Stammesgeschichte* of 1929, Abel later fleshed out this fundamental insight with a real-world example that took him back to his early work on marine mammals. Among the mysticete (baleen whale) families Balaenidae and Balaenopteridae, he found the balaenids to combine a primitive forelimb skeleton with specializations in the cervical vertebrae, whereas the balaenopterids retain the primitive condition of the cervical vertebrae, but show specializations in the skeleton of the forelimb. The conclusion must be that the two families cannot stand in an ancestor–descendant relation to one another, but must instead be derived from a hypothetical common ancestor who retained the primitive condition in both organ systems.[†] It is the neglect of the issue of *Spezialisationskreuzungen* that resulted from the focus on the part of phylogenetic morphologists on a single organ or organ system that, according to Abel, generated the multitude of unwarranted phylogenetic schemes that had so abundantly sprouted in the minds and writings of authors who enthusiastically yet uncritically followed Haeckel's lead: "rash conclusions have, indeed, frequently been drawn. But when continued morphological research discloses an hitherto unknown crossing of specializations, these *Ahnenreihen* will have to be dissolved again … it is easy to combat Haeckel's *Ahnenreihe* of hominids …"[‡] What had at one time been identified as ancestors all too often were later recognized as side-braches of the phylogenetic tree.[§]

So what was, according to Abel, the proper way to conduct phylogenetic research? Like Gegenbaur and any other comparative morphologist, Abel started from the seriability of form conditions of organisms, their organ systems, and organs. If organisms could be arranged in a continuous series, if within that series the ancestral and the descendant condition of form could be identified (i.e., if the series could be polarized), and if that series indeed represented a chain of ancestors and descendants, then Abel would call that series an *Ahnenreihe*. But there are many "ifs"—so how could we know? An *Ahnenreihe*, Abel specified, should never be built on consideration of a single organ or organ system only. Like Gegenbaur before him, Abel requested that phylogeny reconstruction be based on several organs or organ systems, the more the better. The transformation series of a single organ or organ system Abel called a *Stufenreihe*. If, for a set of organisms, the *Stufenreihen* of multiple organs or organ systems run parallel to one another, that is, if there are no crossings of specializations, then each of these *Stufenreihen* would also represent the *Ahnenreihe* for the set of organisms under consideration. If, however, *Spezialisationskreuzungen* were observed, then the *Stufenreihen* could not also represent the *Ahnenreihe*, and the set of organisms under consideration and the species they represent would have to be

---

[*] Abel, 1911, p. 246.
[†] Abel, 1929a, p. 266.
[‡] Abel, 1911, p. 245; see also Abel, 1912, p. 13.
[§] Abel, 1920, p. 20.

grouped in the way they relate to hypothetical common ancestors that are inferred on the basis of the observed character distribution.

Abel's approach to phylogeny reconstruction thus differed in fundamental ways from Haeckel's. Abel did not take a developmental stage, such as the gastrula, to have represented, in the distant geological past, a mature physiological individual, the *Gastraea*, which then could serve as the ancestor of all metazoans. Of course, Abel left the door open for the possibility that a direct ancestor, an ancestral species, could be discovered. But given the ubiquity of Dollo's *chevauchement des spécialisations*, this for him would have been the exception rather than the rule. Quite generally, Abel replaced the Haeckelian concept of "ancestor" with his own concept of "common ancestry": it placed on the branching points of a phylogenetic tree not a specific species, fossil, or extant, but instead a hypothetical common ancestor that instantiates the primitive condition of form relative to all of its putative descendants. Abel thought it is possible that such hypothetical ancestors could at some time or other be identified in the Fossil Record, but given its notorious incompleteness, this would again remain the exception rather than the rule.[*]

Gegenbaur had read a process of evolutionary transformation into his serial arrangement of organ systems or organs, and the same is true for Abel, of course. But the switch from form-relationships to blood-relationships entailed some rather tricky issues, and these started with the species already. Abel himself wanted to ask big questions. In a Haeckelian spirit, Abel deplored the fact that Darwin had lost his good fight against the "species manufacturers" (*Speziesmacherei*), the "priority snoopers," and "[collection] cabinet builders" (*Kistentischler*).[†] But ever since Linnaeus, the species was and remained the basic category in systematics and classification. In contemporary biota, the species—at least of those organisms that are visible to the naked eye—seem to form rather discreet entities, reproductive communities that can be described, diagnosed, identified, and re-identified with some confidence most of the time. But placed into the vertical dimensions of a phylogenetic tree, and barring saltational change that Darwin's theory would leave unexplained, the species becomes a chunk of the genealogical nexus, a segment in an evolutionary lineage with no clear objective boundaries. What about the species recognized by paleontologist? Biostratigraphy, for example, depended crucially on accurate species description and identification. But just as the botanist or zoologist works in a time plane that is the present, so does the stratigrapher describe and identify the species in a certain geological section or horizon, that is, in a time plane that intersects the phylogenetic tree horizontally. Inspecting such a horizontal section through the phylogenetic tree, its individual branches will all appear as discreet cross sections, well separated from one another. So, whereas the species loses its identity along the vertical dimension of the branches of the phylogenetic tree, there is no identity crisis for species if that tree is inspected in successive horizontal, that is, temporal transects. The cardinal question then only is to show how these two perspectives

---

[*] A similar concept of an entirely unspecialized ancestral form was later invoked by the paleontologist Karl Beurlen (1930, p. 537). The zoologist Adolf Remane called this a "zero-value-ancestor" (*Nullwert—Ahne*: Remane, 1948), a concept which he rejected because he considered such an organism, which could have no specific adaptations of its own, not viable.

[†] Ehrenberg, 1975, p. 26; see also Abel, 1914, p. 380; 1929a, p. 130.

could be combined, how to allow species to exist both in a horizontal as well as in a vertical dimension.

Among those who struggled with this thorny issue was, once again, the notable Friedrich Brauer. In his more popular expositions on systematics, presented in a talk held on March 23, 1887, Brauer likened the extant biota to a horizontal transect through the terminal tips of the branches that make up the phylogenetic tree. He cited the botanist Carl Nägeli, who had anchored the reality of taxonomic categories in their history,* but from this insight he drew the conclusion: "Systematic categories exist only in a cross section through time; within time they merge into one another."† There is no standing still of nature, Bauer continued his exposition, everything in nature is subject to constant change, tied into continuous cycles. "Just as the biological individual is composed of a number of developmental stages, so is the species composed of a number of races."‡ A species, according to Brauer, is composed of a number of temporally succeeding races: "the descent of races we can still observe, the descent of species or higher categories we can only infer."§ Higher taxonomic categories (i.e., more inclusive taxa) are composed of a number of temporally succeeding species, but: "the natural system can only represent a certain developmental stage of the genealogy of animals," for which reason "the natural system cannot represent the entire genealogy of animals, but only that time period [that part of the genealogy] which is accessible to direct observation."¶ The consequence of all this reasoning is that the paleontological system must not only complement the zoological system, but paleontologists must also devise as many systems as they recognize geological horizons.** In his earlier, more technical treatment of the same issue, Brauer had insisted that species must objectively exist at least for some duration, for if they did not, "it would be nonsensical to speak about the origin of species."†† Conversely, a species *can* objectively exist only for a limited duration, because through extended periods of time it indiscriminately merges with the succeeding species. A natural system that is based on species as its basic currency thus can capture a time slice, a horizontal transect through time only. And the same holds for more inclusive taxonomic groupings: "the systematic categories are not delimited within the phylogenetic tree; but they

---

* Nägeli, 1865; Brauer's rendition of Nägeli requires some clarification to avoid confusion of the modern reader. Systematic "categories" have no history. The species category is the class of all species; the particular species themselves are taxa. The modern category—taxon distinction was often not drawn in the systematic literature of the 19th and early 20th literature. It would thus be more appropriate to say that Nägeli anchored the reality of systematic groupings in their history.

† Brauer, 1887, p. 583.

‡ Brauer, 1887, p. 590.

§ Brauer, 1887, p. 606.

¶ Brauer, 1887, p. 601.

** Brauer, 1887, p. 604.

†† Brauer, 1885, p. 242. Here resonates a criticism that Agassiz had raised against Darwin. Darwin (1859, p. 52) had characterized species demarcation as an arbitrary convention, whereupon Agassiz claimed it nonsensical to speak about species variation: "How absurd that logical quibble ... As if anyone doubted their [i.e., the species'] temporary existence" (Darwin to Asa Gray, August 11, 1860, quoted from Hull, 1973, p. 429).

are sharply delineated within a certain time period, which represents a horizontal plane that transects the branches of the phylogenetic tree."[*]

For much the same reasons, there existed for Abel a fundamental incompatibility between systematics and phylogeny: "horizontal boundaries between species are sharp, vertical boundaries between genealogically connected species are purely artificial."[†] Referring back to the former Haeckel student Ludwig Plate's (1862–1937) influential definition of species as a reproductive community whose constituent parts recognize each other as conspecific,[‡] Abel found it to apply only to those entities that exactly fit the size of the box the systematist has built for them.[§] The species of paleontologists that are segments of an evolutionary lineage, incessantly subject to variation and natural selection and hence to change, cannot have "real boundaries,"[¶] which explains the difficulty "to combine phylogeny and systematics to form a 'systematic phylogeny' or a 'phylogenetic systematics' … the system is a transect through the phylogenetic tree."[**] The zoological system for sure is limited to one time horizon only, that is, the present. And the question for Abel was how it could be combined with a paleontological system that would represent the phylogenetic tree. Abel tried to bridge this gap between the two-dimensional zoological system and the three-dimensional phylogenetic tree with such concepts as "stem-group,"[††] "ancestral group,"[‡‡] or a "root-group" (Wurzelgruppe),[§§] but found the results awkward and insufficient. As he saw it, there had been no problem to consider fossil forms a mere supplement to a zoological system as long as the latter had not been genealogically interpreted. But as the theory of descent came to underpin systematics, a genetic element was introduced in the diagnoses of taxonomic groupings. The species, genus, family, and so on changed from purely classificatory concepts to concepts that apply to historically conditioned entities. "Are we doomed to reach a point at which it will no longer be possible to portray the results of phylogenetic research in a natural system?"[¶¶] To answer that question, Abel invoked a thought experiment: think of a sheet of glass horizontally transecting a phylogenetic tree. The intersections of the branches of the tree will appear as spots on the glass. As we inspect it from above, we will recognize branches that extend downward from these spots, joining each other as they converge toward the stem, in a maze so complex that, given the methods of contemporary systematics, we will not be able to connect all of these spots to a common point of origin.[***]

---

[*] Brauer, 1885, p. 247.
[†] Abel, 1919, p. 13.
[‡] Plate, 1914, p. 117. Plate was an outspoken anti-Semite, who considered research into race formation (from a hereditary, geographical, etc. perspective) a matter of zoological research (Hutton, 2005, p. 191). This motivated a physiological definition of species, rather than a morphological one.
[§] Abel, 1929a, p. 101.
[¶] Abel, 1929a, p. 102.
[**] Abel, 1919, pp. 1–2; see also Abel, 1913, p. 123.
[††] Abel, 1919, p. 4; 1929a, p. 129.
[‡‡] Abel, 1920, p. 31.
[§§] Abel, 1920, p. 31.
[¶¶] Abel, 1913, p. 123.
[***] Abel, 1913, p. 124.

Handlirsch, the erstwhile student of Brauer, likewise had no solution to offer. Instead, he actually formalized Abel's thought experiment. He again started out by claiming the reality of systematic entities such as species, which he characterized as a sum of individuals of similar constitution and capable to interbreed. Viewed this way, a species is just as real as is a battalion of soldiers.[*] But again, viewed in a time slice such as the present, this reality was unproblematic; with their extension through time such systematic entities lose their individuality as they lose their distinctive boundaries. A natural system thus can capture biodiversity only in a horizontal transect through time, that is, through the phylogenetic tree. Looking at such a transect from above, the twigs and branches appear in cross section as dots, which corresponding to the degree of their phylogenetic relationships form smaller (less inclusive) and larger (more inclusive) circles. The natural system captured in such a transect through the phylogenetic tree can consequently be represented by a nested system of circles within circles. "One could," Handlirsch continued, "directly apply to these circles names of the appropriate categorical rank, and would thus obtain a truly phylogenetic system for that temporal transect,"[†] but since it remains restricted to a time horizon, the system is still not the same as the phylogenetic tree. Handlirsch thus ran two different concepts together. The one is the temporally extended phylogenetic tree, the other a temporally restricted phylogenetic system that did not capture ancestors and descendants, but only expressed relative degrees of phylogenetic relationships. Phylogenetic tree and phylogenetic system: complementary but still essentially different.

Clearly, the attempt to merge phylogenetics and systematics had run into trouble. The continuity of the phylogenetic process clashed with the discontinuity expressed in classification. The presumed reality of species and more inclusive systematic entities, anchored in their history, that is, in their spatiotemporal location, clashed with the conceptual, that is, abstract and logical, nature of systematic categories. Phyletic entities that are subject to Darwinian evolution undergo constant change, forming evolving linages that split and split again with continuous transitions which blur all attempts to establish discreet boundaries. The natural (phylogenetic for Handlirsch) system is expressed in a nested hierarchy of circles, or boxes, each representing a systematic category.[‡] Phylogeny is dynamic, the system is static; phylogeny is historically structured, the system is logically structured. In their critique of the Haeckel student Strasburger, who sketched a phylogenetic systematics for botany in his early career when a professor at Jena,[§] the botanists Alexander Braun and Karl von Goebel both voiced the same criticism[¶]: the natural system of plants is based on morphology, on relations of form. To call it a phylogenetic system changes neither methods

---

[*] Handlirsch, 1925b, p. 67.

[†] Handlirsch, 1925b, p. 75.

[‡] For an analysis of the different underlying ontologies, see Williams, 1992.

[§] Strasburger, 1872, p. iii: "The theory of descent has given a new direction to modern research; the comparative analysis has gained a phylogenetic significance; it has become the tool to discover the real relationships of organized beings, and the *natural* system, which was imagined by earlier researchers in terms of an abstract system, has thus obtained a real basis: it is the *natural* phylogenetic tree (*Stammbaum*) of organisms."

[¶] See the previous chapter.

nor results, but merely infuses the system with a new meaning. As the contorted arguments of Brauer, Abel, and Handlirsch show, such infusion engendered serious confusion and contradictions. It is at this juncture that a young outsider hailing from the *hinterland*, intelligent, aggressive, and self-assured, set out to clear up the conceptual mess in which phylogenetic systematics had become entangled. It was the young man who knocked on Arnold Lang's door at the University of Zurich.

HAILING FROM THE HINTERLAND

Adolf Naef[*] had an unlikely career, and given the fact that, in terms of his institutional affiliations, he remained an outsider throughout his career and without students who would carry his banner forward, he had an equally unlikely yet very strong influence in the field of comparative morphology and phylogenetics, not to mention the international acclaim he earned as a leading capacity in cephalopod research (morphology, embryology, and systematics). Adolf was born in his mother's hometown Herisau (located in the *Hinterland*, Canton Appenzell Ausserrhoden) on May 1, 1883, as the oldest of three children to Martin Näf from Niederhelfenschwil (Canton St. Gallen), and Bertha *née* Rutz. In 1903, Adolf graduated prematurely from the Evangelical Seminary in Zurich-Unterstrass as a certified grade school teacher. With the start of the summer semester 1904, he enrolled at the University of Zurich to study philosophy and literature. Soon attracted to natural sciences, however, biology in particular, he graduated from the university in 1908 with the diploma for High School teacher in natural sciences. On March 30, 1907, Adolf married Elisabeth (Lili) Rosenbaum from Minsk, who gave birth to their daughter Gerda on Christmas Day of 1907. A financially demanding spouse, she enrolled at the medical school of the University of Zurich in 1909. Their marriage was divorced on April 13, 1910, leaving Adolf behind with considerable debt and a daughter to care for.

Throughout his student years, Adolf had to provide for himself and, later, for his young family, with teaching engagements, which is how he might have met his first wife. A most welcome opportunity for part-time employment opened up at the *Institut Tschulok*, a private school that prepared students for the *Matura*, the federal qualifying exam required for admission to Swiss universities (similar to the *Abitur* in Germany). The acquaintance Naef established with Tschulok would prove crucial, as Tschulok greatly influenced Naef's theoretical approach to morphology, systematics, and phylogenetics. Sinai (Samuel) Tschulok[†] was of Jewish descent, born on April 17, 1875, in Konstantinograd (Ukraine). He grew up on an agricultural estate administered

---

[*] The biographical data on Naef are based on documentation deposited in the State Archive Zurich (*Staatsarchiv Zurich*, StAZH U110 e 13, and U 110 d.2, *Mappe* 111), in the Archive of the University of Zurich (UniAZH, *Rektoratsmappe* AB.1.0705), and in the State Archive Basel (*Staatsarchiv Basel*, StABS, ED-REG 1 a 2 1401). Additional information was provided by Naef's daughter Claudia Neuenschwander-Naef, Zurich, who also compiled a biography of Adolf Naef (unpublished) on the basis of his correspondence with his siblings. Further information and analysis is provided by Reif (1998), Boletzky (1999), Breidbach (2003), Williams and Ebach (2008), Rieppel (2012b), and Rieppel et al. (2013).

[†] The biographical data on Tschulok derive from material deposited at the State Archive, Zurich, StAZH U 110d.2, folder 103; see also http://www.matrikel.uzh.ch/active/21776.htm (accessed on August 1, 2012). The original spelling of his family name was Tscholek: see Huser-Bugmann, 1998, p. 151.

by his father. He attended grammar school (*Gymnasium*) in Jekaterinoslaw. As part of the first immigration wave of Eastern Jews[*] in Switzerland, Tschulok arrived with his parents in Zurich in March 1894. In October 1894, he enrolled in the agricultural school at the Swiss Federal Institute of Technology (ETH), "with the goal to teach Russian farmers modern methods."[†] After graduating in the spring of 1897, Tschulok married Rachel Wainstein MD from Riga and Paulograd,[‡] but continued his studies at the natural sciences division of the ETH, where he obtained his certificate for High School teaching in August 1900. From there on he earned his living at the school he founded,[§] all the while pursuing private research in the history and conceptual foundations of natural sciences, especially biology. His first publication[¶] he submitted to the faculty of natural sciences at the University of Zürich, who on that basis granted him the doctoral degree in 1908. In 1911, Tschulok was granted citizenship by the city of Zurich. On recommendation of Arnold Lang,[**] he was awarded the *venia legendi* by the University of Zurich in 1912, on the basis of his dissertation on *Logical and Methodological Investigations: the Position of Morphology in the System of Natural Sciences*. The thesis was published in the same year in the multivolume treatise on the morphology on invertebrates that in its second edition was edited by its initiator Arnold Lang in collaboration with his erstwhile student, then assistant and now professor Karl Hescheler.[††] His *Habilitation* thesis contains many of the ideas Tschulok would later flesh out in his university course on the theory of descent with Naef in attendance; the lecture notes ultimately resulted in his classic textbook of 1922. In recognition of his achievements, the board of the University of Zurich named Tschulok an Adjunct Professor (*Titularprofessor*) in March 1920. Tschulok died on December 6, 1945, in Zurich.

---

[*] Huser-Bugmann, 1998, p. 151.

[†] Huser-Bugmann, 1998, p. 151.

[‡] http://www.matrikel.uzh.ch/active/27360.htm (accessed August 1, 2012).

[§] Adolf Naef's daughter Claudia Neuenschwander-Naef (Zürich), in her privately circulated biography of her father, called the *Institut Tschulok für Maturitätsvorbereitung*, disparagingly *Tschulok's Schnellbleiche*. Other Eastern Jews arriving in Zürich, such as Max (Mejer) Husmann (1888–1965) and Max (Moses) Stern (1889–1943) founded similar schools as Tschulok's, preparing students for the *Matura*: Huser, 2005, p. 290.

[¶] Tschulok, 1908.

[**] "Mr. Tschulok is a very bright man with an independent, brilliant intellect … he recently became a citizen of Zurich." State Archive, Zurich, StAZH U 110d.2, folder 103.

[††] Karl Hescheler (Rübel, 1947, p. 98; http://www.matrikel.uzh.ch/active/9079.htm [accessed August 13, 2012] was born on November 3, 1868, in Schönenwerd, Canton Solothurn. His family was from Bad Schussenried in Upper Swabia, but obtained citizenship of St. Gallen, Switzerland, in 1885. Having passed the *Matura* examination, Hescheler worked in a pharmacy for six months before enrolling at the Federal Institute of Technology, motivated by Arnold Lang's infectious lectures. He obtained his diploma in natural sciences in 1893, then continued as a graduate student of Lang. His thesis on regenerative processes in earthworms (lumbricids) earned him his Ph.D. in 1895. The *venia legendi* he obtained in 1898; in 1903 he was named associate professor, in 1909 full professor of zoology and comparative anatomy as successor of Arnold Lang at the University of Zurich. Encouraged by Lang, Hescheler turned to paleontology, which he pursued with a paleobiological rather than stratigraphic focus. Hescheler became a founding member of the Swiss Archeological as well as the Swiss Paleontological Society, and indeed established paleontology as a research program at the University of Zurich. Karl Hescheler committed suicide on November 10, 1940. A bachelor throughout his life, he bequeathed his estate to the University of Zurich, thus establishing a *Karl Hescheler Stiftung* which today supports the Zoological Institute, as well as the Zoological and Paleontological Museums.

Adolf Naef would be deeply influenced by Tschulok's expositions in theoretical biology. In his critique of Haeckel, Naef would employ a heavy philosophical machinery, and emphasize logic in his attacks on Haeckel's nature mysticism. In 1908, Naef enrolled at the University of Zürich, changing his name from Näf to Naef to pursue graduate studies in zoology under Arnold Lang. During his stay at the German Zoological Station in Naples in the years 1878–1885, Lang–Longus, as he was called there, or also Don Arnoldo[*]—had engaged in a deep friendship with its founder and director, Anton Dohrn.[†] To work at the station naturally required a desk, and it was Dohrn's idea to finance the station through the renting of desk-space to visiting scientists, the rent to be paid by their institutions or governments respectively. Lang's desk was initially sponsored by the *Schweizerische Naturforschende Gesellschaft*, but by the time Naef knocked on his office door, Lang had secured an endowment that would allow one Swiss graduate student a stay of two months per year at the Zoological Station in Naples.[‡] In February and March 1908, it was Naef's turn to travel south on this stipend, an opportunity for him to collect the cephalopod material on which his doctoral thesis would be based.[§] In 1909, Naef obtained his doctoral degree on recommendation by Lang, who found the thesis to qualify for a rather moderate predicate only: "fairly satisfactory (*recht günstig*)."[¶] Lang praised Naef's ingenuity, thoroughness, and his skills pertaining to histological techniques, deemed the writing acceptable, but found the illustrations light ("no more than truly necessary") (Figure 3.2).

Naef returned to the Zoological Station in Naples on a second award in the spring of 1910, on which occasion he managed to obtain a salaried appointment with the mandate to complete the monograph on the cephalopods from the Golf of Naples that had been left orphaned upon Guiseppe Jatta[**] (1860–1903) untimely death. This was a temporary appointment only that left Naef fretting about his future prospects for an academic career. The issue was pressing as Naef drew a modest salary only, yet had engaged in marriage for a second time. The fact that Naef had been admitted to the graduate program at the University of Zürich on the basis of his High School Teacher's degree, rather than passing the *Matura* exam, barred him from seeking the *Habilitation* anywhere else but in Zurich, which made for a somewhat lopsided academic record. After consultations with his PhD advisors Arnold Lang and Karl Hescheler, Naef submitted on April 16, 1914, his *Habilitation* thesis that was based on his research in Naples and titled *On the Individual Development of Organic Forms as Document of Descent with Modification*.[††] In the printed version, Naef added the subtitle: *Critical Comments on the so-called Biogenetic Law*.[‡‡] The thesis ranks among his most important theoretical contributions to his

---

[*] Eisig, 1916, p. 56.
[†] Boletzky, 2000; see also Boletzky, 1999.
[‡] Eisig, 1916, p. 58.
[§] Naef, 1909.
[¶] State Archive of Zurich (*Staatsarchiv Zürich*), StAZH U110 e 13. In a letter to his younger brother Ernst, dated August 6, 1910, Adolf asserted that "there exists in Switzerland no glut of first-rate intellects (*erstklassige Köpfe*), which is what I take myself to be."
[**] Jatta, 1896.
[††] *Über die Individualentwicklung organischer Formen als stammesgeschichtliche Urkunde.*
[‡‡] Naef, 1917.

FIGURE 3.2 Adolf Naef in Naples, spring 1916. (ETH-Bibliothek Zürich, Bildarchiv/ Port_02904; photo by F. Finicelli, Naples.)

"comparative," "idealistic," or as he preferred to call it, "systematic" morphology. Lang retired in 1914, and died later in the same year, which left Hescheler as the only referee for Naef's *Habilitation* thesis. Puzzled by the heavy philosophical slant that characterized the monograph, Hescheler concluded: "Mr. Naef is ... very much inclined toward philosophical speculation and epistemological as well as methodological investigations ... [his writing] it is a bit stodgy, partially long-winded and, as it seems to me, not always very clear."[*] Naef was granted the *venia legendi*, starting with the summer semester 1915. His second marriage was divorced in Munich on February 12, 1915.

---

[*] Archive of the University of Zurich UniAZH, AB.1.0705.

Naef continued his research in Naples under the difficult conditions created by the Great War. In 1916, Reinhard Dohrn (1880–1962), the son of Anton Dohrn and his father's successor as director of the Zoological Station, relocated with his collaborators to Switzerland, where he was accommodated by Hescheler at the Zoological Institute of the University of Zurich.* Naef persevered, however, and repeatedly sought dispensation from his teaching duties in Zurich in order to continue his research in Naples. Hescheler graciously supported his many requests on account of the prevailing difficult times. When, in 1922, Naef requested dispensation from his teaching duties again as he intended to accept the position of associate professor at the University of Zagreb (Agram)—a provisional or transitional arrangement, he thought to himself, in view of difficult conditions of employment[†]—Hescheler lost his patience and Naef his support. Naef subsequently slid into a deep depression, a condition he had hoped that Hescheler would understand as the latter suffered from a similar predisposition.[‡]

Reflecting on the general mood of the time, the professor of zoology at the Swiss Federal Institute of Technology (ETH) Conrad Keller (1848–1930) published in September 1922, in the *Feuilleton* section of the *Neue Zürcher Zeitung*, an article on the *Crisis in Biological Sciences*: "Our times are marked not only by many political, economic, and social crises, but also by severe scientific crises," manifest in nature mysticism and prognosticating metaphysics.[§] Himself a former student of Haeckel, a close friend and colleague of Arnold Lang,[¶] and also a cephalopod expert who had worked at the marine biological stations in Naples and Trieste, Keller opposed the ascent of experimental biology to dominance in life sciences, an approach that in his eyes sought to reduce all biology to physics and chemistry: "A Darwin, a Huxley, a Rütimeyer,[**] an Ernst Haeckel—they all need no longer to be taken seriously, they all [are said to] belong into the junk room of modern science." He then engaged in a critique of idealistic morphology that he traced back to Goethe who, Keller maintained, had he lived to the present day would have immediately abandoned his idealistic for a historically motivated morphology. Keller's polemic triggered a number of responses both from Swiss and German zoologist, among them Adolf Naef, who—writing from Zagreb (Agram)—attacked what he considered to be a misguided and conceptually deeply flawed characterization of comparative morphology and phylogenetics by Keller.[††] Keller retaliated, calling Naef the "newest shooting star [*meteoroid*] at the zoological firmament." Keller's response opens an important

---

* Strohl, 1940, p. 447.
† For details, see Rieppel et al., 2013.
‡ Strohl, 1940, p. 449; aged 72, Hescheler, who remained a bachelor throughout his life, committed suicide through drowning in Lake Zurich on October 11, 1940.
§ Keller, Conrad. Die Krisis in den biologischen Wissenschaften. *Neue Zürcher Zeitung*, #1228, September 21, 1922.
¶ Hescheler, 1916, p. 219.
** See footnote 99, chapter 2.
†† Naef, Adolf. Biologischer Historismus und idealistische Morphologie. *Neue Zürcher Zeitung*, #1307, October 7, 1922.

window on how at least some of the more senior peers in his field perceived
Naef's stature as a junior scientist:

> Naef compensates lack of general knowledge and insufficient mastery of the technical
> literature with audacity. As is often the case with young people, true science he thinks
> starts only with him; what has so far been accomplished in zoology cannot be taken
> seriously. In his writings he derides and ridicules researchers of various disciplines.
> The French have a nice way of describing such a situation: *C'est l'ignorance profonde
> qui parle d'un ton dogmatique.*[*]

Naef snapped back: "In front of a grey head one is supposed to rise and honor the
elderly!"[†] Haeckel's goal, he complained, was above all to prove as much as possible,
and to do so as fast and as clamorously as possible.

Things took a happier turn for Naef when on July 22, 1924, he was finally able
to join his date since 1917, Maria Melitta Emma Bendiner from Munich, in a lasting
marriage. Also in 1924, Naef applied for a full professorship at the King Fuad
University, Abbassia, Cairo. His predecessor there was the German geneticist Victor
Jollos (1887–1941) of Jewish descent, who had arrived in Cairo in 1925, quickly
establishing good relationships in political and scientific circles.[‡] When Jollos threatened to step down from his position in an attempt to leverage a salary increase,[§] Naef
agreed to fill the position at the original scale of remuneration. Jollos consequently
was forced to return to Germany, where he spread the rumor that Naef had obtained
his position in an improper way. Most Europeans taking a job at Cairo University,
which at the time was run on the British model, expected to sooner or later return to
an academic position in Europe. This was no different for Naef. While still on payroll
in Zagreb, Naef had given a seminar at the Zoological Institute of the University of
Basel in January of 1924, which was very well received. The director of the Institute,
Friedrich Zschokke (1860–1936), invited him to come back soon, in a letter in which
he praised Naef's cephalopod monograph as a masterfully executed milestone in
the field. Naef indeed harbored high hopes to one day return to Basel as Zschokke's
successor. Upon Zschokke's retirement in 1930, a search committee was formed in
January 1931. Zschokke had been asked to present a list of possible candidates, of
which Naef was one, his own student and proxy Adolf Portmann another one. In the
search committee's second meeting held on February 7, 1931, one of its members, the
director of the Anthropological Museum Basel, naturalist and anthropologist Fritz (Karl
Friedrich) Sarasin (1859–1942) related rumors he had heard that Naef had not acted
in an entirely fair and professional manner in regards to his appointment at Cairo
University. The committee asked another one of its members, the director of the
Natural History Museum Basel and paleontologist Hans Georg Stehlin (1870–1941),
to further investigate the matter. Stehlin reported in the meeting of March 10, 1931,

---

[*] Keller, Conrad. Antwort an meine Kritiker in Sachen des biologischen Historismus. Neue Zürcher
Zeitung, #1329, 11 October 1922.

[†] Naef, Adolf. Goethe, Haeckel, und historische Biologie. Neue Zürcher Zeitung, #1390, 24 October
1922.

[‡] On Jollos, see Harwood, 1993; Dietrich, 1996; Levit and Olssen, 2006; Rürup and Schüring, 2008.

[§] Jollos was noted for his high financial demands after his emigration to the United States in 1933:
Deichmann, 1996, p. 20.

that Hescheler had confirmed Naef's "somewhat difficult personality," but testimony obtained from Cairo suggested that the rumors concerning Naef's job application in Cairo had been blown out of proportion. An unnamed source from Cairo confirmed: "I heard, his predecessor, Prof. Jollos, is complaining in Germany about how Naef obtained his position. I happen to know the case and am of the opinion that Naef acted correctly." Nevertheless, concerns regarding Naef's personality persisted, and eventually found their way into the final summary report of the search committee. The *Kuratel* (guardianship) of the university promptly decided against Naef, and recommended the young Portmann—intensely lobbied for by Zschokke—as the best choice for appointment instead. The whole affair touched off a heated controversy about academic standards, academic authority and representation in university administration, and nepotism driving academic appointments.[*]

With Portmann's appointment, Naef's hope to return to an academic position in Europe was dashed. He stayed at Cairo University to the end of his life. Naef died on May 11, 1949, in the hospital *Neumünster* (today *Zollikerberg*), Zurich, Switzerland, of pancreatic cancer.

LOGIC MEETS HISTORY

Whoever was asked to review Naef's work invariably commented on its excessive philosophical burden. This certainly revealed some of Naef's personal inclinations, as well as his impatience with the *Haeckelianer* who rode roughshod over foundational philosophical issues when transforming idealistic morphology into phylogenetic morphology, or phylogenetic systematics. Naef declared as his ultimate goal the development of a unified phylogenetic—morphological comprehension of the *type*. He preferred to call his approach "systematic" morphology, but titled his influential 1919 brochure, which he originally had intended to be an introduction to his cephalopod monograph of 1921,[†] "Idealistic Morphology and Phylogenetics." Ever since his student years, Naef said, the theory of descent had been the focus of his research, and he expressed his hope that he would be able to contribute to its justification and improvement.[‡] Commenting on the Scopes Monkey Trial (The State of Tennessee v. John Thomas Scopes) which played out in the small country town of Dayton, Tennessee, in 1925, Naef expressed disbelief that such proceedings could still take place in a modern civilized society: "... when even a man like Henry Ford seriously asserts that Darwinism is just one of those ideas with which malicious Jews had twisted the heads of pious Christians, one or the other would surely throw up his hands in amazement."[§] All the same, Naef opened his seminal *Idealistische Morphologie* not with a reference to Darwin, but with a reference to Goethe, and later specified that the pursuit of comparative morphology in a Goethean spirit to

---

[*] The documentation on Portmann's controversial appointment is archived in the State Archive Basel (*Staatsarchiv Basel*), StABS, ED-REG 1 a 2 1401.
[†] Naef, 1921, p. 5.
[‡] Naef, 1921, p. 1.
[§] "... *so greift sich wohl der oder jener fragend an das Gehäuse seiner Denkdrüse*": Naef, Adolf. Darwinismus. Neue Zürcher Zeitung, #1289, 10 August 1925.

him was akin to learning to empathize with the spirit of a dead language.[*] As he
went about to flesh out his "systematic morphology," Naef certainly developed a
language that few of his philosophically less sophisticated colleagues would fully
comprehend—Hescheler's comments on his *Habilitation* thesis being a case in point.
But at the beginning of his career, he was certainly strongly influenced by Tschulok.
Naef sat in Tschulok's class on the theory of descent, from which he took away
the message that descent with modification "is no longer a hypothesis, but rather
a full-fledged theory, and indeed a unique one of its kind."[†] Following Tschulok's
lead, Naef outlined in the summary section of his doctoral thesis that systematic
biology quite generally is concerned with the conceptual ordering of the organic
world. Systematic morphology more specifically orders organic form conditions on
the basis of *typical similarity*, "whereby the *type* is understood as that condition of
form, either conceptually reconstructed or instantiated in nature, which allows the
derivation from it of similar conditions of form." But since organic form conditions
are realized through embryonic development, "the *type* ... corresponds to a *morphogenetic process* (not to a static, single condition of form)."[‡]

If Naef declared the unified comprehension of the *dynamic type* as the foundational
concept of systematic morphology to be the goal of his research program, he would
obviously have to attempt to bridge the gap between causal relations that are individuating on the one side, and logico-conceptual relations that are spatiotemporally
unconstrained on the other. To not even have recognized this gap was the most central
and fundamental criticism Naef leveled against Haeckel. Only in 1923 would Naef
publish an essay he had written back in 1912, "during a first period of critical reflection on his position by the author, for which reason [this essay] has to be understood as
a premise to his subsequently articulated theoretical expositions."[§] The essay is a perfect reflection of Tschulok's influence on the thinking of the young Naef, derived—as
he wrote in 1912—from insights gained through personal contact, from Tschulok's
published PhD and *Habilitation* thesis, and from Tschulok's well-received book on
The System of Biology in Research and Teaching, published in 1910. This 410-page
volume launches into a historical review of biology that expands into an analysis, from
an empirio-critical (i.e., positivist) point of view, as to how biologists have dealt with,
and should deal with such issues as the relation of description to explanation, of lawful
causality to history, of morphology to ontogeny and phylogeny, of classification to the
phylogenetic tree, that is, the relation of form-relationship to blood-relationship quite
generally. In the preface, Tschulok identified these antinomies as a manifestation of
"two elements that are inherent in every scientific statement: the results of experience
and the requirements of pure reason."[¶] One would be hard pressed to formulate any
better, more succinct characterization of Naef's research program, whereby the reference to the Kantian concept of "pure reason" is particularly revealing. At least in the

---

[*] Naef, 1921, p. 8.
[†] Naef, 1913, p. 359, n. 2.
[‡] Naef, 1913, p. 369.
[§] Naef, 1923, p. 329. Reif (1998, p. 413) emphasized the invariance of Naef's views over the years.
[¶] Tschulok, 1910, p. iii.

German context, positivist philosophy had important Kantian roots* that were either criticized, or adopted and modified, both positions apparent not only in Tschulok's, but also in Naef's theorizing. In his 1908 publication that he submitted as PhD thesis, Tschulok raised the question what this "true essence" could possibly be that is manifest in the *unité dans la variété* as revealed by relations of homology: "This unity in the multiplicity can be understood in two different ways, either in an idealistic, or in a realistic sense."† The idealistic conception considers the type to instantiate a spatiotemporally unconstrained Platonic idea; the realistic conception finds the type to instantiate blood-relationships. Antecedent acceptance of the theory of descent mandates a realistic conception of the type, and a "realistic morphology" must consequently be preferred to an "idealistic morphology."‡ But why accept the theory of descent antecedently? Because, without such preconception, there would be no motivation to search for homology relations that underpin the natural system and with it blood-relationships, argued Tschulok.

As expressed in his 1912 essay, Naef saw things a bit more complicated: for him, the scientific comprehension of the organic world meant to approach this world with the tools of ordering conceptual thought, cleansed of any mechanistic or vitalistic preconceptions. The concepts underlying such systematic research, that is, the concept of the "type," or of "individuality," he correctly recognized as categories of thought. Scientific comprehension of the organic world is achieved if these categories of thought have successfully been applied to the objects of cognition.§ This is not entirely Kant, but pretty close to Kant's *Critique of Pure Reason* nonetheless. It is not Kant because for Naef, categories of thought combined with intuition do not enable the empirical objects of cognition in the first place. All the same there is a Kantian dimension in Naef's account, in that pre-established categories of thought, which are those of logic, are applied to the experienced world of empirical objects and, if that application is successful, enable cognition. In that sense, Naef is closer to the neo-Kantian thought of his own time than to the original Kant.¶

There are a number of fundamental points that Naef adopted from Tschulok, which he then turned into pillars of his own theoretical edifice. One is, again, the logical and epistemological priority of comparative, or systematic morphology over the phylogenetic interpretation of its results. For Tschulok, it is the empirical success in researching the natural system that delivers the proof for phylogeny, and not the other way around. If, conversely, phylogenetic preconceptions would be allowed to influence comparative morphological research and its results, the proof for phylogeny would dissolve into a *circulus viciosus*. The never-ending criticism of phylogenetic

---

* Friedmann, 2000; Carus, 2007.
† Tschulok, 1908, p. 39.
‡ Tschulok, 1908, p. 40.
§ Naef, 1923, pp. 331–332.
¶ See discussion of neo-Kantian schools of thought (the Marburg and Southwest traditions) and their relation to Kant in Friedmann, 2000; Carus, 2007. For an introduction to Kant's Critique of Pure Reason, see Gardner, 1999 (2008).

hypotheses had been, and continued to be, that phylogeny is not directly observable, is not experimentally testable, and thus defies empirical control. Tschulok answered in his PhD thesis:

> The facts that force an acceptance of descent with modification result from *comparative* investigations ... a "mechanistic justification" or an "experimental proof" are neither possible, nor required for the theory of descent, in so far as the latter is based on facts adduced by comparative research.[*]

Not only did Naef adopt Tschulok's position, but again took it a step further. So-called facts generated by comparative analysis can be quite controversial, as the history of comparative morphology shows. Proof of descent with modification for Naef lay not just in comparative analysis capable of generating a hierarchically structured natural system, but in the fact that abstract categories of thought had *successfully* been applied to objects of cognition, that is, that the multiplicity and ambiguity of the experienced world had been mastered (ordered) through the consistent application of categories of logic. Therein, that is, in the very fact that we *can* apply categories of thought based on logic to the experienced world without contradictions, lay for Naef the proof that we have grasped this world correctly and meaningfully.

But debates about proposed phylogenies persisted, of course. Naef was quick to dismiss Haeckel's many phylogenetic trees as spurious, suspect, or plain wrong, but later authors would judge many of Naef's trees just as phantastic.[†] Tschulok, in his *Habilitation* thesis, again tried to improve conceptual clarity by drawing a distinction, fleshed out in his textbook of 1922,[‡] which was not only adopted by Naef, but which proved highly influential in German discussions of phylogeny and evolution in years to come. He saw for historical biology the need to first answer a fundamental question (*Grundfrage*): did descent with modification happen or not? Only if this question is answered affirmatively did two subordinated questions arise: the *Stammbaumfrage*, that is, the problem of the shape of the phylogenetic tree that results from descent with modification; and the *Faktorenfrage*, that is, the causes that drive the process of descent with modification. The affirmative answer to the *Grundfrage* is based, according to Tschulok, not on the proof of any particular phylogenetic tree, but rather on the fact that without acceptance of descent we could not comprehend the graded similarity relations that are expressed in the natural system. Comprehension in this context meant once again consilience of evidence, that is, the fact that we can relate the hierarchy of the natural system with other, independently discovered regularities that prevail in nature, as those apparent in ontogeny, in the Fossil Record, and in biogeography.[§] From there followed, Tschulok's and Naef's interest in Haeckel's Biogenetic Law, and Tschulok's biting critique of Oscar Hertwig's "Ontogenetic Causal Law" that was supposed to render

---

[*] See discussion in Tschulok, 1908, pp. 45–50.
[†] Weber, 1958, p. 10.
[‡] Tschulok, 1912, p. 11 (*Stammbaumproblem, Faktorenproblem*; Tschulok, 1922, pp. 181ff).
[§] Tschulok, 1912, p. 35.

Haeckel's Biogenetic Law superfluous.* Naef, in contrast, replaced Haeckel's Biogenetic Law with his own law of the *Primacy of Ontogenetic Precedence*[†] that was based on von Baer's model of individual development and the principle of generality derived from it (see below).

Naef took an affirmative answer to the *Grundfrage* for granted, even though critics of his "idealistic morphology" would falsely accuse him of rejecting phylogeny. He was not going to waste time on this issue, however. He was also not going to address the *Faktorenfrage* in any detail, as he considered it to be a problem of physiology, not of comparative morphology. For him, comparative, idealistic, or systematic morphology, which are all one and the same thing, was a science of organismic form. The central problem he focused on was how a science of form that underpins the logically structured hierarchy called the natural system can relate to causal processes such as ontogeny and phylogeny, the first epistemically directly accessible, the second leaving its traces in the Fossil Record. That was a question the empirio-critical Tschulok had neither raised nor answered, as for him, the empirical objects of cognition were directly given in perception without the mediating role of matching categories of thought. But before the natural system could be related to any historical processes, its phylogenetic interpretation therewith strengthened, it had to be reconstructed in the first place. It was, again, Tschulok who in his *Habilitation* thesis, and in his textbook of 1922, introduced a crucial distinction, one that was again adopted by Naef. Both Naef and Tschulok, as well as Gegenbaur and all comparative morphologists in between started from the seriability of organisms, their organ systems, and organs in their comparative analysis of morphology. In the context of idealistic morphology, such a series of form conditions could be read in either direction: all it was meant to express was the *unité dans la variété*, the unity of type. When transforming idealistic into phylogenetic morphology, Gegenbaur had read evolutionary transformation into such a series of form conditions. But such a processual interpretation polarized the series of form conditions: it would progress from an evolutionary more primitive, ancestral condition to an evolutionarily more derived, descendant condition of form. To render the natural system amenable to a phylogenetic interpretation, Tschulok consequently requested that "for any sorts of conditions of form, a comprehensive and critical examination of the material must seek to identify the original and the derived type." Later, he issued the same imperative more forcefully: "the *conditio sine qua non* for the reconstruction of phylogenetic trees is the distinction of the primitive and derived condition of form."[‡] The principle that Tschulok used to polarize morphological transformation series, one that Naef also adopted in his own work, was based on logic and was later dubbed

---

* Tschulok, 1912, pp. 40ff. Hertwig's *ontogenetisches Kausalgesetz* was derived from Carl Nägeli's idioplasm theory on the one hand, from August Weismann's separation of the germ-line from the somaticline of cells on the other. Hertwig argued that the soma of organisms had experienced progressive evolution through geological time that led to greater complexity of organization, a similar increase in complexity must have taken place in parallel in the germ cells. That way, species-specific traits of the adult organism are determined by particular parts of the idioplasm of the species-specific germ cell (*Artzelle*) from which it develops.

[†] Naef, 1917, pp. 3, 36; 1919, p. 19.

[‡] Tschulok, 1922, p. 197.

the principle of generality. According to that principle, the more general condition of form is the original, primitive, or ancestral one, the less general is the derived, specialized, or descendant condition of form. Although employed by both Tschulok and Naef, neither of them provided a definition of that principle. Its most succinct definition stems from modern systematics,[*] but it specifies a pattern of reasoning that was entirely characteristic of Tschulok and Naef as well: "character x is more general than character y if and only if all organisms possessing y (at some stage in ontogeny) also possess x and in addition some organisms possessing x do not possess y."[†] The three-fold parallelism that Agassiz had identified in classification, ontogeny, and fossil succession, and that Haeckel had infused with evolutionary meaning, Naef restated in terms of this principle of generality: a parallelism of morphological precedence, ontogenetic precedence, and stratigraphic precedence.[‡]

PLATO'S TURNTABLE

Naef worked with three fundamentally (i.e., ontologically) different, yet closely interconnected theoretical constructs: the transcendental hierarchy of types, the logically structured natural system, and the historically conditioned phylogenetic tree. The ontological glue between these categories is the species, the basic unit of the hierarchy of types, of the natural system, and of the phylogenetic process/tree. In contrast to Tschulok, who—like Haeckel—considered the species to be a conceptual construct just like any other taxonomic group/category,[§] Naef took species to be real, that is, spatiotemporally located entities, complex wholes composed of causally interdependent parts[¶]: "the species, like individuals, are given in nature, forming entities that are tied together through physiological relations, that is, they constitute naturally occurring reproductive communities."[**] To be sure: in an epistemological, that is, methodological context, the systematist discovers species through the recognition of the species-specific type, by the characteristics that guide the systematist in his practice of sorting biodiversity into Linnean species.[††] But in the ontological dimension, the species forms a naturally occurring reproductive community located in space and stretching through time, that is, it forms a species lineage which, as it splits and splits again, gives rise to the phylogenetic tree. To characterize the sexually reproducing species as distinct from all other categories in the hierarchy of types, in the natural system, and in phylogeny, Naef invoked Oscar Hertwig's insight that the causal interrelations that tie together organisms of the same species are not bifurcating, but reticulating, thus forming a "genealogical network" that is

---

[*] On the relevance for modern systematics, see Rieppel, 2010.

[†] de Queiroz, 1985, p. 283, following a suggestion by Wayne P. Maddison.

[‡] Naef, 1917, pp. 3, 36; 1919, p. 19.

[§] Tschulok rejected talk about the "origin of species," "for species are abstract concepts" (Tschulok, 1912, p. 11).

[¶] Naef, 1932, p. 6.

[**] Naef, 1932, p. 6; see also Naef, 1933, p. 12.

[††] Naef, 1921, p. 12. Compare Brigand (2009) for a modern analysis of such context sensitivity of the ontology of biological species as either individuals or natural kinds. Note, however, that Brigand works with two, Naef in contrast with three categories.

instantiated by sexually reproducing organisms.[*] For Naef, species are real, that is, spatiotemporally located relational systems, integrated through reproductive relations that—in contrast to the hierarchy of types,[†] the natural system and the phylogenetic tree—form a reticulate structure. As such, and in contrast to higher categories of the natural system, species are epistemically (empirically) directly accessible, "tied together through testable (*kontrollierbare*) natural factors."[‡] The higher categories, in contrast, correspond to "concepts, which result from a comparative-abstracting contemplation,"[§] although monophyly remained a desideratum:

> With adequate knowledge of the phylogeny, one rule would suffice, "to place in a systematic group always all and only such forms which have derived from a common ancestor" ... groups of a polyphyletic origin are not natural groups.[¶]

The centerpiece of Naef's theoretical edifice was the hierarchy of types, which itself was built on metaphysical premises, two of which were of prime importance. The first is the Aristotelian/Leibnizian Principle of Continuity, expressed in what Darwin called "that old canon in natural history of '*Natura non facit saltum*'."[**] The second is a dynamic conception of nature, ultimately derived from German Idealism. Naef found in Goethe's musings on plant metamorphosis the motivation for a dynamic conception of the type. The type becomes a *Werdetypus*,[††] a becoming type that is variable and capable of metamorphosis, such that one type can morph into a derivative one through an uninterrupted series of intermediates. To understand what Naef is saying here requires more analysis of his concept of the type in the first place. This is perhaps best achieved using the crystal analogy of organic bodies, popular since Georges Buffon's *Histoire Naturelle* of 1749, and Pierre Louis Maupertuis' *Système de la Nature* of 1751, and culminating in Haeckel's *Promorphologie*.[‡‡] Naef was motivated to turn to the crystal analogy not by Haeckel's writings, but by the work of the prominent mineralogist Paul Niggli[§§] (1888–1953) from the University of Zurich instead. Niggli sorted crystals into species just as Naef sorted organisms into types. And just as particular crystals approach their ideal physical structure that defines their species to an imperfect degree only, so do organisms instantiate their type to a variable degree only. Consequently, Niggli subsumed under the same species all those particular crystals that could be arranged in a continuous series of form that would allow their phenomenological transformation one into the other—and the

---

[*] Naef, 1921, p. 31, see O. Hertwig, 1917; 1918a, p. 236.

[†] "There exist no reticulating relationships" of form (Naef, 1931a, p. 13).

[‡] Naef, 1933, p. 44.

[§] Naef, 1933, p. 44.

[¶] Naef, 1911, pp. 152–153; translation from Bieler 1992, p. 311.

[**] Darwin, 1859, p. 194; see Naef, 1931b, p. 97; 1932, p. 4; 1933, pp. 14, 39.

[††] Naef, 1931b, p. 98.

[‡‡] The neo-idealistic morphologist Eduard Jacobshagen (1886–1967) continued to pursue *Promorphologie* (Jacobshagen, 1927), for which he was criticized by Naef: "Jacobshagen's *Promorphologie* and doctrine of symmetries, as well as Haeckel's, is of no central significance, but at best an accessory auxiliary method, which nobody will oppose as long as its inherent limitations are observed" (Naef, 1927, p. 188).

[§§] Laves, 1953; on Niggli's congeniality with idealistic morphology, see Weber, 1958, p. 12.

exact same was true for Naef and his systematic morphology. Organisms not only instantiate their type variably; to the same type belong those organisms that can be arranged in an unbroken series of form that would allow the ideal (abstract, mental) transformation, or metamorphosis, of one into the other.

Naef took token organisms to relate to their type in the same way as its variations relate to a musical scheme: "The relation between the organic forms that group around an archetype (*Urform*) can thus be of a purely ideal nature, comparable to the variations of a musical theme."[*] Significantly, this is not a causal relation, but an entirely ideal relation, one that is not accessible to direct observation, but must be based on a mental representation, a vision generated by the "inner," "spiritual eye."[†] The metaphor of a musical scheme and its variations reveals the type as a *Gestalt*[‡] that is grasped by the intuition of the expert with the primacy of inner necessity.[§] Metascientific concepts such as *Gestalt*, intuition, and visions generated by the spiritual eye are clearly elements of Goethe's synthetic morphology seeking a qualitative, not a quantitative comprehension of nature, one rooted in intuition rather than being based on the principle of causality. But even though the type itself is abstract, its metamorphosis ideal, it must at least in principle be instantiable in nature; it cannot be "an arbitrary, unnatural, imaginary construction," a mere "scheme on paper, without life and colors."[¶] The characteristics that mark out a type must be of a kind and combination that could in principle characterize a living organism.

What is true of any one type is also true of the nested hierarchy of types that results from broader morphological comparison: although ideal, transcendental in nature, that is, transcending the multiplicity of organic appearances in the search for underlying unity (regularity, lawfulness[**]), the hierarchy of types must in principle be instantiable in nature and in that sense can teach us something about nature—if it has been successfully constructed. Given Naef's belief in a lawful determination of all of nature, he found it "possible to predict the existence of as yet unknown forms: This is quite similar to the prediction of as yet unknown chemical elements that was made possible by the Periodic Table."[††] Since "the fossilized remains of lost worlds can only be naturally interpreted in the context of the lawfully determined order that is revealed by systematic morphology,"[‡‡] his systematic morphology, so Naef asserted, could serve as a guide to the "targeted search" for predicted intermediate forms in nature, not only among extinct, but also among extant species.[§§] But in order for this to be possible, Naef had to show that his hierarchy of types could be related to classification, that is, the logically structured natural system on the one hand, and the historically conditioned phylogenetic tree on the other.

---

[*] Naef, 1917, p. 16.

[†] *geistiges Auge*: Naef, 1932, p. 12; see also Naef, 1931b, p. 96.

[‡] A *Gestalt* was most famously compared with a musical theme by Ehrenfels (1890). While not discussed by Naef, Ehrenfels (1890) offers valuable insights into the nature of such a *Gestalt*: more than the sum of the elements it is based upon, the *Gestalt* is immediately apprehended, not rationally analyzed.

[§] Naef, 1932, p. 12.

[¶] Naef, 1921, p. 10.

[**] Naef, 1925, p. 235.

[††] Naef, 1913, p. 348.

[‡‡] Naef, 1925, p. 238.

[§§] Naef, 1933, p. 39.

With his distinction of the hierarchy of types, the natural system, and the phylo-genetic tree, Naef was seeking a clear separation of the ideal from the real, of the conceptually comprehended from the causally determined, of the logical from the historical. On the one side of this ontological divide reside individual organisms, or complex wholes that are located in time and space and form causally integrated relational systems, that is, species and groups of species of common evolutionary origin. On the other side reside spatiotemporally unconstrained conceptual constructs, that is, classifications of the organic world that are integrated not through historical, but through logical relations instead. Naef's third category, the transcendental hierarchy of types, was intended to serve as a bridge across this ontological divide. One of the central concerns of German Idealism was to transcend the divide between Platonic realism, according to which only universals are real, and nominalism, according to which only particulars are real. How is it possible to bridge that gap, to imagine a way in which the universal would be instantiated in particulars without suffer-ing metaphysical headaches?[*] The philosopher Ernst Cassirer characterized Naef as a "modern exponent of the theory of evolution," one who raised questions that could "almost be called 'transcendental' in Kant's sense." Only through thorough conceptual analysis, Cassirer insisted, could "naïve phylogenetics be transcended and the doctrine of evolution [be] raised to the rank of a science that is conscious of its special task and methodological implications."[†] It was Naef who had criticized Haeckel's research program as "naïve phylogenetics"[‡] because Haeckel, as much as Gegenbaur, had failed to adequately deal with this ontological divide that so deeply tore their phylogenetic morphology apart.

For Naef, the "natural system" is a nested hierarchy of intensionally defined classes, or *structural kinds*. As such, the natural system mirrors the hierarchy of type, but *is* not the hierarchy of types. The natural system is not intuited, or appre-hended as is the hierarchy of types, but logically structured instead. Each level of inclusiveness in the natural system forms a systematic category, the corresponding systematic groupings (structural kinds) diagnosed (i.e., defined) by properties that are each necessary and jointly sufficient for inclusion in the kind. All organisms that share the necessary and sufficient membership criteria are included in a systematic, that is, a structural kind, while all other organisms remain excluded: "there is no third alternative."[§] The logical basis for the reconstruction of the natural system is thus revealed to be the Law of Excluded Middle. As the more general condition of form defines the more inclusive, the less general condition of form defines the less inclusive systematic groupings or structural kinds, the diagnoses of more inclusive groupings will be more austere, the diagnoses of the less inclusive groupings more rich in detail. The natural system so conceived affords considerable predictive power with respect to structural characteristics: any snail will be characterized by a shell, unless it is a slug, a sub-group of snails which lost the shell. As one proceeds from more inclusive to less inclusive levels within the hierarchy of the natural system,

---

[*] Naef, 1932, p. 15.
[†] Cassirer, 1950 (1978), p. 170.
[‡] Naef, 1919, p. 3.
[§] Naef, 1919, p. 25.

the predictive power with regards to structural characteristics will increase, whereas the predictive range will correspondingly decrease.

As to the graphic representation of the natural system, Naef realized that its hierarchical structure lends itself most readily to a depiction in terms of a nested system of circles, or of boxes within boxes,* forming a *"Schachtelhierarchie."†* The same type of hierarchy can also be readily represented in terms of a bifurcating tree-like diagram, but in that case, Naef realized, phylogenetic connotations would quickly be associated with the natural system. He therefore took care to speak of a "tree-like construction,"‡ to again emphasize the fact that the natural system is a conceptual construct without genealogical dimension. Reading a tree-like representation of the natural system as a phylogenetic tree would amount to fall victim to the very category mistake that Haeckel committed when dressing up the results of comparative morphology in terms of phylogenetic trees without any further conceptual analysis.

To go from the natural system to the phylogenetic system means to cross the ontological divide that separates logic from history, form-relationship from blood-relationship. Phylogeny is about descent with modification; it is a causally determined process located in space and stretching through time, a process furthermore that is subject to unbroken continuity as *"Natura non facit saltus."§* This clashes with the discontinuity inherent in the conceptual foundation of the natural system. The natural system can for this reason not simply be turned into a phylogenetic one. On the other hand, Naef emphasized that both, natural systematics and phylogenetics deal with the same facts yet from different perspectives, and the facts he had in mind were those expressed in the hierarchy of types. Considered from the point of view of systematics, which seeks a comprehension of form-relationships, the hierarchy of types readily translates into the natural system, a hierarchy of structural kinds. Considered from the perspective of the theory of descent, which seeks a reconstruction of blood-relationships, a phylogenetic tree can again readily be derived from the hierarchy of types. In that sense, the hierarchy of types provides a link between the two worlds involved—the conceptual-logical (*erkenntnistheoretisch-formal*) one of the natural system, and the empirical-historical (*naturtatsächlich*) one of phylogeny.¶

The categories of the natural system, intensionally defined classes or structural kinds, are universals. Species, and groups of species of common descent, are spatiotemporally located particulars or groups of particulars. How could one be understood in terms of the other without committing a category mistake? Impossible, argued Naef, unless in a transcendental realm where—as in Goethe's synthetic science—the universal and the particular are not mutually exclusive branches on the porphyrean tree, but instead are "intimately connected," interpenetrating one another** as they do in Naef's hierarchy of types.†† The hierarchy of types

---

* Naef, 1931b, p. 84.
† Naef, 1931b, pp. 3, 9.
‡ Naef, 1922, p. 302; compared to a "mathematical decision tree" by Breidbach (2003, p. 185).
§ Naef, 1931a, p. 97; 1932, p. 4; 1933, pp. 14, 39.
¶ Naef, 1926, p. 406.
** Cassirer, 1950 (1978), p. 45.
†† Bloch, 1956, pp. 102–103; see also Bloch, 1952.

is Naef's Platonic turntable*: turning it in one direction allows the derivation from it of the logically structured natural system; turning it in the opposite direction allows the derivation from it of the historically conditioned phylogenetic tree. The pivot around which the table turns is the species.

BIOLOGICAL FRAGMENTS CONCERNING AN UNDERSTANDING OF MAN[†]

Whenever he could travel, Naef would spend summer in Switzerland. Arriving there in 1947, he carried with him a frayed leather briefcase with an as yet incomplete manuscript for a three-volume textbook on comparative anatomy, for which he hoped to win a contract with a Swiss publisher. His previous publisher, Gustav Fischer, was now located in Jena in the Eastern (Soviet) Zone of occupied Germany, and hence beyond Naef's reach. Indeed, Naef was unable to obtain a book contract for this project until his untimely death in 1949. Naef's textbook remained incomplete and unpublished.[‡]

Naef's erstwhile contender for the Zschokke succession in zoology at the University of Basel, Adolf Portmann (1897–1982), proved more successful with a more modest project. He published, in 1948, a one-volume textbook on the comparative anatomy of vertebrates through Schwabe & Co. in Basel. Although less drenched in philosophy than Naef's tracts, the book nevertheless carried the spirit of idealistic morphology forward into an increasingly hostile environment. The book is still in print today, albeit in a revised and amended edition. Another project Naef left behind unfinished was the cephalopod volume of the highly regarded, standard setting reference book series *Traité de Zoologie*, edited by Pierre P. Grassé from the Sorbonne University in Paris. After Naef's death, the project was re-assigned to Portmann.[§]

Portmann was tall and lank, with clear-cut features, the eyes set deep in their sockets, friendly but reserved.[¶] As was customary in those days, he wore a white lab coat when at work at the institute. The Zoological Institute, of which he was director, occupied an original building of the University, founded in 1459. Perched high above the Rhine, it was located right at the center of the old town. The *Unteres Kollegium* formed (as it still does) a landmark in a triangle that also included the cathedral (the *Basel Münster*) and the Augustinian Monastry. The latter had served as headquarters for the University (*Oberes Kollegium*) once the monks had abandoned it, but in the 1840s partial demolition and rebuilding transformed the site into the *Museum an der Augustinergasse* with an anthropological, zoological, osteological, and geological section. When a new administrative center for the university was inaugurated on June 10, 1939, the old structure at the *Rheinsprung* was turned over to zoology,[**]

---

* Rieppel, 2011b.
† Portmann, 1944.
‡ An introductory chapter to Naef's textbook was translated and annotated by Rieppel et al. (2013).
§ Roper, 1990.
¶ Illies, 1976. This book contains an authorized biography of Portmann, from which some details of the present account are taken. See also Stamm and Fioroni, 1984; Koechlin, 2004.
** http://www.altbasel.ch/haushof/kolleg.html (accessed March 10, 2016).

except for the uppermost floor that housed the theological faculty. From the window of his office, Portmann could overlook the Rhine flowing below the oldest and central bridge—the *Mittlere Brücke*—that connects greater Basel with lesser Basel situated on the northern bank of the river, the proletarian section of town where Portmann grew up. From there, he could let his gaze wander across the border into German territory where the foothills of the Black Forrest rise behind Lörrach, a town where his grandfather had found employment in the textile industry. In clear weather, a look to the left would reveal the distant Vosges Mountains, located in ever-disputed Alsatian territory that was once again wrenched from German occupation late in 1944 as the war drew to an end. While taking in the view, Portmann pondered remodeling projects that would render the old structure more suitable to his needs (Figure 3.3).

Portmann was still at an early stage of what would turn out to be a brilliant career, scientifically as well as administratively, at the university of his hometown. Along the way, he would collect numerous honors and accolades, among which four honorary doctorates.[*] But back in 1931, he had made a rather unlikely contender for the position he now held. Born on May 27, 1897, in Basel, he came from a blue-collar background steeped in a commitment to socialist reform politics.[†] Never an active

FIGURE 3.3   Adolf Portmann. (UB Basel, Portr BS Portmann A 1897, 2; courtesy of Claire Roessiger. With permission.)

---

[*] 1956: Université d'Aix-Marseille; 1957: University of Freiburg im Breisgau (Portmann had consulted the French Occupation Authorities on the denazification of that university); 1967: University of Heidelberg; 1970: Université de Fribourg (Switzerland).

[†] Simon, 2010, p. 25.

politician himself, Portmann remained supportive of socialist politics throughout his life: "as a young man, I was labeled a bolshevist," Portmann once explained.[*] This is not the sort of provenance that was typical of the Basel professoriate. Portmann earned his PhD in 1921 with research on the systematics and behavior of dragon-flies of the surroundings of Basel, which he pursued under the supervision of his fatherly mentor, the limnologist Friedrich Zschokke, then director of the Zoological Institute. Supported by an anonymous patron, Portmann traveled vastly during his early postdoctoral career, visiting the universities in Geneva, Munich, Berlin, and Paris, and the marine laboratories in Helgoland, Roscoff, Banyuls-sur-Mer, and Villefranche-sur-Mer.[†] While at the University of Geneva, Portmann balanced the Darwinism that pervaded the courses of the experimental biologist and compara-tive anatomist Émile Guyénot (1885–1963), with the anti-Darwinism articulated by Jean-Henri Fabre (1823–1915) in his famous *Souvenirs Entomologiques*. In Munich he heard Germany's leading zoologist Richard Hertwig (1850–1937), and watched Karl von Frisch (1886–1982) decipher the language of bees. Working in Helgoland, Portmann befriended the holistic/organicist biologist and animal psychologist Friedrich Alverdes (1889–1952), at the time at the University of Halle (Saale). In Berlin, he imbued Oscar Hertwig's[‡] (1849–1922; Richards elder brother, both for-mer students of Ernst Haeckel) critique of the purported sufficiency of Darwin's twin factors, random variation, and natural selection, to explain all evolutionary change.[§] The cell biologist Oscar Hertwig was also known for his opposition to reductionist perspectives in experimental biology,[¶] as well as for his biting critique of Social Darwinism,[**] which thrived in Germany well before the *Machtergreifung* by the National Socialists under Hitler in 1933.

In 1925, Zschokke promoted the traveling Portmann *in absentia* from second to first assistant at his institute, where he would earn the *venia legendi*, the right to teach through his *Habilitation* in 1926. To defend his *Habilitation*, Portmann had to deliver a public lecture, which he did on Monday, June 13, 1927, in the auditorium of the Basel Natural History Museum. A summary of his lecture, titled "Current Problems of the Theory of Descent" Portmann published two days later in the local press.[††] Portmann's presentation contained many lines of thought that reflected the anti-Haeckel faction of German biology of the past two or three decades, and that would persist throughout his own later work: the modern mutation theory that was called upon to complement Darwinian selection theory still remained rather incom-plete, insufficient to explain all evolutionary change; the problem of saltational evolutionary change and the acquisition of detrimental features seemingly defy the effectiveness of natural selection; Darwinian selection theory consequently has lim-ited explanatory power; the current fad to favor and support experimental biology at the detriment of comparative anatomy (morphology) is misguided in light of the fact

---

[*] Ritter, 2000, p. 212; see also Illies, 1976, p. 115.
[†] Portmann, 1964; see also Illies, 1976.
[‡] Weindling, 1991.
[§] O. Hertwig, 1918a.
[¶] O. Hertwig, 1897.
[**] O. Hertwig, 1918b.
[††] *Basler Nachrichten*, June 15, 1927.

that the latter alone provides the basis for the reconstruction of the natural system that triggers interest in evolutionary questions in the first place. And finally, but perhaps most importantly, Portmann called for resistance to the threat or temptation to abuse popularized versions of evolutionary theory in misguided yet dangerous political or religious debates.

Per decree of the Education Committee of the government of Basel from January 7, 1929 (renewed September 24, 1929), the young *Privatdozent* took over the management of the Zoological Institute as an ailing Zschokke found himself unable to perform his duties as its director. Eventually, Zschokke retired at age 70, at which point Portmann—among others—applied for the vacant position. In its recommendation[*] to the board of directors (*Kuratel*) of the university dated April 18, 1931, the search committee placed Portmann at the bottom of the short list, on third place together with a young German zoologist from Kiel, Adolf Remane[†] (1898–1976), whose candidacy had been promoted by the director of the Kaiser-Wilhelm Institute of Hydrobiology in Plön near Kiel, the famous limnologist August Thienemann (1882–1960). *Primo loco* was listed Jean Strohl (1886–1942), a former student of August Weismann[‡] and currently full professor of physiological zoology at the University of Zurich.[§]

In its report dated June 10, 1931, the board of directors (*Kuratel*) of the Basel university rejected applicants of German nationality as there were Swiss applicants of equal qualification. Going against the search committee's recommendation, the *Kuratel* placed Portmann ahead of Strohl for a number of reasons, viz. his reputation as an excellent teacher at university and high school level, his successful *interim* management of the Zoological Institute, his youth that promised greater engagement and creativity in research and education, and which also made him a cheaper hire than Strohl. The ensuing election of Portmann as Zschokke's successor by the Executive Committee of the government of *Kanton Basel-Stadt* on September 4, 1931, triggered bitter controversy, because it violated the principle of hiring the best qualified candidate as determined by an academic search committee composed of competent peers. The documentation relating to Portmann's appointment reveals a heavy hand played by Zschokke, himself member of the search committee, in support of his former apt student and assistant. Theodor Beck, member of the Education Committee of *Kanton Basel-Stadt*, and *ex officio* also a member of the search committee, complained about the "favoritism on the part of retired professors in support of their students in the resolution of issues of succession."[¶]

Portmann's brilliance in teaching, education, and administration soon dispelled any doubts that might have lingered since his controversial appointment. He also sought broader exposure to the general public through books and newspaper articles, public lectures, and radio shows. At the end of his career, he had acquired a reputation as an unconventional, if not eccentric scientist, one who through his critique

---

[*] *Staatsarchiv Basel*, StABS, ED-REG 1 a 2 1402.
[†] J. Remane, 2003; Zachos and Hoβfeld, 2001, 2006.
[‡] Harwood, 1993, p. 235.
[§] Peyer, 1942.
[¶] Letter to the Secretary of the Education Council, dated August 25, 1931. *Staatsarchiv Basel*, StABS, ED-REG 1 a 2 1402.

of the rising potential, and threat, of molecular biology had effectively maneuvered himself into the offside. But at the outset of his career, when molecular biology was still the promise of a distant future, his main concern was the ideologization and instrumentalization of biology along social and political dimensions.[*] A classic if distorted analogy that placed biology in the service of social and political indoctrination was the Cell State Theory,[†] one that Portmann never tired to criticize and expose as being vacuous. The Cell State Theory originated with the German anatomist, biologist, and political activist Rudolf Virchow, and was further most famously expounded by the Social Darwinist Ernst Haeckel, and the evolutionary biologist August Weismann (1834–1914),[‡] with a notable anti-Darwinian contribution from Oscar Hertwig.[§] The latter turned against the idea that the Darwinian principle of the struggle for survival could be extended to the parts within the organism,[¶] ultimately determining the relations among the cells that make up an organism, with putative social implications for individuals that collectively make up a state. The earliest critique by Portmann of the "state as an organism" metaphor appears in lectures he delivered in 1943,[**] where he took issue with the claim that a state forms a supraindividual complex whole, a superorganism so to say, whose struggle for survival requires the subordination of all its constituent individuals under the interests and goals that define the greater common good. Portmann returned to such totalitarian portrayal of the German National Community again and again, eventually appealing to the concept of *Diachorese* developed by the Tübingen anatomist and histologist Martin Heidenhain (1864–1949).[††] Arguing against the building block theory of cellular composition, Heidenhain had maintained that it is not cells coming together to form a complex whole, but that instead the complex whole partitions itself through cell division. Friedrich Alverdes,[‡‡] by that time at the Philipps-University of Marburg, took Heidenhain's theory to imply that the complex whole that is an organism provides a preformed matrix—a *Gestalt*—into which the cells are cast as they divide and multiply. For Portmann, cells are never true "elements" of an organism, as individuals are of a state. Instead, there is in nature "a plasmatic whole partitioning itself—that is the central point" which, according to Portman, refutes not only the building block theory of cellular structure, but with it also the Cell State Theory.[§§] These are all concepts that will require careful unpacking in the pages to follow.

The biological discipline on which the heaviest ideological burden was placed was the theory of descent, a process that in German-speaking Europe had started with Ernst Haeckel, and that reached its climax during the National Socialist regime under Hitler. "National-Socialism is politically applied biology" was the famous slogan proclaimed by the "Nazi of the First Hour" Hans Schemm (1891–1935; NSDAP, 1923),

---

[*] Ritter, 2000; Simon, 2010, p. 55.
[†] Reynolds, 2007, 2008.
[‡] See discussion in Richards, 2008.
[§] O. Hertwig, 1922.
[¶] Roux, 1881a.
[**] Ritter, 2000, p. 213.
[††] Heidenhain, 1907.
[‡‡] Alverdes, 1932, p. 103.
[§§] Ritter, 2000, p. 224; compare Portmann, 1960, pp. 36–37.

founder then *Führer* of the notorious National Socialist Teachers' Association (*National-Sozialistischer Lehrerbund*, NSLB) until his death in a plane-crash on March 5, 1935.[*] It was precisely such biologism that pervaded the German National Community (*Volksgemeinschaft*),[†] promoted as it was by Nazi propaganda machine, against which the young Portmann took a stand. But in so doing Portmann attempted to broaden the scope of biology in a philosophical, anthropological, indeed humanistic direction. Portmann's agenda in that respect mirrored the circumstances in which he worked. Colleagues among the faculty at the University of Basel with whom he interacted included such luminaries as the theologian Karl Barth (1886—1968) who resided in the same building as Portmann, and the psychiatrist and philosopher Karl Jaspers (1883–1969). As of 1946, Portmann was invited to participate, and eventually became Chair of the annual *Eranos* workshops held in Ascona, located at the foot of Monte Verità that rises up from the Lago Maggiore in southern Switzerland. There he would meet the psychiatrist C.G. Jung (1875–1961), the philosopher and religious scholar Karl Kerényi (1897–1973), the physicist Erwin Schrödinger (1887–1961), and others of like mind. Also after the war, Portmann became member of a circle that regularly met at the house of the botanist Willhelm Troll (1897–1978) in Mainz to discuss aspects of holistic/organicist biology. Troll had taught a *Gestalt* colloquium at the Martin Luther University of Halle (Saale) and co-founded/co-edited a journal with the same title, prior to his deportation to the Western (US) Sector in the so-called *Abderhalden* transport of June 24, 1945. Having found a new position at the Johannes Gutenberg University of Mainz, he continued discussions of *Gestalt* phenomena in biology at his private residence.[‡] Against such a background, Portmann quite generally rejected a purely mechanistic approach to biology as had been pioneered by the experimental embryologist Wilhelm Roux (1850–1924), who sought to reduce biology to physics, the organism to a system of forces.[§] He also rejected the instrumentalization of biology in the attempt to dominate nature, to engineer life. Portmann infused his approach to biology with his deep appreciation for art, he allowed for ultimate mysteries that governed the shaping of life through geological time, and he emphasized the outward manifestation of an inward essence in the organic *Gestalt* that was not and could not become subject to the forces of natural selection. With respect to the Portmann School, Harvard luminary Stephen Jay Gould famously commented: "These works deserve to be far better known among English-speaking scientists. They have been neglected, in part because the articles are long, difficult, and invariably *auf Deutsch*; in part because Portmann's non-Darwinian perspective evoked little sympathy."[¶] Indeed, it is ironic that his erstwhile competitor for the Zschokke succession, Adolf Remane—who along with his promotor August Thienemann got entangled in ways to be discussed later in the very Nazi biologism that Portmann combatted—would turn against Portmann after the war. At a time when people in Germany still tried to blanket the past and its horrors, motivated as they had been

---

[*] Klee, 2003, p. 530.
[†] Pine, 2007.
[‡] Nickel, 1966, p. 182.
[§] Cassirer, 1950 (1978), p. 189. Portmann gleefully pointed out that even Wilhelm Roux had allowed for an inwardness of organisms which ultimately transcends physiological analysis (Illies, 1976, p. 169).
[¶] Gould, 1977, p. 349.

by a misguided, trivialized and instrumentalized Social Darwinism, the German Zoological Society held its annual meeting in Munich from June 2 to June 6, 1963. Remane, then president of the society, opened the meeting with a speech in which he called the twentieth century unabashedly and without remorse the "century of biology." Taking up the Portmannian concepts of "self-representation" and "inwardness," he rejected such anthropocentric tendencies, which he characterized as "metaphysical thoughts *(metaphysische Schreibtischgedanken)* of an arm-chair biologist" without any empirical foundation.*

During the era of the Third Reich, the *Abstammungsgedanke*, the Theory of Descent, had given rise to a whole series of powerful myths: the struggle for survival as the foundation of all life, the superiority of race that would result from it, and the option of keeping the *Volk* healthy and strong through racial hygiene and eugenics based on the *Lebensgesetze*, the laws of genetics.[†] As will be discussed in detail later, the latter was a perspective that Remane himself had promoted at the first annual meeting *(Reichstagung)* of the scientific academies of the NSDDB, the National Socialist University Lecturers' Association *(National-Sozialistischer Deutscher Dozentenbund)*, that was held in Munich from June 8 to 10, 1939;[‡] and again later in an invited contribution to the journal *Archiv für Rassen- und Gesellschaftsbiologie*.[§] In his accommodation to the ideological matrix of the time, Remane followed the lead of his earlier mentor Thienemann,[¶] who saw in such rhetoric a means to increase funding for his research program under the new regime. As of 1942, if not even earlier,[**] Portmann arranged to set the record straight with respect to the tenets of evolutionary theory, at least as he saw it, using the radio as a means to reach a broader public. Interfering with Goebbels' preferred propaganda medium, the *Volksempfänger*,[††] the Swiss National Radio *Beromünster* transmitted Portmann's voice from its studio in Basel: for him, "the radio [was] one of the most important tools in this fight in support of democracy."[‡‡] In a series of six presentations, Portmann in 1944[§§] outlined the contemporary understanding of evolution, the metes and bounds of evolutionary theory, and its significance for an understanding of human origins.

LIMITS OF SCIENTIFIC KNOWLEDGE

The first and fundamental premise deployed by Portmann was the fact that evolution happened in the past, with the consequence that the evolutionary process is not directly epistemically (empirically) accessible. Evolutionary theory for Portmann was of a highly complex structure, many its parts not based on observation and hence neither verifiable, nor falsifiable. Evolutionary theory is not in nature,

---

* Remane, 1964, p. 36. For the deteriorating relations between Portmann and German zoologists such as Köhler, Stresemann, and Remane, see also Ritter, 2000, p. 249; Illies, 1976, p. 167.
† Voss, 1934.
‡ Remane, 1940, p. 126.
§ Remane, 1941, pp. 91, 119; see also Remane, 1942, p. 80.
¶ Thienemann, 1935a, 1939, 1941.
** Roger Alfred Stamm, Portmann estate manager, personal communication.
†† Pine, 2007, p. 169.
‡‡ Portmann, cited in Ritter, 2000, p. 228.
§§ Portmann [1944?].

it is something biologists say about nature, and what biologists—or any other scientists—say about nature can in principle be either right or wrong. Of course Portmann accepted that evolution did happen in the past in the organic world, and that it would continue to happen into the future, but evolutionary theory he did not take as a fact, but as a construct of the human mind instead. Portmann also did not claim that evolutionary theory as it had been worked out by the time of his radio shows was largely wrong or misguided, but he objected to the misguided popularization of evolutionary theory in the service of a totalitarian regime, and he strongly insisted on the claim that there are phenomena in nature that cannot be explained by the Darwinian mechanisms of variation and natural selection. And most importantly, Portmann insisted on the independence of the cultural and spiritual sphere of human existence from a mechanistic biological basis—as was implied by the Cell State Theory in its Darwinian guise.

In his first radio lecture, Portmann emphasized the fundamental distinction of form-relationship (*Formverwandtschaft*) from blood-relationship (*Blutsverwandtschaft*). Whereas the former is the foundation for systematics, the latter is the stuff of evolutionary theory. While DNA had not yet been discovered, first forays into molecular systematics were at the time pursued using blood and the methods of blood serum diagnostics. Quite generally, however, systematics was still based almost exclusively on comparative anatomy (morphology). It was the comparison of animal (and plant) form that would reveal the natural system. The natural system is thus based on the observation of organismic form (morphology); similarities or dissimilarities of form allow the grouping of organisms into a hierarchically structured natural system that forms the basis for animal and plant classification. The natural system is therefore based on observable form-relationships. In Portmann's words, it is what we can *know*; it is empirically secured knowledge. To interpret the form-relationships that are apparent in the natural system as blood-relationships is not something we can know, but something we *surmise*. It is not empirically secured knowledge, but a hypothesis, a theoretical construct. To explain the natural system on the basis of evolutionary theory implies the transformation of form-relationships from the ancestral to the descendant condition of form. Such transformation of form cannot be observed, however, but only inferred on the basis of the results of comparative morphology. According to Portmann, then, there is a stark contrast between empirically based comparative morphology, and the theory of descent with modification. The consequence is that hypotheses of evolutionary transformation must be based on empirically secured form-relationships, that is, on the results of systematics, and not the other way around. This pattern of argumentation is the same that had been used by Naef, a pattern indeed that still was very widespread in German biology. But given the crisis of the time, Portmann went beyond Naef in his radio talk. He emphasized how in spite of its hypothetical nature, evolutionary theory had been abused in political debate for almost a century, leading up to the catastrophic destruction that Germany was facing during this final phase of the war. Concepts such as race (*Rasse*) and living space (*Lebensraum*)* had been loaded with Social Darwinist connotations of the crudest and most unscientific kind in order to justify German expansion into those eastern territories that Germany had lost under the Versailles treaty, and

---

* The concept of *Lebensraum* was famously introduced by Ratzel (1901).

beyond.* Such instrumentalization of evolutionary theory should have been recognized as deeply flawed, given its theoretical nature: "The perspective of deep time—that is the fundamental fact"† which evolutionary theorists have to deal with. Yet Portmann agreed with Naef and other idealistic morphologists (of whom more later) that evolutionary theory most successfully integrates the results of geology, paleontology, and biology, and for that reason is more than a mere hypothesis, but rather a well-founded scientific theory indeed.

Continuing to expound the biological contributions to evolutionary theory before sketching the Fossil Record of hominids, Portmann landed himself in the muddy waters of genetics and the nature of mutations, again ideologically highly charged issues in contemporary biology. Portmann acknowledged the possibility of saltational evolutionary change, and in that context introduced Jurij Philiptschenko's‡ distinction of micro- from macroevolution. At the bottom line, however, and as Remane had done before,§ Portmann issued an *ignoramus* (we do not know) with respect to the effectiveness of mutations in evolutionary transformations: "nobody knows what far-reaching effects mutations can have … We are completely ignorant about the numbers of mutations it took for the emergence of humans."¶ While microevolutionary phenomena appeared to fall within the scope of contemporary explanatory power of biology, macroevolution for Portmann still remained a mystery. Portmann concluded his discourse on the evolutionary origin of humankind with an appeal to a holistic perspective. Against the "ghost in the machine"** doctrine he proclaimed it impossible to conceptually dissect a human being into mind and body. He implied Jungian psychology when he emphasized the uniqueness of human individuality, and of human nature in general with its light and shadow—capable of good and evil as the war had all-too painfully brought out. The reductionist-mechanistic approach to evolutionary theory that had been developed by Ernst Haeckel and his successors in the 1870s and beyond he declared to be dead. Instead, he proclaimed the future to hold the promise for an expansion of biology that will allow the development of a new perception of human nature.

In sketching Adolf Portmann's reception of contemporary evolutionary theory it has become clear how close his views were in one or another aspect to those of his erstwhile competitors for the chair in zoology at the University of Basel, Adolf Naef and Adolf Remane. Indeed, with his emphasis on holism and the importance of *Gestalt*; with his distinction of form-relationship from blood-relationship; with his insistence on the logical and epistemological primacy of comparative morphology (systematics) over the causal explanation of the natural system on the basis of evolutionary theory; with his distinction of micro- and macroevolution and his caution with respect to mutation theory, Portmann echoes many issues prominent in idealistic morphology as developed by German speaking biologists in the first half of the twentieth century. With a professorship in Basel, Portmann was somewhat on

---

* Blackbourne, 2006, chapter 5.
† Portmann, [1944?], p. 11.
‡ Philiptschenko, 1927.
§ Remane, 1939a.
¶ Portmann [1944?], pp. 22–23.
** The slogan is Ryle's, 1949.

the fringes of the debate that raged particularly in Germany between the idealistic morphologists and the pioneers of phylogenetic systematics. He was able to look in from the outside, so to speak, to synthesize what he found of value, and to reject what he did not agree with. Among the idealistic morphologists of the first half of the twentieth century, he was also somewhat of a latecomer. Not Portmann's "biology in the grand style"* to which he aspired—one that dealt with many issues not treated here—but his specific views on comparative morphology and evolutionary theory were retrospective in many ways. Yet this also highlights that much had happened in German biology between the meetings of Abel's evening discussion group in Vienna, starting in 1908, and the start of the appearance of Portmann's voice on Swiss radio in 1942. The people he talked about, the concepts he endorsed or rejected, and lots more will need to be enriched with more detail and history in the chapters to come. Portmann's synthesis sketched above provides a first approach into the tangled and much misunderstood field of idealistic morphology. But as wide and diverse ideal-istic morphology was in terms of a research program, the field as a whole provided the foil against which the Darwinists among the biological systematists developed phylogenetic systematics. The Darwinists represented the Haeckelian phalanx in German biology of the time, and they developed phylogenetic systematics largely as a critique of, or an antithesis to idealistic morphology. To understand the history and philosophy of phylogenetic systematics, it is necessary to understand not only idealistic morphology in its various manifestations, but also the social and, more importantly, the political context in which the often polemical debate between ideal-istic morphologists and phylogenetic systematists was embedded.

---

* *"Biologie grossen Stils,"* Karl Jaspers in his address for Portmann's 60th birthday (Stamm and Fioroni, 1984, p. 112).

4 The Rise of Holism in German Biology

THE CELL STATE

Portmann allied himself with the organicist-holistic camp prominent in German biology, taking a stand against reductionist biology with the cell state as one of its central metaphors. That metaphor, like any other, had its own, long and convoluted history. On February 12, 1859, Rudolf Virchow delivered a lecture in the *Sing-Akademie zu Berlin* on *Atoms and Individuals*. The issue Virchow wanted to address in that lecture was—once again—the intriguing question of biological individuality. Atomism, he explained, was a materialist branch of ancient Greek philosophy represented primarily by *Leucippus* and *Democritus*, where the term "atom," Greek "átomos," signified "uncut" or "uncuttable." The same philosophical tradition continued in ancient Rome with *Lucretius*, who used the term "atomus" to refer to an indivisible particle, the ultimate building block of material objects. The Latin term "individuum" was introduced by Cicero to translate the Greek term "átomos" (ατομος), and thus means "indivisible" as well. "Atom" and "individual" thus became synonymous; to talk about the divisibility of atoms, or individuals, would thus result in a contradiction of terms, a *contradiction in adjecto*. But just recently important work on polymorphism and division of labor in colonial marine invertebrates had been published, such as the 1851 monograph by Leuckart,[*] or Carl Vogt's 1851 treatise[†] on siphonophores, authors Virchow found to have problematized the notion of biological individuality. Reference to Abraham Trembley's (1710–1784) famous and pioneering experiments of 1740, probing the seemingly limitless regenerative powers of the green hydra (*Hydra viridis*), was *de rigeur* in this context as well.[‡] These studies provided powerful motivation for Virchow to investigate the problem of *biological* individuals, which he contrasted with physical atoms. Biological individuals are composite wholes, but so are crystals. So where is the difference? The botanist Carl Nägeli, in his seminal paper on *Individuality in Nature*, had argued in 1856 that individuals are the indivisible carriers of all life in nature, yet he went on to recognize individuals of higher complexity, comparing crystalline formations with plants that he considered a "conglomerate of many individuals."[§] Not so for Virchow, who proclaimed: "*The individual is alive.* Even the most beautiful crystal remains but an *exemplar*, albeit a prime specimen."[¶] For Virchow, then, an individual is unique,

---

[*] Leuckart, 1851; see also Nyhart and Lidgard, 2011.
[†] Vogt, 1854.
[‡] Virchow, 1862, p. 64; see also Trembley, 1744, 1943.
[§] Nägeli, 1856, pp. 182, 184, 186. Nägeli, 1856, was cited by Virchow, 1862, p. 62.
[¶] Virchow, 1862, p. 49; in calling a crystal an "exemplar" Virchow implied it to be an exemplar of its (natural) kind, where a natural kind is a class concept, not an individual.

a crystal but one of potentially infinite instantiations of its kind, In addressing the problem of biological individuality, Virchow started from the theory that was closest to his heart, as he considered its domain to provide the essence of all understanding of life, of birth, health and death, that is, the Schleiden—Schwann cell theory: "all life is tied to the cell, but the cell is not just a vessel containing the substance of life, it is itself the living part."* Virchow earned immortal fame in the history of medicine for having endowed this cell theory with a clinical dimension, through his *Cellular-Pathologie* (Figure 4.1).†

Goethe, wrote Virchow, had said that "everything living is not a particular, but a plurality; even if it appears to us as an individual, it still is an assembly of independent living entities."‡ The line Virchow quoted is from a text titled *Die Absicht Eingeleitet* (1817), in which Goethe sketched the purpose of his new science of *Morphology*.

FIGURE 4.1 Rudolf Virchow. (Museum für Naturkunde Berlin, Historische Bild- u. Schriftgutsammlungen [Sigl: MfN, HbSb]. Bestand: Zool. Mus., B I/235 [Virchow].)

---

* Virchow, 1862, p. 54.
† Virchow, 1855, 1858.
‡ Goethe, in Kuhn and Wankmüller, 1955, p. 56. Virchow's citation of Goethe has been taken as an indication that he, as well as Haeckel after him, took from Goethe the inspiration for their Cell State theory and metaphor. For a recent and most comprehensive analysis of the Cell State theory in Virchow and Haeckel, and its biological as well as political connotations, see Reynolds, 2007, 2008.

Goethe wanted to go beyond the prevailing scientific attitude, according to which objects of nature, in particular living ones, are best understood through dissecting analysis, through the "separation of [their] parts." Goethe found this problematic, since "the living has been dissected into its elements, yet it is impossible to reconstruct and revive it again from these elements." What his *Morphologie* would instead be aiming at is an understanding of the constituent elements in their interrelations, and thus "to some extent to comprehend the Whole [*das Ganze*] in contemplation": "the German has the term *Gestalt* to denote the complex of the existence of a natural entity." Virchow was less polarizing in his presentation: "Science surely unites, but only after it separated; the first task of research is decomposition ... we seek unity, and we find it in plurality."[*] Contrasting the atom of physics with biological individuals, and forcefully rejecting any possible synonymy of the terms, he asserted: "Individuals are not partial entities, but entities with parts."[†] "So let's move forwards, for community is only revealed through its parts! The community of an individual is composed of ... [interconnected and interdependent] elements. This is why we call it an *Organism*."[‡] The elements, of course, were the cells prominently brought into focus by Schleiden and Schwann.

> What is the organism? It is a community of living cells, a small State, well organized with all the required accessories, with super-ordained and subordinated officials, with servants and masters, leaders and followers.[§]

The intriguing question remains as to whether Virchow took an organism that represents a Cell State to be a *Gestalt*, a complex whole composed of parts, or whether he advocated a more mechanistically motivated "building block" theory of cells. His writings do not offer a conclusive answer to this question. In an early paper on his *Cellular-Pathologie*, Virchow argued that "under the apparatus of a biologist [i.e., the microscope], all living entities decompose into small elements,"[¶] the cells. He followed this observation with the assertion that "*the constituent elements will always only find their full significance in the Whole*,"[**] just as proliferating cancerous cells will ultimately cause the death of the whole organism. But then he continued to consider it "no trouble that we lose the unity of the living organism in the multitude of its constituent epicenters of life," only to further emphasize the mutual physiological interdependence of cells that constitute the organism.[††] In his lecture course on cellular pathology held in 1858, he characterized plants and animals as a summation of cells, the "so-called individual" as a communal structure, a social structure, where a multitude of singular entities are reciprocally dependent on one another.[‡‡] Can bodies be decomposed into cellular territories? Carl Reichert had

---

[*] Virchow, 1862, p. 51.
[†] Virchow, 1862, p. 44.
[‡] Virchow, 1862, p. 52.
[§] Virchow, 1862, p. 55.
[¶] Virchow, 1855, p. 17.
[**] Virchow, 1855, p. 20; emphasis in the original.
[††] Virchow, 1855, p. 25.
[‡‡] Virchow, 1858, pp. 12–13.

issued a histological Law of Continuity where tissues are identified on the basis of continuous cell-lineages which come together to form a communal Whole, "but this so-called *Continuitäts Gesetz* has soon experienced the most severe convulsions," and was rejected by Virchow because of the prevalence of un-sharp boundaries between tissues.[*] So what is an individual, a complex whole? Virchow found himself unable to root biological individuality in nature, especially not considering the complex phenomena described in colonial and polymorphic invertebrates by Leuckart and Vogt. So he turned to human self-consciousness as the origin of the *ich*, and hence as the hallmark of biological individuality at the pinnacle of life's scale of complexity, only to resignedly conclude: "The 'me' of the philosophers only follows from the 'us' of biologists."[†]

> So are the cells therefore the individuals, or are the humans the individuals? Can there be an easy answer to this question? I say: No! ... The problem is that the term "individual" has been used long before there had been a clear understanding of the [complex] nature of all these creatures which have collectively been referred to by this term.[‡]

In the ideal condition, in the healthy state or in the healthy individual, "the health of the Whole is conditional on the well-being and on the intimacy of the relations between its elements," but the whole disintegrates if these constituent elements turn against it. A 1912 state-of-the art review on implantation and transplantation in a clinical context commented on Virchow's conception of the Cell State:

> In view of Virchow's emphasis on the cellular basis of all life, "the organism is not necessarily a unitary entity, but rather a coordination of a multitude of living separate parts, a Cell State. Just as in a well-organized State, every singular element has a function that serves the Whole, and that receives nutriment and stimulation from the center ..."[§]

With Haeckel, the Cell State became a *bona fide* unitary entity, a complex whole composed of parts, a biological individual of higher order. Haeckel recognized several layers of individuality of increasing complexity, both morphological and physiological (as discussed in the second chapter). Commenting on his former advisor and supervisor Virchow, Haeckel accepted the cell as an "elementary organism,"[¶] the human body as an "organized society, a State of cells."[**] In his *Natürliche Schöpfungsgeschichte* of 1868, he devoted an entire section specifically to the "comparison of a multicellular organism with a State."[††] He called the cells the "building materials," which come together to build the body of a vertebrate through

---

[*] Virchow, 1858, pp. 62–63.
[†] Virchow, 1862, pp. 71–72.
[‡] Virchow, 1862, p. 73.
[§] Axhausen, 1912, pp. 306–307.
[¶] *Elementar-Organismus*, a term Haeckel took from Ernst Wilhelm von Brücke (1819–1892), like Haeckel a student of Johannes Müller in Berlin (see Brücke, 1861). Haeckel also deployed Virchow's term *Lebensheerd* in the same context, characterizing the cell as the "hearth of life"; Virchow, 1855, pp. 19, 25; see also Haeckel, 1874, p. 97.
[**] Haeckel, 1866, vol. 1, pp. 264, 270. See also Reynolds, 2008, p. 130, and p. 148, n. 40.
[††] Haeckel, 1868, pp. 246ff.

three steps: cell division, cell differentiation and consequent division of labor, and cell association to form organs and organ systems. This process he characterized as one of "step-wise progression or perfection," and directly compared it with "citizens who set about to establish a State." Multiplication, differentiation, and division of labor characterize the "biological progress" of that State as well, and render it capable to achieve an efficiency and performance that would be impossible for the singular individual, the singular citizen. In the exact same sense, the whole body of a vertebrate is a republican Cell State. Now, everybody would agree, according to Haeckel, that the establishment of a State by humans and its purposeful and goal-directed organization is the work of its citizens and their government, not that of a benevolent Creator. The latter has therefore likewise not to be invoked in an explanation of the origin of the vertebrate body and its adaptations. It is interesting to note that Haeckel calls the multicellular organism a republican state in his *Natürliche Schöpfungsgeschichte* of 1868, as in his *Generelle Morphologie* of 1866 he had characterized—with respect not to the Cell State, but to the states formed by social insects—the republican state (ants) as superior to the monarchic state (bees) both in terms of its overall efficiency, as also in terms of the perfection it allows for the constituent persons: "The same laws that govern the development of human states also cover in the same way the communities and states of all other animals."[*]

The same arguments re-appear in Haeckel's *Anthropogenie* of 1874[†], but in a talk he delivered in the *medizinisch-naturwissenschaftliche Gesellschaft zu Jena* on November 19, 1875, Haeckel placed a more holistic emphasis on the concept of the Cell Sate, and reversed his earlier classification of a Monarchy relative to a Republic. Haeckel explained, with reference to Virchow:

> Just as in any civilized State, where the singular citizens are to a certain degree independent from one another, yet at the same time remain dependent on one another due to the division of labor and consequently are dominated by the law that governs the Whole, so do the microscopic cells in the body of any higher animal or higher plant enjoy a certain independence, and yet they are interdependent due to differentiation and division of labor; at the same time they are more or less governed by the laws of the centralized Whole. This completely accurate and often deployed political comparison is not just a vague metaphor, but can claim validity in reality[‡]; the cells are true citizens of a State. The comparison can even be extended as we may consider the tightly centralized animal body a cell-Monarchy, the less centralized plant organism a cell-Republic. Just as political science documents a long series of stages of ascending perfection in the formation of human States, from the crude hordes of the savages to the highly civilized cultural State [*Kulturstaat*], so does the comparative anatomy of plants and animals demonstrate a long series of steps in the increasing perfection of the Cell State.[§]

Defending the "sensibility or irritability, and motive power or will" of the protoplasm implied by his Cell Soul Theory against Virchow's attack from 1877, Haeckel had

---

[*] Haeckel, 1866, vol. 2, p. 143.
[†] Haeckel, 1874b, pp. 97, 118.
[‡] On Haeckel's belief in a unity of biological and social sciences; see Weindling, 1985, p. 696.
[§] Haeckel, 1879b, pp. 36–37; see also the discussion of political connotations in Reynolds, 2008, p. 133.

likewise maintained the doctrine that in higher animals and plants, the cells, "to a great extent, give up their individual independence, and are subject, like good citizens, to the soul-polity which represents the unity of the will and sensations in the cell community."* Evidently, for Haeckel, there existed a gradation that ranged from plastids or single cells to the loosely integrated amoeboid structures, to the polymorphic invertebrate colonies exhibiting integration through division of labor and so on up to the vertebrate body with its centralized nervous system, the human body finally residing at the top. The Great Chain of Being reflecting increasing levels of integration and perfection continued to underlie Haeckel's vision of the branching phylogenetic tree, a progressionist interpretation of nature inspired by the English philosopher Herbert Spencer (1820–1903).† Wholeness, unity of sensation, and unity of will correspondingly come in degrees as well, dependent as they are on the degree of differentiation and integration of the parts in the complex whole. In Haeckel's monism, the soul is coextensive with matter, wholeness is coextensive with the community or society of cells and its degree of differentiation, integration, and cohesiveness. Rejecting any dualism of soul and matter, the function of the soul becomes tied to bodily organization, differentiation, and integration. "A psychological comparison of different animals reveals to us a long series of developmental stages of the animal soul,"‡ a phylogenetic progress of perfection that is recapitulated in ontogeny: "Aristotle and Plato, Spinoza and Kant were once children too; their superior, far-reaching rational soul (*Denkerseele*) underwent gradual, step-wise development as well."§ Just as there are gradations of soulfulness from ameba to man, so there are corresponding gradations of wholeness. This is a thoroughly externalist conception of holism, attempting to avoid any metaphysical taint, but rooted in nature instead, that is, in the organicist tradition. Cells, the parts, come together to form a complex whole, where the whole is more than the mere sum of its parts to the degree that it exhibits both morphological and physiological integration according to the laws that govern the whole. The community of cells is more than a mere aggregation of individuals, as the cells enter relations of interdependence through differentiation and consequent division of labor. Such abandonment of independence by the constituent individuals results in greater efficiency and potential of the community, and in a unity of will that subordinates the constituent individuals to the laws that govern the complex whole. This is a most powerful *völkisch* metaphor that was to resonate through decades to come, with disastrous consequences, but it is not one that is easily reconciled with the atomistic conception of organisms as is implied by Darwin's theory of variation and natural selection—on which even Haeckel "placed only limited faith."¶

---

* Haeckel, 1879b, p. 58; see also Haeckel, 1878a, p. 48; and Reynolds, 2008, p. 135.
† Weindling, 1985, pp. 694–695. Spencer, in turn, was heavily influenced by Karl Ernst von Baer, and German idealistic morphology, and conceptualized evolutionary change as a progressive development of heterogeneity from homogeneity: Poggi, 1989, p. 108.
‡ Haeckel, 1878b, p. 146.
§ Haeckel, 1878b, p. 147.
¶ Weindling, 1985, p. 688.

THE STRUGGLE AMONG THE PARTS

Ever since Haeckel, Darwinism in German biology was identified not with Tschulok's *Grundfrage*, that is, the fact that species transmutation does occur, for others had claimed so before, most prominently the French naturalist Jean-Baptiste Lamarck (1744–1829), but also Charles' grandfather Erasmus Darwin (1731–1802). It was rather identified with Tschulok's *Faktorenfrage*, the twin mechanisms of random variation and natural selection. And ever since the publication of Bronn's German translation of Darwin's *Origin* of 1860, its reception was ambivalent in German biology. Critics of Darwinism rejected it for its externalist, causal-analytic, and mechanistic approach in the explanation of the origin of organismic diversity and adaptation. However, taking a Darwinian approach to organisms thought to be composed of elementary living building blocks can result in quite a different perspective on plant and animal organization, as was the case with the *Entwicklungsmechaniker* Wilhelm Roux. Roux can justly be seen as the godfather of an unimpeachable reductionist research program in German biology.

Wilhelm Roux[*] was born on June 9, 1850, in Jena, where from 1857–1864 he attended the humanistic *Stoy'sches Institut* (directed by Karl V. Stoy, 1815–1885), before transferring to High School (*Oberrealschule*) in Meiningen (Thuringia, Germany) in 1870. He was somewhat of a loner, a withdrawn child who immersed himself in Johann H. J. Müller's (1809–1875) textbook on physics and meteorology, a liberal German adaptation and updated version of C. S. Pouillet's *Éléments de physique et de météorology* of 1837.[†] His preferred classes at High School were physics, mathematics, and chemistry. In the spring of 1870, he enrolled at the University of Jena to study medicine. Soon thereafter drafted into the army at the outbreak of the Franco-Prussian War, Roux returned to the university in 1871. During the preclinical semesters, he was most impressed, and influenced, by the courses taught by Gegenbaur and Haeckel, "without engaging in personal contact,"[‡] however. He also enjoyed lecture courses in philosophy, particularly on Kant. As Roux himself stated, he eventually "adopted the Kantian mechanistic approach to natural science."[§] Following Jena, Roux continued his studies at the University of Berlin, attracted by Virchow, and at the University of Strasbourg, the city at that time annexed to the newly founded German *Reich*. In 1877, Roux passed his state medical examination, and in 1878 he published his PhD thesis on the branching patterns of blood vessels in humans and some other vertebrates. Given his interests and competence in physics, he sought a causal-analytic explanation for the investigated phenomena and analyzed the branching patterns of blood vessels in terms of hydraulic principles and laws.[¶] His results eventually found their way into the technical literature on the construction of water supply systems (Figure 4.2).

---

[*] The biographical data on Roux were gleaned from Barfurth (1920), and Churchill (1975, 2008).
[†] Kangro, 2008.
[‡] Roux, 1923, p. 3.
[§] Roux, 1923, p. 6.
[¶] Roux, 1878, pp. 242–243.

FIGURE 4.2 Wilhelm Roux. (Museum für Naturkunde Berlin, Historische Bild- u. Schriftgutsammlungen [Sigl: MfN, HbSb]). Bestand: Zool. Mus., B I/146 [Roux].)

Roux kept studying philosophy for another year with Rudolf Eucken (1846–1926), a friend yet critic of Haeckel, while looking for academic employment. Such he found as an assistant first in Leipzig, then in Breslau, where he sought his *Habilitation* by defending a series of theses about the development of the liver on July 31, 1880. The mammalian liver, he claimed, passes through a developmental stage that corresponds to the adult liver of the lamprey; the liver is structured by the action of blood vessels; the development and branching pattern of blood vessels is determined by hemo-dynamic forces. His inaugural speech he published in 1881, as part of his book *Der Kampf der Theile im Organismus.*[*] In his autobiography, Roux defended the priority of his ideas against a brochure published by E. du Bois-Reymond in August of the same year, *On Exercise*[†], which cited the abstract of his *Kampf der Theile* that Roux had first published in the *Biologisches Zentralblatt.*[‡] Having earned the *venia legendi*, Roux embarked on a brilliant academic career that would earn him many

---

[*] Roux, 1881a.
[†] Du Bois-Reymond, 1881.
[‡] Roux, 1923, p. 10.

distinctions and awards: first *Privatdozent*, then associate professor, he was named in the spring of 1889 the director of his own *Institut für Entwicklungsgeschichte* at the University of Breslau. The Breslau years were among the most productive of Roux' career as an empirical scientist. Right from his start in 1879, he found in the lab of Carl Hasse (18841–1913) congenial colleagues, particularly in the *Prosektor* Gustav Born (1851–1900). The latter was a student of Eduard Pflüger (1829–1910) and Rudolf Heidenhain (1834-1897), and hence deeply steeped in physiology.[*] After having served as a professor of anatomy at the University of Innsbruck for six years, Roux accepted an offer from the University of Halle in 1895, a post he held until his retirement in 1921. He died in Halle on September 15, 1924. In 1909, Roux was awarded an honorary doctorate by the University of Jena. The eulogy marking that event celebrated Roux for having been the "first to have addressed the question of the causes of organic development (*Gestaltung*) with appropriate methods, and to successfully pursue this research program with skill, ingenuity, and sharpness of mind."[†]

Haeckel celebrated Roux' 1881 *Der Kampf der Theile im Organismus. Ein Beitrag zur Vervollständigung der mechanischen Zweckmässigkeitslehre* as a brilliant extension of his own ideas, whereas Roux' advisor in Jena, Gustav Adolf Schwalbe (1844–1916) "warned him never again to publish such a 'philosophical' book."[‡] Many developmental and evolutionary biologists of the time, Haeckel included, were left dissatisfied by the Darwinian twin mechanisms of random variation and natural selection, as they believed Darwin had not provided a satisfactory explanation for the perceived purposefulness of animal organization and adaptation. In his *Critique of Judgment* of 1790, Kant had famously argued that a teleological conception of organisms is required in biology for a comprehensive understanding of natural processes as they are observed in plants and animals. Importantly, however, the teleological perspective was to function as a regulative principle only that allowed reasoned judgments about organismic function and adaptation without the implication of an ontological underpinning. To declare the teleological perspective as a regulative principle only that finds its heuristic value in the form of "as-if-statements" allowed it to be combined with a purely causal-mechanistic analysis of organisms without contradiction. Ultimately, the arm of empirical sciences, for Kant, reaches only as far as causal-mechanistic explanations do.[§] This is exactly the perspective that Roux took away from his studies of Kant.

Roux' agenda was to bring the causal-analytic approach back into comparative anatomy. Gegenbaur had argued for the separation of anatomy from physiology, and helped to institutionalize this separation when joining the University of Jena. As one eulogist observed on the occasion of Roux 70th birthday, Gegenbaur had studied the pectoral girdle of turtles in great detail, and compared its components throughout the series of vertebrates as part of a research program that sought to demonstrate the significance of comparative morphology as proof for the theory of descent.[¶] But never had he raised the question of a possible role the pectoral girdle might

---

[*] Mocek, 2001, p. 462.
[†] Roux, 1923, p. 33.
[‡] Churchill, 1975, p. 571.
[§] Höffe, 1981, p. 36.
[¶] Gegenbaur, 1865.

play in respiration in turtles, a question that would seem to naturally impose itself given the fact that the peritoneal cavity is encased in a rigid shell in chelonians. It is true, the well-wisher continued, that the thrust of Roux' research program had been anticipated by the Hertwig brothers' (Oscar and Richard, both students of Haeckel and so close to each other during their student years that they were referred to as "the Siamese twins"[*]) fertilization experiments in sea urchins, even by Haeckel[†] himself, who had taken a needle to early developmental stages of *Staatsquallen*, the siphonophores. But Roux, he judged, deserves ultimate recognition and praise for having revived Johannes Müller's old program by bringing the analytic perspective, that is, physiology back into comparative anatomy, thus forming a comprehensive and integrated research program, a *kausale Morphologie*.[‡] In his inaugural speech for the anatomical institute at the University of Innsbruck that had been founded for him, delivered on November 12, 1889, Roux praised his *Entwickelungsmechanik*—a term that had been suggested by Rudolf Heidenhain[§]—as the new anatomical science of the future. Morphology is a science of form, but what had been missing in morphology, he argued, was the investigation of the immediate causes (*directe Ursachen*) of form generation through experimentation. "We now know that our individual life as a whole is composed, just like the life of a State, of many ... singular, individuals, the so-called cells,"[¶] but the unity of the complex whole that is an organism was for Roux primarily a functional one. The immediate causes of form generation are anchored in function and in that sense it is possible to talk of function as a formative effect, resulting in "functional adaptation."[**] "How does one discover such things," asked the emperor Franz Joseph I of Austria when visiting Roux' institute in Innsbruck? "Your Majesty, this requires one to have a question in one's mind, and then to seek the appropriate means to extort [from nature] an unambiguous answer," replied Roux.[††]

Turning to the potential and limitations of the theory of descent in the thesis submitted in support of his *Habilitation* on August 1, 1880, Roux confronted the problem of purposefulness and teleology head-on. He located in the ancient Greek philosopher *Empedocles* the assertion that apparent usefulness of organismic design can result from purely mechanical forces without any impetus acting according to preconceived goals. Roux acknowledged the mechanistic nature of random variation and natural selection, and called the resulting adaptations purposeful, but:

> The usefulness is not a wanted, but a becoming one, not a teleological, but a natural-historical one; and it is solely in this sense that we want to talk of usefulness.[‡‡]

---

[*] Nyhart, 1995, p. 157.
[†] "The strange amoeboid movements described above ... of the cells, which originated from the cleavage of the egg cell and which compose the larval body of *Crystallodes*, obviously indicate a high degree of phsyiological independence and of relative individuality of these cells." Haeckel, 1869, p. 73.
[‡] Braus, 1920, p. 438.
[§] Mocek, 2001, p. 475.
[¶] Roux, 1889 (1895), p. 32.
[**] Roux, 1883 (1895), p. 23.
[††] Roux, 1923, p. 6, n. 1.
[‡‡] Roux, 1880 (1895), p. 103.

What drives adaptation, as also progressive evolution, that is, "a continuously increasing diversity and perfection," he located in the "struggle for existence" and in "sexual selection,"[*] but beyond those mechanisms he also emphasized the effect of use, or disuse on the evolutionary transformation of organs, or organ systems, a mechanism invoked by both Darwin and Haeckel, and one he wanted in the future to refer to as *"functional adaptation."*[†] Variation he found to be constrained by the laws of growth and the norm of reactions inherent in organisms; natural selection he found to be dependent on variation and hence similarly constrained, a principally sufficient but empirically possibly insufficient force for evolutionary transformation. Worse, the Darwinian twin principles seemed incapable to bring about saltational change as seemed to be required—in Roux' estimation—in certain adaptational changes, such as the transition from water to land (gill breathing to lung breathing) in vertebrates. Such concerted and correlated adaptational changes involving almost all parts of the body require functional adaptation that plays out during the form generating process of development:

> Those considerations lead to the insight that "functional adaptation" not only really occurs, and that its effects must become heritable after several generations, but also that it can, in contrast to natural selection ... affect the entire inner harmonious design of millions of furnishings useful not only relative to one another but also useful relative to the new or modified parts.[‡]

While not (entirely) Darwinian in its conceptual content, this statement nicely documents Roux' atomistic conception of life. Roux comprehended organisms as complex wholes, composed of parts that are individually malleable through variation and natural selection, collectively malleable through functional adaptation, and whose coherence in life processes is functionally, rather than structurally grounded. Such conception of the organism allowed Roux to apply the Darwinian principle of the struggle for existence and the resulting *Zuchtwahl*[§] to the microscopic level within the organism, to the cells of which the organism is composed. If the Darwinian struggle for existence could propel the evolution of the most simple, unicellular forms of life to greater complexity and perfection, then, by analogy, a struggle for existence could also play out between the cellular components of higher organisms.[¶] If the branching pattern of blood vessels was determined by the reaction norm inherent in the tissues relative to the hydrodynamic pressure exerted by the blood stream, the question may be raised as to what conditions that reaction norm? To answer this question, Roux pondered an appeal

---

[*] Roux, 1880 (1895), p. 105.

[†] Roux, 1880 (1895), pp. 114–115. His "functional adaptation" was to provide a mechanism that would bring about useful adaptation much faster than natural selection.

[‡] Roux, 1880 (1895), pp. 125–126.

[§] The term replaced Bronn's translation of natural selection as *Züchtung*, when as of 1866, Victor Carus succeeded Bronn as translator of Darwin's work: Weindling, 1985, p. 688.

[¶] Roux, 1881b, p. 245.

to the *"ultima ratio,"* *"the struggle amongst individuals."* Here, one could think of a struggle among Cell States as among nations, but:

> this would seem as if one wanted to reduce all the positive attributes of a State manifest in its government, legislation, administration, sciences, commerce, trade etc. solely to the battle with other states ... a better explanation would result from the assumption of a struggle amongst similarly functioning parts of an organ, in this case of a *struggle of the cells* that constitute the wall of the blood vessel for nutrients and space ...*

Roux retitled the original publication of 1881 for its inclusion in the 1895 edition of his collected papers as *Der züchtende Kampf der Teile*, meaning the selective struggle for existence among the parts of an organism. In his introduction to the 1895 reprint, he called his first book-length treatise a theoretical essay of my youth, heavily influenced by Haeckel and Gegenbaur, and written in a state of near unconsciousness.[†] The fundamental premise of the *Kampf der Theile* reflects Roux' doubts about the sufficiency of the externalist Darwinian selection theory to explain organismic design and adaptation: "The individual [organism] has not only to prove itself with respect to its external conditions of life, but also and primarily has to maintain itself"[‡] through self-regulation. That self-regulation in turn is driven by the selective struggle among the parts of the organism, its cells, the molecular complexes they contain and the organs they compose. The parts of higher organisms compete primarily for nutrients and space. A change of the environment and with it of the nutrients available for assimilation would result in a changing stimulation of the cells, and thereby would trigger a competitive struggle for existence among them. The consequence is functional adaptation as the result of a process that plays out directly within the individual organism, as opposed to adaptational processes that are driven by a competition between organisms for resources in their external environment. In that sense, the struggle between cells and organs of an organism is comparable to the competition between the representatives of different professions in a State[§], as opposed to war, which is a struggle between States.

In his mature reflections on his youthful writing, Roux acknowledged the dependence of his theory of functional adaptation on the possibility of inheritance of acquired characteristics, but writing in 1895, he asserted that such could even at this time not be ruled out with "absolute certainty."[¶] He expressed his disappointment that pathology had not made any use of his concept of "selection of parts within the organism," and he deplored the neglect of his ideas by orthopedics in particular, given the plasticity of bone and the fact that successful orthopedics could always only be a "functional orthopedics."[**] Roux further elaborated on a letter from Darwin to George J. Romanes (1848–1894) dated April 16, 1881[††], in which Darwin announced

---

* Roux, 1879, p. 336.
† Roux, 1895, p. 139. "This theory formed in my head all by itself, from seeds of thoughts I absorbed during my student years": Roux, 1923, p. 9.
‡ Roux, 1881b, p. 244.
§ Roux, 1881b, p. 248.
¶ Roux, 1895, p. 140.
** Roux, 1895, p. 148.
†† F. Darwin, 1887, vol. 3, pp. 243–244.

the reception of Roux' book sent to him by its author. He found the German language difficult, but in his possibly imperfect judgment Darwin called the book "the most important book on Evolution, which has appeared for some time." Roux agreed with Darwin's criticism that he never reflected on plants in his *Der Kampf der Theile*, but vehemently disagreed with Romanes's review of his book in *Nature*. In this review, which was incited by Darwin, Romanes expressed his surprise about Roux apparent "ignorance of the fact that the doctrine is not original."[*] Romanes called attention to Herbert Spencer, who had introduced the terms "indirect equilibration" for natural selection working from without, and "direct equilibration" for selection working from within the organism, and criticized Roux for having "over-charged" his work "with analogies drawn between the organism physiological and the organism social."[†] Roux rejected Romanes' review as superficial, missing his essential point of a struggle among initially equivalent parts that results in the selection of new tissue properties in a way that differs from Spencer's account.[‡]

GLASS HALF FULL OR HALF EMPTY?

Consider the case of the clubfoot. The orthopedic familiar with Roux doctrine of functional adaptation should have recognized the possibility of its clinical application. Following correctional surgery, sustained and goal-directed exercise should result in a struggle of cells such that those best absorbing the altered physical stimulation should prevail in the competition for nutrients and space, a process that would result in an improved functional adaptation of the previously ill-formed foot. However, it is one thing to talk about the functional adaptation of existing structures, but another to explain the embryonic development of those structures through stages that precede their becoming functional. As was recognized by Hans Driesch—who called himself the grateful methodological pupil but frequent scientific opponent of Roux[§]—the concept of functional adaptation necessarily had to lead up to embryological investigations. It is in the context of these investigations that the concept of Wholeness (*Ganzheit*) assumed a more prominent place in Roux theorizing, trumping the atomism inherent in the *Kampf der Theile*.

Roux is famous for his experiments, first reported at a meeting on pathological anatomy in Wiesbaden on September 22, 1887, where in the two-cell stage of the frog[¶] embryo he would kill one of the two blastomeres with a hot needle to find the other blastomere to develop into a half embryo (*Hemiembryones*).[**] He consequently compared embryonic development to a process that pieces together a mosaic (*Mosaikarbeit*),[††] where cells must be at the right time in the right neighborhood within the embryo for normal (typical in his parlance) development to occur. However,

---

[*] Romanes, 1881, p. 506.
[†] Romanes, 1881, pp. 505–506.
[‡] Roux, 1895, p. 141.
[§] Driesch, 1920, p. 446.
[¶] Roux used the common European green frog, *Rana esculenta*, which has been recognized as a hybrid species of *Rana lessonae* and *Rana ridibunda* (e.g., Uzzell et al., 1976).
[**] Roux, 1888 (1895), p. 437.
[††] Roux, 1888 (1895), p. 455.

Roux also observed that, as an exception to the rule, a hemiembryo might—starting from various developmental stages—regenerate a whole organism, a process Roux referred to as postregeneration (*Postregeneration*).[*] Although a whole organism had reconstituted itself through self-differentiation, Roux called such postregeneration an atypical development.

In his inaugural speech on the newly founded anatomical institute in Innsbruck of 1889, Roux invoked the future Newton of the movements of the parts in the developing organism, comparing the cells to "microscopic building blocks which act as construction workers and at the same time, within a certain domain, possibly also as construction site managers."[†] This would imply reciprocal interaction among the parts in the self-differentiation of the embryo. Indeed:

> ... the effects that bring about and maintain the "Whole" must be rooted in the type
> of its species ... It seems to me that the doctrine of the composition of an organism of
> individual living parts has underestimated *the effects that mediate the typical unity
> of the Whole* ... The *unity* of the Whole is purely *functionally* conditioned; and the
> concerted action of all the parts is mediated by their subordination under a *singular
> will* that dominates the function of the Whole."[‡]

Even the reductionist *Entwicklungsmechaniker* Wilhelm Roux resorted to a rhetoric that invoked a *singular will* that dominates the whole for its own good. Particularly phenomena of atypical development, as are manifest in processes of postregeneration, document the regulative potential inherent in developmental processes, which *"indicate a most intimate combined activity of the parts towards the Whole, as also a greater dependence of the parts on the Whole"*[§] than he had previously assumed. But whereas the deciphering of the mechanical basis of these regulative processes remained reserved for the future Newton of *Entwickelungsmechanik*, Roux insisted that the proper approach would remain causal-analytic coupled with experimentation. This was, and remained, the crucial point in his debate with Driesch, who noted on the occasion of Roux' 70th birthday: "Roux is an analyst through and through. The desire for synthesis is minimal."[¶]

Positioning his *Entwickelungsmechanik des Embryos* within the various branches of biological science in 1885, Roux invoked a key concept that was to guide the design of developmental experiments, one however that for some of his readers would prove to be difficult to incorporate in a purely mechanistic explanation of life: the *self-differentiation* of the egg and the embryo that develops from it, as opposed to the differentiation of the embryo in subordination to causally efficacious external factors.[**] As he set about to separate the blastomeres of the developing frog embryo, especially when faced with postregeneration, Roux invoked another key concept of embryogenesis that for many seemed to elude a strictly causal-analytic,

---

[*] Roux, 1888 (1895), p. 484.
[†] Roux, 1889 (1895), p. 29.
[‡] Roux, 1889 (1895), pp. 39–40.
[§] Roux, 1889 (1895), p. 41.
[¶] Driesch, 1920, p. 450.
[**] Roux, 1885 (1895), p. 14.

that is, mechanistic explanation: the capacity of *morphological self-regulation* of the developing embryo.[*] In his highly influential *Foundations of Theory Construction in Biology* (*Grundzüge der Theoriebildung in der Biologie*) of 1919, the developmental-theoretical biologist and leftist activist Julius Schaxel (1887–1943) clearly identified the dangers of treading such thin ice for someone who seeks a purely mechanistic basis of life. Schaxel was deeply entrenched in the Haeckel school: starting out as a student of Haeckel himself, and continuing his academic education in Munich under Richard Hertwig, a former Haeckel student, Schaxel earned his PhD in 1909 in Jena under Ludwig Plate, another former Haeckel student. In 1918, he was appointed head of the Institute for Experimental Biology at the University of Jena. A left wing social democrat who "harnessed developmental biology to dialectical materialism,"[†] he was dismissed from the university in 1933, following the enactment of the Law for the Restoration of the Career Civil Service (*Gesetz zur Wiederherstellung des Berufsbeamtentums*) of April 7, 1933. Schaxel subsequently emigrated to St. Petersburg via Switzerland[‡], and died in Moscow in 1943 "under unclear circumstances."[§] Reflecting the spirit of the time, that is, a "craving for crises,"[¶] Schaxel opened his 1919 analysis of theory construction in biology with the statement: "contemporary biology is in a state of crisis … the neo-vitalism which opposes biological materialism has significantly contributed to its acute state."[**] Commenting on Roux' concept of self-regulation, Schaxel found that by invoking such a concept the "*Entwicklungsmechanik* … runs the risk to negate itself both in terms of development as well as in terms of mechanics."[††] This nicely captures the point that Driesch had earlier capitalized upon.

In terms of academic *mores*, Hans Driesch was an outsider among contemporary developmental biologists. He was born on October 28, 1867, in Bad Kreuznach, Germany, as the only child of a well-to-do family, the father a merchant from Hamburg where Driesch grew up.[‡‡] In 1886, he enrolled at the University of Freiburg to study under August Weismann, and earned his PhD in 1889 under Haeckel[§§] at the University of Jena, where he also attended courses offered by Oscar Hertwig, and in particular enjoyed the course in comparative embryology offered by Arnold Lang.[¶¶] Independently wealthy, Driesch set out on an extensive travel schedule accompanied by Curt Herbst (1866–1946), a fellow embryologist and lifelong friend Driesch had met in Jena in 1887.[***] Driesch and Curt Herbst were part of a group of four graduate students who met weekly to discuss new publications in the field of zoology.

---

[*] Roux, 1889 (1895, p. 25).
[†] Weindling, 1989b, p. 327.
[‡] Reiß, 2007; Reiß et al., 2007.
[§] Deichmann, 1996, p. 23.
[¶] Forman, 1971, p. 58.
[**] Schaxel, 1919, p. 2.
[††] Schaxel, 1919, p. 45. See also the analysis in Cassirer, 1950 (1978), p. 193.
[‡‡] The biographical data on Driesch are from Herbst, 1941; Ungerer, 1941; Oppenheimer, 1971; and http://www.uni-leipzig.de/unigeschichte/professorenkatalog/leipzig/Driesch_30/ (accessed August 29, 2012); http://vlp.mpiwg-berlin.mpg.de/people/data?id=per63 (accessed August 29, 2012).
[§§] Driesch, 1890a, p. 224.
[¶¶] Driesch, 1951, p. 46.
[***] Herbst, 1941, p. 112. See also Oppenheimer, 1991.

Within that group, Driesch had established particularly close ties to Herbst: "our views of science and of life were almost entirely identical."[*] Driesch identified 1890 as a particularly critical year, as he encountered "by chance" the publication of Roux inaugural speech for the Anatomical Institute at the University of Innsbruck. Previously he had known from Roux' pen only the *Kampf der Theile*, motivated by Darwinism in Driesch's opinion, but here, in Roux' *Die Entwicklungsmechanik, eine anatomische Wissenschaft der Zukunft*, he recognized for the first time a well-designed experimental approach to comparative embryology. His excitement led to the idea to repeat Roux' experiments, not with frog eggs, however, but with abundantly available sea urchin eggs instead (Figure 4.3).[†]

Driesch initiated his crucial experiments on the development of sea urchins in the Marine Biological Station in Trieste in the spring of 1891, and from the fall of 1891 continued his experimental work at Dohrn's Zoological Station in Naples, working at the *"Hamburgischer Tisch"* (a table sponsored by his hometown) during the winter months throughout the 1890s.[‡] His last experiments there he conducted in 1909, the same year he earned the *venia legendi* in Philosophy of Nature at the

FIGURE 4.3   Hans Driesch. (Leopoldina-Archiv [Halle/Saale], BM, Bnd. 08.)

---

[*] Driesch, 1951, p. 39.
[†] Driesch, 1951, pp. 66, 68, 74.
[‡] M. Driesch, 1951.

University of Heidelberg, where he was promoted to associate professor in 1912. In 1919, Driesch became a professor of systematic philosophy at the University of Cologne, and in 1921 accepted a chair in philosophy at the University of Leipzig. During his career as professor of philosophy, Driesch held guest lectureships in Nanjing, Beijing, Madison (Wisconsin), Buenos Aires, and Great Britain. Driesch became the most important proponent of neo-vitalism in modern German biology, but—liberal-democrat, pacifist, and internationalist yet not of Jewish descent, nor married to a Jewish spouse—he was forced to retire from his academic position in October 1933, and forbidden to hold public lectures in 1935.[*] Driesch had supported and signed off on petitions in favor of the German-Jewish philosopher and liberal political publicist Theodor Lessing (1872–1933), and the German-Jewish mathematician, pacifist and political activist Emil J. Gumbel (1891–1966), actions that the *Reichsministerium für Wissenschaft, Erziehung und Volksbildung* (Reich Ministry of Science, Education and Culture)[†] classified as satisfying §4 of the Law for the Restoration of the Career Civil Service. A somewhat Kafkaesque letter from the *Reich* Ministry of Education applauded his "straight political attitude," and recognized his "outstanding scientific reputation," yet still pressed Driesch to seek early retirement in order to avoid a humiliating dismissal.[‡] Driesch stayed in Germany and continued to publish, however, while his vitalism was assimilated into the *völkisch* ideology spiced up by National Socialist demagogy. Driesch died in Leipzig on April 16, 1941.

Driesch was inspired to study zoology by his mother, who was fond of keeping birds and other exotic pets at home, and by Haeckel's books, particularly the *Natürliche Schöpfungsgeschichte*.[§] His PhD thesis dealt with *Tektonische Studien an Hydroidpolypen*, that is, the investigation of levels of individuality in hydrozoans. His main interest in this project concerned the laws governing the branching pattern that results from reproduction through budding. In the first installment of his thesis, he did consider the applicability of the biogenetic law to the growth patterns that characterized the hydrozoan colonies (*Stöcke*), drew phylogenetic conclusions, and even offered a phylogenetic tree.[¶] In the second installment, tackling different taxa, Driesch emphasized the increasing difficulty of demarcating individuals: as a new polyp sprouts—is it a "person," or a mere "appendix" of a (parent) person?[**] While Driesch dwelt extensively on issues of individuality, phylogeny got shorter shrift. Pondering various hypotheses of homology and of phylogenetic derivation, he issued a possible hypothesis of monophyly, which he immediately qualified thusly: "We are dealing here, however, merely with words."[††] The investigation of endogenous laws of growth trumped phylogeny reconstruction: laws of growth are grounded in causal relations; phylogenetic hypotheses are mere words. This indicates that by 1890 already, Driesch had become disenchanted with phylogeny reconstruction in favor of

---

[*] Hutton, 2005, p. 178.
[†] On the history of the Reich Ministry of Science, Education and Culture, see Nagel, 2012.
[‡] Driesch, 1951, p. 272.
[§] M. Driesch, 1951, p. 9.
[¶] Driesch, 1890a, p. 225.
[**] Driesch, 1890b, p. 657.
[††] Driesch, 1890b, p. 685.

the investigation of laws that govern growth and development.[*] In 1893, and drawing on Albert Wiegand (1821–1886)[†], Driesch mounted a full-scale attack on descriptive morphology and phylogeny reconstruction based on it, as it was practiced by Haeckel and his school. What is it, Driesch asked, that a theory of genealogical relationships can in fact deliver?

> It is self-evident that a hypothetical demonstration of a historical connection … must not be confounded with causal analysis: a *portrait gallery of ancestors*[‡] is all that a history of form can deliver. Descent may perhaps contribute something to the understanding of form, but if so, only very little.[§]

There might be, Driesch conceded, some *heuristic* value inherent in the genealogical perspective if applied to low taxonomic levels such as species and below, and related to patterns of geographical distribution, but large-scale phylogeny reconstruction Driesch rejected as lying outside the scope of proper science. Quite in contrast to experimental sciences, which afford predictions, their tests, and mechanistic explanations.

Driesch returned to the topic in a polemic published in 1899: "This thing with the '*Gastraea*' cannot possibly be taken seriously"[¶]—targeting Haeckel of course. Without universal laws, without causally efficacious stimulations, without experimentation phylogeny reconstruction is not only useless, but even impossible: "And yet it persists all the same"! Yes—but not as a natural science, only as a subjective addition to the results of comparative systematics.[**] Homology and phylogenetic derivation—"the old story of the *Baron Münchhausen* … what [does it mean to say that something] 'is' homologous? I do not know, my opponents do not know … Homology is a question of 'assumption', not one of 'knowledge'. How could it be otherwise?"[††] Driesch recognized two types of laws that characterize biology. The first category comprises the laws of succession, as are manifest in developmental processes for example, in terms of which biology is a science of "the lawful becoming" (*Werdegesetzwissenschaft*[‡‡]), that is, physiology. The second category comprises laws of correlation, on the basis of which biology is a science of systematics, but a systematics that is, indeed must be, independent of phylogeny: "the theory of descent has corrupted systematics."[§§] To understand the intricate interrelations among the laws of succession and those of correlation means to understand that biological systematics is—or can be if correctly practiced—much more than a mere catalog of biodiversity, even if it is—as it must be—decoupled from phylogeny. Small wonder that Driesch's relationship with Haeckel deteriorated. With his experimental, mechanistic outlook, the young Driesch felt closer to Roux than to his earlier mentor.[¶¶]

---

[*] Oppenheimer, 1971, p. 186.
[†] Wiegand, 1874–77.
[‡] Translation from Nyhart, 2002, p. 9.
[§] Driesch, 1893a, pp. 46, 48.
[¶] Driesch, 1899a, p. 48.
[**] Driesch, 1899a, p. 37.
[††] Driesch, 1899a, p. 41.
[‡‡] Driesch, 1911, p. 33.
[§§] Driesch, 1911, p. 55.
[¶¶] Ungerer, 1941, p. 458.

When tackling embryonic development on an experimental basis, Driesch used sea urchins abundant in the coastal waters of the Adriatic Sea and the Gulf of Naples. Starting with the two-cell stage, but eventually using embryos as late as the 32-cell stage, Driesch did not use hot needles to kill off cells, but instead used the "shaking method" (*Schüttelmethode*) developed by the Hertwig brothers to separate the cells from one another.* The results of his first classic experiment he published in 1892, under the title, *Entwicklungsmechanische Studien*—"in order to emphasize the relationship of my research program to the one pursued by Wilhelm Roux."† His problem was well defined. August Weismann, under whom Driesch had started out his career, had famously stipulated the distinction of the "germ line" (*Keimplasma*) from the "somatic line" (*Soma*). It is the soma that develops into the reproductively active adult organism guided by the heritable factors preserved in the "germ plasm." To account for the differentiation of the developing embryo into a variety of tissues and organs, Weismann had assumed that cell cleavage results in qualitatively unequal nuclear divisions followed by differential allocation of nuclear material to the resulting cells. The assumption of a qualitative inequality of somatic cell cleavage appeared to be confirmed by the mosaic development, the *Mosaikarbeit* performed by the embryo that was apparent in Roux' experiments. It was this mosaic conception of embryonic development that Driesch set out to test. Shaking the two blastomeres of the two-cell stage of the sea urchin *Echinus* (*Psammechinus*) *microtuberculatus* apart did not result in hemiembryos, as in Roux' frogs, however, but in the development of two individuals (*pluteus* larvae) that were complete but of half the size of normal individuals: "this refutes the principle of organ forming germ domains,"‡ that is, a mosaic mapping of the egg into tissue- and organ-specific domains. Driesch tested the hypothesis of qualitatively unequal distribution of nuclear material during cell cleavage by squeezing a four-cell embryo between glass plates, which resulted in an altered cleavage plane and corresponding subsequent distribution of nuclear material. The prediction would have been that abnormally formed embryos would develop from the compressed eggs, but such was, again, not the case: "this definitively refutes [the] principle of germ domains."§

Driesch continued a long series of experiments, which clearly demonstrated that sea urchin eggs showed a strikingly different potential for regulation during development in comparison to frog eggs, which conformed better to a mosaic model. However, Roux would have none of it. In light of the postregeneration he had observed in frogs, he argued that both types of eggs develop on the basis of the same mechanistic principles, but that in sea urchins, animals of lower complexity than frogs, postregeneration occurs not only regularly but also much earlier during development, whereas in frogs, animals of comparatively higher complexity, such postregeneration—if it occurs at all—occurs at variably later stages.¶ But Driesch

---

* Herbst, 1941, p. 113.
† Driesch, 1892, p. 160.
‡ Driesch, 1892, p. 178.
§ Driesch, 1893b, p. 22. "His' Principle" makes reference to Wilhelm His (1831–1904), another first generation *Entwicklungsmechaniker* who studied first under Johannes Müller in Berlin, then under Kölliker, Virchow and Leydig in Würzburg.
¶ Roux, 1892 (1895).

held the longer end of the stick, as his discoveries and the conclusions he drew from them had much greater predictive force: each blastomere of the two-cell stage of the sea urchin egg *always* developed into a complete *pluteus* of reduced size, whereas Roux' frog embryos might or might not regenerate, and if they did, they did so at variable stages of development. Driesch's observations appeared to reveal true law-fulness, Roux' experiments did not quite match that standard.* Driesch recounts in his autobiographical sketch the "war" he engaged in with Roux during the years 1894–1901. Both he, as well as Curt Herbst, liked to publish their results in Roux' *Archiv für Entwicklungsmechanik*, but eventually Roux—the editor—took to the habit to comment on Driesch's observations in footnotes that he added to Driesch's papers without consulting the author: in spite of his protests, Roux continued to "correct my papers like class essays," Driesch complained. Eventually, Driesch abandoned publishing in Roux' *Archiv*, the editor of which he met for the first time in 1901 at the International Congress for Zoology held in Berlin. The two men imme-diately struck off a great personal relationship, celebrated by Driesch as evidence for the fact that scientists could be personal friends even if they held different views in science: in Driesch's estimation, "Wilhelm Roux must be counted amongst the very best biologists of all times."[†]

At the outset of his engagement in *Entwicklungsmechanik*, Driesch defended a strictly mechanistic position, but that was soon to change:

> Dissatisfied by the indeterminate, vague and unfounded fantasies of our modern biological science, I was first attracted by those research strategies that call themselves exact. I consequently started to analyze these "mechanistic" efforts. I did, however, har-bor doubts about their sufficiency for an understanding of deep biological questions.[‡]

Given the regulative potential in sea urchin eggs that Driesch had discovered, he was able to distinguish the "prospective potential" of germ material as opposed to its "prospective significance," whereby the first he recognized as more encompassing than the latter. The prospective significance of germ material was revealed in normal development: the tissues and organs, the organism that forms through ontogeny. His experiments had shown, however, that development could be disrupted, and yet a normal embryo, a complex whole comprising all the typical parts—if at a smaller size—would be formed. Therein the prospective potential manifested itself. There was, in the sea urchin egg, a potential to fulfill the goal of development, to form the complete and functional complex whole, even if the starting conditions has been per-turbed. The inspiration that Driesch carried away from these observations led him to underwrite the validity of the Aristotelian concept of teleology, of a goal-directed developmental process, if it could be coupled with the Kantian notion of causality.[§]

---

* Karl Heider (1856–1935) reviewed the state of the art of developmental biology at the 1900 annual meeting of the German Zoological Society, especially with respect to developmental determination. He distinguished the "mosaic egg" from the "regulative egg," whereby pure mosaic and regulative development were recognized to be connected by intermediate (mixed) developmental patterns: Kühn, 1935, p. 793.

† Driesch, 1951, pp. 97–98.

‡ Driesch, 1894, pp. v–vi.

§ Driesch, 1894, p. vi.

Driesch's break away from the machine analogy to life[*] was cemented in his seminal book of 1899 on *The Localization of Morphogenetic Processes, a Proof of Vitalistic Manifestations* (*Die Lokalisation morphogenetischer Vorgänge, ein Beweis vitalistischen Geschehens*),[†] which at the same time marks Driesch's forceful invasion of philosophical territory. It would be a mistake, however, to interpret Driesch's vitalism along any metaphysical dimension. It has its root in strictly controlled developmental experiments, instead. The regulative capabilities of sea urchin blastomeres reveal a potential and a lawfulness which, in Driesch's interpretation, could not be further reduced to purely mechanical causes, be those chemical or physical in nature. His experiments, Driesch claimed, delivered empirical proof that organisms are governed by laws that are fundamentally different from, and irreducible to the laws known to govern the anorganic realm. It is this autonomy of life, governed by irreducible elementary laws that are manifest in developmental processes, which Driesch subsumed under his concept of vitalism.[‡] For Driesch, the term "vitalism" did not refer to mysterious forces or causes; it referred instead to the autonomy of biology, manifest in a lawfulness that cannot be reduced to the laws of physics and chemistry.

Decompose a machine into its parts—it will not be able to reconstitute itself. Take the cells apart during the early cleavage phases of a sea urchin egg, and each blastula will develop into a complete complex whole, the *pluteus* larva—albeit of smaller size. There is more to Driesch's vitalism, namely a goal directedness of vital processes, a teleology that reigns over the developing embryo. As he thought to have demonstrated with his experiments, the *goal* of embryonic development is the complete, causally integrated complex whole, the organism. Teleology is a loaded term, however, as it implies the Aristotelian *causa finalis*, a metaphysically tainted causality that implies prescient design and purposefulness. This is precisely the implication that Driesch did not want to have associated with his use of the term. Instead, teleology for him was an elementary causality inherent in the organic system that is the developing embryo.[§] And it is this causality which renders the developing embryo a Whole, a *Ganzheit*, and hence fundamentally different from a machine which is nothing but an aggregate of mechanical parts. To capture this causality and the resulting holistic conception of the organism, Driesch introduced the old Aristotelian notion of entelechy. It is through this entelechy that the organism functions as an integrated whole at every stage of its development, such as to ensure the completeness and wholeness of the fully developed organism.

MECHANICS AND BIOLOGY

Driesch started out as a staunch empiricist, a genuine *Entwicklungsmechaniker*, who even derided phylogeny reconstruction in the name of empiricism. But in the course of his experimentation with sea urchin eggs, he developed views that to many of his

---

[*] Herbst, 1941, p. 114.
[†] Driesch, 1899b; see also Herbst, 1941, p. 122; Ungerer, 1941, p. 458.
[‡] Driesch, 1899b, p. 70.
[§] Driesch, 1899b, pp. 67, 76.

contemporaries seemed to transcend true natural science, while others celebrated the autonomy, the holism, and organicism that Driesch brought to biology. But Driesch was not the only critic of Weismann, Roux, and like-minded developmental biologists. Another one was Oscar Hertwig*, who also distinguished himself as a critic of Darwinism, in particular in its social and political dimensions. In that respect, he set himself clearly off from his former advisor Ernst Haeckel; as a consequence, a deep rift opened between Hertwig and his erstwhile professor. Lawful organization, division of labor, and cooperation were for Hertwig the glue that held organisms, and their societies, together, not the struggle for existence—let alone *random* variation and natural selection. At the same time, however, Hertwig rejected all forms of metaphysically tainted vitalism, albeit he was himself labeled a vitalist by Haeckel. Born into a wealthy family on April 21, 1849, Oscar and his brother Richard grew up in Mühlhausen (Thuringia), where they attended the *Gymnasium* (grammar school). Both brothers enrolled at the University of Jena in 1868 to study under Haeckel. Oscar earned his PhD under Max Schultze (1825–1874), an anatomist at Bonn University. His *Habilitation* thesis, submitted to the University of Jena in 1875, earned young Oscar a lasting place in the history of biology. He was the first to have observed and comprehended fertilization as the consequence of a fusion of the nuclei of the egg cell and sperm cell—using the transparent eggs of sea urchins again. Oscar Hertwig rose through the ranks of associate and full professor in the medical school of Jena University, before he accepted in 1888 the position of professor and director of the Institute of Anatomy that had been created for him at the University of Berlin. He retired from that position in 1921, and died in Berlin on October 25, 1922 (Figure 4.4).

In his dispute with Weismann, Hertwig defended the maxim that "any theory of inheritance has to match cellular theory."[†] The "Weismann barrier" against the inheritance of acquired characteristics required a strict separation of gametes (sexual reproductive cells) from soma (bodily) cells. Weismann thought the exclusive carrier of inherited information to be the *Keimplasma*, the plasma of the gametes. Since these were strictly separated from the soma, there was no possibility that characteristics acquired by the soma through its lifetime could become heritable. Given his studies on fertilization and cell division, Hertwig took the nucleus of the gametes to be carrier of genetic information, an insight that had independently been gained by the plant cytologist Eduard Strasburger. The issue as to whether the cell plasm or the nucleus is predominantly responsible for the transmittance of inherited characteristics would continue to be debated for decades to come.[‡] But Hertwig concurred with Weismann that the cell is the fundamental unit transmitting genetic information, in contrast to the much discussed botanist Carl Nägeli, who saw the cell as only one of several hierarchically nested levels of biological complexity or individuality—intracellular and supracellular—that merits no special predominance in its role in shaping the developing embryo. Not so for Hertwig:

---

* On Oscar Hertwig, see Uschmann, 1969; Weindling, 1991; Nyhart, 1995.
† O. Hertwig, 1894, p. 29.
‡ For a historical account and analysis of that debate, see Harwood, 1993.

FIGURE 4.4 Oscar Hertwig. (Museum für Naturkunde Berlin, Historische Bild- u. Schriftgutsammlungen [Sigl: MfN, HbSb]. Bestand: Zool. Inst., B I/14 [Hertwig].)

> throughout the organic kingdom the cell as a unit both in a morphological and a physiological sense assumes the greatest importance amongst all other elementary units ... The unity of the cell must particularly be recognized in any theories of inheritance, because it has been demonstrated that the units through which species procreate, spores, eggs and sperm of plants and animals, have the status [*Formenwerth*] of cells.[*]

The cell, Hertwig continued, comprises protoplasm and a nucleus and in that sense is a primary organism, which either by itself or in combination with many others forms the basis for all plant and animal organization. He went on to describe the components of the cell body, anorganic and organic, the latter allowing the cell to grow and to subdivide and thus representing living units of lower order from which the cell as a unit of higher order is formed. Turning against Roux' model of mosaic development, which requires a heterogeneity of the protoplasm of the egg cell that corresponds to its subsequent diversification into tissue types and organs, Hertwig once again emphasized the cellular nature of the gametes. Even if the egg differs from all other cells in terms of its size, it is a typical cell nonetheless. After all, the minute mammalian egg that escapes detection by the naked eye has the same developmental potential as the enormous ostrich egg. Furthermore, since Hertwig localized the

---

[*] O. Hertwig, 1894, p. 29.

factors of inheritance in the cell nucleus[*], he concluded that both gametes, egg and sperm, contribute equivalently to the genetic makeup of the fertilized egg. This, he thought, violates the assumption of a heterogeneity of the egg cell protoplasm and the consequent preformation of embryonic development. Instead, Hertwig argued, the developmental fate of the blastomeres is determined in the course of cell cleavage and differentiation that explains regulatory developmental phenomena as those discovered by Driesch.[†] The cells in the developing embryo thus become living entities, individuals, which in reciprocal interaction undergo specification and differentiation in a way that results in normal development: the cells develop in mutual interdependence from one another and in reciprocal relation to the developing complex whole. Roux, of course, objected,[‡] accusing Hertwig, as he had Driesch, of missing similar points he had made himself, and to placate his work in a light that serves to constellate contradictory points of view, when in fact the differences between him and his critics he considered only gradual or, indeed, largely semantic.

Hertwig did not have many students, but exerted a major influence in zoology through his publications, in particular his textbooks on zoology and histology. In his zoology text of 1892, Hertwig put Herbert Spencer's concept of the "survival of the fittest" ahead of the Darwinian notion of the "struggle for survival," but in an anti-Darwinian mode translated Spencer's concept as the *Überleben des Passenden* (survival of the fitting).[§] Against Weismann he defended the inheritance of acquired characteristics, as well as Carl Nägeli's (1884) principle of perfection (*Vervollkommnungsprinzip*), which the latter had claimed to be based on purely mechanistic grounds.[¶] In his two-volume treatise on histology (1893/1898), he called the cell an elementary organism, the ordering of cells into a cellular association a complex whole, an individual of a higher level of complexity. In the second edition of his histology textbook, re-titled *Allgemeine Biologie*, Hertwig equated the multicellular complex whole that is an individual of higher order with a state.[**] A physiological individual, Hertwig defined as a living unit capable of sustaining itself, able to grow and to reproduce—from the single cell to the complex vertebrate. Among multicellular organisms, Hertwig again recognized different levels of morphological individuality of greater or lesser complexity, but whereas the whole multicellular organism (a "person") does represent a physiological individual, its parts may not if they cannot sustain themselves and reproduce (through budding or division) independently of the complex whole. In that sense, then, the complex whole is more than the mere sum of its parts; conversely, and in opposition to Roux' views, the parts collectively contrive to maintain the complex whole through differentiation (division of labor) and collaboration. In contrast to Haeckel, Hertwig recognized as either physiological or morphological individuals exclusively those biological units that originate from other such units either through

---

[*] See O. Hertwig, 1893.
[†] O. Hertwig, 1897, p. 189.
[‡] Roux, 1892 (1895), p. 767.
[§] O. Hertwig, 1892, p. 366.
[¶] O. Hertwig, 1892, pp. 41–42.
[**] O. Hertwig, 1906b, p. 410.

budding, or through division.* The designation of a cell as an "elementary organism" (*Elementarorganismus*)—with reference to Ernst Wilhelm von Brücke (1819–1892)†—appears already in Hertwig's early—1879—review of the history of the cell theory, where he likened the complex whole that results from an agglomeration of cells to form a multicellular organism to a "social association" with new, emergent properties.‡ In metazoans, the cells "have become little builders who construct the most intricate buildings" characterized by a division of labor among them, which renders the cells of the nervous system the "aristocrats of that society."§ In his polemics against Roux, published in 1894, Hertwig drafted a whole section that compares the formation of a human state with the development of an organism.¶ In a speech held in the auditorium of the Kaiser-Wilhelm University of Berlin on January 27, 1899, Hertwig further enriched the analogy of an organism with a state, drawing a parallel between the organization of a state investigated by social science and the organization of organisms investigated by biology: just as the organism, so is the state a complex whole of higher level of complexity. In support of the legitimacy to draw connections and parallels between biological and social sciences, he cited works of Herbert Spencer on sociology and biology.

> Just like an organism, "[t]he state, too, is composed of a multitude of differently functioning organs, it has its own life, it reacts in ways difficult to predict to interferences with is function, it develops and thereby changes its organization, it reaches its prime time after which it ages and finally decays as everything does that is of a finite duration ... And as there is, according to Darwin's theory, a gradual perfection of the organisms in the course of their continuous development [evolution], so there is a slow progression towards higher levels of organization manifest in the historical succession of states."**

Whereas a hydrozoan can be cut into pieces from which a whole organism will regenerate again, the same is no longer possible with a vertebrate. With more complex organization the parts have lost their independence, as is also the case in more complex social organization. The individual contributes to the complex whole that is the state in even more narrowly circumscribed, that is, specialized activities, and as a result becomes more dependent on the state. This, Hertwig concluded, requires the development of just and fair social policies, an expectation that Hertwig coupled with the hope that the "German nation will emerge both refined and strengthened from the natural-scientific—*cum*—social evolutionary process of our current historical epoch."†† As Paul Weindling, biographer of Oscar Hertwig, put it: "The theory of the cell state expressed the sense of social responsibility of the German professor to the nation idealized as a *Kulturstaat*."‡‡

---

* O. Hertwig, 1898, p. 6; 1906b, p. 372.
† Brücke, 1861, p. 381. The concept of a cell state was based on "the theory of the cell as the elementary organism": Weindling, 1981, p. 99.
‡ O. Hertwig, 1879, p. 429; see also Weindling, 1991, p. 262.
§ O. Hertwig, 1879, p. 429; see also Weindling, 1991, p. 262.
¶ O. Hertwig, 1894, p. 133.
** O. Hertwig, 1899, p. 18; see also Weindling, 1991, p. 263.
†† O. Hertwig, 1899, p. 21.
‡‡ Weindling, 1981, p. 116.

Although he was highly critical of Darwinism, at least in the form of its German reception,[*] and of Haeckel, it was Oscar Hertwig who was called upon to represent German biology at the centennial Darwin celebrations held in Cambridge on June 22–24, 1909. Reminiscing his student years in Jena, Hertwig testified to the immense influence Darwin had in Germany. He cited all the prominent German Darwinists of the time, among which Eduard Strasburger, Anton Dohrn, and Arnold Lang, who were all following the lead of Haeckel and Gegenbaur. But, he added, Germany had herself had some influence on the young Darwin through the explorer Alexander von Humboldt (1769–1859) and his books. The days of celebration in Cambridge "will be an incentive for us to continue on the road charted by Darwin that leads us to the forceful investigation of the great secrets of life"—not a word of restraint.[†] But Hertwig's anti-Darwinian, organicist interpretation of society and state hardened after the capitulation of Germany at the end of the Great War in 1918. His critique of Darwinism in his *Das Werden der Organismen* (1916), his *Abwehr des ethischen, des sozialen, des politischen Darwinismus* (1918) that targeted social Darwinism, and his *Der Staat als Organismus* (1922) have been placed in opposition to the Pan-German Imperialism already manifest during World War I, a movement that later sought to regain Eastern German territories lost as a consequence of the Versailles treaty, even to expand Germany beyond its former eastern boundaries in search of new *Lebensraum*. A further motivation for Hertwig's political engagement through book publications was the increasingly shrill propaganda in support of eugenics and racial hygiene.[‡] Hertwig's critique of Darwinism, his rejection of Darwin's selection theory based on randomness of variation, and his defense of the inheritance of acquired characteristics on the basis of the principle of causality in *Das Werden der Organismen* suffered a serious rebuff in the *Zeitschrift für Rassen- und Gesellschaftsbiologie*, at the time in the leading German journal of eugenics and racial hygiene. A 12 page review of Hertwig's book in the first issue of volume 13, published in 1918, stated: "We do not expect any lasting impact of this book on the conceptual/theoretical development of biology. The book lacks originality and, with respect to selection theory, even objectivity, because H[ertwig] battles a caricature, not a historically sound representation of Darwin's principle of natural selection."[§] The bulk of the review focused on a defense of Darwin's and Weismann's doctrines against Hertwig's misunderstandings or misrepresentations, and on a rejection of Carl Nägeli's doctrine of a direct influence of the environment on the organism that had been adopted by Hertwig. God forbid that the Aryan environment could potentially naturalize and assimilate non-Aryans through the inheritance of acquired characteristics!

In the same volume of the *Zeitschrift für Rassen-und Gesellschaftsbiologie*, the co-editor Fritz Lenz, then still *Privatdozent* in Munich but destined to become the leading racial biologist during the time of the Weimar Republic and the Third Reich, mounted an extended refutation of Hertwig's "attacks against 'Darwinism' and

---

[*] Darwinism—a doctrine of "*laissez-faire* liberalism" and "Malthusian competitive individualism threatening social disintegration": Weindling, 1981, p. 116.

[†] O. Hertwig, 1909, p. 958.

[‡] For detailed analysis, see Weindling, 1991, pp. 265, 274, 280.

[§] H. Thiem, in *Zeitschrift für Rassen-und Gesellschaftsbiologie*, 1921, vol. 13, p. 81.

Racial Hygiene." Fritz Lenz was born on September 3, 1887, the son of a farmer in Pflugrade (Pomerania). He attended the *Schiller Gymnasium* in Stettin, from which he graduated in 1905. He went on to study medicine in Freiburg i.Br., where he also attended courses taught by August Weismann and the anthropologist Eugen Fischer (1874–1967), who would later become the director of the *Kaiser-Wilhelm-Institut* of Anthropology, Human Genetics and Eugenics in Berlin-Dahlem. Fusing Mendelian theory of inheritance with *völkisch* idealism, Lenz became convinced of the necessity of racial hygiene by 1907. Later in his life he would praise the Nazi call for large-scale sterilization as a measure to maintain and improve the *Volksgesundheit.*[*] Lenz passed the medical state examination in 1912, and in 1919 earned the *venia legendi* at the University of Munich with a thesis on the effects of hybridization on wing patterns in butterflies. On April 1, 1923, Lenz became Germany's first professor for racial hygiene, appointed at the University of Munich at the rank of associate professor.[†] In 1933, Lenz was named head of the section for racial hygiene and deputy director at Eugen Fischer's Institute in Berlin, a position that was coupled with a full professorship at the University of Berlin.[‡] In his critique published in 1921, Lenz did not target Hertwig's *Das Werden der Organismen*, the failure of which had already been revealed in the aforementioned review, but instead chose to focus on Hertwig's *Abwehr des ethischen, des sozialen, des politischen Darwinismus.* Hertwig had chastised Darwin's doctrine of a relentless struggle for existence because of its ethical and moral implications, which ultimately must result in the suspension of legislation developed in support of a social community. In view of his later involvement in drafting euthanasia legislation,[§] Lenz's position is hardly surprising: "a scientific theory, which deals with the laws of being and becoming in nature has absolutely nothing to do with value principles and hence cannot have ethical implications."[¶] Lenz backed up his rebuttal of Hertwig's position with a reference to Heinrich Rickert (1863–1936), a leading neo-Kantian of the so-called South West School, whom Lenz called "the greatest German philosopher of present times." "Kant had already argued," Lenz continued, that "moral forces cannot be demonstrated in the experienced world, indeed, that there simply are no moral *facts.*"[**] Hertwig's ethics he consequently denounced as naturalistic, not realizing that his own position was guilty of the same fallacy. "Does he [Hertwig] really want," Lenz asked rhetorically, "that all those weaklings, hysterics, idiots, and hereditarily predisposed criminals will continue to be born in the future as well?"[††] Hertwig had called for the equality of all human beings, but, argued Lenz, if we instead prioritize "the life and future of the race, then there simply can be no question of an equality of individuals."[‡‡] By taking a stand against racial hygiene, Hertwig was said to

---

[*] Weindling, 1989b, pp. 143, 454.
[†] Weindling, 1989b, p. 336.
[‡] The biographical data on Lenz are from his vita, University Archive Göttinen, *Rektoratsmappe*, Fritz Lenz.
[§] Klee, 2003, p. 367.
[¶] Lenz, 1921, p. 192.
[**] "… *dass es sittliche Tatsachen einfach nicht gibt*": Lenz, 1921, p. 193.
[††] Lenz, 1921, p. 199.
[‡‡] Lenz, 1921, p. 201.

implicitly promote just such a chanciness of reproduction among the German people that he accuses Darwinism of endorsing. Lenz was proud of the *Archiv* that carried his papers, and which he co-edited with Alfred Ploetz (1860–1940), the founder of the German Society for Racial Hygiene and its journal: "the most important journal in eugenics, as Prof. Corr-Sauders [*sic*] (Edinburgh) attested in a review in the English periodical *Eugenics Review*."*

Hertwig followed the lead of Virchow, Herbert Spencer, and Ernst Kapp (1808–1896)[†] as he drew analogies between biology and sociology, offering an alternative to Haeckel who supported eugenics and expansionist foreign politics.[‡] In his *Der Staat als Organismus* (1922), Hertwig once again emphasized that the organism, just as the state, are complex wholes with emergent properties and hence more than the sum of their respective parts. "This is the great trick performed by nature,"[§] namely to form not merely aggregates, but complex wholes composed of parts. Complex wholes form on the basis of six laws on nature: (1) the Law of Association; (2) the Law of Differentiation and Division of Labor; (3) the Law of Physiological Integration; (4) the Law of the Equality of the Parts; (5) the Law of the Reciprocal Interdependence or Correlation of Parts; and (6) the Law of the Multiplicity of Causes and their Effects. The same laws, according to Hertwig, govern the formation of states. In his review of Hertwig's book, the "father of German Sociology," Ferdinand Tönnies (1855–1936), found it "strange" that such an "eminent biologist" as Oscar Hertwig should revive an organicist approach to sociology, which he claimed had by that time already been completely abandoned by sociologists.[¶] He found the book—largely based on Herbert Spencer's ideas—to give evidence "of sincere, broad studies, and of careful, deep thought. It is also inspired by a noble attitude. But I cannot recognize in it any merit for sociology."[**] In particular he noted that Hertwig missed the distinction of objective and subjective interpretation of social constructs: Hertwig "runs *Volk* (community) and *Staat* (state) together."[††] For Tönnies, the *Volk* is a *Gemeinschaft*, a naturally grown entity, a living organism, a complex whole, an individual of higher order. The *Staat*, in contrast, is a *Gesellschaft*, a conceptual–legal construct. The ageing Tönnies would find himself forced to defend his conception of the *Volk* as a *Gemeinschaft*, based on his "longstanding interest in evolutionary theory,"[‡‡] against its abuse by National Socialist ideology.[§§]

In summary, then, the cell state metaphor encouraged an analogy between the organism and the *Volk*. In its original, reductionist, and mechanistic interpretation,

---

* Sir Alexander Morris Carr-Saunders (1886–1966); the quote is from Lenz's appendix to the questionnaire issued by the Military Government of Germany, University Archives Göttingen, *Kuratoriumsakte Fritz Lenz*.
† Kapp had published, in 1877, a thesis of *Organprojection*, the "projection of organs," which sated that in building their societies, humans had implemented organizations that mirror their own body parts: the hammer is the equivalent to the fist: see O. Hertwig, 1922, p. 3; Weindling, 1991, p. 273.
‡ Weindling, 1991, p. 288.
§ O. Hertwig, 1922, p. 47.
¶ Tönnies, 1923, p. 303.
** Tönnies, 1923, p. 304.
†† Tönnies, 1923, p. 304.
‡‡ Weindling, 1989b, p. 121.
§§ Rammstedt, 1985, p. 304, n. 100; Weindling, 1989b, p. 484.

the organism is a cell state, composed of parts (cells) that compete with each other in a Darwinian selectionist sense for the good of the whole. But the *Volk qua* organism metaphor was also appropriated by the representatives of an organicist-holist biology, who comprehended both the organism and the *Volk* as a hierarchically structured complex whole subject to upward and downward causation in a system of reciprocally interdependent parts. Significantly, the Darwinian selectionists would waste no time to confront, and critique such an organicist–holistic comprehension of the *Volk qua* organism. The conflict between the organicist–holistic and the Darwinian selectionist camp in German biology would take on ever-shriller tones as the ideological polarization accentuated and ideological pressure increased. But a full understanding of the organicist–holistic tradition in German biology of the first half of the twentieth century requires the introduction of yet another founding father, equal to Hans Driesch in his importance.

ANIMAL PSYCHOLOGY AND *UMWELT*

Driesch was not the only one, but one of the major forces who defended holism and the correlated organicism that came to dominate large parts of German biology during the first half of the twentieth century. Another one was Jakob von Uexküll (1864–1944). While a student of zoology at the University of Dorpat (Tartu), Uexküll had taken an interest in marine biology while on excursion in Lesina (Dalmatia), research he would later continue at Dohrn's Zoological Station in Naples. As of 1891, Uexküll became a regular winter guest at the station, where he pioneered physiological investigations, and of course met Hans Driesch as well as Curt Herbst. The three men formed a discussion group, the conversation between Driesch and Uexküll turning ever more on philosophical issues, which marginalized Herbst to some extent.[*] Eventually, Driesch and Uexküll would work "together aggressively at scientific meetings to undermine mechanistic principles in life sciences."[†]

Jakob Johann von Uexküll[‡] was born into an aristocratic family on September 8, 1864, on the family estate Keblas in Estonia. His education instilled in him a pride of place he would retain throughout his life. Aristocratic pride was also prominent among the faculty and students of the University of Dorpat (Tartu), where in the past anti-Darwinian sentiments had prevailed. Darwinism had traditionally been sketched at this institution as committed to an egalitarian bourgeois philosophy. The co-founder of cell theory, Matthias Jacob Schleiden, for example, was offered an honorary professorship for plant physiology and anthropology at the University of Dorpat in March 1863. He held his well-received inaugural lecture on October 16, 1863, subsequent to which he launched his course in anthropology. In December, the Russian government named him Imperial Councilor and Full Professor, which immediately triggered opposition of the faculty to the creation of an off-budget position. The affair escalated, the university faculty, the Russian Orthodox clergy, and the press turning against Schleiden, accusing him of having

---

[*] Bethe, 1940, p. 821.
[†] Harrington, 1996, p. 39.
[‡] The biographical sketch for J.v. Uexküll is based on Mildenberger, 2007, 2010, and Rüting, 2004.

defended Darwinism in his lectures. Schleiden left Dorpat in the summer of 1864.[*] In opposition to Darwinism, the clergy encouraged the students to take classes taught by the philosopher Gustav Teichmüller (1832–1888), who defended a vitalistically tainted teleology, a goal-directed and purposeful development of nature. This, of course, matched the philosophy of the ageing Karl Ernst von Baer, who had left St. Petersburg in 1867 to retire to the hometown of his *alma mater*, Dorpat (Figure 4.5).

The year of von Baer's death, 1876, saw the publication in St. Petersburg of the second volume of his collected presentations delivered at scientific meetings and essays of varied content, the *Studien auf dem Gebiete der Naturwissenschaften* (reprinted in 1886). The book contains Baer's testimony on Darwin's theory, prefaced by four essays that presented a forceful defense of teleology in natural

FIGURE 4.5 Karl Ernst von Baer. (Smithsonian Institution Archives. Image 85-4432.)

---

[*] Jahn, 2001, p. 317.

processes. In an essay on the goal-directedness (*Zielstrebigkeit*) inherent in organic bodies, Baer asserted:

> ... the elaboration of Darwin's hypothesis by his successors shows an increasing tendency ... to deny with respect to natural processes all relations to the future, to what is to develop, all goal-directedness and purposefulness ... Should scientific warrant be accorded to Darwin's hypothesis, it will have to accommodate this universal goal-directedness. If it is incapable thereof, then one will have to reject it.[*]

His conclusions were unequivocal: "All organic development is through and through goal-directed ... the result of development is therefore predetermined."[†] His essay *Ueber Darwin's Lehre* was not motivated, von Baer explained, by outright opposition to Darwin, but should rather be seen as an explication of Darwin's work from his own perspective, which must include, however, critique where critique is due. One of the major striking points was Darwin's replacement of lawful goal-directedness with a collection of accidents as expressed in random variation.[‡] In a letter to Anton Dohrn from the year 1875, von Baer wrote: "I became a partial Darwinist, or better transformationist, but a full devotee I can be no more than a full opponent."[§] Many of these themes and concerns would become apparent in Uexküll's work as well, who when starting out with his studies in 1884 in history, mineralogy, and zoology was still impressed by materialism and Darwinism, the latter at this time revived again among the faculty of Dorpat University. Socio-political developments instigated by the central Russian government that targeted the Baltic aristocracy, and that were underpinned with social-Darwinist propaganda, drove Uexküll away from both Darwinism and Dorpat, as also did the Haeckel-style phylogenetics propounded by the professor of zoology at Dorpat University, the entomologist Julius von Kennel (1854–1939). Independently wealthy, Uexküll was able to divide his time between the University of Heidelberg in the summer months and Dohrn's station in Naples during the winter in pursuit of his physiological studies on octopuses, sea urchins, and peanut worms. His thinking during these years was dominated by a tension between vitalism and mechanism, a conflict that fueled his discussions with Driesch. Ultimately, however, Uexküll's lack of sensitivity and diplomacy in the political arena of science led to his estrangement from Anton Dohrn and from the management of the Naples station, as also to his loss of laboratory space in Heidelberg. He was, however, awarded an honorary PhD degree by the University of Heidelberg in 1907 for his work in muscle physiology.

After marriage, Uexküll settled in Heidelberg in a house close to Driesch's. It is at Uexküll's house that Driesch first met the philosopher Hermann v. Keyserling (1880–1946), also living in Heidelberg and a distant relative of Uexküll's. In Driesch's recollection, "Uexküll and Keyserling were both ... typical representatives of the Baltic aristocracy; highly intelligent, highly educated, very energetic, also very much the 'distinguished gentlemen' ... despotic at times but good-humored at heart, more

---

[*] von Baer, 1876a, p. 173.
[†] von Baer, 1876a, p. 235.
[‡] von Baer, 1876b, p. 240.
[§] Quoted from Kuhn-Schnyder, 1976, p. 16.

'Russian' than German."[*] It is through Keyserling that Uexküll became acquainted with Houston Stewart Chamberlain (1855–1927), an Englishman living in Bayreuth, who was married to Richard Wagner's daughter Eva. Chamberlain was an emphatic Wagnerian, a Pan-German who was critical of the Weimar Republic and of democracy in general. In 1899, Chamberlain published his notorious *Foundations of the Nineteenth Century*, which constituted a foundational contribution to anti-Semitism and contributed to the myth of the Aryan race, a book that would later influence the Nazi Party "philosopher" and deputy of the Führer for the surveillance of the entire intellectual and ideological schooling in the NSDAP, Alfred Rosenberg (1893–1946),[†] as well as Hitler himself. Chamberlain was personally acquainted with the emperor Wilhelm II, and later with Hitler. "Flirt[ing] with Darwinism as a 'proof' of the importance of race" when a young man, Chamberlain's "mystical concept of race"[‡] led him to reject Darwinism later in life, and to turn instead to Goethe's organicist approach to natural sciences and the intuitive apprehension of *Gestalt* as a legitimate source of knowledge. In his intellectual autobiography, written as a letter to Jakob von Uexküll, Chamberlain praised the innocence of the pure, searching vision.[§] Indeed, Uexküll and Chamberlain engaged in an extensive correspondence concerning biological and political issues, including aspects of race ideology.[¶]

During the planning phase of the *Kaiser-Wilhelm Institute for Biology*, Uexküll presented himself in the summer of 1912 as the ideal person to spearhead that effort. However, Darwinians prevailed in the competition for leading positions in this institute, while the neo-vitalist Uexküll was appointed head of a mobile aquarium. In this capacity he traveled to France in 1914 to study lobsters, accompanied by the Viennese Social Darwinist Lothar Gottlieb Tirala (1886–1974), whom he had met through Chamberlain. Tirala would later, in 1933, succeed Fritz Lenz—against the latter's advice[**]—as professor of racial hygiene at the University of Munich on the recommendation from Chamberlain's wife Eva Wagner, the *völkisch*-racist Munich publisher J. F. Lehmann (1864–1935), and the co-founder of Aryan Physics and Nobel laureate Philipp Lenard. Tirala was removed from his post on April 18, 1936, amidst complaints from both students and assistants about Tirala's manifest incompetence as a researcher and teacher.[††] An abstract of his inaugural lecture for his Munich professorship was published in the outlet for Aryan Biology *Der Biologe* in 1934, where Tirala praised Chamberlain as a "friend and teacher" and concluded that the task of

---

[*] Driesch, 1951, p. 206.

[†] On January 24, 1934, Hitler named Rosenberg the *Führer*'s "deputy for the supervision of the entire spiritual and ideological schooling and education of the NSDAP": Nagel, 2012, p. 123. Rosenberg developed his "blood myth" (*Blutsmythos*) of the race following Chamberlain: Weingart et al., 1992, p. 377.

[‡] Kelly, 1981, p. 109.

[§] Hutton, 2005, p. 187.

[¶] J. Schmidt, 1975.

[**] *"Ich habe den in Aussicht genommenen Parteikandidaten Tirala eindeutig als wissenschaftlich unfähig begutachtet. Er ist trotzdem berufen worden, allerdings nach einigen Jahren wegen Unfähigkeit wieder entfernt worden"*: Lenz, statement from April, 21, 1947, to the chair of the committee for the political review of the faculty of the University of Göttingen. Lenz presented his opposition to Tirala's appointment as an opposition to party authority. (University Archive Göttingen, Fritz Lenz – Rektorats-Akte.)

[††] Weindling, 1989b, p. 510.

racial hygiene is not just the investigation of the laws of life (i.e., the laws of inheritance), but also to prompt their practical application, even if compulsorily*: "Racial hygiene represents the same in biology as national-socialism does in economics."† In another entry in the same journal, Tirala praised Haeckel, Chamberlain, and Uexküll as pioneers of politically applied biology.‡ On the occasion of Uexküll's 70th birthday, Tirala arranged for a photograph of the honoree to be published with a congratulatory note on the opening page of the November 1934 issue of *Der Biologe*, followed by an imaginary dialog *pro* and *contra* Uexküll's doctrines, which lent the concluding words to Uexküll's voice: ".... that there is something Higher, that we do not confront the world as singular and forlorn individuals, but that we are nurtured and sheltered by the higher unity of the *Volk*, which alone gives meaning and *Gestalt* to our own life."§

As a consequence of the Russian October Revolution of 1917, Uexküll lost his estate and investments in Estonia. The November Revolution that broke out in German cities in the following year reinforced Uexküll's belief in a conspiracy of the imaginary world Jewry and Bolsheviks¶, which resulted in virulent anti-Semitic statements issued by Uexküll, from which he distanced himself by 1923, however. Yet in the meantime, in 1920, he had published his ominous *Staatsbiologie*,** which once again drew an analogy between the state and the organism. But in contrast to Hertwig's organicist analysis of the state, Uexküll's opus reveals a "feudalistic and capitalistic"†† motivation. Uexküll squarely laid the blame for the contemporary crisis on the inefficiency of the Weimar Republic's form of government, and, in agreement with Chamberlain, he blamed the Jews who had already infiltrated all organs of the German state.‡‡ In the second edition of his *Staatsbiologie*, published in 1933, he married his hope in Hitler's leadership to a vision of cleansing the state that is an organism from its illnesses and parasites, yet pleaded for tolerance and restraint when dealing with the "Jewish question,"§§ only to be bitterly disappointed as years went by.¶¶ Ernst Lehmann, the self-proclaimed spokesman for Aryan Biology and founder of *Der Biologe*, reviewed Uexküll's *Staatsbiologie* (the second edition of 1933) in his journal: although he found himself not in complete agreement, he nevertheless found it to be "a pleasure to accompany the author in his analysis of the biological foundations of the state. An even greater pleasure is it to see that the leaders of the state today are implementing the biological foundations as guiding principles of their actions."***

On the recommendation of a former student of his, Otto Cohnheim (1873–1953), who changed his Jewish name to Kestner in 1917, and became full professor for physiology at the University of Hamburg in 1919, Uexküll was appointed as

---

* Tirala, 1934a, p. 53.
† Tirala, 1934a, p. 52.
‡ Tirala, 1934b; see also Stella and Kleisner, 2010, p. 47.
§ Tirala, 1934b, p. 284.
¶ Mildenberger, 2007, p. 157.
** Uexküll, 1920.
†† Harrington, 1996, p. 61.
‡‡ Harrington, 1996, p. 6; Mildenberger, 2007, p. 157.
§§ Harrington, 1996, p. 62.
¶¶ Rüting, 2004, p. 42; Stella and Kleisner, 2010, pp. 41–42.
*** Lehmann, 1934a, p. 25.

scientific assistant at the University of Hamburg in 1925 with the remit to revamp the much neglected aquarium. The following year, Uexküll founded his *Institut für Umweltforschung*, and in 1927/28 was promoted to full professor and director of that institute. At the University of Hamburg, Uexküll made the acquaintance of the neo-Kantian philosopher Ernst Cassirer (1874–1945), who emigrated to the United States in 1933, and more importantly perhaps he befriended the bio-philosopher and historian of biology Adolf Meyer (1893–1917), who after 1945 named himself Adolf Meyer(-Abich).[*] Adolf Meyer(-Abich) had studied philosophy in Göttingen and Jena, and came to Hamburg as a university librarian. Meyer(-Abich) and Uexküll co-taught seminars on the philosophy of science. With his publications, Meyer(-Abich) established himself as one of the foremost representatives of holism in German biology,[†] recognized as having "played an active role in defining the significance of holism for National Socialism."[‡]

With the escalating radicalization of the Nazi regime, vehemently rejected by Uexküll, and the increasing preparations for war, Uexküll's science fell into disregard and neglect. To finance the research activities at his institute proved increasingly difficult, and pressure on him to retire increased. A disillusioned Uexküll withdrew to the island Capri in 1940, where throughout the vagaries of turbulent times he had managed to maintain a villa. He died in Capri on July 24, 1944. In his attack on Uexküll's holism and neo-vitalism of 1929, Rudolf Ehrenberg (1884–1969)—an experimental physiologist and bio-philosopher at the University of Göttingen—characterized Uexküll as the contemporary incarnation of his compatriot Karl Ernst von Baer. Expressing his disbelief that biology could afford such an inconsequential debate as the one that pitches mechanism against vitalism, he concluded: "Today, in the age of nuclear physics and quantum mechanics ... every natural science must be atomistic. Biology cannot make an exception, or else it is not a natural science."[§] Uexküll in contrast had re-interpreted Bohr's model of the atom as one that was not based on causal relations between mass particles, but rather on systematic (*planmässig*) relations between the whole and its parts.[¶] The title of Ehrenberg's 1929 critique in the widely circulated interdisciplinary science journal *Die Naturwissenschaften* reveals that the target of his review was the second edition of Uexküll's book *Theoretische Biologie*, published in 1928. The first edition of the same book, published in 1920, had received a far better reception in *Die Naturwissenschaften*. Uexküll was celebrated as the Einstein of biology, "the most articulate leader towards a new biological world view [*Weltauffassung*],"[**] who proclaimed biology to be an autonomous science independent of, and irreducible to physics: Karl Ernst von Baer had already defended the goal-directedness of developmental processes, "subject to the reign of a rule acting according to plan,"[††] as Uexküll aptly put it in his treatise.

---

[*] Rüting, 2004, p. 47.
[†] Amidon, 2009.
[‡] Harrington, 1996, p. 190; see also Amidon, 2009.
[§] Ehrenberg, 1929, p. 780.
[¶] Uexküll, 1931, p. 387.
[**] Asher, 1922, p. 477.
[††] Asher, 1922, p. 475.

Concerning his *Theoretische Biologie,* Uexküll wrote to Chamberlain: "It is not my ambition to compete with the truly great men, for I have no other merit than to have pointed to the fruit which have slowly ripened on the tree of Kantian wisdom."[*] Indeed, Uexküll applied Kant's *a priori* categories of cognition to animal psychology, thus construing a very relativistic relation of animals to their *Umwelt.* Uexküll wanted to emphasize the importance of the subject that meets the outer world with its sensory apparatus, and in so doing constitutes its world according to its sensory capacities, perceptions, and potential for reaction: "the living cell," he wrote, "possesses a specific energy with which it reacts individually to external stimulation. The living cell relates to external objects not as an object, but as a *subject.*"[†]

For Uexküll, scientific theories relate to the outer world as does the scaffolding to the house being built within it: "a new scaffolding is needed for biology; the old scaffolding, borrowed from chemistry and physics, will suffice no longer."[‡] And that new scaffolding had to accommodate holism, neo-vitalism, and conformity to plan (*Planmässigkeit*). The latter was (and continues to be) a particularly controversial issue, as in the Aristotelian tradition it seems to imply a conscious, willful entity. As von Baer had put it:

> For the whole of nature I want to use the concept of purposefulness (*Zweckbegriff*) in its strong sense, although I have to admit that this implies the thought of a conscious and willing entity.[§]

In his obituary for Hans Driesch, Curt Herbst sketched Driesch's influence on Uexküll, and claimed to have immediately recognized that the latter's doctrine of conformity to plan is nothing else but a rewording of Driesch's doctrine of entelechy, "for design implies its goal."[¶] Uexküll did not want to mince words:

> Instead of conformity with plan, we might just as well speak of conformity with function, or of harmony, or of wisdom. The name does not matter; what does matter is that we should recognize the existence of a natural force which binds according to rules.[**]

What Uexküll wanted to highlight—with Karl Ernst von Baer and Hans Driesch—is that organisms, as much as the species, indeed the whole tree of life, is a complex whole composed of interdependent parts, where "somehow the parts determine the whole and the whole [determines] the parts."[††] The complex whole develops, and behaves, according to an inherent plan with an immanent goal-directedness, driven by endogenous forces that Uexküll referred to as "will." For him, willfulness, purposefulness, and goal-directedness, were not "mere formulas" as those that physics is dealing with, nor were they supranatural agents, but instead they were natural factors. Darwinism, for Uexküll, was a religion rather than a science, "nothing but

---

[*] J. Schmidt, 1975, p. 122. The letter is dated December 27, 1920.
[†] Uexküll, 1931, p. 386.
[‡] Uexküll, 1920 (1926), p. xi.
[§] von Baer, 1876c, p. 82.
[¶] Herbst, 1941, p. 146.
[**] Uexküll, 1920 (1926), p. 176.
[††] Uexküll, 1920 (1926), p. 241.

the embodiment of the impulse by the human will to get rid ... of plan in nature."[*] For Darwinians, the genealogical tree is nothing but "the result of the influence of external forces," Uexküll complained; instead, he sided with "the Lamarckians," which allowed him to see the phylogenetic tree as the result of inner growth,"[†] striving to even greater complexity of organization according to plan. Organization in that context meant a "unity in which the different parts are combined into a whole through the agency of a common activity,"[‡] and that agency is the will of the organism to act. It is through its actions driven by its will that the organism confronts the exterior world in its struggle for existence, but very much in contrast to Darwinism the organism is not thereby merely subject to the influence of external factors, but actively creates its *Umwelt* instead through its interactions with it. This allowed Uexküll to put the brakes on Darwinian progressionism as heralded by Haeckel: perfection obtains on every rung of the ladder of life, since it "merely means the correct and complete exercise of all the means available"[§] to any one organism. But since the means of perception, and reaction of a paramecium differs from those of a horse, the experienced *Umwelt* will differ accordingly: "By its resources, limits are set to the achievements of every animal. The sum of all resources at an animal's disposal ... the sum-total of all its properties and capacities—these make the organism."[¶]

This is a profoundly relativistic understanding of the world, motivated by a psychological interpretation of Kant's concept of *a priori* categories of cognition: "All reality is subjective appearance. This must be the great fundamental admission even of biology."[**] Impressions of the *Umwelt* are received by the organism, and used by it in its reactions, all according to its species-specific capacities, which in turn "divide into two halves—a receptor half, corresponding to the world-as-sensed, and an effector half, corresponding to the world of action."[††] In contrast to Darwinism, in Uexküll's interpretation anyway, the struggle for existence is not a one-sided affair, where the external world shapes the organism in a series of accidents. Instead, it is a reciprocal interaction between the organism and its *Umwelt*, and in that sense "an essential part of the general plan of life ... where there is a weapon, there is also an enemy."[‡‡]

"The proof adduced by Driesch," in Uexküll's estimation, "is as simple as it is enlightening"—it is a "proof" against a "mechanical framework."[§§] Driesch and Uexküll were unquestionably the most influential authors bringing neo-vitalism to early twentieth-century German biology, to be built upon and expanded by such authors as Adolf Meyer(-Abich), the post-World War I Richard Woltereck (1877–1944)[¶¶], and others. The organism for Driesch, the organism and its *Umwelt* for Uexküll, formed a whole, a *Gestalt* that reigns over and organizes its part. There is a goal-directedness in nature, an *entelechy* in Driesch's term, a non-mechanistic force

---

[*] Uexküll, 1920 (1926), p. 265.
[†] Uexküll, 1920 (1926), p. 261.
[‡] Uexküll, 1920 (1926), p. 17.
[§] Uexküll, 1920 (1926), p. 164.
[¶] Uexküll, 1920 (1926), p. 164.
[**] Uexküll, 1920 (1926), p. xv.
[††] Uexküll, 1920 (1926), p. 81.
[‡‡] Uexküll, 1920 (1926), p. 131.
[§§] Uexküll, 1920 (1926), pp. 181, 195.
[¶¶] Harwood, 1996.

inherent in living matter that urges the actualization of the whole. Driesch recognized the danger of misunderstandings he might promote with his choice of terms. The term *entelechy* derives from Aristotle, in whose writing it stands for the realized, the accomplished whole. Not so for Driesch, for whom entelechy is inherent in the beginning of the living process that serves the actualization of the whole, it is the potential in the sea urchin blastomeres to develop into the whole that is the pluteus larva.[*] Ontogeny for Driesch, phylogeny for Uexküll cannot be reduced to purely mechanical causes, nor is evolution sufficiently explained in terms of the externalist doctrine of natural selection. There is, for Driesch and Uexküll, an irreducible internal regulative principle, a "force which binds according to rules,"[†] active in living substance, one that renders the reduction of life to purely physico-chemical mechanisms impossible. Driesch acknowledged that Adolf Meyer(-Abich)—a self-declared disciple of Driesch and Uexküll[‡]—had established himself as the most forceful promoter of holism[§] in Germany, but he rejected Meyer's attempt to reduce both mechanism and vitalism to one and the same elementary causality, that is, to expand the concept of wholeness to inanimate objects, and therewith to cleanse holism of any vitalistic connotations.[¶] Driesch's vitalistic grounding of the autonomy of biology would clash, however, with the *neue Sachlichkeit*[**] that was disseminated by philosophers and scientists who called themselves members of the Vienna Circle.

THE CONGRESS IN PRAGUE

With a solid background in experimental biology, Driesch became one of the most prominent bio-philosophers defending a holistic–organicist conception of nature. "Natural science without philosophy is blind, philosophy without natural science is empty," he proclaimed, paraphrasing Kant.[††] The search for reciprocal relations and illumination between biology and philosophy pursued by Driesch matched similar efforts by philosophers of science, in particular those that have become known as the members of the Vienna Circle; their philosophy in turn became known as logical positivism or logical empiricism. As such, it stood in the tradition of the great British empiricist philosophers John Locke (1632–1704), David Hume (1711–1776), and John Stuart Mill (1806–1873).

---

[*] Driesch, 1935, p. 198.
[†] Uexküll, 1920 (1926), p. 176.
[‡] Mildenberger, 2007, p. 176.
[§] The term "holism" was introduced into biology by the South-African General and later Prime Minister Jan C. Smuts (1870–1950) in his 1927 book on *Holism and Evolution*. Equally influential in the articulation of holism was John Scott Haldane's *The Philosophical Basis of Biology* (1931), in which resonated many themes that were also articulated by Driesch and Uexküll. These are, among other, the irreducible wholeness of organic structures and/or organisms (including supra-organismal structures such as colonies and biocoenoses) where the whole organizes its parts; the renunciation of a separation of morphology from physiology, as also of the distinction of an internal and external environment of an organism; more fundamentally the autonomy of biology, the impossibility to fully reduce biology to the causality revealed by chemistry and physics, or, more generally, the impossibility "to apply mathematical reasoning to life" (Haldane, 1931, p. 14).
[¶] Driesch, 1935.
[**] Galison, 1990, p. 725; Friedmann, 2000, p. 158.
[††] Driesch, 1911, p. v; Kant's famous line (from *The Critique of Pure Reason*) reads: *Thoughts without content are empty, intuitions without concepts are blind.*

The renaissance of empiricist philosophy is generally traced back to the Austrian physicist and philosopher Ernst Mach (1838–1916), who was identified as "the godfather of logical positivism."[*] Mach was appointed to the chair for the history and theory of inductive sciences at the University of Vienna in 1895.[†] Successors of Mach in that position were the physicist turned philosopher Ludwig Boltzmann (1844–1906) and, later, the German philosopher and physicist Moritz Schlick (1882–1936), a former doctoral student of Max Planck (1858–1947). Given this succession, it is hardly surprising that the philosophy articulated by the members of the Vienna Circle was primarily a philosophy of science, which, as a distinct branch of inquiry, had not existed before. Schlick started to teach at the University of Vienna in 1922. The Vienna Circle, a Thursday night discussion group that as of 1925 regularly met at Schlick's house, had its ultimate origins in a seminar that Schlick taught in 1923.[‡] The circle included men representing a variety of professions, among whom most prominently the mathematician Hans Hahn (1879–1934), the sociologist Otto Neurath (1882–1945), and the logician Rudolf Carnap (1891–1970), who had come to Vienna in 1926. Other philosophers such as Ludwig Wittgenstein (1889–1951) and Karl R. Popper (1902–1994) from Vienna, Hans Reichenbach (1891–1953) and Carl G. Hempel (1905–1997) from Berlin, were not actual members of the Circle, but affiliated with it or with some of its members in one form or another.[§] Among the younger participants was Kurt Gödel (1906–1978), who would become famous for his Incompleteness Theorem concerning mathematics, while the British philosophers Alfred J. Ayer (1910–1989),[¶] a visitor to the Circle, earned lasting fame with his tract that summarized the logical positivist position in an engaging polemic style.[**] The Circle itself published a manifesto under the name of its public branch, the *Verein Ernst Mach*, which was founded in 1928 on the initiative of Otto Neurath, with Schlick as its president and with the ambition to gain influence in politics and adult education.[††] In the recollection of one of its younger members, Herbert Feigl (1902–1988), one of Schlick's students, "the discussions of the Circle centered on the foundations of logic and mathematics, the logic of empirical knowledge, with only occasional excursions into the philosophy of the social sciences and mathematics."[‡‡] The programmatic pamphlet published in 1929 under the auspices of the *Verein Ernst Mach* was titled *Wissenschaftliche Weltauffassung. Der Wiener Kreis* (The Scientific World View: The Vienna Circle), and contained an introduction signed by Otto Neurath, Hans Hahn, and Rudolf Carnap. The tract opens with an expression of concern about the growing popularity of metaphysical and theological thinking, which must be opposed by the spirit of enlightenment and an anti-metaphysical research of facts.[§§] The Weimar Culture was in part characterized by a deep-seated

---

[*] Janik and Toulmin, 1973, p. 133.
[†] Joergensen, 1970, p. 853.
[‡] Joergensen, 1970, p. 848.
[§] Stadler, 1997.
[¶] Rogers, 1999.
[**] Ayer, 1936.
[††] Joergensen, 1970, p. 850; Janik and Toulmin, 1973, p. 132.
[‡‡] H. Feigl, cited in Joergensen, 1970, p. 849.
[§§] Neurath et al., 1929, p. 301; see also Carus, 2007.

cultural pessimism, a sense of crisis pervading arts, humanities, and sciences, with a consequent anti-scientific intellectual movement appealing to irrationalism and nature mysticism*: Neurath "(rightly) saw these movements—which were very popular with students—as broadly sympathetic to the authoritarian fascism that was gaining ground throughout Europe."† At least some of its members thus understood the *Verein Ernst Mach* as a tool to promote a political agenda in opposition to such societal tendencies, articulating a program that sought to spread the gospel of a demystified, objective, and rationally structured science.‡ As far as the logic of scientific inquiry was concerned, the young Carnap in particular, influenced as he was by Wittgenstein and others such as the British philosopher Bertrand Russell (1872–1970), sought to confront the perceptually immediately given with logic, the latter providing analysis that proves propositions to be either true or false. If psychology, social sciences, and biology could be reduced to physics, if physics in turn could be reduced to mathematics, and if mathematics could be reduced to logic, then science would have reached unassailable bedrock on which to erect its theoretical edifice. The task at hand for the Vienna Circle was thus primarily the demarcation of science from metaphysics, and further to analyze in ever greater depth the meaning of scientifically tenable statements, that is, of statements that can be shown to be either true, or false, in short, statements that are testable against the experienced material world. This latter program has become known as the "linguistic turn" in empiricist philosophy, as the demarcation of science from metaphysics was pegged to the meaning of the statements, propositions, or theories. The scientific status of a statement obtained from it being meaningful, that is, testable, whereas metaphysical statements were recognized as being meaningless, that is, untestable. Schlick in particular is famous for his example that involves the description of Vienna's Cathedral in a travel guide: "I found, for instance, in my Baedeker the statement: 'This cathedral has two spires', I was able to compare it with 'reality' by looking at the cathedral, and this comparison convinced me that Baedeker's assertion was true."§ Although this type of argument was soon to be recognized as highly problematical, it is nevertheless instructive to compare the statement analyzed by Schlick in this example with statements such as "There is a God," or "There is an entelechy which is the leading principle in living beings."¶ The first of those two latter statements asserting the existence of God was dismissed outright as unscientific by the members of the Vienna Circle, but with respect to the latter statement they conceded in their manifesto from 1929 that *if* vitalism could be thoroughly cleansed from all metaphysical connotations, and consequently could be recognized as an integral element of a science of biology, then biology could indeed not be fully reduced to physics, a unification of science could therefore not be achieved.**

To disseminate its philosophy, expose it to discussion and criticism, and to further an international exchange of ideas, the Vienna Circle started to organize congresses,

---

* Forman, 1971; Peuckert, 1987; Galison, 1990.
† Carus, 2007, p. 208.
‡ Stadler, 1997; Reisch, 2007.
§ Cited from Oberdan, 1993, p. 61.
¶ Joergensen, 1970, p. 850.
** Neurath et al., 1929, p. 312.

the first one of which was held in Prague in 1929. The second such meeting was held in Prague again in 1934, a time at which Carnap was teaching at the German University in Prague. The meeting was called a preparatory one, that is, in preparation of an International Congress of Philosophy of Science (*Congrès International de Philosophie Scientifique*) to be held in Pars in the following year. The Vienna Circle meeting immediately preceded the Eighth International Congress of Philosophy, also held in Prague, at which Neurath presented a paper on the unity of science, Schlick talked about the concept of wholeness, and Carnap on logical analysis.[*] One of the two plenary speakers at that Congress was, however, not a member of the Vienna Circle but Hans Driesch instead. Allowing for a disunity of science, Driesch called for a special kind of non-mechanistic causality in biology, and defended his thesis of a "vital factor" or "entelechy" as a constitutive explanatory principle that is required in life sciences. Adolf Meyer(-Abich), in his talk on holism, took issue with Driesch's concept of teleology, declaring it not a constitutive principle but a kind of causality in need of investigation and scientific explanation. Driesch's talk was followed by a heated discussion, in which Schlick remained noticeably silent, as he would later have his own talk on the concept of wholeness to deliver. Hans Reichenbach, however, called Driesch's concept of the organism a mystical one, while Carnap found Driesch's organicism to lack stringent lawful structure. Strict lawfulness and consequent testability alone would, according to Carnap's comments, render biology a true science. This "vehement and well organized attack" on Driesch by members and affiliates of the Vienna Circle was followed, the next day, by Schlick's presentation, in which he argued that in terms of an ontology of the outer world, there is "no whole over and beyond the sum of parts."[†]

A commentator on the 1934 philosophy congress in Prague saw the debate around Driesch's talk to reflect the philosophy of the "so-called Vienna Circle around Moritz Schlick," which considers all metaphysics to be meaningless: The logicians (*Logistiker*), he noted, might claim to have succeeded at this congress, whereas Driesch—who was right in his holistic approach to nature—should nevertheless be encouraged to "further develop his philosophy in a way that invites less criticism."[‡] This comment, as indeed the whole dispute at the congress in Prague, nicely announced the developments to come. On the one hand there was a strengthening organicist–holistic comprehension of the organism, its ecosystem, and of the *Volk*, increasingly tainted by an appeal to vitalistic, in some cases even mystic or at least irrational principles or forces, thus imbuing German biology with a *völkisch* spirit. This tendency was opposed by social Darwinists, who emphasized the rational, lawful structure of the natural world, including the world of living organisms and its analog, the world of the National Community, both being subject to the immutable *Lebensgesetze*—the laws of inheritance and of selection—that shape race and through it culture. In the words of one of the leading German idealistic morphologists of the first half of the twentieth century, Wilhelm Troll, the tension that developed during

---

[*] Joergensen, 1970, p. 891.
[†] Galison, 1990, p. 744.
[‡] Sauter, 1934/35, pp. 441–442.

the times of crisis that characterized the Weimar Republic was one between *Logos* and *Bios*.* Whereas the members of the Vienna Circle and their followers stood for a rational comprehension of the world around us, the temptations inherent in a neo-vitalistic, even mystical comprehension of the biological foundations of life were perhaps most glaringly articulated in Oswald Spengler's highly influential two-volume tome *The Decline of the West (Der Untergang des Abendlandes)*.

---

* Troll, 1937, p. 8.

5 The Rise of German ("Aryan") Biology

A TIME OF CRISIS

The cultural pessimism so characteristic of the Weimar Republic has been characterized as a "craving for crises."[*] A perceived "crisis in physics,"[†] precipitated by a "crisis of the principle of causality,"[‡] indeed a "crisis of reality,"[§] eventually motivated a "crisis in biology."[¶] Even talk of a "crisis in morphology"[**] became fashionable. Life scientists did not want to be sidelined, but rather wanted to participate in the discussion of the scientific revolution that was triggered by the burgeoning quantum mechanics. One of the books that most critically influenced the intellectual currents of the era was Oswald Spengler's (1880–1936) *The Decline of the West* (published in three installments in 1918, 1922, and 1923). Spengler took his inspiration once again from Goethe, who on his reading had contrasted the "world-as-mechanism" with the "world-as-organism," "dead nature" with "living nature," "law" with "form."[††] Spengler again adopted an organicist understanding of cultures, that is, he saw "organic states" as "higher individuals," subject to the eternal cycle of "youth," "growth," "maturity," and "decay." That cycle, which governs all life on earth at all levels of complexity, he rooted in *"the Destiny in nature and not* [in the] *causality"* of classical physics, that is, Newtonian mechanics.[‡‡] Of Schopenhauer's concept of a "Will to Life" Nietzsche had noted that it begged the question, since living is a prerequisite for willing. Spengler thus turned to Nietzsche's concept of a "Will to Power," an "energy"[§§] which he found to "lie deep in the essence of Western civilization," and hence to be of "decisive importance" in the analysis of that civilization at its successive life stages.[¶¶] Einsteinian relativity theory, with "its specific tendency to destroy the notion of absolute time" he called a "ruthlessly cynical hypothesis,"[***] while the Darwinian vision of a slow, steady, and gradual progressive development in nature he found conclusively refuted by paleontology, which documented in his view not only the fact of extinction, but also the sudden appearance of new forms "without transition types."

---

[*] Forman, 1971, p. 58.
[†] Schottky, 1921, p. 492.
[‡] Petzold, 1922.
[§] Ash, 1995, p. 286.
[¶] Bertalanffy, 1927.
[**] Hammarsten and Runnström, 1926, pp. 53–54; Schuster, 1929, pp. 203, 206.
[††] Spengler, 2006, p. 20.
[‡‡] Spengler, 2006, p. 21.
[§§] Spengler, 2006, p. 193.
[¶¶] Spengler, 2006, p. 19.
[***] Spengler, 2006, p. 215.

Darwinism, he claimed

with its talk of adaptation and of inheritance ... sets up a soulless causal concatenation of superficial characters, and blots out the fact that here the blood and there the power of the land over the blood are expressing themselves ...*

By putting the soul into the blood, and placing the blood into the landscape, Spengler massively influenced the racist *völkisch* spirit of the time. "A race has roots. Race and landscape belong together ... A race does not migrate."[†] Blood relationships determine speech communities,[‡] "the 'people' is a *unit of the soul*"[§] that inspires a "sense of duty and sacrifice."[¶] In the struggle for perseverance, "[t]he powers of the blood ... resume their ancient lordship. 'Race' springs forth, pure and irresistible— the strongest win and the residue is their spoil."[**] The course of history, according to Spengler, is both willed and destined—"[w]e have not the freedom to reach to this or that, but the freedom to do the necessary or to do nothing."[††] This, of course, is rhetoric that signals worse to come.

The concept of *destiny*, so central to Spengler, he characterized as an "indescribable inward certainty,"[‡‡] a description that would certainly have been classified as untestable, meaningless, and hence metaphysical by members of the Vienna Circle. The Weimar period of crisis was, indeed, permeated by an appeal to irrationality and nature mysticism. In the words of the historian Paul Weindling, Germany's defeat in the Great War promoted "mystic irrationalism, occultism, and theosophy," as well as an "attack on narrowly restrictive positivism," tendencies which could accommodate vitalistic biology far better than the strictures of the "rationalism of physical sciences."[§§] Holism trumped atomism, intuition trumped mechanistic explanation, as the emphasis shifted to "*völkisch* tropes that spoke of the German people (*Volk*) as a mystical, pseudobiological whole and the state as an 'organism' in which the individual was subsumed in the whole."[¶¶] A united *Volksgemeinschaft* was the avowed political goal of Gustav Stresemann (1878–1929) who, in 1923—the year of crises— was appointed chancellor and foreign minister of the Weimar Republic.[***] The concurrent militarization of the German society during the years 1923–1930 imbued the *Volksgemeinschaft* with military values: obedience, willingness to self-sacrifice, and self-discipline.[†††] It is these very values, which National Socialism was to claim and amplify in the name of biology. The politicization of biology was not a Nazi invention, however. During the first months of the Weimar Republic, the secondary-school

---

* Spengler, 2006, p. 255.
† Spengler, 2006, p. 254.
‡ Spengler, 2006, p. 259.
§ Spengler, 2006, p. 264.
¶ Spengler, 2006, p. 378.
** Spengler, 2006, p. 379.
†† Spengler, 2006, p. 415.
‡‡ Spengler, 2006, p. 76.
§§ Weindling, 1989b, p. 321.
¶¶ Harrington, 1996, p. 175.
*** Elz, 2009, p. 61.
††† Mulligan, 2009, p. 94.

principal Otto Rabes (b. 1873) wrote: "In the new state and the new school, it is a self-evident duty of biology instruction to aid in the political education of our youth … Biology, more than any other science, establishes a foundation for political understanding."[*] To translate in that quote "biology" as "Mother Nature," one who rules with an iron fist according to the laws of life means to capture its true spirit. The Haeckel student and later leading Nazi educationist for biology, Paul Brohmer[†] (1885–1965; NSDAP 1933, NSLB, 1933), whose writings "omitted none of the *völkisch*-racist stereotypes,"[‡] praised in a 1927 publication the fact that the concept of the homeland (*Heimatgedanke*) had finally fully permeated the training of elementary school teachers: "great emphasis is justifiably placed on the idea of nature conservation."[§] In 1934, Rabes would review Brohmer's notorious *The Teaching of Biology and Völkisch Education* (*Biologieunterricht und völkische Erziehung*, 1933), noting the author's emphasis on holism, coupled with racial hygiene: "the apprehension of holism is naturally given, as for example in the relations of tissues to organs, organs to organisms, and organisms to their environment, where in each case the particular stands in a most intimate relationship to the next higher unit."[¶] In his 1933 *opus*, Brohmer railed against Darwinism for being mechanistic and simplistic, rejected logical positivism for the same reasons, and praised *Heimatkunde* as the highest educational value as it furthers the love of the homeland. The whole organism he claimed, implicating H. Driesch and J.v. Uexküll, is "conditioned and regulated by a meaningful plan," something that would also create in the human biotic community a supraindividual attitude that is ultimately grounded in the blood relationships of all Germans. Harking back on Oswald Spengler, Brohmer appealed to the "will to struggle, body and soul, for the growth and health of this [German] biotic community."[**]

Analyzing the life and work of the biologist Richard Woltereck, the historian of biology Jonathan Harwood raised the question "in what sense might it be possible to speak of a characteristically 'Weimar biology'?"[††] He found in the German biology of that time a predominance of vitalism and neo-romantic holism, coupled with a critique of a purely mechanistic–materialist approach to nature, primarily directed against what was perceived as the "paragon of early twentieth-century mechanism: the theory of natural selection."[‡‡] Another concept, central to the Weimar culture and coupled with holism, was that of *Gestalt*, alive not only in biology, but also in physics and psychology, always with "at least indirect connections with broader ideological concerns and debates of that time."[§§] Derived from Goethe's *oeuvre*, the concept of *Gestalt* became an important integrating principle for German culture (*Kultur*), as opposed to "Western civilizations," which were characterized by a sterile

---

[*] Quoted in Weiss, 1994, p. 184.
[†] Harten et al., 2006, pp. 345–346.
[‡] Harten et al., 2006, p. 186.
[§] Brohmer, 1927, p. 338.
[¶] Rabes, 1934, p. 112. See the discussion below of the then widespread and much discussed concept of enkapsis, of a nested hierarchy of complex wholes with emergent properties at each level of complexity.
[**] Brohmer in Mosse, 1966, pp. 83, 88.
[††] Harwood, 1996, p. 365.
[‡‡] Harwood, 1996, p. 361.
[§§] Ash, 1995, p. 284.

utilitarianism and materialism. The *Gestalt* is not analyzed, nor rationally analyz-able; it is, instead, "an ordering principle in partnership with wholeness"[*] that can only be intuitively apprehended. Intuition thus became not only a valid form of discovery, but also a valid basis for the justification of (scientific, i.e., biological) knowledge claims—certainly an understanding of natural science that clashed with the demarcation criteria for scientific as opposed to metaphysical statements prof-fered by the logical positivist. One German biologist who, among others, paradig-matically impersonated such a "Weimar biology" undoubtedly was the botanist and idealistic morphologist Wilhelm Troll: "As for my scientific attitude, I always tried to avoid … narrow specialization. In that spirit … I was determined to revive mor-phological research. The morphology and biology of plants are those areas of botany that are closest to my heart."[†] The stage for a discussion of Troll's work, its sig-nificance, and the tensions it created, was nicely set by Max Hartmann (1876–1962) in his postwar review of German bio-philosophical essays published between 1939 and 1945. In 1941 (second edition in 1942), Troll had collected four older essays of his in a book under the title *Gestalt und Urbild*. In the preface to this book, Troll characterized the Darwinian twin mechanisms of variation and natural selection as "crass externality of the English perception of things," against which he pitched his essays in the hope to bring about a "rebirth of morphology from the spirit of German science."[‡] Commenting on those essays, Hartmann noted:

> His rejection of the possibility of any causal explanation of *Gestalten*, their innovation and their transformation in the course of evolution is a purely dogmatic point of view, which is refuted by the results obtained by experimental developmental physiology.[§]

THE REBIRTH OF SCIENCE IN A GOETHEAN SPIRIT

The life and work of Wilhelm Troll, an influential yet controversial figure in German botany, was recounted by several of his students, most comprehensively by Gisela Nickel.[¶] Troll was born in Munich on November 3, 1897, the son of the psychiatrist Theodor Troll and his wife Elizabeth, *née* Hufnagel. He attended the elementary and Latin schools in Wasserburg am Inn in Upper Bavaria, then the *Gymnasium* (gram-mar school) first in Rosenheim, later in Munich. Following his graduation he was drafted into the army in 1916. Dismissed from the army as a decorated lieutenant at the end of the First World War, Wilhelm Troll took up the study of natural sciences and especially botany at the University of Munich in 1918. His major professor in

---

[*] Harrington, 1996, p. 104.

[†] Troll, October 27, 1932; archive of the Martin Luther University of Halle, UA Halle Rep.II, PA 16126 (Wilhelm Troll).

[‡] Translation by Thomas Dunlap, cited from Deichmann, 1996, p. 150.

[§] M. Hartmann, 1950, p. 137.

[¶] Nickel, 1996; see also Rauh, 1979; Weberling 1981, 1999; Meister 2005a. Further information was gathered in the archive of the Martin Luther University of Halle (UA Halle Rep.II, PA 16126 [Wilhelm Troll]), and in the archive of the Akademie der Wissenschaften Leopoldina (Wilhelm Troll, Matrikel Nr. MM 4178).

botany was Karl von Goebel (1855–1932), but as was noted by one of his former students, Troll "from the very beginning placed his studies on a broad scientific and philosophical base."[*] Troll drew his main philosophical inspiration from the study of Goethe and Kant. On June 30, 1921, Troll was awarded his doctoral degree with the highest predicate available, *summa cum laude*, and in 1922 he obtained the diploma for High School Teacher in chemistry, physics, botany, zoology, geology, paleontology, and geography. In 1923 Troll was appointed scientific assistant to Goebel at the Botanical Institute of the University of Munich. Troll obtained his *Habilitation* on February 27, 1925, offering a public inaugural lecture on *The Problem of Teleology in Biology*,[†] and continued to pursue his academic career at the University of Munich until he was offered a full professorship at the University of Halle in 1932, where he became head of the Botanical Institute and Garden. Among the few people with whom Troll forged a close friendship during his years in Halle was the physical chemist Karl Lothar Wolf (1901–1969; NSDAP, 1933),[‡] who had joined the University of Halle in 1937, leaving behind a troubled and ideologically tainted administrative legacy at the University of Kiel.[§] Wolf was a fellow holist who sought to apply the concepts of idealistic morphology to chemical compounds and thus to establish—without much success—a "German (Aryan) Chemistry,"[¶]

---

[*] Weberling, 1999, p. 9.

[†] "Das Problem der Zweckmässigkeit in der Biologie."

[‡] The personal relationship between Troll and Wolf, and their respective families, was sketched by Nickel (1996, pp. 94ff).

[§] Grüttner, 2004, pp. 185–186. Wolf had been appointed as head of the Chemical Institute and professor at the University of Kiel effective October 1, 1930. With Hitler's ascent to power, Wolf was named president of the University of Kiel, in which function he served from the summer semester 1933 through January 1935. Rumors started to circulate that he had been removed from this office against his will, which prompted the dean to intervene with the Ministry of Science, Arts and Education in Berlin with a letter dated February 16, 1935: "During those difficult times … Wolf has pushed back all his scientific and personal interests and committed a huge personal sacrifice in the name of the new University." The Ministry promptly replied with a decree dated March 27, 1935, officially acknowledging Wolf's much appreciated achievements during most difficult times. By early 1936, however, Wolf had managed to raise the opposition of the university president, the *Dozentenbundführer*, the dean, and several faculty members against him, who all requested in a meeting with the Ministry Director Theodor Vahlen (1869–1945) in Berlin on February 15, 1936, that a formal investigation be opened against Wolf with the goal of his dismissal. Wolf was characterized as extremely combatant, and accused of being obstructive, and spreading false accusations as well as insulting remarks about colleagues. In a letter to the Reich Ministry of Education (REM) dated February 19, 1936, the president of the Kiel University diagnosed the root of the problem in Wolf's adherence "to a specific spiritual and ideological position that characterizes the admirers of the poet Stefan George. For Mr. Wolf, the spiritual rejuvenation of the German people is an 'aristocratic' movement, restricted to a few special minds, not a movement of the whole *Volk*, let alone of the whole university." Whitewashing his chicanery against faculty, Wolf portrayed these initiatives as an elaborate intrigue against himself, fueled by misgivings about hiring decisions he had made as president of the university, and animosities that arouse from his duty to cleanse the University of Kiel from its many Jewish faculty, following the legislation of spring 1933. Wolf was eventually ordered by the REM to teach physical chemistry at the University of Würzburg, which he did in 1936/37, before moving to Halle. It is hardly surprising that similar animosities soon sprung up at the University of Halle between himself and the president, Johannes Weigelt. Archive of the University of Halle, UA Halle, Rep. II, PA 17240 (K.L. Wolf); on Wolff's intermezzo in Würzburg see Nagel, 2012, p. 278.

[¶] Nickel, 1996, p. 103.

analogous to the "German (Aryan) Physics."[*] Wolf also took an interest in botany, he accompanied Troll on his botanical excursions, the two co-taught interdisciplinary seminars,[†] and they co-authored a study on Goethe's plant morphology which, in a footnote most probably penned by Wolf, contained an assault on "Jewish Science." Referring to quantum mechanics, and citing Werner Heisenberg (1901–1976) and Niels Bohr (1885–1962) as its prominent exponents, Wolf pointed to the avalanche of relevant publications of mostly Jewish and American authors:

> quantum mechanics is often related to Einstein's relativity theory ... which in essence negates the concept of time. This may explain the popularity that such theories enjoy with Americans and Jews, for the first have no history, the latter have no understanding of history.[‡]

If Troll did not condone such statements, he also did not distance himself from them, nor did he ask his friend and co-author to refrain from putting them into print in a joint paper. As sketched by Gisela Nickel, the two couples, Wolf and Troll, also entertained a close private relationship, spending evenings and weekends, even the summer vacation in 1938, together (Figure 5.1).

At the University of Halle, Troll and Wolf inaugurated a *Gestalt Colloquium* that ran from 1942 through the winter semester of 1944/45,[§] and founded—together with the anti-Semitic historian Wilhelm Pinder (1878–1947)—the associated publication series *"Die Gestalt."* The attendees of the *Gestalt Colloquium*, the *"hallische Gestaltler"*[¶]— distinguished representatives of a variety of academic disciplines—were motivated by a common ideal of scientificity. They opposed the dominance of a purely causal–analytic, positivistic and mechanistic, ultimately mathematical approach in science,

---

[*] Deichmann, 1996, p. 80.

[†] In his capacity as president of the University of Kiel Wolf wrote to the Minister for Science, Arts and Education (Bernhard Rust) on June 18, 1934: "One of the most important and difficult problems in natural sciences is to once again emphasize the importance of creative forces as an element of productive science." This is why in his lectures he sought to imbed physical chemistry in its general historio-cultural context. To formalize that commitment, Wolf asked for—and obtained—an official expansion of his teaching duties beyond general and physical chemistry to cover the history of natural sciences. After his transfer to Halle, Wolf issued the same request, specifying that he would be collaborating in these efforts with Troll. Both the dean and the president of the university, Johannes Weigelt, supported these initiatives, heeding Wolf's request that history of natural sciences should also qualify as a doctoral program (letters of December 4 and 8, 1937). Rust's Ministry replied (letter dated April 20, 1938) that although Wolf would be officially sanctioned to teach history of natural sciences, that program would not be accepted as one granting a doctoral degree. Archive of the University of Halle, UA Halle, Rep. II, PA 17240 (K.L. Wolf). Weigelt's support for Wolf and Troll in that respect is remarkable, as he would later virulently oppose both in these and similar efforts.

[‡] Troll and Wolf, 1940, p. 62, n. 18. This passage was highlighted in his review of Troll and Wolf (1940) by Waaser (1940a, p. 330) in the notorious *Zeitschrift für die gesamte Naturwissenschaft*; Waaser found this essay to provide an excellent introduction to the history of thought which ultimately led to an abandonment of the "creative realism" (*"schöpferischer Realismus,"* a notion apparently first introduced by Wolf and Ramsauer [1935/36, p. 139]) that characterizes the German idea of *Gestalt*, and which led to the type of "objective" and "modern" conception of nature that ignores all of its aspects that cannot be quantified.

[§] Nickel, 1996, p. 97.

[¶] Kaasch and Kaasch, 2003, p. 1047.

FIGURE 5.1 Wilhelm Troll. (Courtesy of Botanischer Garten der Johannes Gutenberg Universität Mainz [Prof. Dr. Joachim W. Kadereit, Director].)

seeking a holistic, synthetic, and integrating understanding of nature instead. Among them were Troll's brother, the geographer Carl Troll (1899–1975); the anti-Nazi philosopher Hans-Georg Gadamer (1900–2002); the pro-Nazi psychiatrist and philosopher Kurt Hildebrandt (whom Wolf as president of the University of Kiel had appointed at that university in 1934, Kiel having been one of several German universities to aspire to the status of an SS model institution),[*] the Zurich mineralogist Paul Niggli; and, among others such as *Gestalt* psychologists, historians, and musicians, the paleontologist Otto H. Schindewolf, whose saltational evolutionary model (typostrophism) Troll would eventually adopt.[†] The Troll biographer Gisela Nickel was told by acquaintances or wives of some of the *Gestaltler*, such as Wolf's wife Anneliese, that at least some of its members considered the *Gestalt Colloquium* as a welcome opportunity to retreat into inner emigration during politically and ideologically most challenging times.[‡]

---

[*] Strübel, 1984, p. 169.
[†] Nickel, 1996, p. 99.
[‡] Nickel, 1996, p. 103.

Following denouncements by the president of the University of Halle, Johannes Weigelt (1890–1948; NSDAP, 1933),[*] the *Gestalt Colloquium* was secretly spied upon, as it found itself caught between conflicting interests of the pluricentric national–socialist administration.[†] As its president from 1936 through 1944, the paleontologist Weigelt sought to reorganize the University of Halle as a national–socialist institution that would focus on ideological education and applied research.[‡] His predecessor in the office of the president of the University Halle-Wittenberg praised Weigelt as "one of the most active members of our faculty, who in a dignified way represents the new Germany and its science in every respect."[§] Weigelt himself rose to that position as of September 30, 1936. In print he boasted:

> It is not a matter of pure chance, nor one of blind action, that it was this university and none other who sought the leadership of Alfred Rosenberg, the *Führer's* deputy for all matters that concern the spiritual and ideological education of the party members. It is him with whom we have sought an alliance such that he may show us the path along which we can carry forward the new spiritual revolution … he calls upon us to instill the German youth with the spirit of national-socialist ideology, and he will help us to select the right men for vacant faculty positions.[¶]

Weigelt was lauded for his leadership all through to the very end. Receiving a military medal—his third, the *Ritterkreuz zum Kriegsverdienstkreuz*—on March 22, 1945, the *Gauleiter* Joachim Eggeling (1884–1945) commended him for his tenure in the presidential office: "you were a true leader of our university in the ideological battle—even if our current military and political situation appears nearly hopeless, history has taught us again and again that victories are not won with strong weapons, but with strong hearts instead. The flower of victory can only be picked at the abyss of death."[**]

Weigelt denounced the *Gestalt Colloqium* for its *weltanschauliche* commitments, and Troll in particular for his alleged Catholicism, his Platonism, and his rejection

---

[*] Johannes Weigelt was born into an evangelical family on July 24, 1890, in Reppen near Frankfurt a.d. Oder. He obtained higher education in the schools (*Gymnasium*) of Halle a.d. Saale, then in Blankenburg a. Harz, graduating in 1909. He subsequently studied geology, paleontology, biology, and geography at the University of Halle, where he obtained his PhD on May 3, 1914. During First World War, in January 1915, Weigelt was severely wounded in the battle around Soisson near Nouvron in France, causing ailment throughout his later life. On July 27, 1918, he was awarded the *venia legendi* for geology and paleontology at the University of Halle, where—after a stint in Greifswald—he was appointed Full Professor of Paleontology and Director of the Geological Institute as of April 1, 1929. He was elected a member of the Academy of Sciences Leopoldina in 1927. Weigelt joined the NSDAP on May 1, 1933. After his deportation to the Western Sector in 1945, Weigelt was remembered in Halle as a "Nazi-Activist." Archive of the University of Halle, UA Halle, Rep. II, PA 16768 (Johannes Weigelt); Archive of the Akademie der Wissenschaften Leopoldina (Johannes Weigelt, Matrikel Nr. MNr. 3736).
[†] For more details see Kaasch and Kaasch, 2003.
[‡] http://www.catalogus-professorum-halensis.de/weigeltjohannes.html (accessed March 10, 2016). See also Klee, 2003, pp. 661–662; Gerstengarbe et al., 1995, p. 171.
[§] Archive of the University of Halle, UA Halle, Rep. II, PA 16768 (Johannes Weigelt).
[¶] Leaflet, Archive of the University of Halle, UA Halle, Rep. II, PA 16768 (Johannes Weigelt).
[**] Archive of the University of Halle, UA Halle, Rep. II, PA 16768 (Johannes Weigelt).

of Darwinism.* Troll had opted out of the Catholic Church in 1919, but he retained a deeply religious spirit, which earned him attacks not only from Weigelt, but also from members of the National Socialist German Students' Association and the National Socialist German University Lecturers' Association at the University of Halle, from the office of Alfred Rosenberg, and from the "terrorist group"† around the eugeneticist Karl Astel (1898–1945; NSDAP, 1930; SS, 1934) at the University of Jena, among which the zoologist turned anthropologist Gerhard Heberer (1901–1973; NSDAP, 1937; SS, 1937)—of whom much more later. As most university lecturers and professors, Troll was a member of the National Socialist German Teacher's League, but his application for membership in the NSDAP was rejected.‡

After the collapse of the Third Reich, the Swiss born Emil Abderhalden (1877–1950), biochemist, physiologist, and eugeneticist at the University of Halle,§ and president of the Academy of Sciences Leopoldina, along with Eastern German intellectuals, scientists, engineers, and industrials was deported under American command by train to the Western Sector of occupied Germany. The so-called *Abderhalden Transport* left Halle for Oberramstedt near Darmstadt on June 24, 1945, with a distressed Wilhelm Troll, a deeply pessimistic and hurting Johannes Weigelt, and Karl

---

* Nickel, 1996, p. 103; see also Kaasch and Kaasch 2003. Troll notified the Kurator of the University in a letter dated November 18, 1941, that he had repeatedly been informed of derogatory remarks made about him and his science by Weigelt (who labeled Troll a *Kathol* and identified catholic tendencies in his science). Emil Abderhalden commented in a letter dated November 20, 1941, that it was him who had informed Troll, feeling that the young colleague deserved to know of such intrigue. On January 30, 1942, Troll requested a formal investigation against Weigelt as well as against himself, in order to clear his name (Archive, University of Hale, UA Halle Rep. II, PA 16126 [Wilhelm Troll]; Archiv, Wissenschaftliche Akademie Leopoldina, Wilhelm Troll, Matrikel Nr. MM 4178). Weigelt is renowned for the role he played as Vice President of the Academy of Sciences Leopoldina in Halle (Saale). Against the President, Emil Abderhalden's support, he opposed the publication of Goethe's writings on natural sciences under the editorship of Günther Schmid, Wilhelm Troll, and Karl Lothar Wolf, as well as—and especially—the theological writings of Paracelsus in the name of the Academy: he feared "that the commentaries added by the editors to those writings could not stand up to scientific scrutiny and standards" (Gerstengarbe et al., 1995, p. 195).
† Nickel, 1996, p. 105.
‡ Nickel, 1996, pp. 109–110. The Halle office of the NSDAP informed the President of the University Halle of the rejection of Troll's application for membership on August 17, 1934. Troll had apparently designated himself as a member of the NSDAP in 1933 (a rumor Troll himself denied), and subsequently became subject of investigations by the NSDAP judicial court for derogatory remarks he had made about the party and its representatives. Since he was, in fact, no party member, the procedures against him were dropped on September 3, 1934. Troll himself claimed that his application of 1933 was part of a collective effort initiated by the leadership of the student association, and that he purposefully jeopardized his admittance to the party by dropping the occasional incriminating remarks. Archive, University of Halle, UA Halle Rep. II, PA 16126 (Wilhelm Troll); Archiv, Wissenschaftliche Akademie Leopoldina, Wilhelm Troll, Matrikel Nr. MM 4178.
§ According to Weindling (1989b, p. 579), Abderhalden "provides a classic example of the transition of science from being a means of social emancipation to one of racial persecution." Himself committed to "temperance, family welfare and medical ethics," his science "ultimately provided the scientific rationale for human experiments" conducted by the Nazis. During the time of his tenure at the University of Halle, Emil Abderhalden, who also served as the president of the prestigious Academy of Sciences Leopoldina, was a towering presence. His nomination, as a Swiss citizen, for the Goethe-Medal for Arts and Sciences noted his selfless efforts for "his 2. Homeland Germany" in the Great War, during which he worked for the Red Cross. "Even if before 1933 he was no National Socialist, he always adhered to a German spirit, and worked for German science." Bundesarchiv, DS, formerly Berlin Documentation Center, microfilm signature DS B 026 (Emil Abderhalden).

Lothar Wolf on board.* Neither of them cheerfully accepted their fate. In a letter to the University of Halle, dated March 1, 1940, Troll explained: "A full professor of Botany at the University of Halle, I was deported on June 22, 1945, to Western Germany following the order of the American military administration. My forceful attempts to resist these measures remained unsuccessful ... I have wowed to return to Halle whenever the first opportunity arises."† But a job offer from Halle failed to materialize. Settling in the newly founded Federal Republic of Germany, Troll became headmaster at the High School of Kirchheimbolanden, the town in which K. Lothar Wolf was born. Indeed, the close personal relation between Troll and Wolf continued in the years after the war, both relocated to western Germany.‡ In 1946, Troll was appointed professor of botany at the University of Mainz, a post he held until his retirement in 1966. The deputy of the Secretary of Education informed the dean for natural sciences at the University of Halle in a note dated June 26, 1946, that Troll had been hired at the newly founded University of Mainz: "Should those rumors prove true, I would like to ask for recommendations with whom to fill the Chair for Botany."§ But in spite of several inquires on his part, Troll was never called back to Halle. The relocated Weigelt succumbed to his ailments on April 22, 1948, already. In western "exile" he felt as if his thread of life had been torn.¶ Nonetheless, in April 1946, Weigelt tried to obtain from his *alma mater* a certificate that even if he had implemented "politically dubious" measures during his tenure as president of the university, he had done so exclusively in order to keep the Theological Seminary from being shut down.**

In Mainz, Troll continued to host a circle of *Gestalt* devotees, but now privately rather than in an official university function. As he grew older, Troll's publications became increasingly colored by religious overtones. Harking back on Karl Ernst von Baer, Troll eventually tied the teleology and consequent meaningfulness that govern the *Gestalt* to creative thought. Philosophical realism, Troll argued, finds the extramental world to be "governed by meaning"; meaning must therefore exist objectively, that is, independent of human thought, and it reveals itself in the usefulness and function-ality of organic structures brought forth through development: "meaningfulness is a principle that pervades all of nature ... the Gestaltung of natural entities according to plan thus reveals 'Divine thoughts' (Goethe) ... K.E. von Baer called the living beings 'thoughts of Creation' ..."†† Troll died on December 28, 1978, in Mainz.

On the occasion of the 70th birthday of his doctoral advisor, Karl Ritter von Goebel, on March 8, 1925, Troll contributed to the *Festschrift* dedicated to the hon-oree an essay that serves as a perfect introduction to Troll's early thinking. Goebel, although critical of the countless phylogenetic trees mushrooming in the gardens

* Nickel, 1996, p. 131; Kaasch and Kaasch, 2003, p. 1052.
† Archive, University of Halle, UA Halle Rep. II, PA 16126 (Wilhelm Troll).
‡ Letter from Troll to the President of the Academy of Natural Sciences, dated April 29, 1955. Archiv, Wissenschaftliche Akademie Leopoldina, Wilhelm Troll, Matrikel Nr. MM 4178.
§ Archive, University of Halle, UA Halle Rep. II, PA 16126 (Wilhelm Troll).
¶ Weigelt, letter to the University of Halle, August 5, 1946. Archive of the University of Halle, UA Halle, Rep. II, PA 16768 (Johannes Weigelt).
** Letter from Oskar Schürer to the President of the University of Halle, April 3, 1946. Archive of the University of Halle, UA Halle, Rep. II, PA 16768 (Johannes Weigelt).
†† Troll, 1952, pp. 11ff.

of many botanists, was much more sympathetic to Darwinian mechanisms of evolution than Troll, and had proclaimed in an earlier paper: "morphological is what cannot yet be understood physiologically."* This would provide for many years to come the foil against which Troll pitched his idealistic morphology of plants: "morphological is what will never be explicable physiologically,"† thereby allowing for a disunity of biology. In his contribution to his former supervisor's *Festschrift*, Troll motivated such disunity of biology, indeed of natural science in general, with reference to Kant and Goethe respectively,‡ which he portrayed as impersonating once again the old contrast of the analytic versus the synthetic mindset. In Troll's analysis, Kant stands for the analytic, that is, the causal–reductionist approach to science, and for the requirement to formulate universal laws of nature in the rigorous and unambiguous language of mathematics. Natural science, according to Kant, must firmly be anchored in the principle of causality, in the sense that the relation of succession between cause and effect must be both necessary and irreversible.§ Scientific theories, Kant claimed, serve the expansion of empirical knowledge only if they are formulated in hypothetico-deductive form,¶ for rendered in such form, theories allow the derivation of testable, that is, logically falsifiable predictions according to the *modus tollens* form of argument.** For Troll, Goethe represented the antithesis to such a Kantian, ultimately Newtonian understanding of natural sciences. Goethe does not seek to decompose the whole into its parts, but instead seeks to comprehend the whole as something that is more than the sum of its parts. Comprehension of nature for him is synthetic, apprehensive (*anschaulich*), and intuitive. Causality as invoked in Newtonian mechanics he declared unable to elucidate the profound secrets of life, such as development and differentiation, that is, the metamorphosis of a plant according to plan, for example. Kant had criticized any strong rendition of empiricism, which bases all cognition of the world on what is immediately given in sensual perception. For Kant, the multiplicity of sensual appearances had to be sorted into genera and their species, thus revealing regularity and hence lawfulness in nature. But what are those laws, asked Goethe, that govern plant development in such a way that not only perpetuates the species-specific morphology, but also sorts species into genera, genera into families, thus revealing an order of nature that can be expressed as a hierarchy of types? It is easy to comprehend that the movement of billiard balls is subject to cause and effect according to Newtonian, indeed Cartesian mechanics. But the parts of a plant, dynamically developing one from the other rather than being mechanically juxtaposed like building blocks, combine to form a whole that as such is more than the mere sum of its parts. That is what according to Goethe renders biology different from physics, and what according to Troll renders morphology different from physiology.

---

* Goebel, 1905, p. 82; cited by Troll, 1925, p. 554.
† Troll, 1925, p. 565.
‡ See also Meister, 2005a.
§ Gardner, 1999 (2008), p. 175; see also Reiser, 2002, pp. 184–185.
¶ Gardner, 1999 (2008), p. 222.
** Kant, 1919, p. 656. Kant's position was famously adopted and exploited by the philosopher Karl R. Popper, a satellite of the Vienna Circle.

In his contribution to the Goebel *Festschrift*, Troll introduced Kant as having articulated the theoretical foundations for an exact science, the paradigm of which would be "mathematical physics." In contrast, he argued, there are "products of material nature" that require for their proper understanding causes other than purely mechanistic ones, namely, final causes. The concept of the "type" he took from Goethe, as one that captures what is "significant, exemplary, as also what is common and characteristic" for a group. But, he hastened to add, such a type is not a merely abstract concept, one that exemplifies the Kantian imperative to seek unity within the multiplicity. Teleology for Troll is also not a merely heuristic device as it was for Kant's biology, cashed out in the form of the so-called as-if-statements. The type, Troll insisted, was called an "idea" by Goethe, where this type *qua* idea is an "inner," that is, endogenous formative principle that generates the *Gestalt* and therewith wholeness from the inside out. Such a conception of the type is certainly a vitalistic one:[*] Goethean morphology, Troll maintained, is independent from the Kantian principle of causality. It is not a causal–analytic, but instead a depicting science. The type revealed by comparative morphology is not an organism that can be comprehended as the sum of its parts, but instead is a *Gestalt*, a complex whole that is "composed of parts that are united according to plan ... ."[†] Physiology, with its search for mechanistic laws of biological organization, reduces the organism to a lifeless machine, Troll insisted. Morphology, in contrast, is not a science seeking to understand the becoming of organic form in a causal–mechanistic sense, but instead seeks the ideal derivation of form-*Gestalten* from one another. "Biology is thus not a unified science. Morphological is not what cannot as yet be explained physiologically, but rather what can in no possible way be explained physiologically."[‡]

Reflecting on such an understanding of morphology, the botanist turned historian of biology Julius Schuster[§] (1886–1949)—at that time scientific assistant at the Geological-Paleontological Institute and Museum of the University of Berlin—called it a "Goethean Morphology," analogous to "Euclidean Geometry," and noted that while it is "in its theoretical foundation a historical throw-back, [it is] not a historical accident ... it seems that an era has begun, which has to solve its own particular problems" (*die eine eigene, nur ihr eigentümliche Aufgabe zu erfüllen hat*)."[¶]

---

[*] While he himself took the type as being real, Troll allowed that such Platonic realism might not be palatable for some of his readers who, he suggested, might take it as a "methodological-heuristic principle" (Troll, 1928, p. ix), one that would be called a "model" in modern science (Weberling, 1981, p. 313): "In a report from Palermo GOETHE calls the *Urpflanze* a model" (Troll, 1949, p. 493).

[†] Troll, 1925, pp. 556–561.

[‡] Troll, 1925, p. 565. Troll's friend, the botanist August Seybold pointedly characterized the polarity between Troll's and Goebel's thought in a letter to Troll, dated June 8, 1931: "If one were to pursue *Gestalt* physiology again, it would be taken as proof that morphology can be reduced to physiology, and yet physiology would get lost in a terminology that is no more than empty verbiage, completely neglecting fundamental facts. Goebel's experimental morphology has done more harm than it has helped. I do not wish to pretend that these arguments don't deserve interest, it only seems to be too early for investigations along these lines. We need the extreme morphology of Troll, we need an extreme physiology à la Kaspar Hauser; a welcome spirit will integrate the two when the time is right." Archiv, Akademie der Wissenschaften Leopoldina (Wilhelm Troll, Matrikel Nr. MM 4178).

[§] Herter and Bickerich, 1973, p. 128. For a biographical sketch of Julius Schuster, see Junker and Landsberg, 1994.

[¶] Schuster, 1929, p. 196.

The "crisis-ridden exclamation … 'back to Goethe'," which had been invoked to express a "desire for morphology'"* Schuster wanted to replace with the exclamation: "onwards beyond Goethe!"† More in line than Troll with the demarcation criteria for science set up by the Vienna Circle, Schuster accepted intuition as a pathway to discovery, with a proviso however: "we will refrain from publishing our intuitions prior to their examination in the light of logical reasoning as the ultimate deciding factor"‡ in the justification of knowledge claims.

A "desire for morphology" in a Goethean sense was again expressed by Troll§ in 1928, in his first major monograph, a book that dealt with organization and *Gestalt* of flowers. Unlike his later co-author Karl Lothar Wolf, Troll in that treatise turned to quantum mechanics to use it in support of his claim for the legitimacy of a disunity of biology. The pointer in that direction he had been handed by Ludwig von Bertalanffy, who in 1927 published a paper on the contemporary crisis in biology.

Ludwig von Bertalanffy (1901–1972) started his studies in natural sciences and philosophy at the University of Innsbruck, but after the first year transferred to Vienna. He earned his PhD in 1926 with a thesis on the physicist and philosopher Gustav Theodor Fechner (1801–1887), written under the supervision of Moritz Schlick, founder of the Vienna Circle and president of the *Verein Ernst Mach*. Other philosophers who influenced the young Bertalanffy during his student years included the neo-Kantians Robert Reininger (1869–1955; University of Vienna) and Hans Vaihinger (1852–1933; University of Halle/Saale). Supported by the Viennese botanist Richard von Wettstein, Bertalanffy published the first part of his important *Theoretische Biologie* in 1932, a book that outlined an organicist foundation for biology that was developed independently and thus differs quite substantially from Uexküll's *Theoretical Biology* of 1920. On recommendation from Schlick and Reininger among others, the book earned Bertalanffy the *Habilitation* in 1934. Bertalanffy also joined the *Studiengruppe für wissenschaftliche Zusammenarbeit*, a section of the *Verein Ernst Mach* founded in 1930 and chaired by Rudolf Carnap. The idea was to bring together representatives of various scientific disciplines (physics, biology, sociology, economics, psychology, and psychoanalysis) in a discussion as to how these disciplines could be integrated into a unified science. Upon his promotion to associate professor in 1941, Bertalanffy offered courses in theoretical biology at the University of Vienna through the winter semester 1944/45, before emigrating to Canada via London after the war.¶

In his 1927 paper concerning the crisis in biology, Bertalanffy sketched the profound changes physics had undergone due to the rise of relativity theory and quantum mechanics. Classical Newtonian mechanics, based as it was on universal and

---

* Ernst, 1926, p. 1080. The "desire" for morphology, Ernst (p. 1075), motivated with an appeal to a holistic worldview: "*keine Weltanschauung ohne Anschauung, keine Anschauung ohne Schauen, ohne das Schauen des äusseren, dann des inneren Auges.*" "Science," he continued (p. 1079), "must necessarily be understood as an art if we expect from it any insight into wholeness."
† Schuster, 1929, p. 196.
‡ Schuster, 1929, p. 205.
§ Troll, 1928, p. 16, n. 1.
¶ Salvini-Plawen and Mizzaro, 1999, pp. 33, 65, 70; Seising, 2005, pp. 105–106, and p. 105, n. 1; Stadler, 1997, p. 380; Deichmann, 1996, p. 339.

hence deterministic laws of nature and as such the model for a mechanistic program in biology, had been enriched by quantum mechanics, which at its core is stochastic in nature, that is, based on statistical laws and hence indeterministic. "The infiltration of what once was an exact science by statistical methods results in a renunciation of universally valid natural laws,"* noted Bertalanffy. In other words, quantum mechanics pulled the carpet from below a Kantian interpretation of the principle of causality, which implied a necessitation of the relation between cause and effect: "Modern Physics has thrown the principle of an all-pervading causal determination of nature to the scrap heap."[†] In Bertalanffy's view, this had far-reaching consequences: on the strong, Kantian interpretation of the principle of causality, scientific theories could be cast as universal statements, which allowed for a hypothetico-deductive structure of science—hence the applicability of the *modus tollens* form of argumentation to empirical sciences. If successful, scientific theories would thus capture a logically, that is, rationally structured world. The indeterministic nature of quantum mechanics would in contrast seem to render the structure of the world an "irrational" one—"irrational" in this context not meaning incomprehensible, but rather the impossibility of a hypothetico-deductive structure of science. The nuclear physicist Arnold Sommerfeld (1868–1951) from the University of Munich did not want to throw in the towel that quickly: "every fundamental physical theory must proceed deductively ... One can say that quantum theory may not yet be mature enough to be cast in a purely deductive structure. That is well possible ...," but at the time this needed not to be the final assessment of quantum mechanics.[‡] Other physicists more readily accepted the consequences of Niels Bohr's atomic model. Hermann Weyl (1885–1955), in his widely read textbook *Raum. Zeit. Materie*, proclaimed: "[i]t has once and for all to be stated that modern physics no longer supports the notion of a complete causal determination of material nature according to stringently formulated precise laws."[§]

The consequence of these developments, Bertalanffy argued, was the fact that "ideas are spreading in physics, which in the realm of biology would be called vitalistic."[¶] It almost seemed that modern developments in physics might revitalize the debate about a controversial point made by Oswald Spengler, according to whom the belief in strict causality and exact laws of nature would eventually be replaced by a new mysticism. Vitalism in biology had been criticized for its rejection of a reductionist–mechanistic approach, for its violation of the imperviousness of causal relations, but all of this had now become true of physics as well; even teleological argumentation patterns had found their way into physics, as for example in Sommerfeld's teleological rather than causal interpretation of the principle of least action.[**] Given those developments in physics, some people had already conjured up

---

* Bertalanffy, 1927, p. 654.
† Bertalanffy, 1927, p. 655.
‡ Sommerfeld, 1924, p. 1048.
§ Weyl, 1921, p. 283; Troll (1928, p. 13), cited from Weyl (1923, p. 286): "Modern physics, this needs to be stated once and for all ... no longer supports the principle of causality."
¶ Bertalanffy, 1927, p. 653.
** Sommerfeld, 1924, p. 1049; cited by Troll, 1928, p. 12.

the immanent collapse of science,[*] but he, Bertalanffy, would rather like to see in them the promising seeds of interesting future developments. One conclusion he found unavoidable in the light of modern research into nuclear physics and its significance for biology was, however, that Darwin's mechanistic principle of natural selection was to be considered an insufficient explanation for the becoming of organic form.

Troll got all the keys to these developments in physics from Bertalanffy's paper,[†] as he proceeded to use this literature in support of his claim that a disunity had emerged in physics that was brought about by the rise of relativity theory and quantum mechanics. In spite of his roots in the *völkisch* tradition, and in contrast to his later friend and collaborator Karl Lothar Wolf, Troll did not reject modern theoretical physics, but instead used it in support for his claim that a similar disunity must also be accepted in biology.[‡] In the introduction to his 1928 monograph, Troll emphasized that his goal was to oppose the current "one-sidedness and impoverishment of biological thinking" that resulted from an exclusive emphasis on the "causal-quantitative approach and experimental method."[§] In this context Troll cited a sentence coined by the physiological botanist Jacques Loeb (1859–1924), which he lifted from a paper by Hans Driesch's friend Curt Herbst: "Biology will be scientific only to the extent that it succeeds in reducing life phenomena to quantitative laws." Like Troll, Herbst himself opposed this attitude as well, as he found Loeb's notion of what is "scientific" to be too restrictive: "Biology deals first and foremost with qualities, such as the specific form of organisms, its ontogenetic and phylogenetic development, and its adaptation to the respective environment,"[¶] he asserted.

Morphology, according to Troll, deals with *Gestalt*, which cannot be conceptually analyzed, and in that sense is irrational: *Gestalt* is an apprehended, or intuited (*anschaulich*) whole that transcends causality, but that requires a holistic perspective (*Ganzheitsbetrachtung*) instead. Yet even if physics invokes concepts that originated within biology, the problematic unit of physics remains the atom, whereas that of biology is the organism, and in that sense the processes of life play out on an entirely different level than subatomic processes. Teleology in physics, and in the realm of the organic, can therefore only represent analogies, not any essential correspondence. In a much later analysis, published in 1951, Troll would take recourse to Nicolai Hartmann's (1882–1950) ontology of "levels of reality" to back up his claim of the autonomy and irreducibility of the organic relative to the anorganic realm.[**] Biology, or at least morphology, requires an "integral concept of *Gestalt*": *Gestalt* is not just an unanalyzed aggregate of parts; it is, instead, an unanalyzable unity, a wholeness which is superordinate to the phenomena investigated by physics, yet

---

[*] See also Ash, 1995, p. 286: "... a self-styled avant-garde, including [Herman] Weyl, Hans Reichenbach, Max Born, and Werner Heisenberg, confidently proclaimed, in Heisenberg's words, that 'quantum mechanics establishes definitively that the law of causality is not valid'."

[†] For a detailed analysis, see Rieppel, 2011e.

[‡] Troll would continue to do so even in his late publications. In a paper published in 1952, Troll (1952, p. 16) quoted from the preface of Sommerfeld's *Atombau und Spektrallinien*: "What today we discern from the language of spectra is the real spherical music of atoms, a consonance of whole numbers, an increasing order and harmony amidst all multiplicity" (the quote is from Sommerfeld, 1921, p. vii).

[§] Troll, 1928, p. x.

[¶] Herbst, 1924, p. 398.

[**] Troll, 1951a.

irrespective of this fact is unquestionably suitable to serve as a foundation for the autonomous science of morphology.* Going from there, Troll launched into a critique of Darwin's theory of natural selection, which he characterized as pseudoteleological and utilitarian doctrine, backing up his attack with a quote from Nietzsche: "As NIETZSCHE so aptly said: 'one should not confuse Malthus with nature. The entire English Darwinism is shrouded in the stuffy air of overpopulation, enveloped in the smell of low-class people, poverty and crampedness'."† Turning from the Darwinian mechanisms of evolution to the issue of phylogeny, Troll called the theory of descent a highly fruitful working hypothesis, the renunciation of which would, however, in no way affect the science of morphology. Although still finding naïve phylogenetics alive and well among contemporary biologists, Troll maintained that

> The times of naïve phylogenetics (NAEF) have long past ... if [Eduard] Strasburger claims that nothing but phylogenetic origin adjudicates any morphological questions, he seeks answers in an essentially unknown state of affairs which itself can only be approached through morphological research.‡

Once again, Troll pitched descent with modification, which is epistemically not directly accessible and hence necessarily hypothetical in nature against morphology, the latter based on observation and hence empirical in nature. He did not, by that, mean to deny the fact of descent with modification, which he considered to have been established, but he did deny the sufficiency of the Darwinian twin mechanisms of random variation and natural selection as an explanation of this fact, as it completely neglects the possibility of endogenous laws of form generation. It is these laws that condition the type, which in turn is not a conceptual construct, but—in Schopenhauer's terms—an *"unitas ante rem,"* an apprehended or intuited wholeness that precedes the multiplicity of appearances.§ Apprehension of the wholeness that is manifest in the type is based on an intuitive function that enables the structures that exist in the living world to become manifest and thus to become objects of preconceptual cognition. But, as a function of their unity and wholeness, the types must also be discontinuous (Platonic) entities, which—given the fact of descent with modification—must result in gaps in the Fossil Record: "these gaps are in no way closed even under the assumption of a very large number of intermediate forms."¶ Later, Troll adopted the views of his fellow *Gestaltler* Otto Schindewolf—"a phylogeneticist of strictest observance"**—who had bridged the gaps in the Fossil Record with a theory of saltational evolutionary change (typostrophism).†† Ultimately, however, Troll kept higher forces firmly in sight: "Natural events are never, especially not in the organic realm, the result of blind coercion by chance, but instead express the phenomenal richness of cosmic reason (*Weltvernunft*)."‡‡

---

* Troll, 1928, p. 16.
† Troll, 1928, p. 33, n. 1.
‡ Troll, 1928, p. 23, n. 5.
§ Troll, 1928, p. 24.
¶ Troll, 1928, p. 34.
** Troll, 1943a, p. 435. Also in Troll, 1944, 1951b; and Troll and Meister, 1951.
†† Schindewolf, 1936, 1937.
‡‡ Troll, 1928, p. 51.

The becoming of organic *Gestalt* through ontogeny, according to Troll, can be analyzed using causal–analytical methods as was done by the *Entwicklungsmechaniker*, but cannot be comprehended other than teleologically, since the ontogeny of *Gestalt* reveals its meaningfulness only relative to its final goal. Roux had correctly understood Kant's appeal to teleology as a heuristic device, but this was not Troll's reading of the immortal philosopher, citing the philosopher Emil Ungerer* (1888–1976) in support of his own interpretation: "Final causes … according to Kant, are not real, but ideal causes."[†] This in turn requires the invocation, in biology, of special "factors," which recall their "older brother that was called '*vital force*' (*Lebenskraft*)."[‡] It is the *Gestalt*, the type which cannot be comprehended in a purely causal–analytical sense. Darwin himself had characterized morphology as the science of homologies, but homologies are revealed exclusively through topological correspondences of parts within complex wholes that represent the same type. When Strasburger said that morphological comparison implies common descent of the compared organisms, he put the cart before the horse seeking salvation in historicism. According to Troll, "it is not descent which decides morphological issues; instead, it is morphology which informs about the possibility of descent."[§] It is for this reason that there cannot be a proper phylogenetic method, Troll maintained. Phylogeny, instead, is a reinterpretation of the results of comparative morphology from a historical perspective, one that can be enriched only insofar as the incompleteness of the Fossil Record permits.[¶]

Come winter semester 1935/36, ideological pressure had significantly increased at German universities, especially (but not only) in Halle (Saale) with Johannes Weigelt as university president. By that time, Troll had been working on his *magnum opus*, his multivolume textbook on the comparative morphology of higher plants, which was to be delivered in a series of irregularly appearing fascicles, the first published in 1935. Volume 1, Part 1, was completed by 1937, published by Bornträger in Berlin.[**] It opened with an introductory section titled *The rebirth of morphology in the spirit of a German science*, which underpinned comparative morphology with a *völkisch* perspective amplified as it was by National Socialist ideology.[††] The same section, written back in 1935, was independently published in the first volume (1935/36) of the newly founded *Zeitschrift für die gesamte Naturwissenschaft* (ZGN), with Troll's later companion Karl Lothar Wolf as one of the founding editors. Troll himself was appointed a member of the editorial board, along with Jakob v. Uexküll, Adolf Meyer(-Abich), and others. With his piece, Troll aligned himself with the organicist–holistic *Ganzheitsbiologie*, which by that time had become a predominant pillar of National Socialist ideology. The philosopher Ernst Ungerer found Troll's text to match contemporary efforts "towards a renewal of our understanding of cognition, and of the relation of science to our world-view (*Weltauffassung*), which is a

---

* Ungerer, 1922.
† Troll, 1929, p. 45.
‡ Troll, 1929, p. 48.
§ Troll, 1929, p. 50.
¶ Troll, 1932, p. 3.
** Troll, 1937.
†† Bäumer, 1989, p. 76, n. 2; see also Deichmann, 1996, pp. 81, 149–150, 269.

central concern of this journal, and also one of the most important tasks of our time."[*] The Munich botanist Ernst Bergdolt (1902–1948; NSDAP, 1922), a Nazi of the first hour, called Troll's Morphology "part of the best *German* science."[†] Troll's essay fitted well into the general context of the first volume of the ZGN, the contributions to which collectively pushed an organicist/holistic agenda, a commitment to *Gestalt*, an engagement against a positivistic philosophy of science, and an anti-mechanism fueled by the "German spirit." Himself (together with Wolf) an editor of Goethe's scientific writings,[‡] Troll insisted that a true rebirth of morphology could only be achieved in a Goethean spirit, which pitches morphology "against the positivistic ideal of science" that prevails among the "western people (*Westvölker*)."[§] The "western people" were, of course, France and England, the Allies who dictated the Treaty of Versailles, and whose liberal democracies were characterized as *undeutsch*, as unsuitable for the German *Volk*, which by that time had been brought into line (*gleichgeschaltet*) with National Socialist ideology under the *Führer*-principle.[¶] "Western science" was dismissed as positivistic, atomistic, and mechanistic, characterized by arbitrary thinking that was content with indeterministic statistical laws, and unfavorable to natural order governed as it is by the *Lebensgesetze*, the laws of life.[**] Morphology in the German spirit meant for Troll the celebration of Goethe and Schelling, the deification of nature, and a firm commitment to Platonic realism. In opening his text, Troll celebrated the "excellent speech" in support of a "desire for morphology" that Paul Ernst had delivered at the 89th Assembly of German Natural Scientists and Physicians, held in Düsseldorf in September 1926. In that speech, Ernst had identified Goethe as "the greatest spirit of the German *Volk*, the greatest educator of the German spirit."[††] Yet even within Germany, there was no unity of opinion regarding Troll's position on morphology, the mechanisms of *Gestalt* generation, and Goethe's significance for contemporary science, biology in particular. Troll's position was fiercely attacked by Darwinists and geneticists alike, authors who called themselves phylogenetic systematists, as will be detailed later. The tension surrounding Troll's morphology is perhaps best expressed in a short note written to him by his friend, the botanist August Seybold (1901–1965), on the occasion of the latter's inaugural speech following his appointment at the University of Heidelberg, which he delivered on December 7, 1934:

> Since you cannot come personally, send your *Daimon*! I shall dare to talk about Paracelsus, Haller, Goethe, and Schleiden, I shall talk against Darwin and his theory of descent, against Americanism, and against modern sensationalism in biology. Will we be understood? Many will shake their head, when the "freemason Goethe" will be introduced as the principal advocate of true German science. But I shall be seeking shelter behind Paracelsus: "May a venomous fart escape from your arse"![‡‡]

---

[*] Ungerer, 1936/37, p. 88.
[†] Bergdolt, 1937/38a, p. 181.
[‡] Troll, 1926.
[§] Troll, 1935/36, p. 350.
[¶] Noakes, 2008, p. 75: *Führerprinzip*—"authority of every leader downwards and responsibility upwards."
[**] Prinz, 1985, p. 72; see also Kühnl, 1984, p. 98.
[††] Ernst, 1926, p. 1075.
[‡‡] Seybold to Troll, December 4, 1934; archive of the Akademie der Wissenschaften Leopoldina (Wilhelm Troll, Matrikel Nr. MM 4178). Freemasons and other such lodges were discriminated against under Nazi rule.

In his review of Troll's textbook, the plant geneticist Otto Renner (1883–1960), professor at the University of Jena, the hotbed of racially motivated social Darwinism with Renner himself more on the sideline in that respect, called Troll's position "consciously anachronistic," and noted of him: "he can remain a good researcher as long as he does not take metaphysical dreams as substitutes for scientific knowledge." The motivation for Troll's rejection of the Darwinian mechanisms of evolution Renner identified in the "calmative of pure metaphysics."[*] As noted above, the philosopher Ungerer read Troll in a much more positive light, as he placed him in the tradition of Adolf Naef, Sinai Tschulok, and Adolf Meyer(-Abich).[†] But whereas Meyer(-Abich) did, in fact, comment on idealistic morphology as expounded by Troll, Naef and Tschulok, both cut from rather different wood (see Chapter 3), did not.

Troll found in Goethe and Schelling the best avenue to overcome the English positivism and mechanism, a strategy which his former student Gisela Nickel[‡] identified as underlying Troll's refutation of the New Morphology[§] propagated by the British paleobotanist Hugh H. Thomas. Basing comparative morphology on nothing but intuitive apprehension for Thomas meant to render it not only subjective, but also—and for that reason—uninteresting. "Most scientists, however, aim at an objective approach to their objects of study. In modern morphology we attempt to compare structures with a view of determining their genetical relationships ... Here the connection of the objects does not depend on the sense of the observer,"[¶] Thomas maintained. Troll, in contrast, invoked Goethe's concept of "Divine Nature" (*Gott-Natur*), and quoted Schelling as saying: "The rebirth of religion through the highest of all sciences, this is in essence the task of the German spirit, the true goal of all his aspirations."[**] In search for universal laws, Troll continued, the exact sciences completely abstract from the "thing in itself," revealing nature only insofar as it is a mechanism. Exact sciences thus deal with nature only in terms of a *natura naturata*, whereas Goethe's science targets the *natura naturans*, the creative nature, confident to find divinity in it. Nature, according to Schelling, is not "only the product of its activity (*natura naturata*)," but instead is "nothing less than living activity, or productivity itself (*natura naturans*),"[††] which results in a dynamic conception of matter. Nature is thus not only fundamentally processual, but natural processes are also goal-directed, and it is through this goal-directedness that nature, in Troll's

---

[*] Renner, 1938/39, p. 527.

[†] Ungerer, 1936/37, p. 87.

[‡] Nickel, 1996, p. 119.

[§] The New Morphology propagated by Thomas (1932/33, pp. 31–32), one that translates observations "into curves by means of formulae," and where "[a] good fit is regarded as evidence that a particular set of assumptions provide a possible explanation of the observed facts" reflects the rise of quantitative ("biometric": Weingart et al., 1992, p. 337) methods in English biology, which led to the first steps toward a numerical taxonomy (see also Gilmour, 1940).

[¶] Thomas, 1932/33, p. 19. Not that Thomas's expositions in the rooms of the Linnean Society in London on November 10, 1932, remained unopposed. As the discussion continued on November 24, Dunkinfield H. Scott (1854–1934) opined: "The rejection of phylogeny by some modern morphologists is, in a sense, justified. We can infer nothing from phylogeny, because phylogeny is unknown, a 'product of fantastic speculations' as [Johannes P.] Lotsy [1867–1931] called it. To infer morphology from phylogeny is to put the cart before the horse" (in Thomas, 1932/33, p. 39).

[**] Troll, 1937, p. 3.

[††] Beiser, 2002, p. 530.

thought, acquires its higher meaning, one that cannot possibly be captured through the Darwinian mechanisms of random variation and natural selection. "What most importantly," according to Troll, "distinguishes the German thinking is not so much its thoroughness but rather its depth, the fact that—to speak in Faustian terms—it descends to the 'mothers',"* the nurturing earth of the homeland.

SCIENCE IN THE NAME OF THE *VOLK*

Troll's appeal to *völkisch* values in his defense of idealistic morphology against the "new morphology" of British authors landed him in the orbit of National Socialist ideology. As recounted above, the introduction to his textbook was separately published during the academic year 1935/36 in the first volume of the newly founded *Zeitschrift für die gesamte Naturwissenschaft, einschliesslich Naturphilosophie und Geschichte der Naturwissenschaft und Medizin* (ZGN). The founding members of the journal were Karl Lothar Wolf,[†] president of the Christian-Albrechts University in Kiel from April 1933 through January 1935, together with three faculty members from the same university, the philosopher Kurt Hildebrandt (1881–1966; NSDAP, 1933), the paleontologist Karl Beurlen (1901–1985; SS, 1931; NSDAP, 1933), and the renowned medical professor and functional anatomist Alfred Benninghoff[‡] (1890–1953; NSDAP, 1941). Both Beurlen and Hildebrandt were members of the so-called Wolf circle ("*Gruppe Wolf*"[§]), fellow party members as of 1933 who had been appointed during Wolf's tenure as president of the University of Kiel in his efforts to bring the University of Kiel into line with National Socialist ideology. The collegial and personal relationships of Wolf with Hildebrandt and Beurlen,[¶] respectively, were close, fueled by a shared ideology. Following his appointment as president, Wolf pursued an aggressive hiring policy designed to "make it clear," as fast as

---

* Troll, 1937, p. 3.

† Karl Ludwig *Lothar* Wolf was born on February 14, 1901, in Kirchheimbolanden, Rhineland-Palatinate, Germany. Following Grade School in Kirchheimbolanden he graduated from the *Humanistisches Gymnasium* in Neustadt and der Weinstrasse (formerly Neustadt an der Haart) in 1920. He studied chemistry and physics in Bonn, Giessen, Heidelberg, and Munich, where he earned his PhD in 1925 with a thesis in theoretical physics supervised by Arnold Sommerfeld. In 1925/1927 he was assistant at the astrophysical observatory in Potsdam. In 1927/1929 he was assistant at the Institute for Chemistry at the University of Königsberg, where he earned his *Habilitation* in physical chemistry in 1928. In 1929 he was appointed Professor and Director of the Institute for Physical Chemistry at the University of Kiel. He served as President of the University of Kiel from spring 1933 through January 1935, and as a senator of the city Kiel during the years 1934–1935. In 1936 he was removed from Kiel University, and taught physical chemistry in Würzburg. In 1937 he joined the faculty at the University of Halle. In 1933 he declined an offer from the University of Karlsruhe on the basis of promises made by the Reich Education Ministry that remained a matter of disputed interpretation throughout Wolf's career until his deportation to the Western Sector in 1945. And again in 1936, Wolf was led to believe he would be offered a position in Munich, a promise that went unfulfilled: "I was later told that Munich is a special case, as it is the 'town of the movement'. This remark I could not understand, since I was a party member [since May 1933], whereas the appointee was not a party member, or at least did not achieve any relevant profile in that respect" (Archive of the University of Halle, UA Halle, Rep. II, PA 17240 [K.L. Wolf]).

‡ Klee, 2003, p. 38; see also Aumüller and Gundmann, 2002, p. 300.

§ Heiber, 1994, p. 393.

¶ Letter from *Dozentenschaftsleiter* Prof. Dr. Holzlöhner of the University of Kiel to Wolf, 1935. Archive University Halle, UA Halle Rep. II, PA 17240 (K.L. Wolf).

possible, "that the old-style university is gone."* The more senior Benninghoff was in that respect somewhat of an outlier among the founders of the ZGN, probably brought in to lend legitimization to the new publication through his eminent status in medical sciences. Troll was elected a member of the new journal's editorial board, along with such biologists as J. v. Uexküll, Adolf Meyer(-Abich), the hydrobiologist and limnologist August Friedrich Thienemann from Kiel and Plön (of whom more later), and the entomologist and "old Nazi"† Hermann Weber (1899–1956) from Danzig, all of whom representatives of the holistic/organicist camp among contemporary German biologists. Other members on the editorial board included the leading racial hygienist Otmar von Verschuer (1896–1969; NSDAP, 1940) from Frankfurt a.M., Troll's fellow *Gestaltler* Georg Gadamer, the *Gestaltler* and physicist Viktor von Weizsäcker (1886–1957) from Heidelberg, and the pro-Hitler philosopher Martin Heidegger (1889–1976; NSDAP, 1933) from Freiburg i.Br. Embedded in the *völkisch* tradition, Heidegger during the early years of the Third Reich indulged in the treacherous illusion that Plato's dream would finally become reality in the New Germany: a philosopher (meaning himself) would consult the political *Führer* in his quest to revitalize and unify the German soul through its roots in the native soil.‡ Through his work in metaphysics, Heidegger had provided the proverbial foil against which the positivist philosophers of the Vienna Circle—antagonists of the organicist–holistic neo-vitalism in biology—developed their demarcation criteria for natural science as opposed to metaphysics. The general direction the ZGN would be headed in its publicist efforts were sketched by Wolf in a letter to the *Reichsministerium für Wissenschaft, Erziehung, und Volksbildung* (REM; Reich Ministry for Science, Education, and People's Education; formerly Ministry for Sciences, Arts, and Education) headed by Bernhardt Rust, which is well worth quoting at length:

> A crucial precondition for the impact of scientific training in the *völkisch* education in natural sciences is the ideological framing of the natural sciences coupled with the insight that scientific research will always proceed in the light of prior ideological commitments. It is amazing to which degree most natural scientists at our universities have lost any understanding of these fundamental facts. The modern name for this separation [from the *völkisch* spirit] is "Positivism," a camouflage of the old Materialism ... *Die Naturwissenschaften* (Springer) ... [a journal] which serves simultaneously as the official outlet for the *Kaiser Wilhelm Gesellschaft* and for the Association of [German] Natural Scientists and Physicians, has for a long time been known to those in the knowing to be the banner bearer of an international and a-racial positivism ... This is all based on a glorification of Einstein. The latter's scientific achievements may be judged one way or another, but it is certainly undeniable that a lot of advertising of his name has been pursued by both Jewish and communist quarters ... to the extent that his achievements have been stylized as "Copernican" ... No reader of *Die Naturwissenschaften* will be able to deny that this journal exerts a detrimental and subversive influence, because it jeopardizes all *völkisch* culture with its soul-less intellectualism.§

---

* Heiber, 1994, p. 409.
† Heiber, 1994, p. 673; Klee, 2003, p. 657. An *Altnazi* was one who joined the NSDAP before 1933.
‡ Rohrkrämer, 2005, pp. 180–181.
§ Letter from Wolf to the Ministry of Sciences, Arts and Education, dated March 23, 1935. University Archives Halle, UA Halle Rep. II, PA 17240 (K.L. Wolf).

In a postscript, Wolf communicated the proofs of an article penned by his colleague Hildebrandt, which criticized *Die Naturwissenschaften* in many respects, and which was to open the first volume of the newly founded *Zeitschrift für die gesamte Natur-wissenschaft*. In a letter dated February 23, 1937, to the German National Science Foundation (*Deutsche Forschungsgemeinschaft*, DFG), the co-editor Karl Beurlen characterized the ZGN as a journal that would seek "[t]o overcome the positivist and specialist scientific thinking, and in so doing eventually to rebuild the natural sciences from the perspective of a worldview that a German would see as truly alive, and that is more than the sum of specific individual facts."[*] The newly confirmed president of the DFG, Rudolf Mentzel[†] (1900–1987; NSDAP, 1925 / 1928; SS, 1932), a high ranking official at the REM, had hoped that the ZGN would replace the more liberal *Die Naturwissenschaften*, as well as the science magazine *Nature* (London), at German universities and research institutions, but by that time had realized already that this was an unattainable goal.[‡]

The very first entry in the ZGN, penned by its founding member Kurt Hildebrandt, defended an anti-positivist stance in natural sciences. He chose Pascual Jordan's 1932 paper in *Die Naturwissenschaften* on quantum mechanics and its significance for biology as his target, accusing Jordan for not having understood even a single sentence of Kant's critique: "the creative cosmic force (*Weltkraft*) invoked by Fichte, Schelling and Hegel has never been understood [by the positivists[§]]. Positivism inherited this poverty from mechanism."[¶] Right in his introductory article, Hildebrandt thus set a tone that would mark the volumes of the ZGN to come as the unofficial outlet for Aryan Physics (*Deutsche Physik*).[**] Hildebrand proceeded to designate as the principal roots of (logical) positivism the replacement of matter by energy, and of the three-dimensional apprehension of material objects by mathematical formalisms. Positivism, he explained, is the total resignation from any notion of a creative force, it is the expression of a tiredness of the world, of a sinking life instinct. The theory of descent, he claimed, is the discharge of French rationalism (Lamarck), and English empiricism (Darwin), both of which he recognized as outright enemies of a creative spirit, one that is paradigmatically exemplified by the German philosophers Leibniz, Herder, and Goethe. Creative nature must be understood in the sense of pantheism, a perspective that unifies matter and mind, body and soul. Against the mechanization of nature as sketched by Darwin, Hildebrandt emphasized that without any creative force at work, no new species could ever evolve.

The path and the task of "German Biology in the German Present"[††] were charted by Hermann Weber[‡‡] (1899–1956); Figure 5.2, again in the first volume of the ZGN.

---

[*] Beurlen, quoted from Deichmann, 1996, p. 81.

[†] Rasch, 1994; see also Nagel, 2012, pp. 112ff.

[‡] Deichmann, 1996, p. 390, n. 55.

[§] Hildebrand specified the positivists as those that had taken the verification principle as the benchmark in demarcation of science from metaphysics, that is, the philosophy of the Vienna Circle. On German Romanticism (Fichte, Schelling, and Hegel), see Beiser, 2002; Richards, 2002.

[¶] Hildebrandt, 1935/36, p. 7.

[**] Deichmann, 1996, p. 81.

[††] Weber, 1935/36, p. 95.

[‡‡] Wenk, 2008/09.

FIGURE 5.2   Hermann Weber. (Universitätsbibliothek Tübingen, Bilddatenbank.)

Weber was a so-called *alt*-Nazi,[*] which implies that he joined the NSDAP prior to 1933. He earned his PhD in 1922 at the University of Tübingen with a thesis involving the morphology of insects. Following his *Habilitation* at the University of Bonn in 1928, and stints at various academic institutions, Weber was appointed professor of zoology at the University of Münster in 1936 by the REM: the search had been on for a good national socialist.[†] Weber was among the many signatories of the "Commitment of the Professors at German Universities and Colleges to Adolf Hitler" (*Bekenntnis der Professoren an den deutschen Universitäten und Hochschulen zu Adolf Hitler*), issued on November 11, 1933, in celebration of the National Socialist Revolution. The Commitment was carried by "the spirit of National Socialism [that] promotes a politicization of German universities to strengthen the national will, that is, to train the will through the training of knowledge."[‡] While at the University of Münster, Weber publicly advocated the formation of a totalitarian *völkisch* state that endorsed policies of racial hygiene and eugenics.[§] In his autobiography, the politically un-implicated zoologist Bernhard Rensch, who arrived at Münster University in March of 1938, described Weber as "pleasant ... pale, stricken by a serious heart

[*] Klee, 2003, p. 657.
[†] Deichmann, 1996, p. 74.
[‡] Reimann, 1984, p. 44.
[§] http://www.uni-muenster.de/unizeitung/2009/6-30.html (accessed March 10, 2016).

condition, and somewhat isolated from his colleagues due to hardness of hearing"*—
the latter two impairments due to a severe bout of influenza he contracted when returning from service in the Great War. His approach to insect morphology was holistic,[†]
which explains his admiration for J. v. Uexküll, H. St. Chamberlain, H. Driesch,
A. Meyer(-Abich), and other like-minded biologists and philosophers.[‡] Following an
appointment at the University of Vienna in 1939, Weber joined the newly founded
"NS-*Kampfuniversität* Straßburg" in 1941, where in his inaugural speech he cast
the relation of organisms to their environment in terms of blood and soil ideology.[§]
Weber was arrested by the French troops liberating Strasbourg in November 1944,
and interned for a year. Upon his release he went to Tübingen, where he became full
professor of zoology in 1954. In that same year, he was awarded, together with Willi
Hennig of whom more later—the founder of modern phylogenetic systematics—the
Fabricius medal of the German Entomological Society.[¶] Weber died two years later
in Tübingen.

In his 1935/36 contribution to the first volume of the ZNG, Weber inferred from
the writings of contemporary holistic and organicist biologists and bio-philosophers
the dawn of a new biology that would turn its back on "materialism and naked
causal-analytic mechanism": For Weber it was Adolf Hitler who led biology's march
in a new, German direction. Hitler had recognized the importance of biology for the
movement he initiated, through his experiences in the trenches of the Great War,
where he had learnt to weigh death against life, the strength of race and of *Volk*
against the individual:

> Biology teaches anyone who wishes to learn a *ganzheitlich*, organismic understanding
> of life in a community; as has to be requested of all natural sciences, biology therefore
> is eminently and in an entirely novel sense also a highly political science.[**]

Ascertaining one more time that for Germans, biology must provide the foundation
of a new political consciousness, and singling out J. v. Uexküll for special applause,
Weber launched into a tirade against the adoption of Darwinism by Marxists
("heinous red seducers of the *Volk*"), praising the anchoring of science in the soul
of the *Volk*, the race, and in the soil of the homeland (*Heimatboden*). All of this ultimately converged on the importance of the modern science of genetics that provides
the foundation for racial hygiene and eugenics. Later, and writing for Lehmann's
Der Biologe (see below), Weber explained: "The twin concepts of 'organism and
environment' ... mean in biology the same as the slogan of 'blood and soil' means
in politics—an expression of a tight connection, of an engagement of two highly
complex systems with one another according to the necessities and laws of nature."[††]

---

* Rensch, 1979, p. 89.
[†] Maier, 2008/09.
[‡] Weber, 1935/36, p. 96.
[§] Klee, 2003, p. 657.
[¶] Wenk, 2008/09, p. 112; the medal is named after Johann Christian Frabricius (1745–1808), a former
    student of Linné, and author of a system of insects published in 1775 (Hoffmann, 1959).
[**] Weber, 1935/36, pp. 96, 106; see also Weiss, 1994.
[††] Weber, 1942, p. 57; on Weber see also Stella and Kleisner, 2010, p. 43.

Hitler himself is well known for his anti-intellectualism, and his penchant for a holistic, intuitive approach to nature: "In certain areas, professorial science is devastating. It leads away from instincts ... if the world were left to the German Herr Professor for a few centuries, nothing but cretins would be seen walking around after a million years—gigantic heads on a diminutive body," he is reported to have once said.[*] Hitler's chief ideologist Alfred Rosenberg likewise adopted organicism, tended to nature mysticism, and rejected mathematical formalism as well as atomistic mechanism in natural sciences.[†] None other than the *Reichsminister für Wissenschaft, Erziehung und Volksbildung*, Bernhard Rust (1883–1945; NSDAP, 1922) considered the *Volk* as a living organism that is protected by the state according to a doctrine that places the interests of the *Volk* ahead of the interests of the individual.[‡] Rust rejected positivism and its claim for an objective science free of moral, ethical, and political values. Such a claim Rust maintained, "is not the mark of a free science, but instead marks the estrangement of the human spirit from the eternal forces of nature and from history, and hence is the mark of pathological degeneration (*krankhafte Entartung*)."[§] All of these topoi found their way into the first, and subsequent volumes of the ZGN.

With volume 3, 1937/38, the editorship for the ZGN shifted from Kiel to Munich, the "city of the movement" (*die Stadt der Bewegung*), which led to a concomitant radicalization of its contents. The new editors were the botanist Ernst Bergdolt (1902–1938; NSDAP, 1922), the mathematician and natural scientist Fritz Kubach (1912–1945; NSDAP, 1933; SS, 1936), and the astronomer Bruno Thüring (1905–1989; NSDAP, 1930; SA, 1933). They all belonged to the anti-Semitic "Munich circle"[¶] that also included the philosopher Hugo Dingler[**] (1881–1954; NSDAP, 1940), a frequent author in the ZGN, writing predominantly on "modern," that is, German (Aryan) Physics. The journal was now called "Organ of the *Reichsfachgruppe* for Natural Sciences and of the *Reichsstudentenführung*," and branded as "a platform (*Kampforgan*) for all active scientists engaged in the fight against all influences alien to the German race and for a science that is rooted in the German soul"[††] In March 1938, it carried a polemic against the "horror-tabloid *Nature* [London]," the epitome of "Western Science."[‡‡] Other notable entries in the third volume include titles such as "*Biology and National Socialism*" ("National Socialism ... an ideology that proclaims the genetically determined and racially conditioned unity of body, mind and soul"[§§]), "*Racial—Völkisch Actuality and Exact Natural Sciences*," "*Physics and Astronomy in Jewish Hands*," or a dedication to one of the founders of Aryan Physics, "*Philip Lenard—Model and Obligation*." Göring's Four Year Plan that was proclaimed at the *Reichsparteitag* in September 1936 had evident reverberations for

---

[*] Tröger, 1984, p. 8.
[†] Cornwell, 2003, pp. 35, 134.
[‡] Nagel, 2012, pp. 48–49.
[§] Hopster, 1985, p. 117.
[¶] Deichmann, 1996, p. 390, n. 56.
[**] On Dingler, see Rieppel, 2012c. On account of his earlier membership in the Masonic Lodge, Dingler's admission to the NSDAP in 1940 required a pardon issued by Hitler (Klee, 2003, p. 112).
[††] Anonymous, 1937a, p. 603.
[‡‡] Rügemer, 1937/38; see also Hoβfeld and Olsson, 2006.
[§§] Hecht, 1937/38, p. 280.

basic research and academic freedom* that were reflected in the ZGN: "The state leadership supports the researcher, but requires of him that he not lead an isolated existence, but instead locates himself at the center of the national community (*Volksgemeinschaft*)."[†] Similar thoughts had been voiced by the philosopher Martin Heidegger—editorial board member of the newly founded ZGN—in his inaugural speech he delivered upon his appointment as president of the University of Freiburg i.Br. in 1933. Academic freedom, he proclaimed, was nothing but a "smokescreen for indifference, for arbitrariness of intentions and preferences, for unboundedness in one's actions," and should be replaced by a commitment to research understood as a service for the *Volk*.[‡] German Science as a national task and a service to the *Volk* was indeed a central concern of the Four Year Plan, with important consequences for *völkisch* science to be discussed later.[§] Its proclamation in September of 1936 was followed by the founding of the Reich Research Council (RFR, *Reichsforschungsrat*) on May 25, 1937, with Hitler, Göring, and Bernhard Rust among others in attendance. The Reich Research Council was conceived by the *Alt*-Nazi Rudolf Mentzel and his allies at the REM as a separate entity within the broader umbrella of the DFG, organized according to the *Führerprinzip* and tasked to bring research funding into line with the goals of the Four Year Plan.[¶] Members of the RFR were directly nominated by the minister of education, Bernhard Rust, on the basis of professional qualifications as well as political allegiance to the party. Most were surprisingly young, many of the generation that had served in the First World War, the so-called front-line generation.[**] In his inaugural speech, Rust celebrated the RFR as a means through which "the Nazi Revolution summoned science to the decisive battle," and again hammered home his point that "complete freedom of opinion and judgment are not marks of a truly free science, but rather an estrangement of the spirit from the eternal forces of nature and history ... ."[††] Academic freedom was nonetheless proclaimed for basic research, but in typical anticipatory obedience the ZGN cautioned that if a scientist relates his work to the broader public in a way that is loaded with political connotations, then the state has the obligation to interfere, that is, confiscate such works of pseudoscience: "the state and the movement must act and obviate in the light of their political responsibility ... It is not the scientist who takes *völkisch* responsibility for the future, but the movement instead, for which the *Führer* alone is accountable."[‡‡] With volume 5, published in 1939, the ZGN was brought into line (*gleichgeschaltet*) with the *Stiftung Ahnenerbe* (Ancestral Heritage Foundation) in

---

* Deichmann and Müller-Hill, 1994, p. 166; see also Nagel, 2012, p. 287. In a circular, dated March 23, 1940, Göring proclaimed: "The tasks set by the war and the Four Year Plan require that universities collaborate more extensively in the search for solutions of technical problems; I therefore intend with the consent of the Reich Education Ministry to foster a close collaboration between universities and the industry" (Archive of the University of Halle, UA Halle, Rep. II, PA 17240 [K.L. Wolf]).

† Passing, 1937/38, p. 162.

‡ Heidegger, cited after Schreiner, 1985, p. 182.

§ Lundgreen, 1985, p. 14.

¶ Flachowsky, 2010, p. 59; see also Nagel, 2012, pp. 112ff.

** Flachowsky, 2010, p. 61.

†† Rust, cited after Fowler, 1945, p. 32.

‡‡ Hecht, 1937/38, pp. 289–290.

Berlin-Dahlem, a research-funding agency* administered by the SS, and thus fell under the ultimate control and leadership of *Reichsführer*-SS Heinrich Himmler. The mission of the *Ahnenerbe* was to research the history, geography, and legacy of the Nordic-racial Indo-Germanic culture; to communicate the findings of such research to the German *Volk* in an engaging language; and to encourage the *Volk* to participate in these activities.† To accomplish its mission, the *Ahnenerbe* awarded research grants, funded expeditions, and sponsored publications.‡ The *Ahnenerbe* was founded in 1937, but held its first annual meeting in 1939 in Kiel under the banner *Germanic History as Political Science* (*Germanenkunde als politische Wissenschaft*).§ The Curator of the *Ahnenerbe*, *SS-Obersturmbannführer* Walther Wüst (1901–1993), expert in Indo-Germanic studies, characterized the core values of the organization as embodied by a *völkisch* science, the study of Germanic and Indo-Germanic history, a voluntariness that is motivated by camaraderie, and a commitment to high levels of performance. The meeting in Kiel was prefaced by a leaflet composed by Dr. Wüst that celebrated *völkisch* values: "Incorruptible and lucid science may be allowed to take the lead [in our investigations], but its work must resonate throughout the *Volk*, allowing every German to participate in its results."¶ This was to highlight the need to tie research to "political life, political will and order."**

The German ("Aryan") Physics rejected relativity theory and quantum mechanics for being counterintuitive (*unanschaulich*) for the German soul and intellect. The movement was initiated in the 1920s by the Nobel laureate Philipp Lenard (1862–1947), the first scientist of high standing to publicly pledge allegiance to Hitler;†† the movement was carried forward into the National Socialist era by another Nobel laureate, Johannes Stark (1874–1957; NSDAP, 1930), the latter through Hitler's designation president of the DFG since 1934.‡‡ Stark appointed Alfred Rosenberg as the honorary president of the DFG, but in 1936 had to yield under pressure and vacate his position, to be replaced by Metzler. The high level of mathematical abstraction characteristic of nuclear physics and relativity theory Stark deemed to be characteristic of "Jewish science." Bruno Thüring, member of the Munich circle, and Munich co-editor of the ZGN, published in 1941 his classic book on *Albert Einstein's Attempted Coups in Physics*.§§ A review in the ZGN praised the text for "the clarity and excellence with which the racial determinedness of the postulates and principles of relativity theory as formulated by the Jew EINSTEIN is documented."¶¶ The second edition of the book (1943) was showered with equal praise by another member

---

* Kater, 1974.
† Wüst, 1939, p. 242; see also Walker, 2003, p. 999: "In 1939, an Ancestral Heritage official described its goals as follows: '[…] to research the space, spirit, deed, and legacy of the racially pure Indogermanic peoples, to mold research results in a lively way, and to present them to the people'."
‡ Walker, 2003, p. 999; see also Kater, 1974.
§ Kaiser, 1939.
¶ Wüst, in Kaiser, 1939.
** Kaiser, 1939, p. 6.
†† Hermann, 1984, p. 161.
‡‡ Nagel, 2012, p. 107.
§§ Thüring, 1941.
¶¶ Stubbe, 1942, p. 306.

of the Munich circle, the philosopher Hugo Dingler,[*] in his reply to Heisenberg. Called a "white Jew" by Stark in view of his involvement with quantum mechanics, Heisenberg had penned a refutation of "German Physics" in 1940, and submitted it to the ZGN on May 20, 1943:

> In contrast to the distinctions that have been drawn in the debates of the past few years concerning the two kinds of physics such as: pragmatic-dogmatic, realistic-irrational, intuitive-formalistic, I would like to put another distinction at the top of my agenda: the distinction between true and false.[†]

Parallel to a "German Physics" there were attempts to "Germanize" other sciences such as mathematics, where intuition should again trump abstract formalisms. Karl Lothar Wolf led the unsuccessful attempt to establish a "German Chemistry," just as the "German Biology" should likewise remain stuck in the mud of ideology and intrigue.

DEUTSCHE BIOLOGIE

The renowned anatomist Benninghoff, cofounder of the ZGN, in its first volume sketched functional anatomy in holistic/organicist terms, relating the functional analysis of organisms to an Uexküllian conception of *Umwelt*.[‡] He adopted Uexküll's notion that there are no greater or lesser degrees of perfection in animal adaptations, but that each organism at every level of complexity of morphological and physiological differentiation lives in perfect congruity (*Einpassung*) rather that adaptedness with its *Umwelt*. Benninghoff concluded his expositions by raising Driesch's question, which is whether function can determine form—at the time a controversial thesis of neo-Lamarckian tinge that will require more discussion.[§] Uexküll himself penned an entry for the first volume of the ZGN, postulating a sharp demarcation of biology from physics: "We leave to the physicists their field of inquiry, but request that the physicists leave our field of inquiry [biology] to us. Any blending is evil."[¶] But the vehicle to spread the gospel of "German Biology" was not so much the ZGN, although it too played a role, but more so the monthly periodical *Der Biologe*[**] that was founded by the Tübingen botanist and geneticist Ernst Lehmann (1880–1957) in 1931.

At its 35th annual meeting, held in Erfurt on April 9 through 13, 1933, the Association for the Advancement of Instruction in Mathematics and Natural Sciences (*Verein zur Förderung des mathematischen und naturwissenschaftlichen Unterrichts*) celebrated a very special "German Spring," a renewed struggle for freedom. On that occasion, it passed three resolutions. The first required that the formulation of means and goals of German education and cultural formation (*Erziehung und Bildung*) be rooted in the spirit of the German Spring:

---

[*] Dingler, 1943, p. 220.
[†] Heisenberg, 1943, p. 201.
[‡] Benninghoff, 1935/36, p. 158.
[§] Bennninghoff, 1935/36, p. 160.
[¶] Uexküll, 1935/36, p. 257.
[**] Bäumer, 1990a.

"The appreciation of the German landscape and its relation to the life and work of its inhabitants, the spreading of the consciousness of race through the *Volk*, the general physical education, and lessons on general public health care are only possible if based on a scientific foundation." The second resolution called for a recognition of the biological foundations of folklore and culture: "The German Association for the Advancement of Instruction in Mathematics and Natural Sciences ... is paying close attention to the efforts of eugenics, and supports the accelerated implementation of eugenic measures, which serve the preservation of a healthy public genetic environment (*Erbgut*), and the liberation of the national body (*Volkskörper*) from heritable defects." The third resolution concerned defensive measures such as protection from poison gas attacks and air raids.[*] The first two resolutions nicely capture the biologism that was to become the centerpiece of National Socialist ideology and propaganda, most famously captured by the incisive slogan "blood and soil" (*Blut und Boden*) that invoked a "geo-cultural linkage of race and place."[†] Oswald Spengler, in his *The Decline of the West*, had already rooted the blood, that is, the race in its landscape, the soil, but the ideological amplification of that metaphor was spurred by Hitler's Minister of Agriculture (*Reichsbauernführer*) Walther Darré (1895–1953; NSDAP, 1930; SS, 1930) through his book titled *Neuadel aus Blut und Boden*, published in 1930 by the racist-*völkisch* Munich publisher Julius F. Lehmann (1864–1935). It was J.F. Lehmann who first introduced Darré to Hitler in 1930, a meeting that lay the foundations of Darré's meteoritic ascent through the Nazi hierarchy until his later demise, triggered by the pressures of war on agriculture. Darré, an ideologically committed Social Darwinist, was an expert in animal breeding who went on to apply corresponding principles of racial selection to humans, promoting racial purity of German peasants. The head of the REM, Bernhard Rust, adopted the slogan of blood and soil as a guideline for education in the new Germany. In agreement with Darré, he praised National Socialism for having managed "to stop the march of the pristine people from the farms and the rural homeland into the asphalt deserts of big cities. It seeks a path out from the sites causing the death of the *Volk* towards the eternal fountains of blood and soul."[‡] One-sided intellectualism is to be balanced by handiwork; a poor grade in Latin can be compensated for by a good performance in sports, Rust thought.

The race in National Socialist ideology is defined by its blood, the latter in turn rooted in the landscape, in the homeland, in its *Lebensraum*.[§] The *Volk* that constitutes a race is tied together by blood relationships that need to be preserved and kept clean. Just like the blood vascular system ties together the organs of an organism, so do the blood relationships tie together the individuals of a *Volk*. The *Volk* itself is conceived of as an organism, a body that needs to be kept clean and healthy. The individual embedded in its *Volk* is exhorted to keep her house clean and well aired; she

---

[*] Anonymous, 1933, p. 113.
[†] Gerhard, 2005, p. 131.
[‡] Rust, cited in Nagel, 2012, p. 162.
[§] *Lebensraum* is a highly loaded term that goes back to an essay published in 1901 by the zoologist turned geographer Friedrich Ratzel (1844–1904). Subtitles Ratzel used were "Conquest or Colonization?," or "The Battle for Space." The concept of *Lebensraum* acquired its notoriety in the context of the eastward colonization of Europe by Nazi Germany.

should observe personal hygiene and stay healthy through the consumption of quality food, through corporal exercise in the open natural landscape, and through the avoidance of alcohol, nicotine, and sexual diseases; particular diligence is required in the choice of a healthy spouse from the same *Volk*. While all of these motives and values gained momentum and intensified after Hitler's ascend to power in 1933, they can be traced back in German culture through the time of the Weimar Republic all the way into the late nineteenth century.[*] Eugenics was called racial hygiene (*Rassenhygiene*) by Alfred Ploetz (1860–1940) in the subtitle of a book he published in 1895.[†] In 1904, Ploetz founded the *Archiv für Rassen- und Gesellschaftsbiologie* (Archive of Racial and Social Biology), which would become a leading journal in its field.[‡] In 1905, Ploetz founded the German Society for Racial Hygiene, an endeavor that was supported—among others—by his friend Ernst Haeckel.[§] For eugenic measures to gain traction in an effort to not only maintain, but even to improve (*Aufartung* or *Aufordnung*) the health and vigor of the race requires the subordination of the interests of the individual to the collective interests of the *Volk*: "*Du bist nichts, Dein Volk ist alles*" (You are nothing, Your *Volk* is everything) was the celebrated motto of Hitler's "national community."[¶] In Hitler's national community, individuality was to dissolve into group consciousness,[**] the individual was reduced to a mere carrier of biological properties,[††] embedded as it is in its race and *Volk*. Arthur Gütt (1891–1949; NSDAP, 19432; SS, 1933), Director (*Ministerialdirektor*) in the Ministry of the Interior and head of the Office for Population Politics and Hereditary Health of the SS, who had helped draft corresponding legislation,[‡‡] explained in an entry in Lehmann's journal *Der Biologe*:

> [The] family [is] the germ cell of the *Volk* ... In National Socialist terms, the individual in itself means nothing; his life acquires meaning and purpose only through the fact that he is born into a community, to which he is connected through natural laws of life.[§§]

In 1933, the professor of zoology Karl Leopold Escherich (1871–1951; NSDAP, 1921) was appointed as the new president of the University of Munich, tasked to bring it into line with National Socialist ideology, that is, "to make the German Universities truly National Socialist—not just ... to 'paint them brown'."[¶¶] In his inaugural speech, delivered on November 25, 1933, and titled *Termiten wahn. Eine Münchner Rektoratsrede über die Erziehung zum politischen Menschen* (Termite Mania—on

---

[*] Weingart et al., 1992.

[†] Weingart et al., 1992, p. 91.

[‡] Weingart et al., 1992, p. 99.

[§] Cornwell, 2003, p. 79. For more context on Alfred Ploetz, his journal and his society see also Weingart et al., 1992, pp. 192–208. Both Haeckel and August Weismann were named honorary presidents of the German Society for Racial Hygiene (Weingart et al., 1992, p. 191).

[¶] In *Mein Kampf*, Hitler wrote of the *völkisch* state: "Thereby the state has to appear as the guardian of a thousand years' future, in the face of which the wish and the egoism of the individual appears as nothing and has to submit" (Hitler, 1941, p. 608; see also Weingart et al., 1992, p. 367; Pine, 2007).

[**] Cornwell, 2003, p. 163; see also Bäumer, 1989, 1990b.

[††] Preuß, 1984, p. 123.

[‡‡] L. Mertens, 2004, p. 277.

[§§] Gütt, 1937, pp. 378–379.

[¶¶] Schultze, 1939 (1966), p. 315.

the Education of the Political Person), Escherich rejected the crass individualism that he found to have characterized the bourgeois–liberal culture of the Weimar Republic. Bolshevist ideology was just as bad, however, as it reduced through the dictatorial rule of the communist society the human individual to a senseless automaton functioning like a social insect. What Escherich promoted instead was the education of the individual to become a "political person," that is, someone who willfully as much as cheerfully adopts National-Socialist ideology (*Weltanschauung*) and its core values, which were obedience to authority (the *Führerprinzip*), strong work ethics, sexual restraint, sobriety, frugality, and the readiness for self-sacrifice in the service of the greater good of the national community.[*] The corresponding slogan read: "The common good trumps self-interest" (*Gemeinnutz geht vor Eigennutz*).[†] In the context of Nazi "biopolitics,"[‡] Escherich followed the party line when arguing that National Socialism is through and through biological: "National Socialism is so as to say the biological will of the German *Volk*."[§] Such biology then is a *Volksbiologie*, a racially conditioned biology, that is, a "German (Aryan) Biology." Reviewing the inaugural speeches of Karl Escherich in Munich, and of the racial hygienist Eugen Fischer (1874–1967)[¶] in Berlin for *Der Biologe*, Ernst Lehmann exclaimed: "Is it a happenstance that the first university presidents to be named in the National Socialist state … are biologists? No doubt the most careful selection" has been made.[**] On the occasion of Escherich's 70th birthday, his inaugural speech *Termite Mania*[††] was hailed as a contribution to the "fight for an applied science," delivered by an entomologist who had left narrow specialization behind as he recognized early on the ideological dimension of biology.[‡‡]

The most vociferous proponent of a German Biology was Ernst Lehmann, botanist, plant geneticist, and director of the Botanical Institute and Garden of the Karl Eberhard University of Tübingen. In 1931, Lehmann founded the journal *Der Biologe*, in the first issue of which he published an appeal for the foundation of a German Association of Biologists (*Deutscher Biologen-Verband*), to serve as a platform to lobby for the interests of biologists both in education and in research.[§§] Full of praise, he used his journal in support of Escherich's program, which called for "an enhancement of the individual through political education to a public personality who voluntarily submits to the service of the communal whole." In Lehmann's view, both teachers and students will follow this call "in the conviction that our *Volk* is not in decline, but instead stands at the beginning of a new ascent."[¶¶] Starting with volume 6, 1935, of *Der Biologe*, Lehmann introduced a rubric portraying biologists with large families in support of the Nazi population policy: "A stormy increase

---

[*] Stephenson, 2008, p. 100; see also Pine, 2007.
[†] Diemberger, 1941, p. 219.
[‡] NS "biopolitics" combined biologism with the concept of *Lebensraum* in a vision of a racially pure national community seeking spatial expansion: Weingart et al., 1992, p. 370; see also Caplan, 2008, p. 19.
[§] Escherich, cited in Kirchner, 1984, p. 85.
[¶] On Eugen Fischer, see Weingart et al., 1992; Hutton, 2005.
[**] Lehmann, 1934b, p. 59.
[††] Escherich, 1934.
[‡‡] Hofmann, 1941, p. 416.
[§§] Bäumer, 1990a; see also Deichmann, 1996.
[¶¶] Lehmann, 1934b, p. 60.

in birth rates amongst the high-valued hereditarily healthy is a prerequisite for the strengthening of our *Volk*."[*]

Lehmann was unquestionably one of the most iridescent figures in the orbit of Nazi ideology and German Biology. Bernhard Johannes Ernst Lehmann[†] was born on June 24, 1880, in Dresden, the son of the geometrician Bernhard Alexis and his wife Sidonie *née* Künzel. His family was evangelical. He passed his *Abitur* at the *Gymnasium zum heiligen Kreuz* in Dresden, and then went on to study in Berne (Switzerland), Kiel, Munich, and Strasbourg, where he obtained his PhD in 1906. There followed assistantships in Bonn-Poppelsdorf (1907–1908) and Kiel, where he defended his *Habilitation* thesis in the winter semester of 1908/09. The summer semester 1911 he studied in England. On October 1, 1911, he started his job as an assistant at the Botanical Institute of the University of Tübingen, where he was named associate professor on April 3, 1913, and was promoted to full professor on May 10, 1922. During those years, he declined offers for employment at the rank of full professor at the Universities of Bonn and Greifswald, and was considered for such appointments at the Universities of Marburg and Erlangen. Lehmann was married and had four children, three sons and one daughter. He died in Tübingen on December 1, 1957 (Figure 5.3).

Prior to Hitler's ascent to power in 1933, Lehmann was for a period of about two years member of the *Deutsche Volkspartei*. Following the wish his father expressed shortly before his death, as Lehmann later explained, he was member of the Masonic Lodge "The Three Sisters" in Dresden "from about 1911 through about 1921," as he recalled. This membership would later thwart his efforts to become a member of the NSDAP.[‡] His deep entrenchment in the *völkisch* tradition is also revealed by his activities in the *Völkisch* Protection and Defense League (*Völkischer Schutz- und Trutzbund*) and in his membership—since 1920, and again following his father's lead—in the Pan-German League (*Alldeutscher Verband*).[§] Against rumors that had been spread in 1932 by members of the National Socialist Teachers' League (NSLB, *Nationalsozialistischer Lehrerbund*) in Kiel,[¶] he successfully documented his Aryan descent following the decree of the Law for the Restoration of the Career Civil Service on April 7, 1933. Lehmann had been denounced of being a Jew, born Levi, from Dresden, who married a Christian woman while at the University of

---

[*] Lehmann, 1936a, p. 416.

[†] *Universitätsarchiv Tübingen*, UTA 126/373, 1; Personalakte Lehmann.

[‡] In support of his renewed application for membership in the NSDAP, Lehmann wrote on October 10, 1937: "My father was a member of the Lodge of the Three Sisters in Dresden. Before his death he expressed his desire that I join the Loge as well. He died in 1909; I became a member in my father's Lodge in 1910 (not 1911). The evening of my admission was the only one I ever attended, I have never attended any other assembly of the Lodge ... [this] shows that I have never had any inner ties to the Lodge" (*Universitätsarchiv Tübingen*, UTA 126 / 373, 13, Bl. 5).

[§] Deichmann, 1996, p. 76.

[¶] Lehmann, it was claimed, was born a Jew named Levi, who either during his time in Kiel, or perhaps during his study visit in England, married a Christian wife—hence his conversion to Christianity (*Universitätsarchiv Tübingen*: UTA 126/373, 10; see also letters in UTA 126/373, 1). The *Gauleitung Württemberg-Hohenzollern* requested in a letter to the party administration dated May 13, 1933, that Lehmann be barred from admission to the NSDAP. His name was consequently flagged at party headquarters (Bundesarchiv, DS [formerly Berlin Documentation Center], microfilm signature DS B 035, file "Ernst Lehmann").

FIGURE 5.3 Ernst Lehmann. (Universitätsbibliothek Tübingen, Bilddatenbank.)

Kiel, and hence converted to Christianity.* The rumors, however, still managed to help jeopardize his application for NSDAP membership in 1933.† He became himself a member of the NSLB on January 1, 1934, and took the oath of allegiance to the *Führer* on September 26, 1934. Lehmann did not sign the "Commitment of the Professors at German Universities and Colleges to Adolf Hitler" from November 1933, because he found it more effective to initiate an independent yet parallel initiative at the University of Tübingen, collecting signatures from his fellow faculty members, an activity that earned him praise from the leadership of the National Socialist German Student Organization (NSDStB, *Nationalsozlialistischer Deutscher Studentenbund*). A memo from the NSDStB, dated March 2, 1933, "cordially" thanked Lehmann for his "letter accompanied by the declaration of Tübingen professors and academics in support of the current *Reich* government. With this deed

---

* *Universitätsarchiv Tübingen*: UTA 126/373, 10.
† "The serious injustice that was done to me has until today not been rectified" (*Universitätsarchiv Tübingen*, UTA 126/373, 13, Bl. 5. A declaration of Lehmann in support of his application to become member of the NSDAP in 1937). Lehmann had been encouraged to apply by the physiologist Rupprecht Matthaei (1895–1976), who also supported Lehmann's initiative to collect signatures from Tübingen faculty in support of Hitler and Hindenburg. According to Matthaei, Lehmann "suffered extraordinarily" from being rejected by the NSDAP (*Universitätsarchiv Tübingen*, UTA 126/373, 17).

Tübingen has completely fulfilled its obligation ... the declaration will be published in the *Völkischer Beobachter,*" the major newspaper published by the NSDAP.

According to his own recollection, Lehmann started a movement to unify German biologists in 1920, which earned him (in his words) the "honor" to be called "the first German biologist to have put himself at Hitler's disposal" by the "communist [Julius] Schaxel," who had denounced Lehmann at the Russian Academy of Sciences in Moscow after his emigration.[*] Lehmann would consent that "Schaxel recognized sooner than most contemporary biologists in Germany that biology is to be called upon to play a most important role in the solution of fundamental questions concerning public life, as well as the private life of everyone," but he rejected Schaxel's analysis because it was based on the communist manifesto.[†] The first issue of Lehmann's *Der Biologe,* which contained the editor's call for "the formation of a German Association of Biologists,"[‡] appeared in October 1931. It was announced by a press release from the University of Tübingen, which proclaimed: "The appearance of this monthly periodical is to be enthusiastically welcomed, as it promises to become the expression during most difficult times of the unwavering will to pursue a unified German biology."[§] The German Association of Biologists (*Deutscher Biologen-Verband*) was called upon "while respecting academic freedom to continuously integrate the results of [biological] research into the organization of the National Socialist state."[¶] In support of his application for membership in the NSDAP in 1937, Lehmann sketched with pride how he could win Julius Friedrich Lehmann (with whom he was not related), Munich, as publisher for *Der Biologe.* Later, Lehmann reminisced: "At the time, the Jew [Victor] Jollos, now in Madison (America) ... told me that this thing with the German Association of Biologists was quite nice, although its journal should have been published not by Lehmanns, but by Springer instead."[**] But, Lehmann claimed: "For many years I have been fighting a battle against the Jewish publishing house Springer."[††] Deemed by E. Lehmann to be the only adequate publisher for his journal,[‡‡] J.F. Lehmann immediately accepted: "I [E. Lehmann] was happy to have been able to show him my gratitude for this, as well as for his entire work [as a publisher] in the *völkisch* spirit as well as in racial studies,

---

[*] *Universitätsarchiv Tübingen,* UTA 126/373, 13, Bl. 5. A declaration of Lehmann in support of his application to become member of the NSDAP in 1937.

[†] Lehmann, in *Heimatbiologie,* the opening lecture of his course of the winter semester 1944/45, manuscript sent to Mentzel at the REM. Bundesarchiv, DS (formerly Berlin Documentation Center), microfilm signature DS G 207, file "Ernst Lehmann" (see also Weindling, 1989a, p. 327). Schaxel, a critic of National Socialist racial doctrine, had urged in 1936 to have that doctrine discussed at the international congress of genetics. Lehmann (1936b, p. 160) retorted: "German biologists are ready at all times to defend the biological foundations of National Socialist ideology ... in front of the scientific world."

[‡] Lehmann, 1931/32, p. 1.

[§] *Universitätsarchiv Tübingen,* UTA 126/373, 1.

[¶] Lehmann, 1936c; in *Universitätsarchiv Tübingen,* UTA 126/373, 10_2.

[**] Lehmann, Stellungnahme; *Universitätsarchiv Tübingen,* UTA 126/373, 14B.

[††] Lehmann, October 23, 1937, "Mitgliedschaft bei der NSDAP"; *Universitätsarchiv Tübingen,* UTA 126/373, 13, Bl. 5.

[‡‡] "When I decided in 1931 to found a German Association of Biologists, there could be no doubt for me that I would ask this "most German publisher" (*deutschester Verleger*) to publish our journal (E. Lehmann, 1934c, p. 305).

by motivating my faculty to award him [J.F. Lehmann] an honorary doctorate."*
J. F. Lehmann was notorious for his role as publisher, providing an outlet for
publications in the *völkisch* spirit and spreading National Socialist racist ideology:
"[J.F.] Lehmann used his publishing outreach and personal wealth to spread
the Nazi message among medical students and practicing physicists."† In 1935,
E. Lehmann published an obituary for the deceased J.F. Lehmann in *Der Biologe*,
stating that: "[J.F.] Lehmann was the first amongst German publishers who clearly
understood that the *völkisch* renewal of our *Volk* would receive decisive support from
biology."‡ This was in line with his pronouncement in the *Völkischer Beobachter*
from September 18, 1936, where in an essay on "The Mission of a German Biology"
E. Lehmann had written: "With unique intuition our *Führer* Adolf Hitler grasped
the need of the hour, which was to proclaim the Laws of Life (*Lebensgesetze*) the
foundation of the German worldview (*Weltanschauung*)."§ When formulating his
views on the mission of German Biology, Lehmann claimed to follow the lead of
Hans Schemm, who in 1929 founded the NSLB under the umbrella of the NSDAP,
and with whom Lehmann discussed the integration of his German Association of
Biologists into the NSLB. Lehmann first met Schemm, a Grade School teacher who
had advanced to the office of the Bavarian Secretary of Education, in December
1933 in Munich. On February 11, 1934, Schemm wrote to Lehmann that "in view of
your excellent work in biology, in racial research, [and] in agricultural fertilization, it
is only natural that your Association [of German Biologists] should be affiliated with
a large National Socialist Organization."¶ The *Anschluss* was executed in the same
year "with the consent of the *Führer*."** Schemm, who signed on to the editorial board
of *Der Biologe* with the appearance of the third volume in 1934, died in a plane
crash on March 5, 1935, subsequent to which he was glorified in Nazi ideology.††
Schemm's spirit transpired in the second, that is, February issue of the third volume
(1934) of *Der Biologe*, which was dedicated to Ernst Haeckel's 100th birthday, in a
laudatory eulogy penned by his grandnephew Werner Haeckel, who applauded the
new regime's Social Darwinist policies as a fulfillment of Haeckel's vision. Chiming
in on the praise of Haeckel, Lehmann recognized in his work the deepest essence
of National Socialist ideology (*Weltanschauung*). The German youth should follow Haeckel—this Nordic giant figure—in gymnastics, in swimming and climbing,
quite generally "in the perseverance in corporeal exercise."‡‡

---

* *Universitätsarchiv Tübingen*, UTA 126/373, 13, Bl. 5. A declaration of Lehmann in support of his
application to become member of the NSDAP in 1937. To award J.F. Lehmann an honorary doctorate
degree met with respectable resistance in the faculty of the University of Tübingen (Heiber, 1991).
During the denazification hearings, Lehmann asserted that after the death of its founder in 1935,
the *J.F. Lehmanns Verlag* was increasingly caught up in National Socialist racial politics, for which
reason he abandoned the publishing house. While this is true for his self-serving book of 1947, *Irrweg
der Biologie*, *Der Biologe* continued to be published by J.F. Lehmanns for as long as it was edited by
E. Lehmann and beyond (*Universitätsarchiv Tübingen*, UTA 126/373, 1).
† Cornwell, 2003, p. 153.
‡ Lehmann, 1935a, p. 143.
§ *Universitätsarchiv Tübingen*, UTA 126/373, 10_2.
¶ *Universitätsarchiv Tübingen*, UTA 126/373, 5.
** Bäumer, 1990a, p. 43.
†† Klee, 2003, p. 530.
‡‡ Lehmann, 1934d, p. 132.

Fleshing out the content of a "German Biology," Lehmann drafted a memorandum[*] dated December 3, 1936, which once again emphasized the fundamental importance of biology, and the need to carry biology into the *Volk* through his journal and at schools in order to strengthen the "biological will": "today, the ideological and political significance of biology is thrown into ever sharper relief." He proposed the establishment of an Institute for German Biology, which should complement Karl Astel's (of whom more later) Institute for Human Genetic Research and Racial Politics at the University of Jena—a plan that went nowhere.[†] In opposition to Political Catholicism,[‡] the institute should have promoted the theory of descent, and "educate about such important issues as the concept of natural *wholeness*." It should have provided a counterweight to a purely mechanistic biology, which is devoid of National Socialist interests and indeed "supported by Jewish science." An important aspect was to teach biology through practice, for example, in school gardens, work which would provide insight into the laws of inheritance as well as into the beneficial nature of medicinal herbs. This latter concern Lehmann shared with the officials at the REM, who were advancing a syllabus reform at the level of both elementary schools and Medical School.[§] Lehmann closed his memorandum with a call to enter into a dialogue with other people (*Völker*)—"not their Jewish exponents," however— with respect to shared "biological aspirations." In his essay on "The Mission of a German Biology," published in the *Völkischer Beobachter* from September 18, 1936, Lehmann referred to a speech Hitler had held in the *Lustgarten* in Berlin on May 1, 1934. In front of "German students" (in fact Hitler Youth), Hitler had emphasized the impossibility to lead a *Volk* or a state unless there prevails a general unanimity in the knowledge and willful acceptance of the laws of life (*Lebensgesetze*) that underlie that community. Lehmann took this as a command to German biologists to seek an ever-deeper understanding of the laws of inheritance, and to communicate them to the broader public.[¶] During those same years, Lehmann published in his journal an article that declared the need for biology to be the central part of German Public Education. Its author, a certain Heinrich Jacob Feuerborn, proclaimed "National Socialism ... [to be] ... nothing else than the comprehensive and targeted application of the laws of life to the individual as well as to the entire Volk. It is the fulfillment of an ideology (*Weltanschauung*),"[**] which itself is based on a scientific comprehension of the world, and consequently is as eternal and immutable as the laws of nature (i.e., of inheritance) on which it is based.

Rust, Schemm, and Lehmann—all of them shared a holistic/organicist perspective of the *Volk*, its *Heimat* (homeland), and the biology on which these are based. This is also the perspective that dominates, although not exclusively, the first volumes of *Der Biologe*, which were published under Lehmann's editorship. In the introductory article to his journal, Lehmann adjured the unity of biology, the unity of anatomy and physiology, the unity of botanical and zoological research, as "today we start

---

[*] *Universitätsarchiv Tübingen*, UTA 126/373, 5.
[†] Deichmann, 1996, p. 78.
[‡] Schreiner, 1985, p. 202.
[§] Nagel, 2012, pp. 185, 218–219; see also Pine, 2010.
[¶] *Universitätsarchiv Tübingen*, UTA 126/373, 10_2.
[**] Feuerborn, 1935, p. 99.

to clearly appreciate to which extent biological insights have to be assimilated to ideological issues."* Looking through the programmatic entries in the first volume of *Der Biologe*, recurrent themes are knowledge of the laws of life (i.e., of inheritance) as they relate to racial hygiene, knowledge of the homeland (*Heimatkunde*), and nature conservation, coupled with an appeal to school reform in order not only to firmly anchor these topics as priorities in the curricula, but also to render Darwinism a staple of school education in a struggle against Marxists who tried to appropriate Darwin for their own agenda.[†]

The Marxists sympathizer Julius Schaxel did not rest his case, however. In a circular distributed in 1936 he requested that the National Socialist racial doctrines be critically discussed at the upcoming international congress of geneticists to be held in Moscow in the second half of August 1937.[‡] Lehmann remained unimpressed: "German biologists are at all times ready to defend the biological foundations of National Socialism in front of the scientific world."[§] To a deeper discussion of these biological foundations of National Socialism we now turn.

---

* Lehmann, 1931/32, p. 4.
† *Der Biologe*, 1931/32, vol. 1, p. 97.
‡ International Congress of Genetics. *Nature*, 137 (1936), p. 941; see also Soyfer, 2003.
§ Lehmann, 1936b, p. 160.

6 Ganzheitsbiologie

ENKAPSIS: HIERARCHICALLY STRUCTURED COMPLEX WHOLES

Testifying on behalf of Ernst Lehmann, who was embroiled in a virulent intrigue of which more later, the Munich dermatologist Franz Wirz[*] (1899–1969; NSDAP, 1933) wrote from the *Hauptamt für Volksgesundheit* (Main Office of Public Health) to the senate of the University of Tübingen on May 10, 1940. In his letter, he praised Lehmann for his efforts in the attempt to establish a "German Biology," a legacy of Hans Schemm who, like Lehmann himself, had sought to expand the significance of biology far beyond narrow subject boundaries:

> [Lehmann] is aware that the entire political thinking of National Socialism is biological, as the *Führer* himself has repeatedly expressed … At this point, however, I want to emphasize that the goals set by Hans Schemm were more far-reaching. Schemm was an enemy of so-called *materialistic biology*. He instead embraced those ideas, which today are subsumed under the slogan of '*Ganzheitsbiologie*'. In terms of its content this [*Ganzheitsbiologie*] most closely corresponds to those thoughts that have more recently been articulated by Meyer [Adolf Meyer (-Abich)], Hamburg, Kötschau [Karl Kötschau (1892–1982)], Nuremberg, and others. But Schemm was right when he recognized that in order to establish such a biology and render it useful especially in schools and at universities, the foundations of the *real material* biology cannot be completely abandoned. It is exactly in that sense that Schemm recognized in Lehmann the ideal person …[†] to carry this plan forward.

The Tübingen philosopher Theodor Haering (1884–1964, NSDAP, 1937) reflected on *Ganzheitsbiologie* in volume 5, 1935, of *Der Biologe*. Haering had been a member of Lehmann's *Vererbungskranz*, another evening discussion circle comprising faculty from various disciplines of the University of Tübingen that met on a monthly basis from 1919 to 1925.[‡] His musings in *Der Biologe* were based on notes he had taken during sessions of the *Vererbungskranz*, characterized by Lehmann as an effort "to forge the weapons for the day when they would be requested from us."[§] Haering confirmed the irreducibility of biology to chemistry and physics, yet warned of an all too literal understanding of the slogan of an "organic state": "Every state is not only an organic-natural whole … but also requires a conscious and leading intellect," who posits unifying goals and ideals: "The race principle is complemented by the *Führer* principle."[¶]

---

[*] Grüttner, 2004, p. 184

[†] *Universitätsarchiv Tübingen*, UTA 126/373, 17.

[‡] Lehmann, 1935b, p. 377; see also Weindling, 1989b, p. 471; Heiber, 1991, p. 435.

[§] *Süddeutsche Apotheker-Zeitung #105, Jahrgang* 1935; unpaginated reprint in *Universitätsarchiv Tübingen*, UTA 126/373, 10_2.

[¶] Haering, 1935, p. 396: "*Neben dem Passenprinzip steht das Führerprinzip;*" see also Harrington, 1996, p. 179.

The trend toward an organicist/holistic biology that started during the years of the Weimar Republic and gathered steam after the ascent of Hitler to power spread well beyond the reach of the *Zeitschrift für die gesamte Naturwissenschaft* (ZGN), or the German biology teachers' monthly periodical *Der Biologe*. It also incorporated some unexpected if remote Jewish roots. In his contribution to the first volume of the ZGN, its cofounder, the comparative and functional anatomist Alfred Benninghoff, had tied the concept of organic wholeness to Martin Heidenhain's concept of enkapsis, that is, a nested hierarchy of complex wholes with emergent properties at every level of inclusiveness. This concept was widespread in German biology at the time, a cornerstone of *Ganzheitsbiologie*, implicit for example in Paul Brohmer's influential book *The Teaching of Biology and Völkisch Education* of 1933.[*]

Martin Heidenhain[†] was born on December 7, 1864, in Breslau, Germany (now Wroclaw, Poland). He was the son of the renowned physiologist Rudolf (Peter Heinrich) Heidenhain (1834–1897), who had coined the term *Entwickelungsmechanik*. Martin Heidenhain was a typical representative of the "Mandarin" tradition within the German professoriate,[‡] born into a family of distant Eastern Jewish descent[§] that on both parental sides counted many well-known physicians and university professors among themselves. His grandfather was a prominent physician from Marienwerder (West Prussia). His mother was the daughter of the anatomist, physiologist, and philosopher Alfred Wilhelm Volkmann (1801–1877) from Halle (Saale), and a sister of the famous surgeon Richard von Volkmann-Leander (1830–1889) also from Halle (Saale), the latter an occasional author of poetry and fiction. Heidenhain himself professed[¶] to have inherited both his scientific and his artistic talent from his ancestors. Mandarin tradition was rooted in the ideal of *Bildung*, as opposed to mere *Ausbildung*. *Bildung* implied *Kultur*, *Ausbildung* merely professional skills. Mandarins generally came from families with an academic background, steeped in an appreciation of humanities, arts, and sciences, indeed of scholarship quite generally. Children would acquire their classic education at the *Humanistisches Gymnasium* with emphasis on Latin, classic Greek, history, and philosophy, before pursuing further studies at university level. Narrow specialization that would be reflected in fragmented university curricula was frowned upon, avoided, indeed opposed by members of the Mandarin professoriate. Mandarins likewise rejected analytic reductionism, barren rationalism, superficial materialism, or shallow utilitarianism. The goal instead was a broad-based humanistic perspective that would provide the backbone for synthesis, a vision of integration and of wholeness, often spiced up with a dose of romanticism or mysticism and proffered with social and/or political connotations. Holism was the overarching theme, Goethe the patron saint. The historian Jonathan Harwood identified a "hunger for wholeness"[**] among Mandarins who rejected an atomistic conception of nature and society, countering it with holistic–organicist arguments (Figure 6.1).

---

[*] See Chapter 5 for details and discussion.
[†] Jacobj, 1952/53; Alfert, 1972.
[‡] Ringer, 1969 (1990); Harwood, 1993.
[§] Adam, 1977, p. 123, n. 18; Grün, 2010, p. 250.
[¶] Jacobj, 1952/53, p. 81.
[**] Harwood, 1993, p. 283.

FIGURE 6.1   Martin Heidenhain. (Universitätsbibliothek Tübingen, Bilddatenbank.)

During his *Gymnasium* (grammar school) years in Breslau, Heidenhain pursued his interests in earth sciences (geology and paleontology), but went on to study biology as he entered the University of Breslau in 1883. Through his father, who held the chair for physiology and histology at Breslau University, he made the acquaintance of Wilhelm Roux, who taught at that university from 1879 through 1889. Moving on to the University of Würzburg, Heidenhain studied zoology under Karl Semper (1832–1893). After having transferred to the University of Freiburg i.Br. in the spring of 1886, he fell back into the family tradition and took up the study of medicine, earning his MD under Robert Wiedersheim (1848–1923) in 1890. While in Freiburg, Heidenhain visited August Weismann's institute and attended seminars offered there. In the spring of 1891, Heidenhain returned to Würzburg as an assistant to the anatomist and physiologist Albert von Kölliker (1817–1905). In 1892 Heidenhain married Anna Hesse, daughter of a judiciary. In 1899, Heidenhain was named associate professor of anatomy at the University of Tübingen. After the death of August von Froriep (1849–1917), he was promoted to full professor and director of the anatomical institute. Following the enactment of the Law for the Restoration of the Career Civil Service on April 7, 1933, the Ministry of Culture of Württemberg alerted the president of the University of Tübingen in a letter dated June 7, 1933, of a new law that would lower the retirement age for professors from 70 to 68 years.[*] This provided additional leverage to get rid of professors who remained guarded with respect to the new National Socialist regime.[†] The new law targeted Heidenhain among others, known as a staunch republican, who was accordingly retired on October 1, 1933. Heidenhain died in Tübingen on February 14, 1949. Today, Martin Heidenhain is most widely honored for his innovative histological techniques,[‡] but at the time his *Synthesioloy* was equally widely received. Beyond his research, Heidenhain was known as receptive to all kinds of cultural activities; he collected antique German craft such as woodcarvings and ceramics, and—pleasant and congenial man that he was[§]—actively participated in the social life of the university and of scientific associations.[¶] One of his students described him as a great aesthetician.[**]

Inspired by the second German edition of Driesch's *Philosophy of the Organism*[††] (yet rejecting Driesch's vitalism[‡‡]), Heindenhain's *Synthesiology* sought an integrative approach to anatomy, where structure, development, and function would all come together. Heidenhain formulated his concept of enkapsis as an alternative to reductionist interpretations of cell theory, which understood tissues, organs, and organisms as aggregates of cells. Heidenhain opposed such an atomistic interpretation of cell theory, one that allowed Roux's extrapolation of the Darwinian struggle for existence to intercellular relations within the organism. In a talk held in front of the Physical-Medical Society (*Physikalisch-Medizinische Gesellschaft*) of Würzburg in January 1899, Heidenhain for the first time publicly called for a "synthetic anatomy," and announced his intention to work up a "synthetic theory" of cells, organs, and organisms which would render the organism as a complex whole, a harmoniously integrated "living cosmos."[§§] Much later, in 1937, when he returned to the same topic, Heidenhain branded "Schwann's cell-theory" as "purely analytic/atomistic (*zersetzend*), lacking the potential of a synthetic science, which alone allows the comprehension of true laws of nature."[¶¶] However, Heidenhain's *Synthesiology* aimed not just at transcending a reductionist interpretation of cell theory, but even more importantly sought to overcome the disunity in biology that was reflected in the philosophical and methodological demarcation of anatomy from physiology, a distinction that had been cemented in Gegenbaur's classic text of comparative anatomy of 1859. That distinction was still very much alive at the time of Heidenhain's writing, as for example in the mind of his colleague, the German anatomist Wilhelm Lubosch (1875–1938). Explicating the science of "biological morphology" in the journal founded by Gegenbaur, Lubosch emphasized the contrast between the "architectural" versus the "technical" approach to comparative anatomy, which

---

[*] *Universitätsarchiv Tübingen*, UTA 117c/16.
[†] Mörike, 1988, p. 72.
[‡] Volkmann, 1935.
[§] Alfert, 1972, p. 224.
[¶] Jacobj, 1952/53, p. 88.
[**] Mörike, 1988, p. 73.
[††] Driesch, 1921.
[‡‡] Jacobj, 1952/53, p. 86.
[§§] Jacobj, 1952/53, p. 83.
[¶¶] Heidenhain, 1937, p. xiv.

he equated with the formal as opposed to the etiological (causal) approach.[*] In his survey of the state of the art in comparative anatomy, published in 1931 in a classic seven volume compendium on the comparative anatomy of vertebrates co-edited by Lubosch, the latter insisted that comparative anatomy does not "explain" the human body, but instead teaches us to "comprehend it meaningfully" (*sinnvoll verstehen*).[†]

Reflecting on that tradition Heidenhain admitted that morphological form can, and has been approached from two different perspectives, which are in its development on the one hand, and in its function on the other.[‡] With his *Synthesiology*,[§] Heidenhain sought an understanding of the body "in its totality, in its being as an instantiation of form (*in seiner Wesenheit als Formerscheinung*)."[¶] Opening his discourse on "a structural theory of living matter,"[**] Heidenhain referred back to Haeckel's distinction of a nested hierarchy of morphological individuals of increasing complexity. He followed Oscar Hertwig[††] when he called cellular associations and colonial organisms (*Tierstöcke*) "persons" of first, second, and third order, a terminology that again had first been used by Haeckel in his *Generelle Morphologie* of 1866, yet cautioned: "it is possible to obtain in this way a *phylogenetic* theory of the body, but not a *histophysiological* theory. In order to obtain the latter, it is necessary to generalize,"[‡‡] that is, to synthesize histology with physiology. In morphological, histological, and physiological terms, the body for Heidenhain constituted an inclusive, that is, nested or *enkaptic*[§§] hierarchy of parts of increasing complexity. The parts Heidenhain called *histomeres*, the systems the parts collectively form he called *histosystems*. Both, histomeres and histosystems, he conceived of as complex wholes, or individuals of different degrees of complexity. The fundamental property of living matter, which Heidenhain found instantiated by the histomeres, is their replication through either division, or budding and segmentation, analogous to asexual reproduction in invertebrates.[¶¶] If the divisible histomeres proliferate yet maintain physical (and hence physiological) integration, the result is "the formation of tissue colonies" (*gewebliche Stockbildungen*) called *histocormi*,[***] analogous to colonies of hydrozoans (*Polypenstock*).[†††] The fundamental histomere in organic bodies is the cell. The organs thus represent a histosystem composed of histomeres, built up by cell lineages splitting and splitting again: "The animal body can be decomposed into structural systems of lesser or higher order, which either effectively, or in their origin, are either divisible or have arisen through division of precursor systems of the same kind."[‡‡‡]

---

[*] Lubosch, 1926, p. 655.
[†] Lubosch, 1931, p. 60.
[‡] Heidenhain, 1920, p. 330.
[§] Heidenhain, 1923, p. 43.
[¶] Heidenhain, 1921, p. 3.
[**] Heidenhain, 1907, p. 84.
[††] O. Hertwig, 1906b, p. 378.
[‡‡] Heidenhain, 1907, p. 85.
[§§] Heidenhain, 1907, p. 92.
[¶¶] Heidenhain, 1923, p. 42.
[***] Sing. *histocormus*: Heidenhain (1921, p. 6; 1923, p. 42); *cormus* is again a term borrowed from Haeckel (1866, p. 285), applied to histological structures by Heidenhain.
[†††] Heidenhain, 1921, p. 66.
[‡‡‡] Heidenhain, 1907, p. 100.

Histomeres and histosystems are thus relative concepts: what is a histosystem at a lower level of inclusiveness becomes a histomere at a higher level of inclusiveness.

Dissecting the structure of a muscle fiber,[*] Heidenhain found histosystems of decreasing degrees of complexity encapsulated within one another, thus forming a hierarchical system, which he called *enkaptic*.[†] By analogy, cells divide to form organs, organs combine to form organisms, such that the organism itself constitutes an enkaptic system, a hierarchical "socialization of individuals" (*Vergesellschaftung von Einzelpersonen*) that represents a "living cosmos."[‡] It is important to understand that the developmental differentiation of this living cosmos is governed not only by the histomeres in an upward chain of causation, but also, and in addition, in a chain of downward causation by the histosystems that form complex wholes with emergent properties at all levels of inclusiveness. At one point, Heidenhain compared the "synthesis" of developing organisms to form harmoniously structured complex wholes across all domains of life to the variation of a musical theme, all renditions structured by the same rhythm.[§] In other words the histosystem that is a complex whole instantiates a *Gestalt*. Such appeal to *Gestalt*, an eminently holistic concept, he thought to transcend the old contrast between morphology and physiology. His synthesiology, Heidenhain consequently claimed,[¶] contributed to a "colossal change [in comparative anatomy] in favor of holism (*Ganzheitslehre*)."

In summary, Heidenhain's synthesiology takes living systems to form inclusive (enkaptic) hierarchies built up from divisible units (*histomeres*), the most fundamental one of which is the cell. The histomeres form lineages that split and split again as they develop into organs, organ systems, eventually whole organisms. The resulting hierarchy is not one of boxes within boxes (sets within sets), but one of lineages splitting and splitting again. It is, as Ludwig von Bertalanffy called it, a *Teilungshierarchie*[**] (division hierarchy), a dynamic, that is, processual and hence *relational* system. Organs, organ systems, indeed the organism form histosystems of increasing complexity; each histosystem at each level of inclusiveness forms a complex whole which is more than the sum of its parts, as it instantiates emergent properties. The developmental differentiation of an organism is governed not only in an upward causation from histomere to histosystem, but also in a downward causation from histosystem to histomere, such that the parts constitute the whole, while the whole integrates the parts. The comparative and functional anatomist Alfred Benninghoff from Kiel, later Marburg University, called Heidenhain's synthesiology "the most important achievement in morphology of the last few decades."[††] Adopting Heidenhain's synthesiology for his textbook *Beiträge zur Anatomie funktioneller Systeme* (1930), Benninghoff emphasized the coming together of parts in a collective function that serves the superordinated tasks of differentiating and sustaining the

---

[*] Heidenhain, 1907, p. 92; 1923, p. 22.
[†] On Heindenhain's concept of enkapsis see also Hueck (1926).
[‡] Heidenhain, 1923, p. 4.
[§] Heidenhain, 1932, p. 2.
[¶] Heidenhain, 1937, p. xv.
[**] Bertalanffy, 1932, p. 265.
[††] Benninghoff, 1938, p. 1378.

complex whole.* Other, less holistically and more analytically inclined minds saw things differently. The Tübingen botanist Walter Zimmermann (1892–1980), at the time the leading phylogenetic systematist, a Darwinian, and (at the time) a supporter of eugenic legislation,[†] considered Heidenhain's expositions on the branching pattern in plant leaves an exercise in idealistic morphology: "he unquestionably is to be counted amongst the 'idealistic' morphological researchers, who are particularly influenced by Goethe's writings."[‡]

The core concepts of Heidenhain's synthesiology nevertheless resonated well beyond the confines of comparative, developmental, and functional anatomy. It is in particular the concept of a nested hierarchy of complex wholes, subject to upward and downward causation, where parts build up the whole and the whole integrates the parts under a unifying superordinate goal, that became absorbed into contemporary thought reaching far beyond biological concerns. As early as 1929, the holistic bio-philosopher Adolf Meyer(-Abich) praised Heidenhain synthesiology as a successful integration of morphology with function in the spirit of the new times.[§] In 1935, the developmental biologist Bernhard Dürken (1881–1944), since 1921 at the Friedrich Wilhelm University of Breslau and later named director of the Institute of Developmental Biology and Heredity at that university, published an essay on the relation of the parts to the whole in the organism in a journal addressing teachers of mathematics and natural sciences in all school grades. "Wholeness, holism, uniformity, these are concepts and words which in the contemporary biological literature increasingly come to the fore," he wrote, citing the usual suspects such as Hans Driesch, Jakob v. Uexküll, Ludwig von Bertalanffy, Oscar Hertwig, and others: "A truly organicist biology must be the goal for the future."[¶] The classical cell theory, "in so far as it is a building block theory," he declared a thing of the past. Calling Heidenhain's critique "excellent," he contrasted the defunct atomistic interpretation of cell theory with the fact that "in reality, the organism is not built up from cells; instead, the cells are from the outset subordinated to the whole."[**] The following year, Dürken continued his project to contribute to the future of biology with the publication of a book on *Developmental Biology and Wholeness* (*Entwicklungsbiologie und Ganzheit*) with the subtitle: "A contribution to the renewal of the world view" (*Ein Beitrag zur Neugestaltung des Weltbildes*).[††] The book was Dürken's answer to the publisher's suggestion to write an introduction to developmental biology and its relation to the problem of wholeness for a broader audience.[‡‡] He introduced the term *Ganzheitsbiologie* as one that implies a new contemporary challenge: *Ganzheitsbiologie* "is not an empty word, but a content-heavy concept expressing nothing less than that the entire current biology is at a crossroads." It is true, he conceded, that several authors had in the past struggled with the notion of the wholeness

---

* Dabelow, 1953/54, p. 159.
[†] Junker, 2001a, p. 289.
[‡] Zimmermann, 1935, p. 26.
[§] Meyer[-Abich], 1929, p. 169.
[¶] Dürken, 1935, p. 57.
[**] Dürken, 1935, p. 63.
[††] Dürken, 1936.
[‡‡] Weber, 1937, p. 590.

of an organism, but "what is new is that this discussion has overflown the narrow banks of the professional scientific world, in order to moisten and inseminate the vast field of the entire intellectual culture and the overall spiritual attitude of our present time."* Insights of contemporary biology have rightly brought to bear on our national and political life, Dürken continued, such that concepts such as wholeness, holistic vision, and totality have come to pervade not just natural sciences, but also arts, humanities, indeed the broader public domain: "the significance of biology for the intellectual culture and for critical contemporary thought is thus clearly underscored."†

Holism, he concluded, is a foundational concept for an all-encompassing biology, putting the emphasis on the whole rather than on its parts.‡ The Tübingen botanist Ernst Lehmann noted in a review of Dürken's book he published in his journal that "few books have ever achieved such widespread attention."§ While endorsing contemporary holism and organicism and its scientific, ideological, and political implications, he still criticized Dürken for glossing *Ganzheitsbiologie* in a "philosophical-scholastic" jargon instead of presenting it in terms of a sober empirical research program. The entomologist Hermann Weber from Münster reviewed Dürken's book in *Die Naturwissenschaften*.¶ He complimented the author for his ability to expound complex issues in an understandable language, carefully charting middle grounds between the Scylla of crass materialism and the Charybdis of neovitalism. In contrast to Lehmann, Weber found Dürken's arguments to be firmly rooted in scientific causal research and, writing for a liberal science magazine, he avoided all allusion to political or ideological connotations for which he was otherwise known. Writing for the popular eugeneticist and racist magazine *Volk und Rasse* (People and Race) published by J.F. Lehmann's house in Munich, the zoologist-turned-anthropologist Gerhard Heberer—of whom more later—had less favorable things to say about Dürken: "It has to be said that in its essential parts the book is to be rejected."** To try and use holism as a vehicle to transcend the philosophically flawed mechanism–vitalism debate he declared a futile enterprise, since the Darwinian twin principles of variation and natural selection expose the primary wholeness of organisms for what it is, a *Deus ex machina*. Going from there, Heberer's review devolved into a polemic against Dürken's fundamental misunderstandings of the principles of genetics and Darwinian evolution, which in Heberer's views alone provide the scientific foundations for racial hygiene. There resonates in these contrasting reviews of Dürken's widely read treatise a difference of opinion and ideology between the holists–organicists on the one hand, and the Darwinian phylogeneticists and eugeneticists on the other, which within a year would erupt into a bitter and ideologically highly charged debate among German biologists (sketched in Chapter 7).

---

* Dürken, 1936, p. 1.
† Dürken, 1936, p. 2.
‡ Dürken, 1936, p. 17.
§ Lehmann, 1937, p. 396.
¶ Weber, 1937.
** Heberer, 1937a, p. 272.

BRIDGING FROM ANATOMY TO ECOLOGY

The animal psychologist Friedrich Alverdes (1889–1952; NSDAP, 1937) was a widely read author who invoked holism–organicism in an attempt to transcend the materialism–vitalism debate, turning to Heidenhain's synthesiology in the course of his argumentation. Like Dürken, he tied the concept of the wholeness of organisms to "what we have learnt in our fatherland through the rise of the idea of totality."[*] Alverdes was among the signatories of the "Commitment of the Professors at German Universities and Colleges to Adolf Hitler" from November 11, 1933. He was labeled an "outspoken activist,"[†] who published in *Der Biologe* an exchange on *völkisch*-political anthropology[‡] that he conducted with Ernst Krieck (1882–1947), a leading National Socialist pedagogue and ideologue who—again—understood the German *Volksgemeinschaft* as an organic totality. For Krieck, it was not just anthropology that had to be pursued from a *völkisch*-political perspective. That perspective had to dominate the entire University, offering training not only in military sciences, but also in racial biology and racial psychology (*Rassenseelenkunde*).[§] The latter issues in particular were dear to Alverdes' heart as well. When he turned to Heidenhain's synthesiology, he did so in an attempt to illustrate levels of individuality, where subordinated parts are (in-)formed through the causal influence of a superordinated whole. An organ cannot be comprehended as the sum of its parts; rather, and in accordance with Heidenhain, "the organ forms so to say a matrix into which the cells are cast" as they divide and multiply.[¶] In explicating his thought, Alverdes employed not only Driesch's entelechy, but also *Gestalt* theory. "The embryonic development of sea urchins demonstrates," as had been shown by Driesch, "how the parts subordinate themselves to the whole, how the particular events sort themselves into the totality of events."[**] Wholeness for Alverdes implied individuality, but as is exemplified by the regenerative powers of a worm, a hydrozoan, or a planaria, biological individuality cannot be anchored in indivisibility. Instead, biological individuality can only be conceptualized, Alverdes argued, as the holistic cooperation of all functionally interrelated parts.[††] Going further, and extrapolating from organs and organisms to supraindividual wholes, Alverdes employed the Uexküllian notion of "congruity" (*Einpassung* rather than adaptedness) to invoke a unity and wholeness of the organism with its environment (*Umwelt*). With this, Alverdes took a major step beyond what Heidenhain and Dürken had argued, as it takes the concept of hierarchically structured individuality beyond the organism and its development into the realm of ecology.

The term *Oecologie* had been introduced by Ernst Haeckel[‡‡] for the science that investigates the "conditions of existence" of organisms in their natural environment. In so doing, Haeckel also invoked the "economy of the whole of

---

[*] Alverdes, 1935a, p. vii; cited in Deichmann, 1996, p. 75.
[†] Klee, 1003, p. 1; see also Harten et al., 2006, p. 341.
[‡] Alverdes and Krieck, 1937.
[§] Rammstedt, 1985, p. 288f, n. 8.
[¶] Alverdes, 1932, p. 103.
[**] Alverdes, 1932, p. 91.
[††] Alverdes, 1932, p. 97.
[‡‡] Haeckel, 1866, vol. 2, p. 286.

nature (*des Natur-Ganzen*)," as also of the "economy of animal organisms"[*]: "Ecology investigates the totality of the relations of an animal to its anorganic and organic environment, in particular the friendly or adversarial relations to those animals and plants with which it comes into direct contact; or, in one word, those intricate reciprocal relations which Darwin characterized as the struggle for existence."[†] It is precisely through such a web of reciprocal functional relations that Alverdes sought to anchor individuality in nature. In 1936, Alverdes published a paper on the biological conception of wholeness in the *Zeitschrift für Rassenkunde und ihre Nachbargebiete*, a journal that was founded in 1935 by Egon Freiherr von Eickstedt (1892–1965), an anthropologist then at the University of Breslau, who although having been refused membership in the NSDAP collaborated with the National Socialist regime in racial programs (for more see below).[‡]

In this paper,[§] Alverdes welcomed the rise of *Ganzheitsbetrachtungen* in biology, a movement he applauded as an effective means to oppose the atomism, mechanism, and reductionism taught by positivist philosophy. Life is dynamic, the organism a system of functional relations, which impart wholeness on it in a way that the whole cannot be reduced to any one of its parts. In the living world, the investigation of efficient causes has always to be followed by an investigation of final causes, which alone can lead to a full understanding of the whole, as had so brilliantly been argued by Driesch and J. von Uexküll. Modern racial science applied to humans has to be commended, he wrote, for having adopted a holistic approach, which forms the basis for an understanding not only of physiological, but also of psychological distinctions between races. However, Alverdes contended, the holistic approach must not end with the investigation of a particular individual, but must reach out beyond it in an attempt to grasp the essence of supraindividual wholes. Just as the races and species of biologists form complex wholes, so are races and people (*Völker*) of humans true wholes of common destiny. Ecology, Alverdes found, had already and successfully implemented the concept of supraindividual wholes: "Thienemann in particular had emphatically made the point that a biocoenosis can only be successfully investigated from a holistic perspective."[¶]

Alverdes returned to his pet topic in a talk he delivered in the Marburg working group on holism, and which he published in 1939 in the by then radicalized ZGN. Discussing the contributions made by Driesch and von Bertalanffy, Alverdes claimed priority for the introduction of the term *Ganzheitsbetrachtung* in 1932 in the context of his animal psychology.[**] "Supraindividual wholeness is nothing mystical," he continued: "as in an individual organism, here also is the whole determined by its parts, the parts determined by the whole ... subject to integrative reciprocal

---

[*] Haeckel, 1866, vol. 2, p. 287.

[†] Haeckel, 1870b, p. 365.

[‡] Eickstedt, 1935, p. 1.

[§] Alverdes, 1936a.

[¶] "... mit grossem Nachdruck betont Thienemann in seinen Schriften, dass Biozönotik nur alss Ganzheitsbetrachtung betrieben werden kann": Alverdes (1936a, p. 7). The terms *Biocoenose* or *Lebensgemeinschaft* were first introduced by the limnologist Karl August Möbius (Möbius, 1877; 1886, p. 247, n. 1; see also Nyhart, 2009a).

[**] Mildenberger (2007, p. 177) called Alverdes the "leading holist" (*federführender Ganzheitler*).

relations."* Societies of bees, ants, and termites form such supraindividual wholes, he claimed, as also entire biocoenoses—citing Thienemann again, in particular his treatise on the significance of limnology for the culture of the present time.† Holistic thought has also taken hold in forestry and nature conservation he noted with approval, emphasizing that all biological individuality comes in the form of a nested hierarchy: "Every individual is always also part of a more inclusive whole-ness, i.e., part not only of a particular living space (*Lebensraum*) ... but also of a species, a race ...."‡ In the first volume of the ZGN already, its cofounder, the anatomist Alfred Benninghoff, had already explicitly applauded the extrapolation of Heindenhain's concept of enkapsis to supraorganismic levels of complexity in ecology. Appealing to complex wholes that include individuals of plants or animals as their parts, Benninghoff welcomed "the attempt ... to capture the whole living world in such an enkaptic system (Friederichs), an attempt which from the perspective of biological organicism is nothing but logical."§

SINGING THE PRAISES OF FORESTS AND LAKES

> Every living being, be that a plant, an animal or a human, is a whole, which is again a part of a whole of higher order. All layers of a forest serve each other as well as the whole – they are interdependent amongst themselves as each one is dependent on the whole.¶

It is this interdependence of all parts of the forest, and the animals living in it, as well as their subordination to the welfare of the complex whole of which they are parts, that maintains the biological equilibrium of the *Lebensgemeinschaft*. In an article published in the philosophical journal *Acta Biotheoretica*, addressing the issues of causality, finality, and wholeness, Alverdes listed Ludwig von Bertalanffy, Hans Böker, Bernhard Dürken, and Adolf Meyer(-Abich) as his allies in the defense of *Ganzheitsbetrachtung* in biology.** In a contribution to *Der Biologe*, Alverdes welcomed holism and organicism as new theoretical perspectives in biology. Organicism in particular, as formulated by various British and German authors including Bertalanffy and Meyer(-Abich), he characterized as the attempt to understand the lawfulness of order and organization of living entities.†† Bertalanffy later contributed himself a piece on the organismic perspective to *Der Biologe*, proclaiming himself the "father of organicism."‡‡ In this context he boasted that even *Reichsforstmeister* (Reich Master of Forestry) Hermann Göring had adapted and implemented his guidelines in forest management.§§

---

* Alverdes, 1939, p. 64.
† Thienemann, 1935b.
‡ Alverdes, 1939, p. 66.
§ Benninghoff, 1935/36, p. 158.
¶ Otten, 1943, p. 77.
** Alverdes, 1935b, p. 167.
†† Alverdes, 1936b, p. 124.
‡‡ Bertalanffy, 1941, p. 337.
§§ Bertalanffy, 1941, p. 257.

   A top ranking Nazi, head of the German Luftwaffe and as of September 9, 1936, the plenipotentiary of the Four Year Plan that sought the rearmament of Germany, Hermann Göring provides a good example of how National Socialism capitalized on *völkisch* ideology[*] that had roots reaching well into the late nineteenth century. As *Reichsforstmeister*, Göring campaigned for sustainable forestry captured by the slogan of the *Dauerwald* (eternal forest). The *Dauerwald* doctrine was based on an organicist understanding of nature, demanding a "holistic form of silviculture" in contrast to the "mechanistic tenets of 'scientific forestry' pursued in past times to maximize yields through tightly managed monocultures."[†] The forest for Göring was a complex whole, an organism, whose health had to be maintained by forestry measures such as growing a mixed community of native species, selective culling to maintain overall strength—all the while drawing an ideologically motivated analogy between the forest and the German *Volk*: "just as the eternal *Dauerwald* tolerated only healthy specimens of native, site-adapted species, the *Volksgenossen* were encouraged to take up the axes and 'excise what is of foreign race and sick' from society."[‡] On Göring's interpretation, National Socialist ideology allowed for explicit parallels between the forest and the *Volk*, where the complete subordination of the individual to the purpose and goal of the supeordinated whole is the only way to assure long-lasting existence: "Eternal forest and eternal nation are ideas that are indissolubly linked."[§]

   The *völkisch* tradition tuned to a Social Darwinist song[¶] through Nazi propaganda, one that found its way into and was derived from Paul Brohmer's school reform calling for a "New Biology,"[**] was deeply entrenched in a holistic–organicist understanding of the *Volksgemeinschaft*. The cohesion of the community, the strength of will that emerges from the home soil in which the *Volk* is rooted though bloodlines was said to have reached its zenith under the National Socialist Regime. Social classes were claimed to have been abandoned; all that remain are individuals who in turn become fully immersed in the community which itself is exclusively focused on the *Führer*—this is the magic power of the blood that guarantees the superiority of the German (Nordic) Race.[††] It is thus only natural that the strength it draws from its homeland naturally mandates the protection of this homeland, its resources both anorganic and organic, as well as its culture. The cornerstone of such propaganda was a community ideology, which would increasingly manifest itself throughout the biological literature of the time.

   August Friedrich Thienemann was born on September 7, 1882, in Gotha, son of the renowned publisher and book dealer Friedrich Thienemann.[‡‡] After graduating

---

[*] Mosse (1998). The grotesque debauchery of Göring's love for German history, customs, and landscape, tainted by romanticism and nature mysticism, is beautifully narrated by Larson (2011, pp. 278ff).
[†] Imort, 2005, p. 43.
[‡] Imort, 2005, p. 56.
[§] Göring, cited from Imort, 2005, p. 54.
[¶] Imort, 2005, p. 56.
[**] Mosse, 1966, pp. 81ff.
[††] Scharfe, 1984, p. 110.
[‡‡] Biographical data for A.F. Thienemann derive from the Archive of the Max Planck Society, Berlin-Dahlem (II. Abt., Rep. 1A, PA Thienemann, August); and Archiv, Akademie der Wissenschaften Leopoldina, August Thienemann, Matrikel Nr. MM 4061; see also Thienemann (1959).

from the *Humanistisches Gymnasium Ernestinum* in Gotha, he studied natural sciences, in particular zoology, at the Universities of Greifswald, Innsbruck, and Heidelberg. Thienemann obtained his PhD in 1905 at the University of Greifswald with the highest honors possible, *summa cum laude*, following a qualifying examination that included examination in philosophy. His thesis advisor was Wilhelm Müller, the stepbrother of the Brazilian immigrant Fritz Müller (see Chapter 2). Throughout his later career, which earned him a prominent place in the history of biology as a pioneer of limnology, Thienemann campaigned for the necessity that every student of biology should take courses in philosophy, especially in epistemology. On October 1, 1907, he started his new job as head of the section for hydrobiology at the Agricultural Institute of the University of Münster. This was the time when Thienemann initiated his research seeking the development of a typology of lakes based on studies in the pre-Alpine and lowland regions in Germany. He obtained his *Habilitation* in zoology at the University of Münster on January 21, 1910. When returning from the Great War, where on the western front he was severely wounded on September 13, 1914, the University of Münster bestowed on him the title of professor on October 5, 1915. Rejecting the offer of a full professorship at the University of Munich early in 1917, Thienemann accepted the position of the director of the Hydrobiological Institute (*Hydrobiologische Anstalt*) in Plön near Kiel, an institution of the *Kaiser-Wilhelm-Gesellschaft*, effective July 1, 1917. He remained at the helm of that institution until 1957, during which time he wielded an enormous influence as editor (since 1916) of the *Archiv für Hydrobiologie*. From 1922 through 1939 he served as president, later as honorary president, of the International Association for Theoretical and Applied Limnology, which he himself had helped bring to life. Decorated with many national and international honors, Thienemann was a very prolific author himself, leaving behind an international following of students, colleagues, and admirers. Thienemann died in Plön on April 22, 1960 (Figure 6.2).

Thienemann was again one of those German scholars and professors deeply entrenched in the Mandarin tradition. One of his collaborators characterized him as "urbane, balanced and tolerant … much influenced by Goethe, deeply rooted in western humanity"[*]: "Thienemann was careful to anchor the programmatic element of 'his limnology' within a bourgeois educational canon, which included as a matter of course a Goethe-inspired natural philosophy and a set of Christian motifs."[†] The eulogy held on the occasion of his funeral on June 4, 1960, characterized him as the son

> of an old, traditional publishing family, deeply entrenched in the culture of Europe … this literary-artistic background he preserved and kept alive. Thienemann would never have become a scientific technocrat or narrow-minded specialist, but always remained a visionary observer (*betrachtender Beobachter*).[‡]

Some applauded Thienemann's mission, which was characterized as a service in the name of the Big and the Whole (*das Grosse und Ganze*) that is ultimately anchored

---

[*] Moss et al., 1994, p. 2.
[†] Schwartz and Jax, 2011, p. 239.
[‡] Eulogy by Harald Sioli, manuscript kept in the archive of the Akademie der Wissenschaften Leopoldina, August Thienemann, Matrikel Nr. MM 4061.

FIGURE 6.2   August Thienemann. (Archiv der Max-Planck-Gesellschaft, Berlin-Dahlem.)

in Devotion to Detail (*Andacht zum Kleinen*).[*] Other more progressive authors found his writings to be characterized by belletristic tendencies, and to reveal a cloistered and romantic mindset.[†] When asked in 1931 to review the shortlisted candidates for the Zschokke succession at the University of Basel (see Chapter 3), Thienemann prefaced his evaluations with some more general comments that outlined his perspective on where zoology should be heading in the future:

> I believe that in contemporary zoology, genetics and physiology (including *Entwicklungsmechanik*) have passed their peak ... Considering the contemporary intellectual life as a whole, it seems to me that a period of analysis is now again being succeeded by a period of synthesis, and that – if all the augury is not deceptive – the predominantly synthetic branches of biology will again become the focus of scientific interest.[‡]

---

[*] Schwabe und Brundin, 1961, p. 310.
[†] Schwabe und Brundin, 1961, p. 311.
[‡] Letter from A. Thienemann to the Dean of the Philosophical Faculty of the University of Basel, dated February 3, 1931. Staatsarchiv Basel, StABS, ED-REG 1a 2 1402.

The ecologist and historian Kurt Jax* identified Thienemann† as one of the first authors to have explicitly treated an ecosystem as a complex whole, that is, an individual. Thienemann fully endorsed the holistic–organicist perspective prevalent in ecology during the 1920s and 1930s on both sides of the Atlantic,‡ one that portrayed nature as an "interconnected whole, in which every part was in tune with everything else."§ In so doing, Thienemann¶ followed the lead of his friend and colleague, the entomologist and ecologist Karl Friederichs (1878–1969) from the University of Rostock, "one of the most prominent ecologists of the Third Reich, [who] studied among other things the tree-destroying insects in occupied Poland."** Both Friederichs and Thienemann were acutely aware of the "interrelationship between politics and science,"†† and hence of the significance of ecology in the context of Nazi "blood and soil" ideology. In his 1937 textbook, Friederichs notoriously characterized ecology as "the science of blood and soil,"‡‡ a deed that earned him applause in the ZGN: Friederichs was praised for having recognized the "intimate relationship" that ecology shares "with National-Socialist ideology."§§ Thienemann would later similarly succumb to the mounting ideological pressures exerted by the Nazi regime on academics.

Extrapolating from Heidenhain and hence from anatomy/histology to ecology, both Thienemann and Friederichs conceptualized the ecosystem in terms of an enkaptic hierarchy,¶¶ that is, a nested hierarchy of complex wholes. In an important, widely cited paper published in *Die Naturwissenschaften* in 1927, Friederichs emphasized the potential for self-regulation at every level of inclusiveness of the hierarchically structured ecosystem, a potential for self-regulation that maintains the living system in a state of homeostatic equilibrium. "The true living community (*Lebensgemeinschaft*) is the populational system of a biotope,"*** a system that forms a complex whole, an *Organization* in the sense of Wilhelm Roux, that is, a "living unit of higher complexity": "but there are not only such units within nature, the *whole* of nature, the entire cosmos, forms a [living] unit,"††† an organization. While Friederichs built his argument on a broad review of the relevant contemporary literature, Thienemann looked back on the primary sources that have helped shaping German, indeed *völkisch* biology: Karl August Möbius (1825–1908), who first introduced the term biocoenosis or *Lebensgemeinschaft* in 1877‡‡‡; Friedrich Junge's (1832–1905) influential *Der Dorfteich als Lebensgemeinschaft*, published in 1885 as a text that should infuse the youth with love for—and understanding of—the homeland and its natural

---

* Jax, 1998, p. 113.
† Thienemann and Kieffer, 1916.
‡ Jax, 1998; Potthast, 1999; Schwarz and Jax, 2011; McIntosh, 2011.
§ Zeller, 2005, p. 154.
¶ Thienemann, 1935a, p. 338; see also Thienemann, 1941, p. 112.
** Stella and Kleisner, 2010, p. 42.
†† Schwarz and Jax, 2011, p. 236.
‡‡ Friederichs, 1937, p. 91; see also Deichmann, 1996, p. 160; Stella and Kleisner, 2010, p. 43.
§§ E. May, 1937/38, p. 486.
¶¶ Thienemannn, 1939, p. 278; 1941, p. 112.
*** Friederichs, 1927, p. 154.
††† Friederichs, 1927, p. 156.
‡‡‡ Möbius, 1877; see especially Möbius, 1886, p. 247, n. 1.

resources through the management of a school pond*; and of course Ernst Haeckel's musings on hierarchically structured biological individuals.

In his *Lebensgemeinschaft und Lebensraum*,† Thienemann outlined his hierarchical perception of ecology, characterizing ecological systems as a nested hierarchy of biocoenoses of first, second, and third order. As in Friederichs' views, such systems of biocoenoses maintain a homeostatic equilibrium through self-regulation. Each biocoenosis at each level of inclusiveness (a river, a forest, the system that includes both the forest and the river winding though it) can thus be seen as an organism of higher complexity,‡ that is, a complex whole or individual of higher order, the causal integration of which Thienemann curiously grounded in Georges Cuvier's Law of the Functional Correlation of Parts. Cuvier invoked this law, which ultimately can be followed all the way back to Aristotle, in opposition to any theory of gradual species transformation. Evolution cannot, according to this law, tinker with any one part of an organism without rendering the whole machinery dysfunctional. The same held in Thienemann's holistic–organicist perspective of ecology: no part of a complex ecosystem could be changed, for example, by human impact, without jeopardizing the health of the whole. The organism for Thienemann was an individual of first order, the biocoenosis or *Lebensgemeinschaft* of organisms is an individual of second order, biocoenosis and biotope together form an individual of third order—a nested hierarchy where complex, reciprocal causal relations govern not only horizontally at each level of complexity, but also in an upward as much as in a downward chain of causation. The same would, of course, hold of human *Lebensgemeinschaften*, a human *Volk* and its natural environment. Thienemann reiterated similar ideas in his *Der See als Lebensgemeinschaft*, published in 1925: "Morphology speaks of different levels of individuality," that is of individuals of first, second, and third order, he explained.§ When talking of morphological individuals, Thienemann looked back on Haeckel's *Generelle Morphologie*, and the enkaptic system Heidenhain derived from it. His discussion of ecological individuals, or organisms of first, second, and third order, Thienemann concluded with a reference to Hans Driesch: "it is possible that Driesch's notion of 'the whole' may play a role here…."¶

The increasing ideological pressure exerted on academics by the National Socialist regime first became apparent in a talk Thienemann delivered at the 37th annual meeting of the German Association for the Promotion of Teaching Mathematics and Natural Sciences (*Deutscher Verein zur Förderung des mathematischen und naturwissenschaftlichen Unterrichts*), held in Kiel from April 14 through 18, 1935. The summary report on the conference was delivered by headmaster Dr. Erich Günther from Dresden, and published in the *Unterrichtsblätter für Mathematik und Naturwissenschaft*.** The conference opened with an evening lecture that celebrated mathematics and the natural sciences as providing the tools with which to shape the unifying will of the *Volk*, and through which the *Volk* becomes a true organism.

---

* F. Junge, 1885; for more detail see also Jax (1998, p. 114, n. 4), and the in-depth analysis of Nyhart (2009a).
† Thienemann, 1918, p. 282; see also Thienemann, 1935a.
‡ Thienemann, 1918, p. 300.
§ Thienemann, 1925, p. 595.
¶ Thienemann, 1925, p. 595.
** *Unterrichtsblätter für Mathematik und Naturwissenschaft*, 41, 1935, pp. 146–153, 178–184.

But, the keynote speaker hastened to add, the racial roots of the German culture and its homesteadiness (*Bodenständigkeit*)* must always be remembered. It is for this reason that pictures of rune stones and Viking art had been chosen to adorn this year's program—a testimony to the Nordic spirit and Germanic culture that were so aptly immortalized by the great past biologists of Kiel: Karl August Möbius, and Friedrich Junge. A public lecture in the auditorium of the University on the evening of April 16, 1935, reinforced the important role played by mathematics and natural sciences in the education of the German youth in a *völkisch* culture. Biology, again, was singled out as "the principal voice in education in the New Germany," because National Socialist doctrine mandated the cultivation of a healthy *Volk*. The insights generated by biological research in the eternal laws of life would not only motivate a joyful communal spirit, but also sustain targeted measures against racial decay and racial death.[†]

In his talk, Thienemann[‡] acknowledged the fact that the slogan "blood and soil" had invited political interpretation and exploitation of ecology, and that the holistic–organicist conception of a *Lebensgemeinschaft* or biocoenosis as a hierarchically structured complex whole had enticed the drawing of analogies to the *völkisch* society, but hastened to add that in his talk he meant to offer a strictly biological perspective, not a political one.[§] While trying to keep his distance, Thienemann nonetheless welcomed the newly gained prominence of biology in the Third Reich, as it promised increased funding levels for both research and teaching. Thienemann indeed ranked fifth among German zoologists in terms of research funding between 1934 and 1945. The researchers topping him were radiobiologists and zoological geneticists.[¶]

In his highly original and influential exposition of the theoretical foundations of ecology published in 1939,[**] Thienemann cited Friedrich Ratzel[††] in support of his claim that spatial expansion, spurred by the overproduction of offspring, is the hallmark of all life: for Thienemann, *Lebensraum* and *Lebensgemeinschaft* are continuously subject to intricate reciprocal relations that create wholeness. Ratzel—of *Lebensraum* fame—had identified the relation of life to space on earth[‡‡] as the most fundamental geographical problem: "it is impossible to imagine a human being, let alone a human society, in isolation of its soil."[§§] It was, in his opinion, Charles Darwin who had most thoroughly dealt with the consequences of overproduction of offspring: geographical expansion driven by the struggle for survival. Among the other authors mentioned by Thienemann figure the usual suspects in the camp of German holist–organicist biologists and philosophers: Friederichs is cited on almost every page;

* On the historical and ideological implications of *Bodenständigkeit* see Bassin (2005, p. 206).
† *Unterrichtsblätter für Mathematik und Naturwissenschaft*, 41, 1935, p. 150.
‡ Thienemann, 1935a, p. 337.
§ Thienemann, 1935a, p. 337.
¶ Deichmannn, 1996, p. 116, Table 2.13; see also Deichmann and Müller-Hill (1994).
** Thienemann, 1939; Reprinted separately by Schweizerbart in Stuttgart, 1939, and in Thienemann (1941, pp. 101–118).
†† Ratzel, 1901, cited in Thienemann, 1939, p. 272; 1941, p. 106. On Ratzel and his theory of causal (ecological) relations tying a society to its natural world see Bassin (2005, pp. 210ff).
‡‡ Ratzel, 1897, p. 363.
§§ Ratzel, 1897, p. 366.

Adolf Meyer(-Abich)'s views on holism and Goethe are discussed; Alverdes is cited in conjunction with the philosopher Wilhelm Burkamp (1879–1939), whose concept of wholeness (*Ganzheit*)[*] Thienemann deemed superior to that of Driesch.[†] Driesch had opposed the application of the concepts of individuality and wholeness to supra-organismic entities, whereas Burkamp conceptualized wholeness more generally, that is, as the togetherness of a multitude of parts.[‡]

The last section of his 1939 *exposé* on the foundational principles of ecology Thienemann titled "*Menschliches – Allzumenschliches (Angewandte Ökologie)*," setting the stage for a discussion of the implications of the principle that "man is part of the whole of nature." To refer to "applied ecology" in that context mirrored Hans Schemm's dictum that "National Socialism is politically applied biology."[§] Thienemann[¶] opened this section with a quotation from the Tübingen paleontologist Edwin Hennig (1882–1977, NSDAP, 1937) who had blamed the detrimental separation of humanities from natural sciences in the curricula of German universities for undermining the fundamental insight that the highest intellectual achievements will only be brought about if the human spirit consciously and willfully submits itself to the eternal laws of life and nature. Rejecting possible loose analogies between *Völker* of social insects and the German *Volksgemeinschaft*, Thienemann referred to the famous *Termitenwahn* lecture delivered by Karl Leopold Escherich on November 25, 1933, when he claimed that "Nature knows only one value judgment," namely, the preservation of the whole, of the universe, the cosmos, whatever that might mean in terms of sacrifice endured by its parts.[**]

In 1941, Thienemann collected a selection of his earlier publications under the title *Leben und Umwelt* as volume 12 of the publication series *Bios*. The series, starting in 1934, had been founded by Adolf Meyer(-Abich), who in the introduction presented himself as a student of H. Driesch and J. v. Uexküll, whose work he was to continue, and whose spirit should pervade the contents of the new series.[††] Co-editors included F. Alverdes, L.v. Bertalanffy, K. Beurlen (of whom more below), H. Driesch, and J. v. Uexküll among others. Thienemann's volume had been prompted by Meyer(-Abich) and was dedicated by its author to Karl Friederichs "in friendship." It opens with a newly penned preface, in which Thienemann[‡‡] renewed his earlier "call for synthesis," and in the spirit of the Mandarin tradition encouraged the younger generation of academics to avoid narrow professional specialization, and to seek instead a holistic comprehension of the world, which he declared to be the "command of our time (*Befehl der Zeit*)."[§§] The included reprint of his *Unser Bild der lebenden Natur*[¶¶] shows that by 1940, Thienemann had abandoned his earlier reservations regarding analogies that were drawn between enkaptically structured biocoenoses and the new

---

[*] Burkamp, 1929.
[†] See also Schwartz and Jax, 2011: 236.
[‡] See also Sapper, 1938, p. 111.
[§] Lehmann, 1935c.
[¶] Thienemann, 1939, p. 282.
[**] Thienemann, 1939, p. 282; 1941, p. 117.
[††] For more detail see Mildenberger (2007, p. 176f).
[‡‡] Thienemann, 1941, p. vii.
[§§] Thienemann, 1941, p. x.
[¶¶] Thienemann, 1940.

German *Volksgemeinschaft*. Contrasting the genetic constitution of a *Volk* with the requirement of adaptation to its environment, Thienemann invoked "the twin concepts of 'blood and soil'," the two inseparably tied to one another. The implication for Thienemann was a reciprocal interdependence between the genetic constitution of an individual that determines his potential, and the living conditions of an individual that determines how far his potential can be realized.[*]

Referring back to the American ecologist Stephen A. Forbes (1844–1930), who in 1887 already had labeled the lake a "microcosm," Thienemann generalized that motive over all of nature, which he claimed forms an ordered, dynamic, homeostatic, and harmoniously structured whole embedded in cosmic relations: this, according to him, is the "essence of the new biological world view (*Weltbild*)."[†] Ecology understood in its broadest sense, that is, as a science of the all-encompassing economy of nature, Thienemann characterized as the most important one of all natural sciences,[‡] concerned as it is with the eternal laws of life (*Lebensgesetze*) that govern nature as much as the German *Volksgemeinschaft*. Biocoenoses as much as the *Volk* form enkaptic hierarchies; to sustain the stability and functionality of such complex systems requires of the included individuals that they subordinate themselves to the goals and greater good of the whole. "As part of his *Volk*, the individual will sacrifice his life if this is required for the survival of the whole ... self-sacrifice is senseless for the hermit, but a—moral—obligation for any individual embedded in a community!"[§] This is because the German *Volk* in particular had become acutely aware of the dependence of an organism on its *Lebensraum*, a truism that Thienemann found succinctly expressed in the slogan of "blood and soil."[¶] A review in the ZGN called Thienemann's collection of essays an authoritative statement "of our modern conception of nature."[**] The *Ahnenerbe* anthropologist Gerhard Heberer (1901–1973, NSDAP, 1937)[††] reviewed Thienemann's book for *Der Biologe*. He praised the author as an important leader in the field of community ecology or *Biocoenotik*, a field of research that "plays a central role in biology; its significance for human racial theory is obvious."[‡‡] When Thienemann's *Leben und Umwelt* was to be reprinted in 1947, the publisher, Johann Ambrosius Barth in Leipzig, hesitated and requested from the *Kaiser-Wilhelm-Gesellschaft* a clearance certificate for Thienemann, which was promptly delivered, dated October 16, 1947: "Thienemann was neither member of

---

[*] Thienemann, 1941, p. 12f.
[†] Thienemann, 1941, p. 4
[‡] Thienemann, 1941, p. 22.
[§] Thienemann, 1941, p. 12. Thienemann's colleague at the Plön Hydrobiological Station, Friedrich Lenz (1889–1972), who had published an essay on *Lebensraum und Lebensgemeinschaft* in 1931, edited a *Festschrift* celebrating Thienemann's 60th birthday that was published in the *Archiv für Hydrobiologie* in 1943. In his introduction, Lenz commemorated two students of Thienemann that had lost their lives on the battlefield: "They suffered a heroic death in the fight for the freedom and right to live (*Lebensrecht*) of our Volk" (Lenz, 1943, p. viii).
[¶] Thienemann, 1941, p. 12.
[**] Vareschi, 1942, p. 261.
[††] Bäumer, 1990a,b; Deichmann, 1996; Hoßfeld, 1997; Klee, 2003, p. 234.
[‡‡] Heberer, 1942, p. 168.

the NSDAP, nor of any of its affiliated organizations."* Which only goes to show how close the affinities were between the *völkisch* heritage and Nazi ideology.

It is interesting to follow the gradual escalation of Thienemann's rhetoric, as the same is apparent in the writings of one of his younger *protégés*, Adolf Remane, considered one of the most influential German zoologists of the twentieth century.[†] When asked to submit names of additional potential candidates for the Zschokke succession at the University of Basel in 1931 (see Chapter 3), Thienemann recommended Remane, characterizing him as an "extremely versatile zoologist" in a letter dated January 2, 1931.[‡] When Ulrich Gebhardt nominated Remane for election into the Academy of Natural Sciences Leopoldina in Halle in December 1935, he again characterized him as the "most broadly interested" among the "younger zoologists."[§] Although much younger than Thienemann, Remane would come to incorporate the same Mandarin philosophy that also characterized his mentor.

Robert Gustav Adolf Remane[¶] was born on August 10, 1898, in Krotoschin in the Prussian province of Posen, since 1918 part of southern Poland. He was the fifth of six children; two of his siblings died at an early age; his brother fell in France in 1917. Under the tutelage of his grandfather, Remane developed an early love for botany, but also became an avid collector of butterflies. While at grammar school, he came across—through pure chance, he said—Gegenbaur's 1898 textbook on the *Comparative Anatomy of Vertebrates, Including Reference to Invertebrates*: "it made the biggest impression on me, and became my favorite book." Going through the school system of higher education in Krotoschin, he graduated early in 1916 (*Notreifeprüfung*) in order to be drafted into the army. He served in France, until he was wounded by shrapnel on April 28, 1918.

With his home country now occupied by Poland, Remane enrolled to study biology, geography, and chemistry at the Humboldt University in Berlin. In December 1918, Remane was actively involved in the suppression of the accelerating Spartacist uprising,[**] a movement carried by the political left and the proletariat, which implicitly places Remane on the right side of the political spectrum: "it is … no coincidence that many young academics participated in … the suppression of the left labor movement."[††] As for his doctoral thesis, Remane sought the agreement from the zoologist and comparative anatomist Willy Kükenthal (1861–1922), director of the Humboldt Museum in Berlin, to be able to pursue a topic of his own choice. He was promoted *summa cum laude* on March 15, 1921, based on his thesis on the morphology of the anthropoid dentition. Kükenthal promised him the next opening among the research assistant positions at the Humboldt Museum in Berlin, but with Kükenthal's death in 1922

---

* Archive of the Max Planck Society, Berlin-Dahlem (II. Abt., Rep. 1A, PA Thienemann, August).
† Junker, 2004, p. 217.
‡ State Archive, Basel, StABS ED-REG 1a 2 1402. Remane was placed *tertio loco*, tied with Portmann, but was not considered for employment because an equally qualified Swiss candidate was available.
§ Archiv, Akademie der Wissenschaften Leopoldina, Adolf Remane, Matrikel Nr. MM 4297.
¶ Archiv, Akademie der Wissenschaften Leopoldina, Adolf Remane, Matrikel Nr. MM 4297. See also Weigmann (1973); Heydemann (1977); Siewing (1977); Junker (2000, 2004); Zachos and Hoβfeld (2001, 2006); Remane (2003); Rieppel (2013a).
** Zachos and Hoβfeld, 2001, p. 314. On the Spartacist uprising see also Henig (1998), Kühnl (1984, p. 95), and Mulligan (2009, p. 81).
†† Kühnl, 1984, p. 97.

also died Remane's dream to obtain such distinguished and desired employment. He therefore stayed on the payroll of the Prussian Academy of Sciences until he was offered an assistantship at the Zoological Institute of the University of Kiel in 1923. He obtained his *Habilitation* in zoology at the University of Kiel on October 25, 1925, and in the spring of 1926 he worked for seven weeks at the German Zoological Station in Naples. He was promoted to associate professor at Kiel university in 1929, and named full professor of zoology at the University of Halle in 1934. In 1936, Remane returned to the University of Kiel as director of the Zoological Institute, in which function he was also charged to develop a marine biological station that opened in 1937 in Kitzeberg near Kiel (Figure 6.3). When asked to outline his research interests in the context of his election to membership in the Academy Leopoldina, he wrote:

1. Clarification of the theoretical foundations of morphology, phylogenetics, and systematics in general (species and race concepts).
2. Expansion of biocoenosis research in the sense of Möbius, which provides a necessary link between the ecological investigation of the particular organism, and the investigation of larger living systems (*Lebenseinheiten*) as pursued by limnology for example.[*]

FIGURE 6.3 Adolf Remane. (Leopoldina-Archiv (Halle/Saale), BM, Bnd. 10a.)

---

[*] Remane, in his Vita; Archiv, Akademie der Wissenschaften Leopoldina, Adolf Remane, Matrikel Nr. MM 4297.

Following Hitler's *Machtergreifung* on January 30, 1933, Remane later in that same year joined the National Socialist Teachers' League (*Nationalsozialistischer Lehrerbund*, NSLB), and the SA. His admission as a member to the NSDAP followed in 1937, after the ban on new memberships had been relaxed.[*] In that same year, Erich Wasmund (1902–1945), adjunct (*ausserplanmässig*) professor of hydrogeology at the University of Kiel, denounced Remane for having allegedly criticized Göring's Four Year Plan that sought the economic independence and rearmament of Germany. The "Four Year Plan," Remane was accused of having said, "ruins natural sciences."[†] Issues very much at stake at that time in German science were academic freedom in general, and the value of basic research in particular.[‡] During the Nazi regime, freedom in general, as also academic freedom, meant the willful submission to the ideological values carried by the party line. In that system, a scientist was not to be a "theoretical person," but a "political person" pursuing science in the *völkisch* spirit committed to an organicist–holistic *Weltanschauung* in opposition to a strictly empiricist–positivistic (i.e., Western) perspective.[§] Director of the Geological Institute, and hence Wasmund's supervisor, was the paleontologist Karl Beurlen, an outspoken supporter of the science policies implemented by the National Socialist regime (more on him below). The fact that Remane retained his position at the University of Kiel indicates that he managed to successfully clear his name from these accusations.[¶] In 1942,[**] the German Zoological Society (*Deutsche Zoologische Gesellschaft*) sent a letter—co-signed by Remane—to the Reich Chancellery in support of Karl von Frisch (1886–1982), the famous interpreter of the bee dance language, who at the University of Munich faced early retirement due to his Jewish ancestry. To make this sensible issue palatable to party officials, the authors chose to introduce their plea with an expression of their general and broad support for the anti-Semitic policies that were introduced at German universities. After the collapse of Nazi Germany, Remane was arrested by British occupation forces in September 1945, and detained in the Civil Internment Camp No.1 in Neumünster-Gadeland for 9 months. In 1947, Remane returned to his academic position at the University of Kiel, but in 1948 he was classified by the denazification tribunal as a "fellow traveller" (*Mitläufer*), a categorization that carried a ban from teaching for 2 years.[††]

In the years to come, Remane continued to serve as the Director of the Zoological Institute, and of the Anthropological Museum of the University of Kiel, pursuing a highly successful career as a researcher until his retirement in 1966. He earned a permanent place in the pantheon of the history of biology through his discovery of the interstitial fauna of sandy beaches. But the primary motivation for his later scientific career he himself traced back to his reading of Gegenbaur's textbook while

---

[*] Zachos and Hoβfeld, 2001, p. 318

[†] Zachos and Hoβfeld, 2001, p. 317.

[‡] Science officials of the Nazi regime considered the declaration of the Four Year Plan in September 1936 a watershed event that pushed basic research on to the backburner: Nagel (2012, p. 287).

[§] Hopster, 1985, p. 116f. On the issue of academic freedom in German science after the proclamation of the Four Year Plan see also Mosse (1966); Pine (2007, 2010); Rieppel (2011f).

[¶] J. Remane, 2003.

[**] Deichmann (1996, p. 46) dates the letter to May 11, 1942; Klee (2003, p. 491) dates the letter to July 7, 1942.

[††] Wolf Herre, quoted in Hoβfeld, 1999a, p. 253; Zachos and Hoβfeld, 2001, p. 318.

still in grammar school.* His own *magnum opus*, a treatise on the principles of comparative anatomy, systematics, and phylogeny reconstruction[†] can in some way be thought of as a continuation of Gegenbaur's program in systematic, or phylogenetic morphology, revealing further influences from Sinai Tschulok and Adolf Naef. His son, Jürgen Remane[‡] (1934–2004), would later—in 1986—attempt to motivate the leading evolutionary biologist Ernst Mayr from Harvard University to arrange for an English translation of his father's book, as he thought it would provide a road block for the rapidly spreading cladistic revolution that had been triggered by the translation of Willi Hennig's work into English. Mayr declined, as he located Adolf Remane's book in the tradition of—in his view—defunct idealistic morphology.[§] But Remane's research reached well beyond the analysis of homology as a clue to common ancestry, as he branched out into functional anatomy, ecology, and biogeography, using exemplars that reached from tiny invertebrates to primates. Remane's zeal in natural sciences was imbued with his "pantheistic religious spirit in the sense of Goethe,"[¶] and balanced by his interest in ancient philosophy and art. Remane's understanding of natural science is reflected in his saying that "most ancient people understood laws of nature not as an antithesis to Divine power, but as one of its many manifestations."[**] He was deeply invested in nature conservation,[††] and sought to inspire an interest for nature and its manifold yet ordered and lawful manifestations among the urban students that took his courses and attended his lectures. Remane passed away in Plön near Kiel on December 22, 1976.

The reason for Remane's detention by the British occupation forces after the collapse of Nazi Germany was his work and publications on animal race, that is, subspecies formation. Race formation as the initial step toward speciation, that is, the formation of a new species, was a hot topic in pre-World War II German biology, both in botany and in zoology. Among Remane's colleagues pursuing similar lines of research was Wolf Herre (1909–1997; NSLB, 1934; SA, 1935; NSDAP, 1937),[‡‡] first assistant (1933), then docent (1936), and finally adjunct (*ausserplanmässiger*) professor for zoology and comparative anatomy at the University of Halle. After World War II, Herre joined the University of Kiel, where he established himself as the leading expert among German zoologists on domestication in mammals. In an early, more general and synthetic paper *On race and speciation* (*Über Rasse und Artbildung*), which for the time of its appearance in 1936 is remarkably free of any ideological taint, Herre explained: "Race and species are relative concepts. They are systematic concepts. They must therefore be applied with the same essentially critical attitude as all other systematic concepts."[§§] Evolution blurs any discrete boundaries

---

* Zachos and Hoβfeld, 2001, p. 314; on Gegebaur and the various editions of his textbooks see Rieppel (2011b).
† Remane, 1952.
‡ Adatte and Stössel-Sittig, 2004.
§ Harvard University Archives, Ernst Mayr Correspondence, HUG(FP) 74.7, Box 34, Folder 1413. For more details, see Rieppel (2013a).
¶ Heydemann, 1977, p. 89; see also Zachos and Hoβfeld, 2006, p. 344.
** Heydemann, 1977, p. 89.
†† Remane, 1935.
‡‡ Klee, 2003, p. 247.
§§ Herre, 1936, p. 217.

between races and species, which cannot even be clearly demarcated on the basis of the "geographical principle" (where geographical isolation of populations is recognized as a prerequisite for speciation). When interviewed by the historian of biology Uwe Hoβfeld on March 3, 1997, in Kiel, Herre explained that Remane had admitted in an interview conducted by the British occupation authorities that he had published on issues of race formation. In Herre's recollection, Remane had also given presentations on the struggle for existence (*Kampf ums Dasein*) in the SA: "After the war, it was our concern to get Remane out [of the detention camp] as quickly as possible."[*] According to Remane's son Jürgen, his father "was accused of publishing on racial politics, whereby a presentation on *Species and Race* from the year 1927 played a certain role. It became apparent, however, that in this talk he [Adolf Remane] had dealt with theoretical issues of zoological systematic."[†] The crux of the matter was, however, that at this time, zoological studies on race and species formation informed anthropological studies aiming in the same direction. A good example demonstrating such a link was provided by the leading (eu-)geneticist Fritz Lenz (1887–1976; NSDAP, 1937), who in 1934 published a paper *On Races and Race Formation*. Capitalizing on recent studies in zoology, Lenz drew from the geographic polymorphism of species the insight that natural selection is the most decisive factor in race formation: "In that respect Darwin's theory is unassailably correct."[‡] Extrapolating to human society, Lenz emphasized, "we must by all means avoid to drive a new wedge into our Volk. All mingling with alien racial elements would certainly be detrimental." But, Lenz went on to explain, interracial procreation is not the only source of racial degeneration. What is required in addition is a healthy and vital communal spiritual life liberated from all detrimental elements. Such cultural selection, Lenz continued, is some sort of environmental selection as well, and selection is, in fact, the only mechanism driving race formation: "and this, ultimately, is the task of Racial Hygiene."[§]

Eugen Fischer (1874–1967),[¶] a leading racial hygienist and cofounder (in 1927), later director of the *Kaiser-Wilhelm-Institute of Anthropology, Human Genetics and Eugenics* in Berlin, praised himself for having been the first to have enriched the previously largely descriptive anthropology with genetics, and thus having founded the discipline of *Anthrobiology*. Its main focus he declared to be "human genetics, the biology of human races, and the effects of bastardization."[**] As was noted by the student of Nazi racism Christopher M. Hutton, many practitioners of *Anthropobiology* who had medical or zoological training believed that humans could be segregated into discrete races, whose mental and physical characteristics were shaped by environmental factors such as climate and landscape.[††] The year 1935 saw the founding of

---

[*] Herre, cited in Zachos and Hoβfeld, 2001, p. 318, n. 21; see also Hoβfeld, 1999a, p. 253.

[†] J. Remane, 2003.

[‡] Lenz, 1934, p. 181. Lenz singled out the recent monograph by Bernhard Rensch (1933) for discussion.

[§] Lenz, 1934, p. 188f.

[¶] Hutton, 2005, p. 219f; see also Hutton (2005, pp. 64ff), on Fischer's work on the *Rehobother Bastards*, from Southwest Africa, a former German colony.

[**] Weingart et al., 1992, p. 355f.

[††] Hutton, 2005, p. 21. As will become clear in discussions to come, the tension between such racial beliefs and a Weismannian conception of hereditary transmission would spark the contemporary debate between neo-Lamarckism and a socially motivated (or distorted) Darwinism.

the journal *Zeitschrift für Rassenkunde und ihre Nachbargebiete* by Egon Freiherr von Eickstedt, a one-time scientific assistant of Eugen Fischer, then professor of anthropology at the University of Breslau. Eickstedt was able to put together a truly international editorial board for his new journal, one that included such diverse personalities as the Swiss Fritz Sarasin, director of the Anthropological Museum in Basel, and Otto Freiherr von Verschuer (1896–1969), director of the Institute for Hereditary Biology and Eugenics in Frankfurt a.M., and as of 1942 successor of Eugen Fischer in Berlin, who supervised Josef Mengele's (1911–1979) doctoral thesis, and who allegedly received "anthropological material" from Auschwitz.[*] The goal of his journal, Eickstedt proclaimed, was to establish an outlet for a broad interdisciplinary approach to racial studies that would put anthropology in contact with its "sister sciences, which include biology, psychology and medicine amongst others."[†] Such reciprocal interaction between zoology and anthropological racial theory is well illustrated in a review, published in the *Zeitschrift für Rassenkunde*, by the zoologist-turned-anthropologist Gerhard Heberer of a paper[‡] that laid out the principles of systematic zoology for younger adepts of the discipline in the *Zoologischer Anzeiger*: "Reading this portrayal of the workings of the modern zoological systematist will be beneficial to the racial theorist as well."[§]

When read against this contemporary background, Remane's writings take on a deeper dimension than a mere technical contribution to speciation theory, one that may explain his internment by the British authorities and subsequent classification as a fellow traveler. The paper which—according to his son Jürgen—got him primarily into trouble was the published version of a talk on *Species and Race* Remane had delivered in front of the Society for Physical Anthropology in Kiel on April 24, 1927.[¶] Introducing his topic, Remane touched on contemporary zoological research on speciation, mainstream but also eclectic, such as the *Formenkreis* theory of the "leading ornithologist, protestant pastor, and sometime Nazi sympathizer"[**] Otto Kleinschmidt (1870–1954). The latter believed each species to have been specially created, and hence to be essentially different from all other species. Yet he recognized at the same time geographical variation within a species, the variable populations of the same species thus constituting a *Formenkreis*.[††] To distinguish the relationships prevailing within a sexually reproducing species that forms a *Formenkreis* from the phylogenetic relationships that prevail between separate species, Remane drew on Oscar Hertwig[‡‡] and Adolf Naef,[§§] who had contrasted the reticulating reproductive relationships within species with the dichotomous relations between species of common origin. On the basis of that distinction Remane concluded that all humans belong to the same species, rendering the concept of human races a biological–ecological rather

[*] Hutton, 2005, p. 222f and elsewhere; see also Weingart et al., 1992; Grüttner, 2004.
[†] Eickstedt, 1935, p. 1.
[‡] Michaelsen, 1935.
[§] Heberer, 1935, p. 321.
[¶] Remane, 1927.
[**] D. Williams (2006, p. 4); this paper analyzes possible influences of Kleinschmidt's *Formenkreis* theory on later theories of speciation and biogeography.
[††] Kleinschmidt, 1926a; see also Kleinschmidt, 1900; 1926b.
[‡‡] O. Hertwig, 1917, 1918a, p. 236.
[§§] Naef, 1921, p. 31.

than a typological–morphological one. His general conclusion was that "Varieties and races do not form [separate, i.e., isolated] genetic entities"[*] as the *bona species* of Haeckel would have to. At this early stage of his career, Remane in his talk touched on all the relevant issues of contemporary research into speciation[†] in an ideologically neutral tone.

His next significant contribution on the problem area of race and species was the published version of a talk he presented in front of the entomological society of Halle: *Population Research, a Necessary Research Program for Systematics* (published in 1937). He elaborated on the difficulties of demarcating Haeckel's *bona species* from geographical or ecological races, a conundrum that Haeckel had already pondered from the perspective of an evolutionary continuum. Turning to the issue of human races in the last paragraph of his paper, Remane judged their status as real entities of nature as still unresolved in spite of intensive research, their demarcation being blurred by extensive and far-reaching migrations of humans.

Things became more serious when Remane was invited to present on the same topic at the first annual meeting (*Reichstagung*) of the scientific academies of the NSDDB, the National Socialist German University Lecturers' Association (*National-Sozialistischer Deutscher Dozentenbund*), held in Munich from June 8 through 10, 1939.[‡] The proceedings of this meeting were published under the auspices of *Reichsdozentenführer* Walter Schultze (1894–1979) through J.F. Lehmann's publishing house in Munich. In his capacity of *Reichsdozentenführer*, Schultze, a Nazi of the first hour and devoted supporter of Hitler's ascent to power, was responsible for the political supervision and indoctrination of university docents, that is, the *Gleichschaltung* of German universities. Charting a remarkably neutral ideological course through much of his paper, Remane mainly reiterated the contents of an earlier contribution of his that had linked the issues of race formation and speciation, as well as questions of morphological evolution at a supra-specific level, to current theories of mutation and inheritance.[§] In that earlier paper, Remane had referred back to Tschulok's distinction (see Chapter 3) of the *Grundsatzfrage* (did evolution occur or not?) from the two subsidiary questions, the *Stammbaumfrage* (what does the tree of life look like?) and the *Faktorenfrage* (what are the causal agents that drive evolution?), relegating systematics and phylogeny to the first, mutation theory to the second of these subsidiary questions. In his talk at the NSDDB meeting in Munich, Remane linked Tschulok's *Faktorenfrage* to the old metaphor of *nature versus nurture*, quickly dismissing all neo-Lamarckian argumentation schemes, as these would have been politically incriminating at this time (a topic of which more later). Instead, he praised the virtues of the Darwinian slogan of the "struggle for survival" (*Kampf ums Dasein*),[¶] which in Nazi ideology had acquired strong sociopolitical connotations. Recognizing the limited explanatory power of current mutation theory with respect to morphological evolution above the species level, Remane nonetheless concluded his presentation with a summation of the merits of modern

---

[*] Remane, 1927, p. 7.
[†] See also Zachos and Hoßfeld, 2001, p. 336: Glaubrecht, 2007.
[‡] *Erste Reichstagung der Wissenschaftlichen Akademien des NSD—Dozentenbundes.*
[§] Remane, 1939a.
[¶] Remane, 1940, p. 118.

theories of inheritance: the fact that current mutation theory cold not satisfactorily explain macroevolutionary change should not be held against the science of genetics in general. Instead, genetics should be celebrated for its contribution in the breeding of superior crops and livestock, and for providing guidance in the struggle against the degeneration of the human race, as well as offering means to preserve our species: "The question whether such means are to be put into practice has for decades no longer been a question in need of scientific warrant, but is only a question of responsible political leadership."*

As the war dragged on and ideological pressure increased, Remane was "happy to oblige"† when he was invited, in 1941, to contribute a paper on the theory of descent to the *Archiv für Rassen- und Gesellschaftsbiologie, einschliesslich Rassen- und Gesellschaftshygiene* (*Archive for Race and Social Biology*), again published by J.F. Lehmann in Munich. The first volume of what was to become the leading German outlet for research on racial hygiene was published in 1904, founded— among others—by Alfred Ploetz (1860–1940), who first introduced the German term *Rassenhygiene* and in 1905 was a founding member of the eponymous society in Berlin, and the zoologist, former Haeckel student and outspoken anti-Semite Ludwig Plate (1862–1937). On October 19, 1903, Ploetz communicated to Ernst Haeckel "that the *Archiv* was founded as a campaigning force on the side of Darwinism and the modern *Weltanschauung*."‡ As of 1908, the *Archiv* was published with the subtitle *A Journal for the Theory of Descent*, a rider that was eventually dropped as the journal became, in 1923, the *Scientific Organon of the German Society for Racial Hygiene*.§ In 1941, when Remane's contribution was solicited, the journal proclaimed to be dedicated to human racial biology, seeking to reach out to all those readers "who share an interest in the fate of our *Volk*," in particular personalities in intellectual leadership positions. Among the editors at the time was the zoologist *cum* anthropologist Gerhard Heberer, and Fritz Lenz. Lenz in particular played an important role in the German science of human heredity and racial hygiene, and helped draft the NS-Euthanasia legislation.¶ Again reflecting back on Hans Schemm's famous slogan, Lenz wrote in a letter to Walter Gross (1904–1945), head of the Race Policy Office (*Leiter des Rassenpolitischen Amtes der NSDAP*) dated April 26, 1937: "Racial hygiene is applied racial theory, that is, applied human genetics."** Although listed as junior author only, Lenz—a former student of Eugen Fischer—claimed to have been the major contributor to the standard textbook on human heredity and racial hygiene of the time,†† the famous *Baur-Fischer-Lenz*‡‡ that went through several revisions and

---

* Remane, 1940, p. 126; see also Remane, 1942, p. 81; Zachos and Hoβfeld, 2001, p. 324.
† Remane, 1941, p. 89.
‡ Translated by and cited from Weindling (1989b, p. 128).
§ Weingart et al., 1992, pp. 199ff; see also Weingart, 1985, p. 326.
¶ Klee, 2003, p. 367.
** A copy of the letter is kept at the University Archive Göttingen (GWDG), Rektoratsmappe Fritz Lenz, Prof. Dr. med.
†† Lenz's *Vita* dated October 13, 1947, written in the third person: "Im Jahre 1921 erschien die erste Auflage der 'Menschlichen Erblehre' von Baur-Fischer-Lenz, von der Lenz den grössten Teil geschrieben hat." University Archive Göttingen (GWDG), Rektoratsmappe Fritz Lenz, Prof. Dr. med.
‡‡ *Grundriss der menschlichen Erblichkeitslehre und Rassenhygiene*, by Erwin Baur, Eugen Fischer, and Fritz Lenz, first edition 1921, fourth edition 1936.

editions in the years to come. After the war and the collapse of the Third Reich, both Lenz and Heberer found controversial employment at the University of Göttingen. In his CV submitted to the University of Göttingen and dated October 13, 1947, Lenz claimed that through his co-authorship in the *Baur-Fischer-Lenz*, and through his role as editor of the *Archiv für Rassen- und Gesellschaftsbiologie*, he had established himself as the most prominent German representative of Sir Francis Galton's (1822–1911) science of Eugenics.* Confronted with denazification efforts, Lenz emphasized in a declaration dated April 21, 1947, that he had highlighted the creative cultural achievements of the Jews in the fourth edition (1936) of the *Baur-Fischer-Lenz*, which not only earned him a reprimand from the *Führer's* chancellery, but also led him to refrain from future editions of "his" book which he did not want to bring into line (*gleichschalten*) with Nazi ideology.† Testifying on Lenz's behalf, one Dr. K.V. Müller from the Ministry of Culture of Lower Saxony, Hannover, confirmed that Lenz did not want to follow up the 1932 [*sic*] edition of his book, as this would have meant to abandon the liberal spirit that still characterized the fourth edition.‡ With respect to the *Archiv*, Lenz specified in explanations he appended to the questionnaire issued by the Military Government of Germany in 1946 (completed by Lenz in July 1947) that he had actively edited the journal together with Alfred Ploetz from 1919 through 1933. During that time, he had always tried to solicit international contributions, including papers from Jews (as well as from catholic priests, as he emphasized elsewhere in 1945§): "Although I was still listed as co-editor on the title page since 1933, I exerted no longer any influence on the ideological orientation of the journal."¶ The denazification committee acquitted Lenz with a notification dated July 15, 1949, classifying him as exonerated (*entlastet*, category V), next to which Lenz scribbled his own self-assessment: "without any qualification."** Interestingly, Lenz could establish credibility for his claim that he never promoted war, as he found—like Haeckel before him—the battlefield to exert negative eugenic effects, killing the best among the youth: "I have always considered war to be the worst enemy of the best genes in our *Volk*."†† Nonetheless, the perceived need to whitewash his own involvement with the *Archiv* after 1933, a journal in which Remane was to publish in 1941, highlights the dubious reputation the journal acquired in a postwar perspective.

---

* Lenz's *Vita* dated October 13, 1947. University Archive Göttingen (GWDG), Rektoratsmappe Fritz Lenz, Prof. Dr. med.

† Declaration to the attention of the committee for the political screening of the faculty of the University Göttingen, dated April 21, 1947. University Archive Göttingen (GWDG), Rektoratsmappe Fritz Lenz, Prof. Dr. med.

‡ Certified copy of testimony by Dr. K.V. Müller, September 23, 1946. University Archive Göttingen (GWDG), Rektoratsmappe Fritz Lenz, Prof. Dr. med.

§ Letter from Lenz to the Education Control Officer of the University of Göttingen, dated August 31, 1945. University Archive Göttingen (GWDG), Kuratoriumakte Lenz.

¶ Appendices to the questionnaire issued by the Military Government, signed by Lenz on July 22, 1947. University Archive Göttingen (GWDG), Kuratoriumakte Lenz.

** Compare Lenz's handwritten entry on his notification of residency in the city of Göttingen, with the official de-nazification notification dated July 14/15, 1949.

†† Declaration to the attention of the committee for the political screening of the faculty of the University Göttingen, dated April 21, 1947. University Archive Göttingen (GWDG), Rektoratsmappe Fritz Lenz, Prof. Dr. med.

It is likely the fellow zoologist and Darwinist Heberer, not Lenz, however, who invited Remane to contribute to the *Archiv*. In his piece, Remane glorified the "struggle for existence" once again, praising the theory of descent for having displaced an anthropocentric worldview in favor of a new "biological *Weltbild* that accepts for all organisms, including humans, the fact of being subject to all-encompassing biological laws."* In his estimation, Darwinian evolutionary theory delivered the tools with which mankind could willfully plan and shape its own biological future, as had likewise been claimed by the future Nobel laureate Konrad Lorenz (1903–1989, NSDAP, 1938) in a recently published paper in *Der Biologe*. In this notorious pamphlet,† Lorenz had identified the opposition of the Catholic Church as the main obstacle to a widespread acceptance of Darwinian evolutionary theory, which in his view provided the only scientific foundation for a planned "purification [*Aufartung*] and improvement of *Volk* and race."‡ "The recognition of our dependence on nature renders us biologists more modest in view of our past, more ambitions with respect to our future," was a sentence that Remane singled out for *verbatim* citation.§ To neglect such rational scientific insight would result, according to Lorenz, in a degeneration of the *Volk*, indeed of humankind, due to insufficient selection pressure. Remane concurred: "relaxation of selection pressure necessarily results in degeneration caused by the fixation of deleterious hereditary factors and mutations ...."¶ Further on he amplified that the theory of descent had become the major driving force for the development of genetics and for the justification of racial hygiene, two fields of research that Remane recognized as indispensable components of the contemporary biological world view."** Recognizing that such staunch materialistic–reductionist (Social) Darwinism, one also forcefully promoted by Heberer and his clique (of whom more later), would clash with the enthusiasm for a *Ganzheitsbiologie* professed by other biologists of the time, Remane—the humanist that he was—tried to build bridges: The *Ganzheitsbiologie*, he proclaimed, seeks at its heart nothing else than to gain an understanding of organic function and organization, governed by the principles of self-preservation, preservation of the species, and preservation of the state.††

Remane's accommodation to the ideological matrix of the time reached its high point in a 1939 paper titled *The Society as a Form of Life in Nature* (*Die Gemeinschaft als Lebensform in der Natur*). The choice of this title revealed Remane's reverence for his mentor August Thienemann, as the latter had chosen the subtitle *Society Is the Form of Life in Nature* (*Gemeinschaft ist die Lebensform der Natur*) for his influential 1939 paper on the foundational principles of ecology. And again like Thienemann, Remane opened his contribution with reference to Escherich's 1933 speech on termite mania (*Termitenwahn*) in support of his plea for a unification of the National Community (*Volksgemeinschaft*) as the latter prepared

---

* Remane, 1941, p. 90; see also Remane, 1942, p. 73.
† On Konrad Lorenz and his relation to National Socialism see Bäumer (1989, 1990b); Deichmann (1996); Burkhardt (2005).
‡ Lorenz, 1940, p. 25.
§ Lorenz, 1940, p. 28; Remane, 1941, p. 91.
¶ Remane, 1941, p. 119.
** Remane, 1942, p. 80.
†† Remane, 1941, p. 92.

to engage in a confrontation with the perils of dangerous times. Remane once again highlighted biology as the only proper scientific foil against which to analyze and identify the natural social dispositions of humans, an endeavor which alone would provide the insights necessary for a proper understanding and structuring of the *Volksgemeinschaft*: "100 years ago biology was already able to show that all that we identify as individuals, i.e., as single organisms in nature, including humans, are in fact communal wholes (*Gemeinschaftsbildungen*)."[*] Reflecting the prevailing bio-sociopolitical spirit of the time, Remane portrayed the family as a complex whole, a nucleus that is nested within a complex whole of greater inclusiveness, the National Community.[†] The survival of the *Volk* in the struggle for existence must be based, according to Remane, on the same motivations as those that drive the defense of the family: a common unified will that overcomes individualism and egotism and so prepares for self-sacrifice should such become necessary. But given the evolutionary emergence of humans from within primates through a prolonged process of a struggle for survival, the disposition for self-sacrifice cannot be part of the human genetic makeup.[‡] This is where the education to a political person has to complement nature along the lines that Escherich had sketched in his *Termitenwahn*.

"National Socialism is so as to say the biological will of the German *Volk*," is one famous line from the Munich professor of zoology, Karl Escherich.[§] The Tübingen plant geneticist Ernst Lehmann fostered the belief in the scientific mission of biology as the foundation of the collective will: "without biological will there cannot be a *völkisch* state."[¶] The love for the homeland (*Heimatliebe*) was to be cultivated: "work in the school garden becomes ever more important."[**] It made possible the experience of the laws of life (i.e., the laws of inheritance), the blessings of medicinal herbs, as well as the virtues of weeding out the weaklings and racially unfit. Evidently, the boundary between science and ideology became highly porous under the pressures of the Nazi regime, not just for such hot-headed southern German ideological zealots like Lehmann, but also in much cooler heads from the north, such as Thienemann and Remane. Small wonder that the constant refrain reeled off by German scientists after the war was that they had sought the inner emigration through a retreat into the apolitical realm of pure science, unimpaired by moral, ethical, or social connotations. And yet "Publications cannot be judged independently from the political context in which they were embedded, one that was plainly evident to the reader. Every author was familiar with those political connotations, and knew what he did."[††] Remane ranked sixth among German zoologists in terms of research funding between 1934 and 1945, right behind Thienemann, although party membership did not in general significantly influence the level of research funding in biology, at least not in genetics.[‡‡]

---

[*] Remane, 1939b, p. 44.
[†] Remane, 1939b, p. 45.
[‡] Remane, 1939b, p. 55.
[§] Kirchner, 1984, p. 85.
[¶] Lengerken, 1934.
[**] Lehmann, E., *Memorandum of the German Association of Biologists*, December 3, 1936. *Universitätsarchiv Tübingen*, UTA 126/373, 5.
[††] Kleβmann, 1985, p. 366.
[‡‡] Deichmannn, 1996, p. 116, Table 2.13, and p. 120; see also Deichmann and Müller-Hill, 1994, p. 167f.

BRINGING FOSSILS TO LIFE

The benefits to be reaped from the newly found prominence of biology in the Third Reich not only tempted biologists, but also stirred the concupiscence of those scientists who worked with the fossilized remains of organisms long extinct, dating back to past eons of earth history. One example among many was the Tübingen paleobiologist Edwin Hennig, well known for his pro-regime stance. Director of the Geological-Paleontological Institute, he justified his entry into the NSDAP in 1937 in a letter to the president of the University of Tübingen dated September 11, 1945, with his experience that a "happy uplifting of the German *Volk* could be observed."[*] Hennig was considered a Nazi sympathizer because he hoped to gain advantages for his field of research in exchange for the alignment of his publications with National Socialist ideology and priorities.[†] A typical example, playing to the anti-intellectualism of the Hitler regime, was his announcement of a new exhibit of Permo-Triassic tetrapods on the premises of the University of Tübingen in *Der Biologe*. He praised the exhibit for the fact that a systematic arrangement of the fossils had been purposefully avoided: intellectualism should not interfere with a compassionate appreciation of the "biological foundations of the ancient *Lebensgemeinschaft* on exhibit".[‡]

But Edwin Hennig was not the first, nor the only one to cast paleontology as a biological discipline at its heart. As mentioned before, it was Henry Fairfield Osborn (1857–1935), an American geologist, paleontologist, and eugenicist who would eventually head the American Museum of Natural History in New York, who was spearheading the effort to base paleontology on a biological foundation in the United States. Another leading paleozoologist of the time was Louis Dollo from the *Musée Royale d'Histoire Naturelle de Belgique*, the Royal Belgian Natural History Museum in Brussels, to whom the Viennese paleobiologist Othenio Abel dedicated the first issue of his new journal, *Palaeobiologica*, complete with a *laudatio* for Dollo signed by 50 colleagues of international status, Edwin Hennig among them.[§] Othenio Abel established a deep personal friendship with both Osborn and Dollo. After the Great War, Osborn encouraged American scientists in a letter to the influential *Science* magazine to renew their relationships with European colleagues. Setting up his own science as a leading example, Osborn wrote "we paleontologists welcome the works of Othenio Abel."[¶] Osborn sent food packages to Abel and his family in those economically most challenging postwar years,[**] and the two corresponded to make sure they had as complete a set of reprints of each other's publications as possible.[††] Abel befriended Dollo during his early studies on marine mammals at the Brussels museum. The two men maintained a close friendship and mutual respect throughout their career. But Osborn and Dollo were not the

---

[*] Adam, 1977, p. 173, n. 117.
[†] Adam, 1977, p. 173.
[‡] E. Hennig, 1938, p. 302.
[§] Abel, 1928a,b, pp. 1ff.
[¶] Osborn, 1920, p. 568.
[**] Letter by Abel to Osborn, Sept. 29, 1920, Folder 5, Box 1, Henry Fairfield Osborn Papers, Archives of the American Museum of Natural History.
[††] For example, letter from Osborn to Abel, July 25, 1920, Folder 5, Box 1, Henry Fairfield Osborn Papers, Archives of the American Museum of Natural History.

only influences Abel acknowledged to have helped shape his views on paleobiology.* Another was the fellow neo-Lamarckian Otto Jaekel (1863–1929), "one of the most brilliant paleontologists"† and a pioneer in the quest to "free paleontology from the shackles of geology ... in which she still remains at most universities in the German Reich."‡ A graduate from the Universities of Breslau and Munich (where he earned his *Habilitation* in 1890), and after stints in Strasbourg and Berlin, Jaekel became professor of geology and paleontology at the University of Greifswald by decree issued on October 18, 1906.§ On November 23, 1915, an application was issued to name Jaeckel "*Geheimer Regierungsrat*"; the initiative was granted on March 24, 1916. Jaeckel's major claim to fame, at least in vertebrate paleontology, was his involvement in the excavation of the Halberstadt (Germany) dinosaur remains, representing the Late Triassic *Plateoaurus*. Yet this very involvement got him caught in a maze of intrigue, deception, and denunciation that culminated in litigation lasting well beyond his death. The Nazi academia became notorious for intrigue, as will be exemplified in Chapter 7 by an episode from the "hot years" of the Third Reich, involving the pioneers phylogenetic systematics. Jaekel's tribulations show that such practices were already in full swing during the years of the Weimar Republic—and at that early time involved one who would become the chief ideologue of the NSDAP, Alfred Rosenberg, the editor in chief of the *Völkischer Beobachter*, the party's main publicist outlet based in Munich (Figure 6.4).

On June 6, and July 26 of the year 1927, the *Völkischer Beobachter* published two articles on *The University Scandal*, and *On the Greifswald University Scandal*, which raised serious allegations with respect to Jaekel's handling of the Halberstadt excavations: perjury in testifying against the assistant fossil preparator Paul Mauersberger accused of theft or embezzlement (Mauersberger had kept a motor for himself that had been made available to Jaekel's field crew by the company Siemens-Schuckert); sale of Halberstadt fossils to museums inside and outside Germany; the embezzlement of fossils for his own private profit; irregular payment of wages to Mauersberger; and having hung a red towel out of his window (*einen roten Lappen aus seinem Fenster gehängt*) during the 1918 revolution, signaling sympathy with the communists. After the first article had appeared, Jaekel wrote a brandishing letter dated July 6, 1927, to *Sehr geehrter Herr Hitler*, identifying Mauersberger—"an associate of the local [i.e., the Halberstadt] communists"—as the informant of the *Völkischer Beobachter*.¶ Having spread the "wildest lies about me," Jaekel demanded a "total satisfaction

---

* Ehrenberg (1975, p. 64) located the first printed appearance of the term "Paläobiologie" in the course catalogue for the winter semester 1908/09. The use of the term was mediated by the engineer Franz Hafferl (1857–1925), who first met Abel at the meetings of the *k.k. Zoologisch-Botanische Gesellschaft* in Vienna, and as of 1907/08 was a regular attendee in Abel's classes for several semesters (personal communication of Hafferl to Ehrenberg, 1975, p. 64).
† Abel, 1929b, p. 144. On Jaekel see also Drevermann (1929); Bather (1930).
‡ Abel, 1929b, p. 153.
§ University Archives Greifswald, Acte des Kratoriums der Universität zu Greifswald betreffend Dr. Jaekel, Heft I. Anfang 1906, Ende 1936. Signatur PA 65.
¶ In a letter dated July 4, 1927, to the Prussian Minister for Science, Arts, and Education, Jaekel suspected Dr. Franz Klinghardt to be the informant, one "who according to my knowledge is still employed as civil servant, and has never been punished nor declared insane on the basis of his doctor's affidavit after his dangerous attack on me on November 9 of last year."

FIGURE 6.4 Otto Jaekel. (Museum für Naturkunde Berlin, Historische Bild- u. Schriftgutsammlungen [Sigl: MfN, HbSb]. Bestand: Zool. Mus., B I/546 [Jaekel].)

(*Wiedergutmachung*) in your newspaper within three days," or else he would involve the state attorney. On July 8, 1927, Jaekel wrote to the *Völkischer Beobachter*, rebutting all allegations raised against him, and explaining that he had experienced a previous altercation with Mauersberger (Jaekel had motivated the Ministry of Culture to sue Mauersberger for insult on August 10, 1922, who was found guilty and fined 100 *Reichsmark*), which might explain his adversarial motivations. The managing editor of the *Völkischer Beobachter*, Wilhelm Weiss, replied to Jaekel on July 12, 1927, acknowledging the receipt of his complaints: "Your letter to *Herr Hitler* was rather insulting and not very considered, and therefore rather useless for us." Weiss further specified that the information contained in the article had not been provided by Mauersberger, but by "a trustworthy man," and that the author of the article had relieved the Editor in Chief Alfred Rosenberg from his duty to keep the author's name confidential. The author, it turned out, was Jacques Rosenbaum-Ehrenbusch, an engineer and architect who owned the house at *Bahnhofstrasse* 46/47 in Greifswald, where Jaekel rented an apartment (the house was later taken over by the University). Rosenbaum-Ehrenbusch and Jaekel again had gone through an earlier altercation, as the owner of the house had obtained a conviction of Jaekel for perjury on April 23, 1923. Pursuing his rehabilitation from the accusations published in the

Völkischer Beoachter, Jaekel motivated the Prussian Minister for Science, Arts, and Education to file a suit against Rosenbaum-Ehrenbusch on August 13, 1927, accusing him of publicly insulting Jaekel. Jaekel himself was investigated for the allegations that Rosenbaum-Ehrenbusch had published, but the proceedings against him were halted on January 14, 1928. In contrast, the suit against Rosenbaum-Ehrenbusch went ahead, the state attorney noting on May 7, 1929, that it would go forward even in spite of Jaekel's death. On September 18, 1928, Jaekel had designated his son Fritz as his legal representative, charged to defend his interests, particularly in court. Rosenbaum-Ehrenbusch was sentenced on October 13, 1930, but appealed the court's decision. The appeal was denied on September 29, 1931, and the earlier sentence upheld: Rosenbaum-Ehrenbusch was fined 600 *Reichsmark* or 16 days in jail for publicly insulting Jaekel, Weiss from the *Völkischer Beobachter* was fined 150 RM or 15 days in jail for complicity. In its overall assessment, the court found Rosenbaum-Ehrenbusch's articles to constitute an effort to discredit Jaekel as a public figure and academic teacher to a degree that would require "to liberate the German youth from him." At the age of 65, Jaeckel retired from his position in Greifswald effective April 1, 1928. Upon his retirement, Jaekel accepted a full professorship at the Sun Yat-sen University in Canton. On March 6, 1929, Jaeckel succumbed to pneumonia at the German Hospital in Beijing.

Abel hailed Jaekel as a brilliant forerunner of paleobiology, who sought to establish paleontology as a subject area independent of geology. During his years in Greifswald, Jaekel pursued several initiatives in that direction.[*] On April 13, 1920, Jaekel wrote to the dean, explaining that methods and research goals in paleontology and geology are sharply divergent, justifying separate chairs for each discipline. At the University of Munich, Jaekel claimed, geology on the one hand, paleontology and zoology on the other, had been separated for 45 years already, and similar initiatives were under way in Berlin, Breslau, and Göttingen. Greifswald should not lag behind only because of financial considerations. Jaekel renewed his initiative with a letter to the dean dated October 10, 1925, at which point it became clear, however, that his campaign had less to do with the autonomy of paleontology as a science, but rather was motivated by his desire to lessen his teaching load: "For this semester I have again announced a course in geology, but only with the understanding that I would be allocated an assistant lecturer." On December 28, 1925, the Prussian Secretary for Science, Arts, and Education permitted the employment of a junior lecturer to teach geology starting with the summer semester 1926. For that task, the Secretary chose Johannes Weigelt from the University of Halle-Wittenberg, who took up teaching geology at the University of Greifswald in the winter semester of 1926/27.[†] In 1928, Weigelt succeeded Jaekel as professor and director of the Geological-Paleontological Institute of the University of Greifswald, before moving back to the University of Halle-Wittenberg in 1929 (for more on Weigelt see Chapter 5, and below).

In 1903, Jaekel had been placed *primo loco* for full professorship in paleontology at the University of Vienna, but—to Abel's later chagrin—his appointment there was

---

[*] University Archives Greifswald, Acte des Kratoriums der Universität zu Greifswald betreffend Dr. Jaekel, Heft I. Anfang 1906, Ende 1936. Signatur PA 65.

[†] Archive of the Martin Luther University, Hale Saale, UA Halle Rep. II, PA 16768 (Johannes Weigelt).

thwarted through intrigues instigated by local dignitaries.[*] In 1912, Jaekel founded the *Palaeontologische Gesellschaft*, a professional society for German speaking paleontologists, which elected him as their first president at their first annual meeting in Budapest.[†] Jaekel's efforts collided with Abel's intention to found in that same year a Paleobiological Society in Vienna, with the consequence that "ill-feelings" developed between the two men; they cleared their differences in later years, however, "as we both valued substantive issues higher than personal quibbles."[‡] Abel succeeded Jaekel as president of the *Palaeontologische Gesellschaft* at its annual meeting in Frankfurt a.M. in 1921, after Jaekel had nursed the society through the chaos of World War I and the immediate postwar years. Abel in turn founded his *Palaeobiologische Gesellschaft* in 1932, an event that was announced in *Der Biologe* as the long awaited creation of an "international society for the care and support of paleobiology in the broadest sense."[§] When serving in a reserve infantry regiment stationed in Belgium during World War I, Jaekel used idle time to write yet another analysis of the biological foundations of state organization, a book he privately published in 1916. From among the few lines Jaekel highlighted in this booklet, the one that stands out reads: "The great benefit of this war is not the booty we capture, but instead the strengthening of the health of our nation (*staatliche Gesundheit*). Who could deny that such was necessary in many respects."[¶] While this might sound like a glorification of warfare in Social Darwinist terms, Jaekel was quick to dismiss Darwinism. Darwin's theory, Jaekel claimed, "invokes a series of contingencies, the links of which are possible in principle, but the causally efficacious concatenation of which is highly improbable."[**] Instead, Jaekel stressed—as he had done before—function as the active, form as the passive principle of life, and countered Darwin's externalist selection theory with his own dictum: "*Omnis transformatio ex vi et anima formae* (form changes as a consequence of its activity)."[††] The father of German sociology, Ferdinand Tönnies (1855–1936), found nothing of value to sociologists in Jaekel's treatise, but praised him for his defense of neo-Lamarckism: "The philosopher has always found this form of the theory of descent more plausible than any other. Lamarck's thinking is more ingenious than Darwin's."[‡‡]

Jaekel was an extraordinarily prolific writer, dealing with topics that ranged from geology, stratigraphy, and hydrogeology to paleontology, human prehistory, arts, and politics. Given the number and breadth of his publications it comes as a small surprise that his publication record received rather mixed reviews in the obituaries penned by Abel, as also by the Frankfurt paleontologist Fritz Drevermann (1875–1932): "who would argue that with so much light, shadows could be missing?"[§§] In Abel's retrospective, Jaekel always tended toward grandiose synthesis to the detriment

---

[*] Abel, 1929b, p. 155.
[†] Jaekel, 1914a.
[‡] Abel, 1929b, p. 154.
[§] Anonymous (1932/33, p. 128). The entry was most probably penned by the editor Ernst Lehmann.
[¶] Jaekel, 1916, p. 190.
[**] Jaekel, 1916, p. 1.
[††] Jaekel, 1916, p. 38.
[‡‡] Tönnies, 1917, p. 136.
[§§] Drevermann, 1929, p. 183.

of detailed analysis.[*] Still, Abel singled out two publications by Jaekel that in his judgment represented milestones in the memorized paleontologist's career: Jaekel's 1902 study of phylogenetic patterns as revealed by the Fossil Record, and Jaekel's 1922 paper on *Function and Form in Organic Development*, which was based on a talk he had delivered at the annual meeting of the *Palaeontologische Gesellschaft* of 1921. Another important paper by Jaekel is the publication of the speech he held at the founding assembly of the *Palaeontologische Gesellschaft* in Budapest.[†] Taking these publications together, a number of core tenets of Jaekel's research program can be identified, many of which would be adopted not only by Abel, but also by other German paleontologists such as Edwin Hennig and Karl Beurlen. Neo-Lamarckism, in Jaekel's estimation, trumps Darwinism in German paleontology; function consequently shapes form and not the other way around; all phylogenetic change results from activities of essentially autonomous organisms; stimuli for evolutionary change can be external as well as endogenous, reflecting the needs (*Bedürfnisse*) of the organism; evolutionary change is goal-directed or orthogenetic; orthogenesis is the phyletic expression of Carl Nägeli's "principle of perfection" that, in Jaekel's interpretation, carried no vitalistic connotations; orthogenesis consequently results in an autonomous perfection of the organism with respect to its bodily functions; the change from one basic plan of organization to another one is saltational, driven by "metakinesis," that is, a "shaking up" or repatterning of ontogeny; each basic plan of organization lays the foundation for the evolution of an orthogenetic lineage subject to developmental constraints, evolutionary inertia, and a lawfully ordered termination (extinction).

Another fellow neo-Lamarckian and friend commended by Abel was the physical anthropologist Franz Weidenreich (1873–1948). Abel found in Weidenreich support for his defense of the importance of comparative anatomy, paleontology, and phylogenetics against the rising influence of genetics. It was Weidenreich who, in Abel's estimation, "in a series of ... excellent ... magnificent publications took position against the nonsense that is currently perpetuated in the distinction of 'genotypes' and 'phenotypes'"[‡]—a distinction that, when cast in Weismannian terms, necessarily had to be rejected from a neo-Lamarckian point of view. Only a neo-Lamarckian conception of morphological evolution could resolve what Weidenreich identified as the central conundrum of Darwin's selection theory: "adaptive modifications [are] first tested, then inherited – the Lamarckian principle – and not the other way around, first inherited and then tested – the selectional principle."[§] Abel enthusiastically explicated Weidenreich's views at the Frankfurt meeting of the *Palaeontologische Gesellschaft* of 1921: "*Everything which has become genotypical must once have been phenotypical.*"[¶] A famous episode played out at the meeting of the *Palaeontologische Gesellschaft* in Tübingen in 1929, which overlapped with the annual meeting of the *Deutsche Gesellschaft für Vererbungskunde*. The two societies organized a joint session,

---

[*] Abel, 1929b, p. 144.
[†] Jaekel, 1914b.
[‡] Abel, 1923, p. 201.
[§] Weidenreich, 1921, p. 106.
[¶] Abel, 1922, p. 112.

in which Weidenreich's theory drew severe criticism from the Finnish (eu-) geneticist Harry Federley (1879–1951)—at the time working in Berlin—who dismissed neo-Lamarckian arguments on grounds of a "complete lack of any experimental proof."[*]

Abel's first major contribution to paleobiology was his *Grundzüge der Palaeobiologie der Wirbeltiere*, published in 1912, which he dedicated to his "honored teacher and dear friend LOUIS DOLLO." The goal of this discipline he defined as the investigation of the adaptations of fossil organisms, of their mode of life, and of their phylogenetic relationships.[†] The centerpiece of Abel's paleobiology was his "reaction theory" (*Reaktionstheorie*), which he[‡] traced back to Osborn,[§] but which in fact is a synthesis of Weidenreich's theories with Jaekel's concept of orthogenesis, rooted as it was in the idea of developmental constraints. Developmental constraints explain why the potential for adaptive morphological evolution is not unlimited, and may even force maladaptations as those that had been identified by the Russian paleontologist Wladimir Kowalevsky (1842–1883) in his classic monograph on the ungulate mammal genus *Anthracotherium*.[¶] Kowalevsky, whom Abel hailed as another pioneer of paleobiology, had dedicated this monograph to his "teacher and close friend" Charles Darwin, noting that "the continued development and dissemination of the principles [that Darwin] formulated ... must necessarily contribute to a reorientation of paleontology towards new goals."[**]

Abel himself identified his biological "Law of Inertia" (*Trägheitsgesetz*) as his major original contribution to paleobiology. Publishing in *Biologia Generalis*, a journal he himself co-edited, Abel[††] referred back to Carl Nägeli again, who in his mechanistic account of evolutionary change of 1884 had in Abel's view provided a perfectly valid and non-vitalistic explanation of directional and progressive phylogenetic change that was based on the inertia manifest in living nature. In order to avoid all possible taint of vitalism, Abel sought to extend the laws of classical physics (i.e., Newtonian mechanics) to the biological realm: "The core of my argument is to compare processes of adaptation with movement in the mechanistic sense."[‡‡] Organisms have long been known to be subject to the laws of chemistry, so why should they not also be subject to the laws of mechanics, asked Abel? In addition to inertia, Abel invoked the "the principle of least resistance."[§§] Jointly, these laws for Abel provided a mechanistic explanation for both, Jaekel's orthogenesis and Dollo's law of the irreversibility of evolutionary processes. That way, Abel claimed to have subsumed two evolutionary laws that had previously been considered to be

---

[*] Federley, 1929, p. 306; see also Schindewolf, 1936, p. 1; Harwood, 1993, pp. 119ff.
[†] See also Kutschera, 2007, p. 172.
[‡] Abel, 1928c, p. 6.
[§] Osborn, 1889, p. 561. In closing this paper, Osborn (p. 566) found himself compelled to "postulate some third, as yet unknown, factor in evolution to replace the Lamarckian Principle," yet retained his belief in the possibility of the inheritance of maternal impressions: "I myself am a firm believer in it, from evidence which I am not free to publish" (Osborn, 1892, p. 563).
[¶] Kowalevsky (1873/74); on Kowalevsky see also Sterelnikov and Hecker (1968).
[**] Kowalevsky (1873), *Widmung*, no page numbers.
[††] Abel, 1928c, p. 2.
[‡‡] Abel, 1928c, p. 95.
[§§] Abel, 1928c, p. 100.

independent under one unifying principle, his own "Law of Inertia." With reference to work pursued by his former student, later research assistant and successor Kurt Ehrenberg (1896–1979) on cave bear fossils from Styria, Abel hinted at the possibility that Haeckel's biogenetic law might be reducible to the same unifying principle as well. Finally, in his opening speech at the 1928 annual meeting of the *Palaeontologische Gesellschaft*, held in Budapest on September 27, Abel[*] subsumed yet another phylogenetic law under his "principle of biological inertia," which was the Italian invertebrate zoologist Daniele Rosa's (1857–1944) "law of progressively diminishing variability".[†] Directional phylogeny (orthogenesis), irreversibility, the progressively diminishing potential for variation within a phylogenetic lineage (i.e., phyletic senescence), and—potentially—the recapitulation of phylogeny during ontogeny Abel reduced to a single unifying principle of "biological inertia," which itself was anchored in laws of classical mechanics. Although neo-Lamarckian at its core, Abel insisted that his *Trägheitsgesetz* is by no means a Lamarckian principle, but corresponds more closely to the ideas Darwin had himself presented in his *"On the Origin of Species"*[‡] on the effect of use and disuse of organs. Quoting extensively from the sixth edition of the *Origin*,[§] Abel drew the paradoxical conclusion that "Darwin has been no Darwinist"[¶]—which was correct when viewed from the perspective of how Darwinism was understood in contemporary German evolutionary biology.

Given that Abel's paleobiology flourished and expanded in German paleontology during the interwar period, it comes as little surprise that it acquired sociopolitical connotations prevalent at the time, fueled in addition by Abel's personal political inclinations. Abel was raised and remained deeply steeped in the *völkisch* spirit of the time, "imbued with history, nationalism, and racial identity."[**] In his autobiography,[††] Abel recounts how under the guidance of his father he became an avid collector of natural history specimens, anything from plants, butterflies, and fossils to crystals, but also of ancient coins. His accounts of the geological–paleontological excursions he undertook during his student years reveal a deep commitment to the homeland paired with a consciousness of race, in particular with respect to Jewry. On a trip through Transylvania Abel acquired for good money an interesting coin from an

---

[*] Abel, 1929c, p. 17.

[†] Rosa, 1899.

[‡] Abel, 1928c, p. 13.

[§] "[This has been effected chiefly through the natural selection of numerous successive, slight, favourable variations]; aided in an important manner by the inherited effects of the use and disuse of parts; and in an unimportant manner, that is in relation to adaptive structures, *whether past or present, by the direct action of external conditions, and by variations which seem to us in our ignorance to arise spontaneously. It appears that I formerly underrated the frequency and value of these latter forms of variation, as leading to permanent modifications of structure independently of natural selection*" (Darwin, 1888, Vol. II, p. 293; emphasis added by Abel). Osborn likewise emphasized Darwin's adoption of "Lamarckism," citing from a letter from Darwin to Moritz Wagner (1813–1887): "In my opinion, the greatest error which I have committed has been not allowing sufficient weight to the direct action of the environment ... independently of natural selection" (Darwin, cited in Osborn, 1891, p. 140. Darwin's letter dates from Oct. 13, 1876: Fr. Darwin, 1959, p. 337f).

[¶] Abel, 1928c, p. 14.

[**] Galison, 1990, p. 717.

[††] Published by Ehrenberg (1975).

obviously dishonest owner of a local tavern. In Abel's words "the owner of the watering hole (*Schnapsbude*), a Jew by the name of Spitz, had obtained the coin from a farmer in exchange for a liter of hooch."[*] Abel's ethos was German Nationalist, rooted in the *völkisch* tradition and spiced up with monarchistic nationalism and anti-Semitism. After failing to establish, with the help and support of faculty colleagues, his own political party in 1927, Abel eventually turned to National Socialism, and joined the NSDAP rather late, in 1938. Always an indefatigable advocate of Greater Germany, Abel rushed from Göttingen to Vienna to celebrate Hitler's arrival there on March 15, 1938.[†] The *Führer* would later, through a decree dated January 17, 1940, recognize Abel for his loyalty and support by awarding him the medal dedicated to the memory of March 13, 1938, the day on which Hitler and representatives of the Austrian government—without parliamentary backing—signed the law that decreed the reunion of Austria with the German Reich.[‡]

Abel's right-wing political leanings and latent anti-Semitism impacted personnel policies at the University of Vienna especially with respect to *Habilitation* promotions and employment at the professorial level. The avenue through which to exercise such influence in promotion committees and at faculty meetings was a coalition of 18 like-minded faculty members that Abel forged in 1919, in a successful effort to thwart social–democratic initiatives for university reform. The way this secret coalition functioned under the code name *Bärenhöhle*[§] was recently sketched by the Viennese journalist and historian Klaus Taschwer.[¶] Abel's *Bärenhöhle* friends were the same clique that unsuccessfully tried to establish themselves in the political arena in 1927. Abel's efforts to influence university policy was particularly palpable during his tenure as dean of the philosophical faculty (1927/28), and later as president of the university (1932/33).[**] Anti-nationalistic politics promoted by Engelbert Dollfuß, who was appointed federal chancellor of Austria in May of 1932, triggered clashes among the students at the university during which American citizens studying in Vienna were injured. The American Ambassador complained to Dollfuß about the growing Nazi violence at the Vienna University later in the same year, an intervention that forced Abel to pronounce an apology. Nazi sympathizers condemned such "humiliation" in the press, conjuring up times soon to come when such diplomatic *démarches* would no longer be honored, let alone acted on.[††] As unrest at the university escalated, Dollfuß dissolved Parliament in March 1933, and banned the Austrian branch of the National Socialist German Workers Party (NSDAP) in June of the same year. Dollfuß fell victim to an attempted coup by the Social

---

[*] Abel, in Ehrenberg (1975, p. 30).

[†] Ehrenberg, 1975, p. 124.

[‡] Universitätsarchiv Göttingen, Kuratorium-Akte, Mathematisch-Naturwissenschaftliche Fakultät, Dr. Abel.

[§] The code name refers to a cave yielding cave bear remains near Mixnitz in Styria (Austria), objects of research not only for Abel, but especially so for his student, successor, friend, and biographer Kurt Ehrenberg.

[¶] Taschwer, 2012.

[**] The dean of the faculty of philosophy at the University of Vienna praised Abel in 1944 for his anti-Semitic activities and supporter of Greater Germany during his tenure first as dean, later as president of that same university: Deichmann (1996, p. 71).

[††] Pauley, 1992, p. 129.

Democrats and Nazis in July 1934. His successor Kurt Schuschnigg forced university professors known for their German Nationalistic or National Socialist ambitions—Abel among them—into early retirement by the start of the academic year 1934/35, under the pretext of budgetary relief.[*] His father's loss of employment prompted his son Wolfgang Abel, a racial hygienist involved with the development and implementation of racial policies in the Third Reich at the *Kaiser Wilhelm Institut für Anthropologie* in Berlin-Dahlem, to contact the fellow eugenicist and paleobiologist Henry Fairfield Osborn, then president of the American Museum of Natural History in New York. Wolfgang Abel pleaded with Osborn to intervene with Nazi officials on his father's behalf, in an attempt to find him a faculty position in a German university.[†] Osborn, since its founding year an honorary member of the *Palaeontologische Gesellschaft*, was happy to oblige, and immediately contacted his acquaintance, the Harvard graduate Ernst "Putzi" Hanfstaengl (1887–1975), who was an early confidant of Hitler and at the time foreign press spokesperson for the NSDAP.[‡] In his letter, Osborn reminisced about a recent road trip he had taken through southern Germany, recounting how "greatly impressed" he was "by the solidarity of the country and enthusiasm for the rebirth of Germany under the new conditions of the Hindenburg-Hitler regime," which promised a "very bright future" if the current financial crisis could be overcome. He was one of many American visitors to the new Germany who "failed to grasp the true character of Hitler's regime."[§] Pitching Abel's case, Osborn concluded with the expression of his belief that the "call of Dr. Abel to Germany would meet with worldwide approval and would do much to offset the criticism aroused by the removal of many university professors of Jewish affinity." Hanfstaengl himself had become acquainted with Abel's work when he was an art student in Vienna: "I believe he was occupied at that very time in getting out his important work *Grundzüge der Palaeobiologie der Wirbeltiere*."[¶] Having been informed of all these activities, Othenio Abel profusely thanked Osborn for his efforts, calling his letter to Hanfstaengl "a most excellent move": "I hope to be able to put the full dimensions of my abilities to the service of my people and my German fatherland."[**] Abel was eventually appointed full professor of paleontology at the University of Göttingen as of April 1, 1935, an appointment through which he, and his wife, were granted official citizenship of the German Reich.[††] With a letter dated

---

[*] Ehrenberg, 1975, p. 106; Deichmann, 1996, p. 71.

[†] W. Abel to Osborn, Oct. 7, 1934, Folder 5, Box 1, Henry Fairfield Osborn Papers, Archives of the American Museum of Natural History.

[‡] Osborn to Hanfstaengl, Oct. 18, 1934, Folder 5, Box 1, Henry Fairfield Osborn Papers, Archives of the American Museum of Natural History. On Putzi see also Heiber (1994, p. 296); Larson (2011).

[§] Larson, 2011, p. 67.

[¶] Hanfstaengl to Osborn, Nov. 2, 1934, Folder 5, Box 1, Henry Fairfield Osborn Papers, Archives of the American Museum of Natural History. For more detail on the Osborn/Hanfstaengl intervention see Rieppel (2012a).

[**] O. Abel to Osborn, Nov. 27, 1934, Folder 5, Box 1, Henry Fairfield Osborn Papers, Archives of the American Museum of Natural History.

[††] Universitätsarchiv Göttingen, Kuratorium-Akte, Mathematisch-Naturwissenschaftliche Fakultät, Dr. Abel. In a letter dated May 26, 1939, the NSDAP communicated to the head of the National Socialist German University Lecturers' Association that Abel had been removed from the faculty of the University of Vienna because of his national socialist activities, but now has a new position in Göttingen: Schleiermacher (2005, p. 86).

June 22, 1940, the president of the University of Vienna Fritz Knoll (1883–1981, NSDAP, 1937) informed Abel that he was named honorary senator of the University of Vienna, an honor that Abel shared with the other colleagues that had been dismissed on Schuschnigg's initiative: "the honorary Diploma will be presented to you in a ceremony to be held in the fall."[*]

Abel's employment at the University of Göttingen did not work out to everybody's satisfaction, however. For one thing, Abel was accused of neglecting his teaching duties through his far-flung travels and his working visits in the collections at the Brussels Museum. More importantly, he was criticized for teaching paleontology in a way that did not cover the interest and needs of students who would later seek employment as field geologists or in the mining industry. With a letter dated December 17, 1939, the dean informed Bernhard Rust's ministry in Berlin: "Prof. Abel teaches paleobiology, not a paleontology of the kind that would be useful to a geologist." Only 10 days later, the dean requested of Rust's REM to initiate procedures that would result in Abel's early retirement, no later than by April 1, 1940 (Abel would have reached the retirement age of 65 years on June 20, 1940). The letter, dated December 27, 1939, cited Abel's self-justification according to minutes taken at a specially convened meeting, in which Abel had insisted that the goal of his research and teaching was life history and biology (of extinct organisms), a direction which left him with little interest in geology: "In that respect it has to be emphasized that Professor Abel pursues no collaboration whatsoever with the other biologists, nor with any other members of the faculty. He therefore stands outside any organic collaborative team (*organische Arbeitsgemeinschaft*) in the faculty." On December 29, the president of the University of Göttingen himself wrote to the REM that Abel had caused conflicts among the faculty in geology, mineralogy, and zoology: "Never has it been possible to convince him of the necessity of collaboration; quite to the contrary, he always expressly refused any such effort."[†] Abel himself turned to the REM in a letter dated January 18, 1940, in which he requested early retirement per April 1 on his own behalf, as he had been called upon by the *Gauleiter* of Salzburg and Carinthia, Friedrich Rainer (1903–1947), to build a Research Institute for the History of Life in the *Haus der Natur* in Salzburg, an endeavor that would be supported by the *Stiftung Ahnenerbe* (Ancestral Heritage Foundation) under the direction of Heinrich Himmler. Himmler, indeed, wanted to make Salzburg the center for German (Aryan) Biology.[‡] Abel made it clear that his extensive private paleontological collections would travel with him[§]: should Abel find it difficult to find space, he should contact the energetic Walter Greite, head of the biological research unit at the *Ahnenerbe*, who would easily be able to provide him with the required rooms, the *Gauleiter* opined.[¶] Following Greite's fall into

---

[*] Universitätsarchiv Göttingen, Rektorats-Akte, O. Abel.

[†] Universitätsarchiv Göttingen, Rektorats-Akte, O. Abel.

[‡] "Besprechung mit Gauleiter und Reichsstatthalter Dr. Scheel über verschiedene Angelegenheiten des 'Ahnenerbes' in Salzburg." Bundesarchiv, DS (formerly Berlin Documentation Center), microfilm signature DS G 119 (file Greite).

[§] See documentation at the Universitätsarchiv Göttingen, Kuratorium-Akte, Mathematisch-Naturwissenschaftliche Fakultät, Dr. Abel, and Rektorats-Akte, O. Abel.

[¶] Bundesarchiv, DS (formerly Berlin Documentation Center), microfilm signature DS G 119 (file Greite).

disgrace in 1942, and due to the chaos of war, the project really never quite got off the ground, however. Abel died of heart failure on July 4, 1946, in his country estate near Mondsee in Upper Austria.

Otto Schindewolf (1896–1971), then president of the German Paleontological Society, congratulated Abel in a letter dated January 22, 1942, for his appointment as honorary member. Abel had, indeed, maintained a high level of engagement in the *Palaeontologische Gesellschaft* throughout his career. In his capacity as its president, Abel had invited the *Palaeontologische Gesellschaft* to hold its annual meeting in Vienna in 1923. After his move to Göttingen, he again invited the Paleontological Society to meet there in 1937, a meeting that took place from October 12 through 16. The president of the society as of January 1, 1937, was the paleontologist Johannes Weigelt, who opened the meeting with a welcoming address held in the main auditorium of the Zoological Institute. During that time, Weigelt served as president of the University of Halle (for more detail see Chapter 5), and as an expert consultant on mineral resources for the Four Year Plan.[*] Accordingly, Weigelt[†] justified his investigation of the Eocene Geiseltal fossils not only with reference to their paleo-*biological* significance for the new German ideology (*Weltanschauung*), but also because of the economically important insights these investigations provided into the Geiseltal geology, an area of brown coal mining located near Halle an der Saale (Saxony-Anhalt). Weigelt was, indeed, celebrated for bridging the perceived gap between basic and applied research: when he was awarded the *Ritterkreuz* in March 1945, *Gauleiter* Eggeling praised Weigelt for his continuous efforts in support of German armament, and especially in the "fight for German iron freedom (*Eisenfreiheit*)." Weigelt's micropaleontological research had led to the discovery of iron ores, which earned Weigelt a position as a consultant for the *Reichswerke Hermann Göring*, a Reich corporation for iron mining and ironworks established in 1937. On the occasion of Weigelt's 50th birthday in 1940, the local press used this example from Weigelt's research to underscore the importance of basic research, as it may generate results with unforeseen applicability (Figure 6.5).

Starting with volume 3, 1930, Weigelt was listed as co-editor of Abel's *Palaeobiologica*. As of its fourth volume, published in 1935, Weigelt also signed on as co-editor of *Der Biologe*, a commitment he continued after the journal had been taken over on Himmler's command by Walter Greite (1907–1984; NSDAP, 1932) of *Das Ahnenerbe*[‡] (more on that unfriendly takeover later). In his opening speech at the Göttingen meeting, Weigelt stressed the fact that "it is precisely paleontology that has greatly gained in significance in most recent times, summoned to play an entirely different role than in the past particularly with regard to ideological issues, indeed facing major new challenges in the biological indoctrination (*biologische Durchdringung*) of the *Volk* in the New Germany."[§] To cast paleontology in the light of Nazi "biopolitics" was a strategy that intensified at the ideologically even more highly charged 1938 annual meeting of the Paleontological Society, which took

---

[*] Klee, 2003, p. 661.
[†] Weigelt, 1939.
[‡] Kater, 1974, p. 413, n. 148; Güttner, 2004, p. 64.
[§] Weigelt's (1937) speech is summarized in the Palaeontologische Zeitschrift, vol. 20 (1938), p. 164.

FIGURE 6.5 Johannes Weigelt. (UA Halle-Wittenberg, Rep. 40 VI, Nr. 2 [Rektoren-Album], Bild 51 [Johannes Weigelt].)

place in Bayreuth from July 11 through 14. Still its president, Weigelt requested that the name of the society be changed to *Deutsche Palaeontologische Gesellschaft*, and that new bylaws should be enacted that better reflect the spirit of the new time. Without concern for a loss of international appeal, Weigelt proposed such a name change in honor of Hans Schemm, the deceased founder and *Führer* of the National-Socialist Teacher's Association that was headquartered in the *Haus der Deutschen Erziehung* where the meeting was held. His plea was seconded by the paleontologist Karl Beurlen (on whom more below), who maintained that science can never be fully separated from ideology (*Weltanschauung*) and consequently requested the alignment of the *Deutsche Palaeontologische Gesellschaft* with National Socialist ideology: "let us go home in recognition of this obligation."[*] Following Weigelt's recommendation, a roundtable discussion on teaching paleontology was organized at the Bayreuth meeting for July 11, 1938, to be chaired by Abel.

---

[*] Anonymous, 1939a.

The introductory commentary that was to stimulate discussion was delivered by Edwin Hennig from Tübingen, who the year before had published a paper in *Der Biologe* in which he stressed the importance of paleobiology as a means to strengthen the German people's ties to the soil in which the German race is rooted, and on whose treasures the German civilization depends.[*] He considered it a heroic achievement to have to enriched paleontology with the insights of biology, rendering paleobiology "a central guiding principle of *völkisch* community life, offering up knowledge of the laws of nature whose willful acceptance turns dominance of nature into freedom of spirit."[†] Hennig went on to characterize paleontology and geology as complementary sciences, which jointly capture the sublime relations that tie life to earth, rooting blood in soil. Rattling on about the heroes of Bayreuth, Richard Wagner, and Houston Stewart Chamberlain, Hennig concluded that paleontology, the science of prehistoric life, offers the best venue through which an ideological foundation for biology can be established.[‡] Beurlen followed up on Hennig's remarks, emphasizing the specific German spirit that in his view must infuse German paleontology, and that should be cultivated through education at schools and universities. Abel urged a broader perspective, as he required every person to be aware of the fundamental facts concerning the history of the soil on which he lives, and to honor the history of his ancestry, one that extends beyond mere family ties all the way into prehistoric times.

German paleontology at the time, and with it paleobiology, underwrote neo-Lamarckian doctrines, motivated by the belief that Darwinian selection theory is an incomplete, or at least an insufficient explanation for large-scale evolutionary change. Certainly conscious of race and accused of anti-Semitism in his actions as university administrator in Vienna, Abel's published writings do not sport eugenicist programs, proclamations, or polemics, as was the case for some contemporary biologists. The same is true of his paleontological colleagues such as Edwin Hennig, who avoided reference to racial hygiene and its sociopolitical consequences fueled as they were by a militant Social Darwinism. Such distance did not stop traditional paleontologists and paleobiologists to jump on the bandwagon that gathered political steam with Hitler's ascent to power, however, and to sketch their discipline in such a way as to be able to cash in on the ascent of biology to foundational importance for National Socialist propaganda.[§]

GROTESQUELY GRANDIOSE: THE EVOLUTIONARY SYNTHESIS IN *VÖLKISCH* SPIRIT

Just as in botany with Wilhelm Troll (see Chapter 5), there was in German paleontology of the time again a tendency to turn to nature mysticism. The most extreme worker in that regard was certainly the paleontologist Edgar Dacqué[¶] (1878–1945),

---

[*] E. Hennig, 1937, p. 1.
[†] E. Hennig, 1937, p. 4. On the German imperative to seek dominance of nature see Blackbourn (2006).
[‡] The proceedings of this roundtable discussion are published in the *Palaeontologische Zeitschrift*, vol. 21 (1939), pp. 5–19.
[§] For example, Adam, 1977, p. 173.
[¶] Meister, 2005b.

who obtained his PhD from the University of Munich in 1902. After a brief stint in Dresden, Dacqué pursued his career at the Bavarian State Collections of Paleontology and Historical Geology first under Karl Alfred von Zittel (1839–1904), then under Ferdinand Broili (1874–1946), and finally under Karl Beurlen (1901–1985). Dacqué earned his *Habilitation* in 1912, and was promoted to associate professor and curator of the Bavarian state collections of paleontology and historical geology in 1915.[*] Author of a series of well-received scientific publications,[†] Dacqué published in 1924 his book *Urwelt, Sage, und Menschheit*, the first in a series of evolutionary treatises loaded with nature mysticism and religious connotations. There is anectodal evidence that Hitler himself was attracted to nature mysticism through Dacqué's *Urwelt, Sage, Menschheit*.[‡] A member of the Theosophical Society, Dacqué was on record for having held—at his own risk—public lectures and courses on mythology, religion, and science. As of 1935, Dacqué's presentations met with increasing resistance, however, until Alfred Rosenberg's office finally intervened, prohibiting Dacqué's public lectures on *Weltanschauung* and religion.[§] In his first book of 1936, the paleontologist Otto Schindewolf, who kept a remarkable distance to the Nazi regime and ideology, spent several pages to refute Dacqué's views on phylogeny and its underlying mechanisms, without touching on his mystical elaborations, however. This contrasts with Beurlen's review of Dacqué's *oeuvre* in the second volume of the ZGN. Beurlen praised Dacqué for his opposition to an understanding of life as a meaningless, soulless mechanism. Instead, life was to be comprehended as a dynamic process (*Geschehen*) with deeper meaning, upon which Beurlen concluded: "A mechanistic explanation of nature will always terminate in the '*ignoramibus*', as it addresses only the superficial, i.e., the meaningless and serendipitous manifestations of life."[¶] Beurlen applauded Dacqué's inspiration that the human soul partakes in the soul of the universe, and insisted: "These are insights that today fall on fertile soil again"(Figure 6.6).[**]

Beurlen, indeed, stood much closer to *völkisch*-motivated nature mysticism than the sober Schindewolf, but being more moderate than Dacqué in his writings, he earned a greater degree of acceptance among his peers. The paleontologist Walter R. Gross (1903–1974) from the Humboldt Museum of Natural History in Berlin for example noted of Beurlen's theorizing that "his special hypotheses are in every respect anchored in natural philosophy (*Naturphilosophie*), holism (*Ganzheitslehre*), and a special theory of knowledge (*Erkenntnistheorie*)."[††] It is hard to tell, however, whether Gross was genuine in his appreciation of Beurlen's work, or whether he was just ridiculing the theoretical foundations on the basis of which Beurlen had acquired prominence in his field under the National Socialist regime.

---

[*] Quenstedt and Schröter, 1957.
[†] For example, Dacqué, 1915, 1921.
[‡] Mosse, 1998, p. 306; Gasman, 2004, p. 178, n. 63.
[§] Meister, 2005b, 202.
[¶] Beurlen, 1936/37, p. 39. The "ignorabimus" was Emil Du Bois-Reymond's (1818–1896) expression for the inexplicable residue in nature that transcends empirical sciences.
[**] Beurlen, 1936/37, pp. 39–40.
[††] Gross, 1943, p. 240.

FIGURE 6.6   Karl Beurlen. (Archiv der Max-Planck-Gesellschaft, Berlin-Dahlem.)

Commenting on Beurlen's *magnum opus*, his *Die stammesgeschichtlichen Grundlagen der Abstammungslehre* of 1937, the Tübingen paleontologist and historian of the German evolutionary synthesis Wolf-Ernst Reif (1945–2009) found it not only drenched in vitalism, but quite generally "overburdened with philosophy."[*] Born in Aalen (Würtemberg) on April 17, 1901, Beurlen earned his PhD at the University of Tübingen, studying fossil invertebrates under the direction of Edwin Hennig. He had chosen a career in natural sciences "with a heavy heart,"[†] as indeed he fostered a deep interest in philosophy and linguistics throughout his life.[‡] The philosophers whom he was most attracted to, and whose work would greatly influence his worldview, were Friedrich Nietzsche, Hans Driesch, and Jakob von Uexküll. Their influence is, however, not yet much noticeable in the published version of his PhD thesis, Beurlen's first book of 1930: an introduction to the theory and practice of phylogenetics, based on the study of fossil decapods. The contents do reflect, however, the influence of his supervisor, Edwin Hennig, who he acknowledged especially for his oral communications.[§] In addition, in an interview conducted with Wolf-Ernst Reif in 1982, Beurlen attested[¶] that both he and Otto Schindewolf had as graduate students been acutely aware of the work of Hans Salfeld,[**] who had

---

[*] Reif, 1983, p. 188.
[†] Beurlen, cited from his unpublished autobiography by Levit and Olsson (2006, p. 114).
[‡] Tollmann, 1986; On Beurlen see also Grüttner (2004); Levit and Olsson (2006); Küppers (2007).
[§] Beurlen, 1930, p. 319; On Beurlen's relations to E. Hennig see also Reif (1986, p. 115, n. 54).
[¶] Reif, 1999, p. 155.
[**] Salfeld, 1921, 1924; cited in Beurlen, 1930; See also Dürken and Salfeld (1921); E. Hennig (1932).

described saltational evolutionary change in fossil cephalopods (ammonoids). But Salfeld not only stipulated saltational evolutionary change, he also insisted that "the propensity to form grooves or keels [on their shell] exists as an endogenous factor in every single ammonite animal."[*] This is a vitalistic, neo-Lamarckian argumentation pattern that would be accentuated in Beurlen's writings.[†]

In his textbook published in 1932, Beurlen's supervisor Edwin Hennig again proclaimed the insufficiency of Darwinian selection theory as an explanation of evolutionary change, and invoked "immanent laws of life" instead that bestow directionality on evolutionary progress.[‡] In an earlier publication,[§] Hennig had recognized in phylogenetic lineages of ammonoids periods of vigorous youth and adolescence that were followed by periods of adulthood, senescence, and death (extinction), that is, a cyclicity of phylogenetic processes that was incorporated in saltational accounts of change by both Schindewolf and Beurlen. Given such parallelism in the life history of an individual organism and the phyletic history of an evolutionary lineage, Edwin Hennig conceptualized phylogenetic lineages as complex wholes, which renders an evolutionary lineage a "phyletic individual of higher order."[¶] The same ideas can be found in Beurlen, who called a phyletic individual a *"phyletische Gestalt."*[**] Beurlen explicitly identified Hans Driesch and Jakob von Uexküll for having played a central role in clarifying the nature of biological individuality, that is, of biological wholeness (*Ganzheit*). And although Beurlen called Driesch "the creator of the concept of wholeness,"[††] he nonetheless criticized the bio-philosopher for having failed to follow the implications of this concept all the way home. Driesch had, indeed, restricted the concept of wholeness to the particular individual organism, denying, or at least questioning the possibility to expand the scope of this concept to cover "supra-personal wholes."[‡‡] This contrasts with Uexküll, who had not only allowed for evolutionary entities to be complex wholes, but who also stipulated a unity of the organism with its *Umwelt*, the two locked into reciprocal causal interdependence. A causal reciprocal interdependence of the biocoenosis and its biotope had likewise been recognized in ecology,[§§] Beurlen noted, whereby Friederichs and Thienemann in widely read papers characterized the nested hierarchy of ecological complex wholes in terms of Heidenhain's concept of enkapsis. Beurlen consequently looked to limnology for an elucidation of a holistic (*ganzheitlich*) comprehension of ecosystems,[¶¶] and in his 1937 book approvingly cited August Thienemann as a source of insight. Heidenhain had applied his concept of an enkaptic hierarchy to organs and organ systems. Friederichs and Thienemann had expanded this concept to ecosystems.

---

[*] Salfeld, 1921, p. 345; emphasis added.
[†] Beurlen, 1930, p. 319.
[‡] E. Hennig, 1932, p. 24.
[§] E. Hennig, 1916, p. 516.
[¶] E. Hennig, 1944, p. 290.
[**] Beurlen, 1937a, p. 102.
[††] Beurlen, 1937a, p. 105.
[‡‡] Driesch, 1910, p. 20.
[§§] See also Stella and Kleisner, 2010.
[¶¶] Beurlen, 1936, p. 36.

Beurlen in turn employed the concept of enkapsis is his characterization of the phylogenetic system:

> The *ganzheitlich* structure of the organism, demonstrated anatomically by the enkapsis of the histosystems, is recognized functionally in the determination of the differentiation of each organ through the functional plan of the whole ... [which] ranges not only over individual development, but also over the supra-individual phyletic development.[*]

Heidenhain employed the enkaptic organization of histosystems to explain the integration and wholeness of the individual organism; Beurlen extended that same argument to the phylogenetic system, where the enkapsis of phyletic *Gestalten* rendered these as integrated complex wholes at different levels of inclusiveness of the phylogenetic system. And again, such complex wholes are not just the sum of their parts, as Driesch's experiments had so aptly documented. Instead, emergent properties characterize the causally integrated and hierarchically structured complex wholes at every level of inclusiveness. In such a system, the co-instantiation of the properties of the parts is a necessary, but not also a sufficient disposition for the instantiation of the emergent properties of the complex whole, because the emergent properties of the complex whole reciprocally condition the nature of its parts. In Beurlen's system, every individual composed of parts at any level of complexity of the phylogenetic system is always also itself part of a superordinated whole, and can be fully comprehended from the perspective of the superordinated whole only: this, Beurlen found, is precisely the conclusion that had eluded Driesch,[†] a conclusion furthermore that Beurlen went on to apply to the *völkisch Lebensordnung*.

In his unpublished autobiography, Beurlen claimed that he was drawn into political activities against his will.[‡] In fact, however, Beurlen enthusiastically embraced the new spirit of the time, actively promoting it during his rapidly advancing career in Nazi Germany, until his sudden fall when the collapse of the Third Reich forced him to leave Germany for Brazil.[§] Still a docent (*Privatdozent*) in Königsberg, Beurlen wrote to the "esteemed Mister Hitler" on November 27, 1931, that he counted himself among those "who believe that the idea of National Socialism is the only avenue towards a great future for Germany."[¶] Associate member (*förderndes Mitglied*) of the SS since 1931, Beurlen joined the NSDAP in 1933.[**] In 1934, Beurlen was offered the position of full professor and director of the Geological-Paleontological Institute of the University of Kiel, where he actively promoted National Socialist ideology under Karl Lothar Wolf, who served as president of the university at the time (see Chapter 5). In 1936, Beurlen was admitted to the board of directors of the German Geological Society,[††] after having been nominated by the Working-Group of National-Socialist University Docents (*Arbeitsgruppe nationalsozialistischer*

---

[*] Beurlen, 1937a, p. 102.
[†] Beurlen, 1937a, p. 105.
[‡] Levit and Olsson, 2006, p. 114.
[§] See also Rieppel, 2011f.
[¶] Heiber, 1991, p. 372.
[**] Grüttner, 2004, p. 22.
[††] Anonymous, 1936, p. 668.

Hochschullehrer) for geology, which Beurlen himself founded and chaired.* Beurlen was elected president at the Society's 1938 annual meeting held in Munich.† This move provided him with the platform to open the annual meeting of the German Geological Society in Berlin the following year, on March 1, 1939, where he used the opportunity to proudly announce that the Viennese Geological Society was all set to join its German counterpart as a junior sister organization under the name *Alpenländischer Geologischer Verein*.‡ In 1940, he was elected an honorary member of the Viennese partner society—"probably for political reasons."§

At the 1938 annual meeting of the *Deutsche Palaeontologische Gesellschaft* in Bayreuth that ran from July 11 through 14, Beurlen chaired the scientific sessions.¶ In his opening remarks, he expressed his joy about the renaming of the society as a specifically German one following Weigelt's suggestion. The year before, in 1937, Beurlen had been admitted to the prestigious German Academy of Sciences Leopoldina in Halle (Saale), the oldest of its kind in Germany (at the time called *Kaiserliche Leopoldinisch-Karolinische Deutsche Akademie der Naturforscher zu Halle*).** In that same year, Beurlen was named *Fachspartenleiter Bodenforschung* (referee for soil sciences) in the *Reichsforschungsrat* (RFR; Reich Research Council), responsible for allocating research funds to projects in geology, geophysics, and mineralogy.†† His nomination as *Reichsforschungsrat* reveals Beurlen's loyalty to the regime, and his willingness to administer geological sciences while aligning them with the regime's guidelines.‡‡ The *Führerprinzip* had been implemented in the DFG by its president Johannes Stark in 1934 already,§§ and naturally carried through into the RFR that was tasked to bring German research into line with the goals of the Four Year Plan. Beurlen certainly fit the profile of a zealous *Reichsforschungsrat*, as he had glossed geology in the new spirit of the time in a piece he submitted to the *Zeitschrift der Deutschen Geologischen Gesellschaft* on October 14, 1936, to be published the following year. Geology, he claimed, deserves the status of *primus inter pares* among all natural sciences, because it alone provides the scientific foundation for the design of the German lifestyle (*Lebensgestaltung*), and for the cultivation of German territory (*Lebensraum*). At German schools, geology must be introduced in the teaching of *Heimatkunde* in such a way as to enlighten the unity of blood and soil (*Einheit von Blut und Boden*). Aligning geological research with the targets set by the Four Year Plan would result in a self-motivated subordination

---

* Range and Mempel, 1936.
† Anonymous, 1938, p. 546.
‡ Anonymous, 1939b, p. 253.
§ Cernajsek and Seidl, 2007, p. 263. Beurlen kept that honorary appointment in what later became the Austrian Geological Society to the end of his life.
¶ According to Tollmann (1986, p. 373) and Küppers (2007, p. 10), Beurlen was elected chairman of the German Paleontological Association at its 1938 meeting. Instead, it was Otto Schindewolf who was elected to succeed Johannes Weigelt in the chair (Anonymous, 1939a).
** Anonymous, 1937b, p. 112.
†† Anonymous, 1937b, p. 292.
‡‡ Beurlen was noted to have been the only member of the Reich Research Council who did not seek profit from his nomination in terms of funding of his own research: "his personal integrity may have played a part here" (L. Mertens, 2004, p. 127).
§§ Deichmann and Müller-Hill, 1994, p. 166.

of individual researchers under the great common task, which is to shape the new Germany. In contrast, geological research pursued in a liberal spirit opens no new perspectives for the new *Reich*, "and will consequently be eliminated."*

In a retrospective of 1939, the Reich Research Council asserted that it had respected, even promoted academic freedom in research, a claim that was backed up with reference to the fact that a significant number of projects in basic research had been funded along those pursuing practical research.† But justified concerns about academic freedom and the status of basic research could never be fully dispelled after the proclamation of the Four Year Plan and the subsequent formation of the Reich Research Council. On the occasion of the 1937 annual meeting of the German Geological Society in Aachen, representatives of both the Geological and Mineralogical Societies presented the *Reichsforschungsrat* in the evening of August 21 with a memorandum that laid out their vision for the future of their science. The paper stressed the virtues of incentivized collaborative research, as opposed to a top–down management of science that mandates common goals, which might suppress the creative initiative of individual researchers.‡ Sketching his vision for geosciences in the Third Reich, Beurlen stressed in his communication of 1937 that he would seek to strengthen research in each of the three disciplines he would represent in the Reich Research Council, in a way that would allow them grow together to form a unified theoretical edifice, a Whole:

> Our task will have to be to foster the three sciences in a way that will enable each to contribute to the Whole accordingly ... straggling individual sprouts will need to be carefully trimmed and trained, however, such that they will grow into the framework of the Whole.§

This was a statement that must have been interpreted by many members of the German Geological Society as a veiled threat to their academic freedom. Beurlen thus followed up with a clarification, explaining that only those researchers who stand apart from *völkisch* life (*völkische Lebensordnung*) have reason to fear any infringement on the freedom of research through the implementation of the Four Year Plan.¶ Reminiscing these troubled times, the Tübingen (later Münster) paleontologist Helmut Hölder (1915–2014) expressed his satisfaction that in spite of transiently changing its name, the Paleontological Society had yielded to ideological pressures of the time to a far lesser degree than the Geological Society.**

In 1941, Beurlen was named successor to Ferdinand Broili at the University of Munich, assuming the position of Professor of Paleontology and Historical Geology, and Director of the Bavarian State Collection of Paleontology and Historical Geology, a post he held until 1945. His declared goal was to align these disciplines

---

* Beurlen, 1937b, p. 54.
† Nagel, 2012, p. 243; see also Deichmann and Müller-Hill, 1994.
‡ Anonymous, 1937c, p. 553.
§ Beurlen, 1937b, p. 360.
¶ Simon, 2008, p. 69.
** Hölder, 1976, p. 7.

as pursued at the University of Munich with the spirit of National Socialism.[*] In 1942, Beurlen accepted his election—by a narrow margin[†]—as co-chairman of the *mathematisch-naturwissenschaftliche Klasse*, which closed the one remaining gap in the National Socialist leadership of the Bavarian Academy of Sciences.[‡] In the academic year 1944/45, Beurlen also joined the board of directors (*Führungskreis*) of the Reich Leadership of University Lecturers (*Reichsdozentenführung*).[§] With the collapse of the Third Reich, Beurlen hurried to rewrite his personal history under the Hitler regime in a new light, seeking written attestation from politically uncompromised colleagues. One such *Persilschein* (named after the German laundry detergent *Persil* that was sold under the slogan: *Persil washes whiter than white*) he requested from the prominent nuclear physicist Arnold Sommerfeld, also from the University of Munich, who declined to oblige. Sommerfeld had for political reasons opposed Beurlen's election as co-chair of the *mathematisch-naturwissenschaftliche Klasse* of the Bavarian Academy of Sciences, and accused Beurlen to have run, in his capacity as dean of the natural sciences from 1942 through 1945, an administration that was partial to sympathizers of the Nazi regime.[¶]

Beurlen's ideological commitments, deeply rooted in the *völkisch* tradition and enriched by Nazi ideology, carried right through into his science, just as he had himself declared to be inevitable. In his major book of 1937, he aimed at a grandiose synthesis of contemporary holistic–organicist biology with the insights gleaned from paleontology in a phylogenetic context. The phylogenetic system forms an enkaptic hierarchy, he claimed, which means that all life forms a complex whole of which the German *Volksgemeinschaft* is an integral part, and hence is governed by the same eternal laws of life that had shaped the entire tree of life. In Beurlen's system, the wholeness of the individual organism mirrors the wholeness of phyletic *Gestalten*, both subject to endogenous forces of development that run parallel to each other. Just as the individual organism passes through a cycle of birth, youth, adolescence, maturity, senescence, and death, so do phyletic lineages. Each major phylogenetic lineage for Beurlen represented a "type," that is, a type of organization or a typical plan of construction, which constrained the ontogenetic development of the particular representatives of each type. New types come into being through *neomorphosis* during the vigorous youth period of a phylogenetic lineage, neomorphosis meaning the initiation of evolutionary innovation during early stages of ontogenetic development ("ontogenetic repatterning" or "ontogenetic deviation" in modern terms). In contrast, the senescence of a phylogenetic lineage Beurlen attributed to Daniele Rosa's principle of the progressive reduction of variability, coupled with increasing

---

[*] Heiber, 1991, p. 372.
[†] Stroemer, 1995, p. 107; Beurlen withdrew from the academy in 1946.
[‡] Hashagen, 2010; Beurlen resigned from the Bavarian Academy of Sciences in 1946.
[§] Grüttner, 2004, p. 22.
[¶] http://sommerfeld.userweb.mwn.de/gif100/05137_01.gif (accessed April 22, 2013). Having lost all his academic positions after World War II, Beurlen had to support his family as a construction worker until he was offered a position by the Brazilian Geological Survey in 1950. In 1959, he was offered a professorial position at the University of Recife, from which he retired in 1969 to return to Tübingen (Germany), where the paleontologist and historian of biology Wolf-Ernst Reif had the opportunity to interview him in 1982 (Reif, 1999, p. 155), before Beurlen's death on December 27, 1985.

developmental constraints.* His concept of *orthogenesis*—the idea that each phyletic lineage goes through a cycle of development that parallels the history of an individual organism—Beurlen anchored in Driesch's concept of *entelechy*,[†] that is, in the "inherent individual lawfulness" (*Eigengesetzlichkeit*;[‡]) of the organism. Driesch's entelechy he declared an ordering principle inherent in the organism, not a rationally positioned goal. The same he found to be true of Jakob von Uexküll's concept of teleology. Beurlen professed to have been deeply influenced by "the entire corpus of work of v. UEXKÜLL's school,"[§] in particular with respect to his account of "organization," and the reciprocity between the organism and its *Umwelt*. On J. v. Uexküll's account, organization implies "conformity with plan," which Beurlen again found to have been confused with premeditated "purposefulness." The rules that determine the "ordered necessity," which according to Driesch governs the developing embryo, according to Beurlen range as well over the temporally successive, irreversible, and unique manifestations of the complex wholes that are phyletic *Gestalten*: orthogenesis means nothing more than the realization of possibilities of form under the constraints imposed by the type.[¶] Against Abel, Beurlen denied that the causes active in organismal development would be the same as those active in classical physics. Driesch, he explicated, had called for a second sort of causality, one he had dubbed a "unifying causality,"[**] rooted in his concept of entelechy. In Beurlen's interpretation, this is

> a causality which DRIESCH dubbed 'the whole' (Ganzheit) and [Oswald] SPENGLER called 'fate' (Schicksal), it is a causality that is irreversible, and characterized by the necessary cycle of birth – youth – maturity – senescence – death. The expression of this 'causality of the individual' in phylogeny is the cyclic developmental process.[††]

With the ascent of the Third Reich, the German *Volk* had irrevocably linked its historical fate to that of its *Führer* and vice versa. Just as National Socialism was understood as politically applied biology, politics was understood as applied history—at least by the Nazi pedagogue and ideologist Ernst Krieck.[‡‡] German politics were defined by the *Führer*, and through it he defined the fate of the *Volk*. But such historical fate is not just passively endured, but rather to be shaped by the will of the individual, the collective will of the National Community, and the supreme will of the *Führer*, all of them aspiring to the same common goal, which is the supremacy of the German *Volk* in its geopolitical struggle for existence. An appeal to will power was already very much in evidence in the revitalization and remilitarization of the German nation during the early years of the Weimar Republic, when only the mobilization of the popular will could throw the allied supremacy into relief and open the door to hopes for a better, greater German future.[§§] With its *völkisch*-racist background,

---

* Beurlen, 1930, p. 552; 1937a, p. 102.
[†] Beurlen, 1935/36, p. 447; 1937a, pp. 131, 139.
[‡] Driesch, 1910, p. 3.
[§] Beurlen, 1937a, p. 257.
[¶] Beurlen, 1937a, p. 170.
[**] Driesch, 1914, p. 52.
[††] Beurlen, 1932, p. 79.
[‡‡] Schreiner, 1985, p. 164.
[§§] Mulligan, 2009, p. 79.

National Socialism understood history as an irrational life-stream carried by a will to power and the conviction that might makes right.[*] Such was also exactly the way Beurlen understood evolutionary history to unfold.

The experienced world, according to Beurlen, is one of an irrational multiplicity of entangled processes.[†] And while he accepted that lawfulness underlies the apparent irrationality of nature, he thought it wrong to believe that history, including evolutionary history, could be reduced to the universal laws of classical physics (mechanics). These laws capture only universal, that is, abstract and eternal relations expressed in mathematical terms, as opposed to history, which plays out in the *hic et nunc*, in the here and now, and hence deals with individuals, indeed with an entire enkaptic hierarchy of individuals. Never missing an opportunity to lash out against Jewish science, Beurlen rejected relativity theory and quantum mechanics as counterintuitive (*unanschaulich*) for the German soul—this being a leitmotif for the ZGN which Beurlen had cofounded and co-edited. To abstract from time was to forfeit any insight into the lawful concatenation of events, which collectively constitute the irreversible history of life and cultures.[‡] To properly understand history requires to turn away from a physicalist and to endorse an organicist worldview instead.[§] This is the juncture at which Beurlen turned to the question as to which kind of causality other than the physicalist's conception of material cause and effect would regiment phylogeny. What exactly is this force, this immaterial factor, or regulative principle that was invoked by Driesch and Uexküll. Could it be that it reigns not only over ontogeny, but also drives phylogeny, generating the enkaptic hierarchy of phyletic *Gestalten* in the process? Beurlen resolved to call this immaterial agent the Will to Be (*Wille zum Dasein*),[¶] citing Nietzsche's *The Will to Power*[**] in that context while noting that this important source also contains a critique of Darwinism that is "worth taking to heart."[††]

In opposition to the externalist Darwinian theory of natural selection, it was the Will to Be, an endogenous force inherent in all living beings and the complex wholes they form at different levels of inclusiveness in the tree of life, which in Beurlen's view propelled the evolutionary process forward. The great German philosopher Arthur Schopenhauer had considered all "evolution" (in an idealist, rather than Darwinian sense) the manifestation of a Will to Life (*Wille zum Leben*).[‡‡] Because the very act of willing presupposes life, Nietzsche objected and introduced his own concept instead, the Will to Power. Rejecting mechanistic explanations based on quantitative correlations, Nietzsche found—much to Beurlen's delight—true explanatory power to be grounded in qualitative relations, as he tied the "fundamental reality of existence"

---

[*] Reimann, 1984, p. 49.
[†] Beurlen, 1942, p. 194.
[‡] Beurlen, 1936, p. 19.
[§] Beurlen, 1936, p. 37.
[¶] Beurlen, 1935/36, p. 455; 1937a, p. 222.
[**] Nietzsche, 1901 (1968); see Beurlen, 1935/36, p. 457. In the spring of 1924 in an article published in *Deutschlands Erneuerung*, Hitler invoked Nietzsche's concept of the Will to Power as a means with which to crush Marxism and break the treaty of Versailles (Horn, 1968, p. 281).
[††] Beurlen, 1937a, p. 20.
[‡‡] For Schopenhauer, the Will to Life is a concept that explains how all forms of life grow, function, and behave: see Janaway (2002, p. 45).

to the Will to Power.* Criticizing Darwinism, Nietzsche argued that "the influence of 'external circumstances' is overestimated by Darwin to a ridiculous extent: the essential thing in the life process is precisely the tremendous shaping, form-creating force from within [i.e., the Will to Power] which *utilizes* and *exploits* 'external circumstances' ... in the struggle of the parts a new form is not left long without being related to a partial usefulness and then, according to its use, develops itself more and more completely."† Nietzsche's Will to Power is not a transcendental concept, spiritual in nature, but a real biological force, comparable to the "will" (i.e., urge) to reproduce. Organisms do not employ power; instead, the Will to Power is a natural force that works through the organism.‡ In Nietzsche's thought, the organism confronts its environment driven by an inner force:

> The will to power can manifest itself only against resistances; therefore it seeks that which resists it – this is the primeval tendency of the protoplasm when it extends pseudopodia and feels about. Appropriation and assimilation are above all a desire to overwhelm, a forming, shaping, and reshaping, until at length that which has been overwhelmed has entirely gone over into the power domain of the aggressor and has increased the same.§

This, in a nutshell, beautifully expresses what Beurlen's concept of the Will to Be (*Wille zum Dasein*) was meant to mean, except that Beurlen specifically tied the Will to Be to J. v. Uexküll's conception of *Umwelt*, where the organism constitutes its environment on the basis of its given perceptional capacities and capabilities.¶ The latter, in turn, are subject to evolutionary progression.

The Will to Be was for Beurlen just as natural a force as the Will to Power was for Nietzsche. It plays out in the phylogenetic process, driving the evolution of organisms toward increasing autonomy and freedom from its expanding *Umwelt*. Beurlen cast adaptation as a struggle against continuously changing environmental conditions, and appealed to neomorphosis (early ontogenetic deviation) as the mechanism through which the organism evolves to gain both control over, and independence from an ever more broadened and enriched *Umwelt*.** The organism, the species, indeed every phyletic *Gestalt* represent a complex whole, driven to self-differentiation (*Selbstgestaltung*), that is, the willing generation of a *Gestalt* by the autonomous, endogenous Will to Be.†† In the formation of a phyletic *Gestalt*, the Will to Be instills an aspiration to freedom that translates into an increasing independence from an expanding *Umwelt*.‡‡ In Beurlen's view, Darwin's externalist theory of natural selection renders the organism a passive entity, exposed and reacting to the impact of the

---

* Tanner, 2000, pp. 56, 65.
† Nietzsche, 1901 (1968, p. 344); cited in Beurlen, 1935/36, p. 457. Wilhelm Roux's struggle of parts within the organism is here related to the typical neo-Lamarckian interpretation of form change as a consequence of change in function.
‡ Hicks, 2010, p. 98.
§ Nietzsche, 1901 (1968, p. 346).
¶ Beurlen, 1935/36, p. 455.
** Beurlen, 1935/36, p. 456.
†† Beurlen, 1937a, p. 222.
‡‡ Beurlen, 1937a, p. 238.

environment. Beurlen found Darwinism to neglect the autonomous creative powers that are active in every living being according to (the Uexküllian concept of) plan.[*] The exposure to the vagaries of the *Umwelt*, the consequent adaptive specializations, and the incorporation of these specializations in the phyletic *Gestalt* through neomorphosis are the means by which living beings conquer the world.[†]

However, the Struggle to Be, driven as it is by the Will to Be, is not just a dull Darwinian one for natural resources and procreation. It is, instead, a Struggle for Power. The Will to Be *qua* Will to Power seeks not just self-preservation, or the preservation of the species, but beyond those proximate goals it also seeks the ultimate dominance of the phyletic *Gestalt* over anything that opposes it: "this is why – and Nietzsche saw this very clearly – the Will to Be is a Will to Power."[‡] However, Beurlen conceded that the potential for independent and free individual development (*Gestaltung*) is not unrestricted, just as any egg cell is likewise not omnipotent.[§] The type, after all, constrains ontogenetic development and through it the potential for phyletic modification, the more so the closer a phylogenetic lineage reaches its senescence.

Accordingly, and for Beurlen, freedom therefore does not mean randomness or chance, but instead to submit to the Will to Be.[¶] This is where Beurlen's Aryan evolutionism meets Nazi ideology: just as phyletic differentiation is limited by the inherited type of organization, so is social differentiation constrained, or enhanced, by the heritable traits of the human races. But the Darwinian mechanisms of variation and natural selection, ultimately grounded in randomness and chance, will not suffice to secure the superiority of the Nordic race. Similarly, social freedom, academic freedom, freedom of research: they all have nothing to do, nor should they have anything to do, with the hedonistic individualism that characterized city life during the Weimar Republic, and continued on in decadent Paris and London where jazz music ringing out the doors of night clubs compete with the hustle and bustle of crowded streets. The university is not to be understood as an institution that provides a carefree space for the pursuit of liberal arts and basic research free of values. Instead, it is to be an institution that continues the educational program implemented in schools at all levels, which is to train and strengthen the national will through the training of body and mind.[**] Freedom, anchored in the Will to Power, is constrained by the willful acceptance of the reign of eternal laws of nature and of history, as Bernhard Rust put it in his memorable speech announcing the foundation of the *Reichsforschungsrat*.

A phyletic *Gestalt* for Beurlen is a supraindividual complex whole shaped by the Will to Be, the latter manifest as a Will to Power. And the same holds true, according to Beurlen, for the German *Volksgemeinschaft*. The phylogenetic systems forms an enkaptic (nested) hierarchy of complex wholes, which is subject to upward and downward causation that results in emergent properties at all levels of inclusiveness. While every part indispensably contributes to the whole in a chain of

---

[*] Beurlen, 1937a, p. 12.
[†] Beurlen, 1937a, p. 338.
[‡] Beurlen, 1937a, p. 223.
[§] Beurlen, 1937a, p. 233.
[¶] Beurlen, 1937a, p. 235.
[**] Reimann, 1984, p. 45.

upward causation, any part cannot be fully understood without a perspective from the whole, as it is embedded in a chain of downward causation. The same, again, holds true for every individual embedded in a *völkisch Lebensordnung*. The German *Volksgemeinschaft*, rooted in its *Lebensraum*, forms an enkaptic (nested) hierarchy of complex wholes: the particular individual, the family, the National Community. And while every individual indispensably contributes to the greater good of the national community, it is necessary to understand that within such an enkaptic social hierarchy, every individual cannot be effective in the best possible way unless it understands, and willfully accepts, its role in a chain of downward causation. The Will to Be, manifest as a Will to Power, requires every individual to shelve its self-interests and to freely submit to the imperative of the emergent properties of the more inclusive complex whole that are manifest in the greater communal goals of the National Community. Freedom is not arbitrariness, nor unconstrained individuality, but rather insight into historical necessity as conditioned by the laws of nature. It would be foolish, Beurlen contended, to think that such a perspective would in any way limit the potential of any individual embedded in the *völkisch Lebensordnung* to live a fulfilling life. Implicitly invoking an Uexküllian value, Beurlen insisted that every spontaneously living human being can experience the world, and think about it, only within the constraints and potential determined by his racial relationships.[*]

Beurlen cast evolution as a process that would bolster "the National-Socialist truth that the *Volk* as a community of blood and fate is the highest of all transindividual units."[†] At the same time, his science embodied the cultural pessimism expressed by Oswald Spengler in the form of orthogenesis, an ontogenetic as well as phylogenetic cyclicity that ends with the death not only of individual organisms, but also of phyletic *Gestalten*. Does that mean that the decline of civilizations (*Kulturvölker*) is a biological necessity as well?[‡] It might have been in the past, but no longer now, at a time when insight into the laws of life, that is, laws of heredity, can be practically applied through racial hygiene. Beurlen appealed to neo-Lamarckism, when he compared the potential of acquired characteristics to become heritable (within the constraints of the type), with education that shapes race (within the constraints of the racial constitution). He appealed to the vitalism of Driesch and Uexküll, when he argued that the whole determines the part as much as the part contributes to the whole. And finally, he appealed to Nietzsche when invoking a Will to Be that would sort out the irrationality of the experienced world. Just as Philipp Lennard and Johannes Stark pushed German Physics as a science commensurable with the characteristics of the blood and soul of the Nordic race, just as Ernst Lehmann pushed for the inauguration of an Institute of German Biology, so did Beurlen formulate a German Phylogenetics, rife with social connotations reflecting the ideological matrix of its time. As the historian Anne C. Nagel succinctly put it: Aryan science was motivated by an "immoderate overconfidence and the grotesque misbelief that a universal understanding of science could be replaced by a *völkisch* one."[§]

---

[*] Beurlen, 1935, p. 520.
[†] Weiss, 1994, p. 193.
[‡] Tirala, 1934a.
[§] Nagel, 2012, p. 282.

7 The Ideological Instrumentalization of Biology

THE RECONSTRUCTION OF ERNST HAECKEL

In 1919, the year of his death and amidst political turmoil, *Die Naturwissenschaften* ran a series of articles commemorating Haeckel and evaluating his life's work from a zoological as well as philosophical perspective. The contributing zoologist was Richard Hertwig from Munich, the philosopher was the erstwhile psychiatrist Theodor Ziehen (1862–1950) from Halle. Hertwig found Haeckel to have maintained a guarded stance with regards to new developments in biology, particularly physiology and *Entwicklungsmechanik*, developments that Haeckel himself had been instrumental in bringing about: he "was a researcher of a kind like Goethe."[*] Of course, Hertwig acknowledged the enormous influence Haeckel had, especially in promoting Darwinism in Germany, but also noted the increasing antagonism Haeckel encountered as he got older—not so much because of what he was saying, but because of the way he would carry his deep convictions into the world. "The impetuousness with which Haeckel challenged many institutions of Church and State, old and venerable doctrines and policies in his *Welträtsel* and *Weltwundern* was the same force it took to help Darwinism to its breakthrough [in Germany]."[†] Ziehen in turn found Haeckel to have been unsuccessful in the development of a *philosophical system*. Haeckel's monism he rejected for the primacy it accorded to the material basis of processes of consciousness, the author he consequently accused of a "naïve materialistic realism."[‡] Worse: Haeckel hypostasized substance supposed to instantiate the laws that govern phenomena and their change, which reveals Haeckel's total failure to grasp the neo-positivism that had been initiated by Ernst Mach for example. But even if Haeckel had found many followers who uncritically adopted his scientific dogmas, he had himself talked to other admirers of Haeckel who had been motivated by Haeckel to take an interest in the deep problems of the world and develop a critical philosophical perspective in that regard. This is where Ziehen located Haeckel's main, and lasting, merit. In a more general commemoration of Haeckel, the zoologist Karl Heider (1856–1935) from Berlin placed him in the tradition of Goethe and Alexander von Humboldt: "Like those, he was an intellectual leader of our nation."[§] Seconding this judgment, the geologist Johannes Walther

---

[*] Hertwig, 1919, p. 953.
[†] Hertwig, 1919, p. 958.
[‡] Ziehen, 1919, p. 959.
[§] Heider, 1919, p. 945.

(1860–1937) from Halle called Haeckel the "most daring champion of freedom of thought"[*]: "[Haeckel] showed little interest for politics and the party system, but wherever any movement would threaten the freedom of research, of thought, or of teaching, he would stand at the frontline of the battle."[†]

To call Ernst Haeckel a trailblazer for National Socialism in Germany certainly entails an element of anachronism. This does not detract from the fact, however, that some German biologists and eugeneticists instrumentalized Haeckel in the service of Nazi ideology. Such glorification of Haeckel in the "new" Germany was nevertheless received controversially even by high ranking Nazi ideologues, pitching the Haeckel supporter Alfred Rosenberg against the Haeckel critic and science pedagogue Ernst Krieck for example.[‡] University of Kiel philosopher and cofounder of the Zeitschrift für die gesamte Naturwissenschaft (ZGN) Kurt Hildebrandt sided with Krieck, whom he quoted as having characterized Haeckel as an "abysmally inflated philistine who portrays himself as the heroic mover and shaker of the world."[§] Although Hildebrandt rejected any theory of inheritance of acquired characteristics, noting though that both Darwin and Haeckel had entertained that possibility, he concluded that the origin of new species could not possibly be explained on purely mechanistic grounds, as Haeckel had claimed: "This [insight] brings the theory of descent back to the tradition of German philosophy of a creative nature."[¶] In Hildebrandt's judgment, Haeckel had committed an incomprehensible error when he tried to unify Goethe's philosophy of nature with a vulgar version of materialism.

The rift between reductionist materialism and organicist holism became glaringly apparent in the intrigues and battles that flared up between the Haeckel supporters (geneticists and phylogeneticists) among German biologists on the one side, the holistic–organicist *Ganzheitsbiologen* on the other. Elements in Haeckel's writings that lent themselves to political interpretation and abuse were his monism, his materialism, and of course his Darwinism, one that was equated—as had been done by Haeckel himself—with selection theory. It was, accordingly, not the holists and organicists who enlisted Haeckel in the name of their science, but rather the Darwinists, propagators of selection theory and racial hygiene, who rewrote Haeckel to reflect their own intensions and goals. Most prominently, it was members of a clique surrounding the racial hygienist Karl Astel at the University of Jena[**] who were most vocal in propagating Haeckel as a precursor of their own ideals and agenda. In the historian of medicine Paul Weindling's assessment, Astel "exemplifies how the spirit of armed aggression infused medical ideals."[††]

Karl Astel[‡‡] was named head of the Thuringian Office for Racial Affairs in Weimar after the Nazi ascent to power, in which capacity he signed off on thousands of forced sterilizations between 1933 and 1945. Without *Habilitation* and without having to

---

[*] Walther, 1919, p. 946.
[†] Walther, 1919, p. 949.
[‡] Hoβfeld, 2007, p. 452; see also Weindling, 1989b, p. 498.
[§] Hildebrandt, 1937/1938, p. 18, n. 1.
[¶] Hildebrandt, 1937/1938, p. 29.
[**] Heiber, 1994, p. 127.
[††] Weindling, 1989b, p. 309.
[‡‡] Klee, 2003, p. 20.

prevail in a competitive job search, Astel was placed,[*] in 1934, on the faculty of the Medical School of the University of Jena by Fritz Sauckel (1894–1946; NSDAP, 1923), *Gauleiter* of Thuringia. Sauckel and Astel were long-time friends, who had first met during their student years in the anti-Semitic *Deutsch-völkischer* Protection and Defense League (*Schutz- und Trutzburg*) in Schweinfurt.[†] Extracurricular appointments like Astel's were part of Sauckel's strategy—allegedly motivated by a decree of Hitler—to transform the University of Jena to a "truly National-Socialist university"[‡]: "during the past years I have endeavored to attract to the University of Jena professors who through their personality provide assurance of their political reliability and National Socialist orientation," Sauckel wrote to Science and Education minister Bernhard Rust in March 1943.[§] Astel was assigned his own Institute for Human Breeding Research and Genetics, later renamed as Institute for Human Genetics and Racial Politics. From 1939 to 1945, Astel served as president of the University of Jena: "Astel was among the most prominent SS figures at German universities."[¶] Throughout his career, Astel acted according to his conviction that "Nazism's significance [lies] in the recognition that the laws of nature [are] the basis of state and society. Only by ridding society of the burden of the weak, infirm, criminal, and alien types can the healthy elements in the *Volk* realize their powers to their full extent."[**] Already on May 3, 1935, Astel had written to Himmler that "the University of Jena must become an SS University!"[††] On July 5, 1935, Astel wrote to Himmler:

> As you know already, the issues dearest to my heart are the consolidation of Thuringia as a leading model in putting racial renovation into practice, the creation of a stronghold of research conscious of race, the development of a science carried by an unambiguous ideology, and a corresponding higher education policy.[‡‡]

With Germany irrevocably on the loosing side of the war, Astel committed suicide on April 3, 1945, in Jena. Those around Astel at the University of Jena who engaged in the National Socialist glorification of Haeckel included most notably the philologist and publicist Hans Friedrich Karl Günther (1891–1968; NSDAP, 1932) also known as the "Race Günther" (*Rassegünther*), a pioneer of the Nordic racial movement; the zoologist turned anthropologist Gerhard Heberer; the ruthless plant geneticist Heinz Brücher[§§] (1915–1991; NSDAP, 1934), graduate student of Ernst Lehmann, then postgraduate student in Astel's institute; the zoomorphologist and racial theorist Victor

---

[*] By decree (*Sondervoranschlag*) of the Thuringian Ministry for Education: Hendel et al., 2007, p. 96.
[†] Weindling, 1989b, p. 309.
[‡] Heiber, 1994, p. 127; see also Harrington, 1996, p. 195; Hoβfeld, 2007, p. 10f.
[§] Hendel et al., 2007, p. 175.
[¶] Deichmann, 1996, p. 258.
[**] Weindling, 1989b, p. 536.
[††] Bundesarchiv, DS (formerly Berlin Documentation Center), microfilm signature DS B 026, file "Karl Astel"; see also Hendel (2007), document Nr. 54, p. 98f.
[‡‡] Bundesarchiv, DS (formerly Berlin Documentation Center), microfilm signature DS B 026, file "Karl Astel."
[§§] Rehm, 1992, p. 1 "Who knew Brücher could imagine that he did not ... proceed diplomatically." On Brücher see Hoβfeld, 2003, p. 571, n. 223.

J. Franz (1883–1950; NSDAP, 1930), a former *"Ritter"*-professor for phylogeny at the University of Jena and from 1935 to 1945 director of the *Ernst-Haeckel-Haus*, an archive (and today an institute for the history of science) located in Haeckel's *Villa Medusa* in Jena; and finally Astel's long-term assistant Lothar Stengel von Rutkowski (1908–1992; NSDAP, 1930).[*] In their reconstruction of Haeckel's ideology, these "Jena scientists drew on Haeckel's monism in an anti-Christian crusade in line with the SS's pantheistic Nordic creed."[†] Astel has famously been reported to have characterized the holists/organicists among German biologists as "subversive hillbillies and pseudo-scientists (*fünfte Kolonne ultramontaner Pseudowissenschaftler*)."[‡] The list of targets of their campaign indeed reads like a "who's who" of *Ganzheitsbiologie*: they included among others Friedrich Alverdes, Karl Beurlen, Bernhard Dürken, Ernst Lehmann, Adolf Meyer(-Abich), Wilhelm Troll, and the *Gestaltler* assembled around him.[§]

Although not directly related to the reconstruction of Haeckel, an exchange in *Der Biologe* paradigmatically exemplifies the schism between a *völkisch* motivated organicist biology as opposed to a materialist–reductionist biology founded on selection theory. This particular controversy revolved around the *völkisch* philosopher Rudolf Steiner (1861–1925), his Anthroposophical Society (that grew out of Madame Blavatsky's Theosophical Society), and the methods of biodynamic farming he propagated. Biodynamic agriculture was (and still is) based on a holistic conception of soil, plants, and animals as a unified, causally interrelated system that was further embedded in cosmic relations and hence regulated by forces emanating from the nightly sky in seasonal rhythms. Controversies around biodynamic farming that played out at highest official rank—prominently involving the Minister of Agriculture Walther Darré and the landscape architect and conservationist Alwin Seifert (1890–1972)[¶]—found their due reflection in *Der Biologe*. Rudolf Steiner's appeal to Goethe's holism, his anthroposophy, and his biodynamic methods of agriculture and gardening were favorably commented upon in *Der Biologe* in 1932/1933,[**] at a time when the journal was still edited by its founder Ernst Lehmann. Come 1941, Astel tried to motivate Himmler to wage a campaign against anthroposophists: "all of us who want to defeat Christianity, Anthroposophy, and all other gospels spread by mentally impaired missionaries" should take a stand against them, recognizing "the perilousness of anthroposophy as an institution of indoctrination contrary to nature, and especially as a melting pot for the half and fully demented."[††] Although in his reply, Himmler pretended to share Astel's assessment of anthroposophy as a dangerous movement, he admitted to be unable to do anything about the school of Rudolf Steiner because Rudolf Hess supported and protected it.[‡‡] Anthrosophical

---

[*] Hoßfeld, 2007, p. 446f.

[†] Weindling, 1989b, p. 536.

[‡] Mildenberger, 2007, p. 191.

[§] Harrington, 1996, p. 177f, 195; see also Hoßfeld, 2003, p. 552.

[¶] Compare Gerhard, 2005, p. 138f, with Zeller, 2005, p. 156f.

[**] Neubauer, 1932/1933.

[††] Letter of Astel to Himmler, November 14, 1941. Bundesarchiv, SS-Führerpersonalakten 19, file "Astel, Karl, Dr."

[‡‡] Letter of Himmler to Astel, August 21, 1941. Bundesarchiv, SS-Führerpersonalakten 19, file "Astel, Karl, Dr."

theories of agriculture were eventually officially banned after Hess had absconded to Scotland on his grotesque peace mission in May of that same year. Yet, in spite of what he wrote to Astel, Himmler himself was quite favorably inclined toward biodynamic farming methods.* At the same time that this exchange took place, and likely instigated by his mentor Astel, the plant geneticist Brücher launched a vitriolic attack against anthroposophical pseudo-biology in *Der Biologe* in 1941, the journal having been taken over by the *Stiftung Ahnenerbe* (Ancestral Heritage Foundation) in 1938/1939: "Holism was a meta-biological doctrine destined to inflict most severe damage to natural sciences ... such jugglery and occultism, along with astrology, the belief in the power of stars ... must decisively be banned from the life of our *Volk*."† The rejection of holism and the nature mysticism associated with it implies no threat of loss of any values, Brücher argued, since only a strictly empirical approach in research will result in a severe test of any proposed new insights.

Quite similar tensions surrounded the reconstruction of Haeckel. Günther Hecht, head of the Department for German People (*Volksdeutsche*) and Minorities in the National Socialist German Workers Party's (NSDAP's) Racial Political Office made it clear through an entry in the third volume (1937/1938) of the *Zeitschrift für die gesamte Naturwissenschaft* that the doctrine of an all-pervading materialistic monism, as was propagated by Haeckel, must be rejected by *völkisch*-biologically motivated National Socialism. Haeckel, he continued, may well remain the subject of strictly scientific debate, but "the Party and its representatives must not only resist any even partial inclusion of Haeckel's thoughts [in contemporary ideology], as has been occasionally called for, but must also abstain from all official discussion of Haeckel's research program and teaching."‡ Co-founder of the journal, the philosopher Kurt Hildebrandt from Kiel, doubled up in the same volume, calling Haeckel's solution to the riddles of the world on the basis of a monistic-mechanistic interpretation of Darwin's theory of evolution an "illusion."§ In contrast, and again in the same volume of the ZGN, a certain H. Froese summarized the discussions that were held among the biologists organized in the Student Association of the Section for Natural Sciences at the University of Munich, on the occasion of the Reich Competition (*Reichsleistungskampf*, RBWK) of the German Student Association (*Deutsche Studentenschaft*, DSt). Complete unanimity, he claimed, prevailed among the biologists in attendance. The theory of descent with modification is by now accepted as a truism. An important consequence of this is the fact that humans form part of an unbroken line of descent reaching into deep time.

> This means that the laws of inheritance govern humanity as much as the entire world of organisms. This of course includes the transmission of intellectual capabilities. From these considerations necessarily follow measures of racial hygiene as well as the eradication of genetically conditioned asocial elements.¶

---

* Longerich, 2012, pp. 483–519.
† Brücher (1941, p. 266); Hoβfeld (2003, p. 543) reads Brücher's tract as a polemic against holism in general.
‡ Hecht, 1937/1938, p. 285; see also Hoβfeld, 2005a, p. 332; Richards, 2007.
§ Hildebrandt, 1937/1938, p. 17; see also Richards, 2007.
¶ Froese, 1937/1938, p. 306.

In the first volume of *Der Biologe*, a reviewer—most likely the editor Ernst Lehmann—of the eugeneticist Fritz Lenz's paper *On the Importance of Education for Racial Hygiene* quoted the latter as having emphasized the need to keep the "fear of Darwinism" away from the German youth. Questioning Darwinian selection theory could only result from a "lack of biological insight."[*] The ease with which arguments were bent to fit the purpose in the Social Darwinist reconstruction of Haeckel's writings is, indeed, remarkable. Consider Heberer, who celebrated the centenary of Haeckel's birthday in 1934 with a lecture on *Ernst Haeckel and His Significance for Science*, in which he concluded:

> Without fear Haeckel built on the despised '*Naturphilosophie*', providing an empirical base for many of its propositions ... [He] transformed the entire system of biology by integrating it under the umbrella of the theory of descent ... Haeckel did not attribute to Darwinism, i.e., to selection theory a universal significance. Instead, like Darwin himself, he was a Lamarckist ... [The inheritance of acquired characteristics] is a problem that is currently the object of intense scrutiny and debate. But whatever the verdict in this respect will be, it will be irrelevant for the significance of eugenics, which Haeckel himself had already called for.[†]

Ernst Lehmann—always eager to dance at every party on offer, and busy to please all sides—celebrated Haeckel's 100th birthday in the third volume (1934) of *Der Biologe* with a series of entries on the famous Jena professor. The volume was introduced by Werner Haeckel, the grandnephew of Ernst Haeckel:

> We celebrate the 100th birthday of the great naturalist from Jena under lucky stars. Finally, his most important demands are being fulfilled: the eradication of inferior beings ... forty years after he made these claims the leaders of the Third Reich fulfill them well beyond his own expectations.[‡]

In addition to his grandnephew's eulogy, there were five entries on the "sun of Jena,"[§] with Heberer sketching his achievements in zoology in ideologically remarkably neutral terms: "he was the one who led the theory of descent to its final victory, he was the brilliant reformer of biology," also contributing to nature, philosophy, and—through speeches and pamphlets—the culture wars. In contrast, the instrumentalization Haeckel's for National Socialist propaganda is particularly apparent in a contribution by the German philosopher Friedrich Lipsius (1873–1934; NSDAP, 1932) from Jena. He elaborated on Haeckel's musings on levels of individuality, which by analogy, he thought, clearly place the human individual in the service of the superordinated complex whole that is the *Lebens-* and *Arbeitsgemeinschaft*:

---

[*] Anonymous, 1931/1932, p. 97.
[†] Heberer, *Vortragsmanuskript*, dated 1934: Universitätsarchiv Tübingen (UAT) 126/256: Rektoramt, Personalakten des Lehrkörpers, Gerhard Heberer, Privatdozent für Zoologie und Vergleichende Anatomie, Lehrberechtigung 11. März 1932.
[‡] Haeckel, 1934, p. 33.
[§] Bölsche, 1934, p. 37.

Nature ... unconditionally sacrifices the individual for the necessities of life of the species, and ties all cultural achievements to the interaction of countless particular forces
and singular individuals. One can, indeed, consider the organization of ever more inclusive *Lebens-* and *Arbeitsgemeinschaften* the secret law of all organic development.[*]

The Haeckel worship in *Der Biologe*, coupled with a rejection of holistic–organicist
interpretations of nature, intensified after Lehmann had been ousted from the editorial office in November 1938 through an initiative of the SS-*Ahnenerbe* and the SS
Security Service (SD). Following a directive from Himmler, Walter Greite, then head
of the Biological Research Section at the *Ahnenerbe* (Ancient Heritage Foundation),
established in May 1939 the Reich Association of Biologists (*Reichsbund für
Biologie*), based on the premise that membership would be mandatory for all
German biologists.[†] Greite, who studied at the University of Göttingen, was part
on an old Göttingen SS alliance that had formed around Rudolf Mentzel, later of
the *Reichsministerium für Wissenschaft, Erziehung, und Volksbildung* (REM) and
successor of Stark as president of the DFG and its Reich Research Council,[‡] where
Greite had acted as a division head before he moved on to the *Ahnenerbe*.[§] With Fritz
Knoll (1883–1981; NSDAP, 1937)—president of the University of Vienna—as its
head, the Reich Association of Biologists succeeded the as of then defunct German
Association of Biologists (*Deutscher Biologen-Verband*) that had been launched by
Lehmann.[¶]

The journal *Der Biologe* was simultaneously taken over by *Das Ahnenerbe*,
with Astel on the editorial board along with other prominent racial politicians
such as Walter Gross (1904–1945; NSDAP, 1925) from the Race Politics Office
(*Rassenpolitisches Amt*) of the NSDAP, Fritz Kubach (1912–1945; NSDAP, 1933),
the *Führer* of the Reich Student Association, and the Curator of the *Ahnenerbe*,
Walther Wüst. The managing editor was Walter Greite, assisted—among others—
by the botanist Walter Zimmermann from Tübingen, Heberer from Jena, the paleontologist Weigelt from Halle (Saale), and the animal psychologist Konrad Lorenz
from Vienna. Greite introduced the first number of *Der Biologe* in its new format
as a "weapon in the fight (*Kampfblatt*) for the research of the Laws of Life and
for the advancement of Biology"[**]: the Darwinian struggle for existence, the eternal
confrontation of hereditary endowment and environment is as old as is the Nordic
race and the blood that sustains it. This, he claimed, must be the base of the German
world view, one that seeks to preserve and maintain the legacy of the forefathers
(*Ahnenerbe*).[††]

His introduction to the "new" mission of the journal reflects the same allegedly
Haeckelian spirit that also pervaded Greite's sketch of the tasks and goals of the
newly founded Reich Association of Biologists, cast as an appeal to the "most able

---

[*] Lipsius, 1934, p. 46.
[†] Grüttner, 2004, p. 63f.
[‡] Walker, 2003, p. 994.
[§] Mertens, 2004, p. 130.
[¶] Kater, 1974, p. 413, n. 148; Grüttner, 2004, p. 63f.
[**] See also Bäumer, 1990a, p. 45.
[††] Greite, 1939a, p. 1f.

ones amongst our ancestors, who in a heroic struggle and through intense work have gained insight into the Laws [of Life], which they recognized as being true and consequently to be applied to their ultimate consequences."[*]

In the 1939 volume of *Der Biologe*, Heberer commented in a similar spirit on the significance of Haeckel's work for phylogenetic systematics in the service of selection theory: "Haeckel is the great classic of phylogenetic systematics, and will remain so in the future ... the theory of descent is a proven theory ... space in this journal is too precious to be wasted on critics of Haeckel, who mostly seek ideological gains from their opposition to Haeckel"[†]—those being mainly neo-Lamarckists and representatives of the Catholic Church. Heberer's plea was seconded by Werner Zündorf[‡] (1911–1943; SA, 1934–1938; NSDAP, 1937), at the time a graduate student of Ernst Lehmann in Tübingen, who articulated a similar appreciation of Haeckel from a botanical perspective. Haeckel, he argued, "is the ultimate biologist and a unifying phylogeneticist,"[§] the one who ultimately secured the breakthrough of Darwin's evolutionary theory in Germany—the latter again equated with selection theory.

Although a graduate student of Lehmann, Zündorf was a *protégé* of Heberer, whose assistant he became after his graduation on December 15, 1938.[¶] Within the National Socialist German Student Association (*Nationalsozialistischer Deutscher Studentenbund*, NSDStB), Zündorf acted as a Section Head (*Fachgruppenleiter*) within Natural Sciences from 1935 to 1938/1939.[**] In this capacity, Zündorf reported in *Der Biologe* in 1938 on the biological activities of the Natural Sciences section of the student leadership at the University of Tübingen.[††] It is the "will of the young student team" (*das Wollen der studentischen Mannschaft*), which leads to success in their collaboration with national-socialist docents, as is documented by the honorable mention their manifesto on *The Theory of Descent in the Ideological Fight of Present Times* had earned in the *Reich* performance competition of students that had been called out in 1935.[‡‡] Heberer, then still a docent in Tübingen, had lent a helping hand in the composition of this ideologically tainted prospectus on the significance of evolutionary theory in the fight against not only the Catholic Phalanx that opposed Darwinian selection theory, but also against neo-Lamarckism and idealistic morphology: the exposition was meant to be a declaration of war against

> biologically tainted ideological currents, which sometimes masked, sometimes in full openness seek to destroy proper biological research with the goal to erect an edifice of mystical speculations on the ruins of Aryan-Nordic natural science, and therewith to deprive the National Socialist ideology (*Weltanschauung*) of its natural foundations.[§§]

---

[*] Greite, 1939b, p. 241.
[†] Heberer, 1939, pp. 264ff.
[‡] Hoβfeld, 2000, p. 261; 2003.
[§] Zündorf, 1939a, pp. 274ff.
[¶] Junker, 2004, p. 251.
[**] Hoβfeld, 2000, p. 262.
[††] Zündorf, 1938, p. 311; see also Potthast and Hoβfeld, 2010, p. 456.
[‡‡] See announcement of the Reichsschaft der Studierenden an den deutschen Hoch- und Fachschulen: "We hope to see a strong participation of students of biology in this performance competition." *Der Biologe*, 1935, 4: 332–333.
[§§] Zündorf, 1938, p. 311.

These were the circumstances, Zündorf claimed, that motivated the format of the paper as a pamphlet (*Kampfschrift*) defending true Nordic science against veiled nature mysticism. Another plant geneticist and graduate student of Ernst Lehmann in Tübingen from 1935/1936 to 1938 was Heinz Brücher,* a fellow party member (he joined the NSDAP in 1934 as a Leader of Hitler Youth) and personal friend of Zündorf, who after his graduation went to work with Astel in Jena. He obtained his *Habilitation* under Astel and the botanist Otto Renner in 1940. Later he became an important plant geneticist working for the *Ahnenerbe* and the SS, joining expeditions into the eastern territories and, as of 1943, headed the newly founded *Ahnenerbe* institute for plant genetics in Lannach, Austria. After the war, Brücher emigrated to South America, and was brutally slain on his farm Condorhuasi in Mendoza, Argentina, on December 17, 1992, probably by the cocaine mafia, who saw in him an adversary as he promoted alternative sources of income for small farms.† Early in the 1960s, Brücher inquired with Walter Zimmermann, who still served as botany professor at the University of Tübingen, about the possibilities to return to Germany.‡ Motivated mostly by concerns regarding the schooling of his children, he asked whether there was any substance to the rumors that German universities kept blacklisting old Nazis in order to exclude them from employment?§ How is the "Barn Owl Lehmann doing," Brücher further inquired (Lehmann was deceased in December 1957), and also—tongue in cheek: "Who was Haeckel? I have almost forgotten. Is it after all permitted to talk of him again in central Germany? I sometimes feel like living in a world full of fools, for which reason I am planting a lot of vines..."¶ Brücher's self-ironic reference to Haeckel reflected back on the notorious book he had published while still a graduate student in Tübingen in 1936 through Julius Lehmann in Munich, titled *Ernst Haeckel's Blut- und Geistes-Erbe*. The book, instigated by Astel and Lothar Stengel von Rutkowski, and its content amplified and critically vetted by Haeckel's grandnephew Werner Haeckel, became a central pillar among three in the effort to cast Haeckel's *Naturphilosophie* in terms of Nazi ideology**: the other two book-length initiatives in support of the same cause were Heinrich Schmidt's *Ernst Haeckel—Denkmal eines grossen Lebens* (1934),†† and Victor Franz's *Ernst Haeckel und das einheitliche Weltbild* (1944 unpublished).‡‡ Western culture, its liberalism and individualism, victorious in the Great War and dictator of the Versailles treaty, rendered a positive Darwin reception problematic in National Socialist Germany—Haeckel, the "German Darwin," was the obvious

---

* Deichmann, 1996, pp. 258ff; see also Hoßfeld, 2005b, p. 179.
† Rehm, 1992, p. 1.
‡ Letter from Brücher to Zimmermann, "*Im September 1961*" (UAT 286/1, Zimmermann Korrespondenz).
§ Zimmermann answered with a letter dated November 30, 1961, assuring Brücher that such blacklisting did not exist (UAT 286/1, Zimmermann Korrespondenz).
¶ Letter from Brücher to Zimmermann, "*Im September 1961*" (UAT 286/1, Zimmermann Korrespondenz).
** Hoßfeld, 2005b, pp. 174–176.
†† Schmidt's, 1934a, Haeckel monograph was reviewed by Ernst Lehmann in *Der Biologe*: "... Haeckel sought closeness to nature (*Naturverbundenheit*) and to biological thinking—the deepest essence of National Socialist ideology is to seek closeness to nature, National Socialism itself is politically applied biology ... Haeckel—the Nordic warrior ... a real man" (Lehmann, 1934d, p. 132).
‡‡ Hoßfeld, 2007, p. 459.

idol to turn to.* Brücher's book was prefaced by Astel, who used the opportunity to sketch the Jena agenda, one that was fully supported by Brücher: to justify the Law, decreed on July 14, 1933, for the Prevention of Hereditarily Diseased Offspring with Haeckelian argumentation patterns, and to use the theory of descent, and in particular selection theory, as a platform for a campaign against Jewry, the Church, and Bolshevism.† On October 17, 1936, Stengel-Rutkowski wrote to Brücher: "And by the way, our ideological front has recently been fully endorsed by the Reich Minister for Science, Rust."‡

In a letter to Himmler, dated April 28, 1939, Astel praised Brücher's Haeckel book as a "culture-biological monograph" that presents "a racial-biological analysis of Haeckel's personality, and highlights the significance of Haeckel's work for modern science."§ Brücher's Haeckel book was reviewed by Heberer in *Der Biologe*: "Ernst Haeckel," Heberer proclaimed,

> the eminent German naturalist, the lucid culture-biologist, the unbending fighter and uncompromising pagan, is today appreciated as a man of Nordic spirit, whose lifework the *völkisch* state uses as a foundation ... [Haeckel] is culture-biologically and politically the forerunner of our state biology, a red-hot *völkisch* nationalist.¶

Heberer praised Brücher as a student of Astel, Head of the Thuringian Office for Racial Affairs, and commended his book as the first attempt ever to sketch Haeckel's eminent personality in the context of Haeckel's own racial and family background, to understand Haeckel's work in the context of his own time, and to derive from it its significance for the present.

Other than through books and articles, it was through the Ernst-Haeckel-Society that the ideological reconstruction of Haeckel was promoted. One year before his death, in the summer of 1918, the company of Carl Zeiβ, manufacturer of optical instruments, bought Haeckel's house, the *Villa Medusa*, with the remit that it would be transformed into a museum and archive after the owner's death.** Victor Franz, its director from 1935 to 1945, engaged Heberer and others in 1941 in discussions revolving around the constitution of an *Ernst-Haeckel-Gesellschaft*. On July 23, 1941, Franz wrote to Sauckel, hoping to enlist him as patron of the new society: "The Ernst-Haeckel-Society is not a political association, but nonetheless aims to direct its members with utmost political clarity toward Adolf Hitler's Germany."†† Members were courted with a letter specifying the mission of the new

---

* Hoβfeld, 2007, p. 454.
† Hoβfeld, 2005b, p. 179; 2007, p. 460.
‡ Hoβfeld, 2005b, p. 177.
§ Hendel et al., 2007, p. 209.
¶ Heberer, 1937b, p. 65; Lehmann characterized Brücher's arrival in his lab: "Mister Brücher was a young man who assessed himself very highly. At the time he was 22 years old. When he came to me, he had just written a book, which he published together with President Astel—Jena (Weimar), and which received a highly laudatory review in the journal *Der Biologe* that I edited." *Universitätsarchiv Tübingen*, UTA 126/373, 18: Lehmann, "Abschliessende Äusserung," 24. Mai 1940, p. 22.
** Schmidt, 1934b, p. 49.
†† Hoβfeld, 2003, p. 545.

society as one that "seeks to preserve the memory of this groundbreaking natural scientist, warrior, and upright professor."[*]

The central message casting Haeckel's reconstruction in the light of National Socialism once again permeated Greite's introduction to the 1940 volume of *Der Biologe*:

> Dualism, the idea of a separation of body and soul, is un-biological ... thinking in terms of the Laws of Life has taught us to recognize this weak and unrealistic idealism, one that weakened not only our personal lives but also undermined and clouded the life of our *Volk*.[†]

THE DECONSTRUCTION OF ERNST LEHMANN

With his efforts as editor of *Der Biologe*, and as leader of the *Deutscher Biologen-Verband*, Lehman had established himself, or at least was acclaimed by some such as Lothar Stengel-von Rutkowski, as the "*Führer* of German biologists."[‡] But by the end of 1938, Lehmann had come under serious attack from members of the SS, who had different ideas about the direction a genuine German Biology should take, one that counters Lehmann's holism and organicism. Pressure on Lehmann built from two sides, one the circle around Walter Greite at the Ancestral Heritage Foundation (SS-*Ahnenerbe*), the other the phalanx around Karl Astel at the University of Jena. The anatomist and anthropologist Robert Wetzel (1898–1962), Vice-President of the University of Tübingen, member of the SS, and an engaged enemy of Lehmann,[§] was later identified by "insiders ... as the man behind the scheme to oust Lehmann"[¶] not only from the University of Tübingen, but also from the German Association of Biologists.

Walter Greite was born on June 13, 1907, in Hannover. He studied forestry in Freiburg i. Br., then biology, physiology, and chemistry in Göttingen, where he obtained his PhD in the autumn of 1932. As of February 1, 1932, Greite was a member in the NSDAP; in 1933 he joined the SS. Also in 1933, Greite became assistant to Ernst Mangold (1879–1961) at the Agricultural College in Berlin, a position from which the REM pulled him out later in the same year with the order to organize an association of Prussian university docents, the leader of which he was to be. As of April 1, 1935, Greite taught genetics and racial theory at the College of Teachers' Education in Frankfurt a.d. Oder, while serving as referee in the DFG, then under the direction of the Nobel Laureate Johannes Stark, cofounder of German Physics. In the fall of 1935, Greite abandoned his teaching job to work full-time for the DFG. When his old friend Mentzel took over as head of the DFG as of November 1936, Greite was tasked to provide "political reviews" of grant applicants.[**] The following year, Greite left the organization and was appointed in the section for biological genetics at the Reich Health Office, effective April 1, 1937. His SS-personnel file

---

[*] Hoβfeld, 2003, p. 546.
[†] Greite, 1940, p. 1.
[‡] Stengel von Rutkowski, 1937, p. 33.
[§] Heiber, 1991, p. 438.
[¶] Harrington, 1996, p. 197.
[**] Mertens, 2004, p. 130.

characterizes him as prevalent Nordic, a focused personality, uncompromising and with a strong will, ideologically steadfast, and well versed in biology.* Greite was listed as a volunteer in Himmler's *Reichssicherheitsdienst des Reichführers-SS* (SD), with whom the SD sought closer contact after Greite's promotion to *Regierungsrat* at the Reich Health Office as of June 1, 1937.

Ernst Lehmann fell into the crosshairs of Greite and Kurt Riedel[†] from the SD in the fall of 1938. At a meeting in Gmünd in Lower Austria on August 11, 1938, Himmler had tasked the Curator of the *Ahnenerbe*, Walther Wüst, with the reorganization and realignment of biology in the Reich. Wüst reported back to Himmler on October 31, that the necessary steps had been taken in collaboration with the SD.[‡] The SD had reported "untenable conditions" prevailing in the German Association of Biologists under Lehmann's leadership. Lehmann was forced to step down as head of the *Deutscher Biologen-Verband* in November 1938, and Fritz Knoll, President of the University of Vienna, was installed as deputy managing director; the position of the managing director was listed as "currently vacant." A Section for Biological Research was established within the *Ahnenerbe*, with Greite—on leave from the Reich Health Office—as its director. *Ex officio*, Greite simultaneously became editor-in-chief of *Der Biologe*, the journal from now on being published under the close auspices of the *Ahnenerbe*. The President of the University of Tübingen, Hermann Hoffmann (1891–1944), in a letter dated November 13, 1938, asked Lehmann—the latter under investigation because of various allegations detailed below—why the German Association of Biologists as well as its journal had been taken out of his hands? The hidden implication of that inquiry was to demonstrate Lehmann's political unreliability. Lehmann answered on June 5, 1939, that his was not the only journal annexed by the *Ahnenerbe-SS*. The same had happened to the Archive for Religious Studies (*Archiv für Religionswissenschaften*), previously edited by his faculty colleagues. There was no reason, therefore, to single him out for inquiry.[§] Greite, however, wrote to Hoffmann on December 15, 1938, that Lehmann had voluntarily stepped down from his positions in the German Association of Biologists and its journal: he understood, certified Greite, that their "affiliation with the *Ahnenerbe* could only be completed following personnel changes," most likely due to the fact that Lehmann had never succeeded to become a member of the NSDAP. There loomed, however, in the background an investigation of Lehmann for embezzlement of funds related to his role as editor of *Der Biologe*.[¶] In his attempt to whitewash himself in a letter dated July 9, 1945, after his dismissal from civil service on command of the French Military Government, Lehmann characterized the unfriendly takeover by the *Ahnenerbe* as an initiative to "turn the [German] Association of Biologists into an organization of racial biology. When I vehemently resisted this attempt, I was removed from my position."[**]

---

* Bundesarchiv Berlin, SS-Führerpersonalakten, signature 31-A, file "Dr. Walter Greite."
† Weindling, 1989b, p. 538.
‡ Bundesarchiv, DS (formerly Berlin Documentation Center), microfilm signature DS GG 119, file "Greite, Walter."
§ *Universitätsarchiv Tübingen*, UAT 126/373.1.
¶ *Universitätsarchiv Tübingen*, UAT 126/373.4.
** Lehmann, Rechtfertigungsschreiben, 9. Juli 1945; *Universitätsarchiv Tübingen*, UAT 126/373.1.

On January 23, 1939, the managing director of the *Ahnenerbe*, Wolfram Sievers (1905–1948; NSDAP, 1929), tasked Fritz Knoll with the organization of a general meeting of the German Association of Biologists in Berlin, in the rooms of the *Ahnenerbe*. At the meeting, which took place on May 5, 1939, Lehmann's *Deutscher Biologen-Verband* was reorganized as Reich Association of Biologists (*Reichsbund für Biologie*) under the patronage of Himmler, with Knoll as its president (*Vorsitzender*), and Greite as its managing director (*Bundesleiter*). Greite had thus taken control of all matters biological in the Reich—or of almost all. There remained the *Zeitschrift für die gesamte Naturwissenschaft*, which was likewise taken over by the *Ahnenerbe* in 1939, and which carried biological entries under the editorship of the Munich botanist Ernst Bergdolt. On December 8, 1939, Greite drafted a memo underscoring the political importance of biology, and the role the ZGN should play in that respect:

> [Biology provides] the strongest pillars for the National Socialist world view ... Biology is represented by the *Reichsbund* and its research section at the *Ahnenerbe*. The platform of *Der Biologe* serves the competent, factually and politically correct representation of biology ... the ZGN should no longer publish biological articles ... Dr. Ernst Bergdolt is a most pleasant person and politically highly trustworthy (decorated with the *Blutorden*), but his science is controversial ... The ZGN was founded at a time when efforts were made to integrate universities and to overcome narrow specialization. I was myself involved with the founding of the journal and have provided the initial financial support through the DFG. But since then, the journal has taken an ill-fated turn given its unfortunate programmatic perspective ... the ZGN is furthermore in competition with *Die Naturwissenschaften* ...[*]

Greite tried to convince the publishing arm of the *Ahnenerbe* to let him restructure the ZGN as a journal that was to treat anorganic nature exclusively, and thus acquire a format for which a "reliable National Socialist magazine" did not already exist. He underscored the blatantly contradictory contributions that had appeared in 1940 in the ZGN and in *Der Biologe*, respectively. The ZGN had carried an entry on idealistic morphology by Friedrich Waaser, who, citing Wilhelm Troll, Karl Beurlen, and of course Goethe, attacked the externalism and mechanism that underlie Darwinian selection theory, the latter labeled a product of the "specific English-Western spirit".[†] *Der Biologe* on the other hand published in the same year a vitriolic attack on idealistic morphology by Werner Zündorf, who defended a strict selectionist point of view against Wilhelm Troll, Edgar Dacqué, and—to a lesser extent—the fellow party member Ernst Bergdolt, citing Ernst Haeckel, Gerhard Heberer, and Walter Zimmermann as his allies.[‡] "Much work remains to be done," urged Greite, to unify—or rather, force into line—biology under the banner of Darwinian selectionism. Greite continued to push this issue throughout 1940, but was brusquely rebuked

---

[*] Bundesarchiv, DS (formerly Berlin Documentation Center), microfilm signature DS GG 119, file "Greite, Walter."
[†] Waaser, 1940b, p. 12.
[‡] Zündorf, 1940.

in January 1941: if he did not like the ZGN, he should simply resign from the editorial board, he was told. The incident foreshadowed Greite's fall from Himmler's grace.

Managing director Sievers informed Greite on December 12, 1938, shortly after he had joined the *Ahnenerbe*, that the *Reichsführer*-SS Himmler had tasked the *Ahnenerbe* on December 8 of the same year to conduct anthropological research on inmates of alien races in concentration camps. Curator Wüst wrote back that such research was customarily conducted by the Reich Health Office, not the *Ahnenerbe*. On February 22, 1939, the DFG awarded Greite RM 10,000 to conduct anthropological research on "alien races and bastards." While aspiring to anthropological research, Greite all the while sought employment as tenured civil servant (*Beamter*) in the SD. This he was not granted, however. Instead, the idea was floated to seek a tenured teaching position for Greite at some German university. On leave from the Reich Health Office to the *Ahnenerbe* until March 16, 1941, Himmler requested a prolongation of Geite's leave of absence by one year, to give him time to write a *Habilitation* thesis. The thesis on Jews of Vienna, which Greite submitted to the University of Munich was rejected, however. Submitting the same thesis at Jena University, Greite had to proactively withdraw it after he failed to get any support from either Astel or Heberer. Greite's *Habilitation* was rejected not only because he had not taken any formal courses in anthropology, but also because the collection of the data for this thesis had been a collaborative effort under the leadership of Christian von Krogh (1909–1992; NSDAP, 1930; SS, 1931) from the University of Munich.[*] It was further alleged that he had not written the thesis himself. Sievers found that Greite's attempt to seek the *Habilitation* on such a fraudulent basis had severely damaged the reputation not only of the *Ahnenerbe*, but also of the SS in general. Sievers further complained that Greite increasingly revealed incompetence in biological matters: it was suggested that the success of *Der Biologe* was not reflecting Greite's competence, but was instead the result of the excellent qualifications of the members of the editorial board. Finally, Greite was accused of having used his position and influence in improper and secret attempts to seek higher level employment in Berlin and Salzburg. The conclusion of the "Greite affair" was that Himmler—with a fiat issued the day before his leave of absence came to an end on March 15, 1942—suspended Greite from all his responsibilities at the *Ahnenerbe* as well as in the Reich Association of Biologists, and ordered him to report back at his previous post at the Reich Health Office on March 17, 1942: "I have reached that decision on the basis of your behavior particularly over the past year, which rendered your employment in any office of the *Reichsführung*-SS or at the *Ahnenerbe* impossible."[†] Caught in all this turmoil and what he considered vile intrigues against him, Greite remarked to Wüst in the fall of 1941 that he would "love to just get rid of my SS uniform and go abroad."[‡] Greite held on to the editorship of the journal

---

[*] Hoßfeld and Junker, 2003.

[†] Himmler to Greite, March 15, 1942. Bundesarchiv, DS (formerly Berlin Documentation Center), microfilm signature DS GG 119, file "Greite, Walter."

[‡] Notes on a conversation with Greite by Wüst, October 10, 1941. Bundesarchiv, DS (formerly Berlin Documentation Center), microfilm signature DS G 119 (file "Walter Greite").

Der Biologe, however, which continued to be published by J.F. Lehmann in Munich but for financial reasons started to decline in 1943.[*]

The second front against Ernst Lehmann was opened by Astel and his *entourage* in Jena, most notably Heberer among them. Born in Schweinfurt on February 26, 1898, into an evangelical family, Karl Astel studied medicine and philosophy after graduating from the *Humanistisches Gymnasium*. In his personnel file of 1934,[†] he declared himself a "specialist for human breeding and genetics, racial theory, and population science and politics." He listed Ploetz, Lenz, Darré, and the race-Günther as authors who had most strongly influenced him, and emphasized the manifold animal breeding projects he had pursued since early childhood. In the spring of 1919, Astel fought against "Bolsheviks" in Würzburg and Munich. In the winter of 1919/1920, he founded the Schweinfurt chapter of the *Deutsch-völkischer* Protection and Defense League, "in which I included the current SS Honorary Group Leader, *Gauleiter* and *Reichsstatthalter* [Fritz] Sauckel." "In the year 1927 [June 4, 1927], I married a healthy woman of the Nordic race," Margarethe Lorenz. On July 1, 1930, Astel became a member of the NSDAP: "I have always only acted in the *völkisch* spirit." He boasted that in the spring of 1932 already, he had founded (together with Fritz Lenz) a public office for hereditary consultation: "I already conducted sterilizations in the year 1932!" "Come January 1, 1934," Astel proudly announced, "21 Hereditary Health Courts will take up their work in Thuringia, which will finally pave the way to the sterilization of carriers of severe hereditary diseases ... Thuringia alone spends 25 million *Reichsmark* per year on [the treatment of] hereditary diseases of all kinds." In June 1934, Sauckel named Astel full professor at the University of Jena, the latter maintaining his leadership position at the Thuringian Office for Racial Affairs in Weimar. On May 8, 1935, Astel communicated to Himmler his goal to "to render Thuringia a fortress at the frontline of the SS-battle against all extra-governmental powers including Christianity, and for the indoctrination of the *Volk* with a mode of thought that conforms to the Laws of Life."[‡]

The most active and influential phylogeneticist and racial theorist around Astel was Gerhard Heberer. Astel had met with Curator Wüst on January 25, 1938, to discuss the integration of the SS squad quarters (*SS Mannschaftshaus*) Jena into the natural sciences division at the University of Jena, in order to enhance the latter's National Socialist reorganization. As a faculty member both at the SS squad quarters and at the University, Astel recommended "*SS-Untersturmführer* Dr. Heberer, currently in Tübingen" for that position: "Since the resident zoologist, whose position he should have taken over, cannot be readily removed, and furthermore is a person who is unsuited for our purposes,[§] [Heberer] will be assigned a parallel chair for 'Theory of Descent and General Biology.' Heberer is one of the most important men in the

---

[*] Bäumer, 1990a.
[†] Bundesarchhiv, PK (formerly Berlin Documentation Center), microfilm signature PK A0089. File "Karl Astel."
[‡] Hendel et al., 2007, *Dok. Nr.* 54, p. 98.
[§] The person in question was the neo-Lamarckist Jürgen W. Harms, "Haeckel-professor" and head of the Zoological Institute of the University of Jena as of 1935, as well as (*ex officio*) Director of the *Phyletisches Museum*: see Potthast and Hoßfeld (2010, p. 442).

field of general genetics, especially in chromosome research. His collaboration is of utmost importance."[*] Not all of Astel's plans materialized in the way he intended.

Gerhard Heberer[†] was born on March 20 in Halle/Saale. He studied biology, general zoology, genetics, and anthropology at the University of Halle. From May 15, 1919 to February 15, 1920, he voluntarily joined the *Freikorps* movement engaged in the suppression of leftist uprisings. In 1924, he assumed a leading role in the *Wandervogel*, a *völkisch* German youth movement that revived German Romanticism in celebrating closeness to nature. On July 23, 1924, he defended his PhD, earning the predicate *summa cum laude*. He described himself as a specialist in chromosome research, hydrobiology, and racial history, declaring a special interest in Indonesian races. The latter he probably derived from his participation, in 1927, in the Sunda Expedition led by the zoologist Bernhard Rensch (1900–1990, of whom more later), and a succeeding 6 months stay at the Zoological Institute in Buitenzorg (Bogor, Java), during which time he traveled throughout Southeast Asia. His interest in Indonesian racial history related intimately to the question of the origin of the Indo-Germanic Race. From May 1, 1928 to October 31, 1935, he was scientific assistant at the Zoological Institute of the University of Tübingen, where he earned the *venia legendi* on November 23, 1931. In 1935/1936, he stood in for Otto zur Strassen (1869–1961) at the University of Frankfurt, hoping he might become his successor. In spite of Sievers' intervention and efforts,[‡] that hope did not materialize, however, and Heberer had to return to Tübingen, where he resumed his position as scientific assistant from April 1, 1936 to October 31, 1939, with stints at the Zoological Stations in Naples and Rovigno d'Istria. When under consideration for a position in Jena, the Dean of the Science Faculty at the University of Tübingen, the paleontologist Edwin Hennig, signed off on a statement lauding Heberer for his publications in theoretical biology, through which he "had actively engaged in various ways in the ideological battle hat was playing out in biology."[§] As of October 1, 1939, Heberer was appointed associate professor at the University of Jena. The SS had tried to find him a professorship at a German University for some time already, and Astel of course wanted him for Jena. In the summer of 1938 already, Heberer was asked whether he would take on the directorship of the *Phyletisches Museum*, a position which through Haeckel's testament was tied to the directorship of the Zoological Institute at the university, and which at the time was held by Jürgen W. Harms (1884–1956). Harms was known as a neo-Lamarckian, and naturally opposed such plans. Heberer wrote to Astel on May 2, 1938, encouraging him to break Harms' opposition to his employment in Jena: "time is of the essence and works against a German Biology as *we* understand it."[¶] Heberer eventually succeed Karl Kötschau at Jena University, as

---

[*] Bundesarchiv, DS (formerly Berlin Documentation Center, microfilm signature DS B 026 file "Karl Astel."

[†] The most complete biography of Heberer was offered by Hoβfeld (1997); see also Potthast and Hoβfeld (2010, pp. 442ff). The present account also draws from Heberer's SS personnel file (Bundesarchiv [formerly Berlin Documentation Center], SS-Führerpersonalakten, 72-A), and Heberer's *Rektoratsakte*, Vol. I, October 1949—December 1959, University Archive Göttingen (GWDG).

[‡] Kater, 1974, p. 137.

[§] Letter from Stolte to E. Hennig, April 13, 1938, and Hennig's consent; *Universitätsarchiv Tübingen*, UAT 117C/16.

[¶] Hoβfeld, 1997, p. 66f; emphasis added.

he became an associate professor for general biology and human evolution, but he was never appointed as director of the *Phyletisches Museum*: Harms had prevailed against Astel.* In the spring of 1943, Heberer was to be transferred to the University of Straßburg, which prompted Sauckel to plead with the Rust ministry to retain the "man who proved to be excellent" in Jena, and to promote him to full professorship (Figure 7.1).†

April 12, 1937, Himmler named Heberer an SS-*Untersturmführer* and member of the board of directors of the SS Race and Settlement Main Office (*Rasse- und Siedlungshauptamt*). As of February 3, 1938, Heberer became a volunteer in the SS Race Office, and was entitled to the title of an adjunct of the *Ahnenerbe*.‡ September 11, 1938, Heberer was promoted to the rank of a SS-Obersturmführer. In a letter dated June 14, 1937, Heberer reported to the SS that the Tübingen office of the NSDAP had sent him the paperwork for application for membership: "My admission as a member is desired in many quarters, and would certainly facilitate my work

FIGURE 7.1   Gerhard Heberer in 1943. (Heberer Estate, Private Property of Uwe Hoβfeld, Jena.)

---

* In a declaration deposited in February 1950 at the University of Göttingen, in the context of an investigation of his political past, Heberer insisted that he would never have aspired to actions that violated Haeckel's will, that he had never taken part in any intrigue against Harms, but always behaved in a most correct manner, a position supported by his biographer Uwe Hoβfeld on the basis of archival material in Heberer's private estate: Hoβfeld, 1997, p. 69. See also Heberer, "Erklärung—hinsichtlich der Frage des Phyletischen Museums—Jena," University Archive Göttingen (GWDG), Rektoratsmappe Gerhard Heberer, Band 1, von Oktober 1949 bis Dezember 1959.
† Letter from Sauckel to Rust, March 8, 1943: Heiber (1994, p. 127). See also Hendel et al. (2007) *Dok. Nr.* 103, p. 176, and Hoβfeld (1997, p. 84f).
‡ See also Kater (1974, p. 66). Sievers to Heberer: "Of course you may also refer to yourself as an adjunct of the *Ahnnenerbe*."

on behalf of a German Biology ... Prof. Dr. Astel, SS-*Obersturmführer* is happy to provide at any time information on me that might be relevant to my membership in the party, and to act as my guarantor."[*] His party membership had taken effect on June 10, 1937. A memorandum from September 9, 1937, noted that the reprints of his publications, which Heberer regularly sent to the SS, showed him to represent his science in a truly National Socialist spirit. In 1941, Heberer was decorated with the SS-Honor Ring (*Totenkopfring*). Effective January 30, 1942, Heberer was promoted to SS-*Hauptsturmführer*. With the collapse of the Third Reich, Heberer became a Czech prisoner of war; he managed to escape,[†] and returned to Germany (to the "civilized world" as he put it)[‡] in 1947. As of October 1, 1949, Heberer headed an Anthropological Research Center at the University of Göttingen. Going through the de-nazification procedures, Heberer had managed to be classified in category V as a non-incriminated person. In the questionnaire for political review, Heberer claimed that his membership in the NSDAP was not entirely voluntary, and that he had never played any active role in the party. Heberer insisted that he had received a letter from the SS following a presentation he delivered at the 1937 Assembly of German Natural Scientists and Physicians, which ordered him to submit reprints of all his publications: "I have not honored that request."[§] The next thing he knew, so he attested, was that Himmler had named him SS-*Untersturmführer* without any prior consultation. Reference letters relating to the de-nazification procedures, as well as those that were solicited when Heberer was under consideration for a lectureship at the University of Göttingen, were mixed in their assessment of Heberer's political past. Some emphasized his enthusiastic alignment with National Socialist ideology, "*ein grosser Nazi*" who was frequently seen sporting his SS uniform. The plant geneticist Otto Renner, at the University of Jena from 1920 to 1948 (later Munich), provided a balanced judgment, pointing to the fact that Heberer had defended him in front of the Gestapo in February of 1945.[¶] Noting that Himmler had wanted Heberer as a professor for human evolution in Jena, he was nevertheless appointed only after proper academic procedures had been followed. He specified that he thought Heberer to have been a resolute party member, and that he had disapproved of Heberer's fight against the Church. On the other hand, Heberer had never participated in anti-Semitic propaganda, and in contrast to Astel had not worn his SS uniform at work. However, "a difference of opinion between him and the fanatical Astel never became apparent."[**] Heberer died in Göttingen on April 13, 1973.

---

[*] Bundesarchiv (formerly Berlin Documentation Center), SS-*Führerpersonalakten*, Gerhard Heberer.

[†] Heberer's *Vita*, undated, in the University Archive Göttingen (GWDG), Rektoratsmappe Heberer, Oktober 1949 through December 1959.

[‡] Heberer, 1956, p. 101.

[§] Heberer, Anlage zum Fragebogen für die politische Überprüfung, University Archive Göttingen (GWDG), Rektoratsmappe Gerhard Heberer, Band 1, von Oktober 1949 bis Dezember 1959.

[¶] Letter of O. Renner to Heberer, January 29, 1950. Renner hinted, however, that the palpable nearness of the end of the war might have made such testimony easier for Heberer. Renner also suspected Heberer's influence on Astel, who as President of the University Jena allowed Renner to speak his mind with unusual latitude. University Archive Göttingen (GWDG), Rektoratsmappe Gerhard Heberer, Band 1, von Oktober 1949 bis Dezember 1959.

[**] Letter by O. Renner, December 10, 1949; University Archive Göttingen (GWDG), Rektoratsmappe Gerhard Heberer, Band 1, von Oktober 1949 bis Dezember 1959.

Clearly Heberer and Astel, as indeed Himmler and the *Ahnenerbe*, had a different idea of what German Biology was to be. Not *Ganzheitsbiologie*, not Lehmann's holism/organicism, but instead an uncompromising commitment to selection theory in its Social Darwinist interpretation, the latter embedded in a theory of descent that left no room for the inheritance of acquired characteristics.* It was to be an externalist selectionist Biology of Race that would not accommodate racial change through the inheritance of characteristics that were acquired as an adaptation to local geographical (*Lebensraum*) as well as socio-political (*Kulturlandschaft*) circumstances. The Astel phalanx against *Ganzheitsbiologie* was another expression of the clash between racism and environmentalism.† In articulating his scientific and political views, Lehmann himself was an utterly ambivalent, non-committal person, trying to wiggle his way through treacherous territory by pleasing everyone. He was keen to let his critics know that he considered it a "crime in today's Germany if the secure, genetic foundations of the legislation regarding racial hygiene and eugenics in the National Socialist state are flippantly called into question."‡ In his review of the classic "Baur-Fischer-Lenz" textbook on eugenics, he similarly applauded the fact that *Erbbiologie* had become the foundation for new legislation.§ And again, in a lecture on "Total War and Biology," he propagated the "selection of vigorous individuals from as large a pool as possible of healthy children" as a precondition for success.¶ But then, in his memorandum on German Biology, he specified that the institute he was aspiring to would complement Astel's institute in Jena: "Beyond those issues of race, which naturally feed into genetics and population politics, there are many more issues discussed in biology today. Let's remember ... such important questions as those that relate to the concept of natural wholeness ..."**

Ultimately, however, it was such ambivalence that raised doubts about his commitment to the new Germany, his political reliability, and his scientific integrity. A former student characterized Lehmann as a radical opportunist,†† while an investigation by the Gestapo following Lehmann's denunciation led the investigators to characterize him as a man who outwardly plays the role of a perfect member of the *Volk*, while inwardly harboring a critical attitude toward the NSDAP and the *Führer*.‡‡ In the midst of disciplinary proceedings against Lehmann, the president of the University of Tübingen, the psychiatrist Hermann Hoffmann (1891–1944), characterized Lehmann—whom he desperately wanted to get rid of—as a man marked by an excessive craving for attention, egocentric, yet insecure. His character, he said, was labile, unpredictable, his political and ideological reliability consequently

---

* For an in-depth account see Harrington (1996) and Weindling (1989b, pp. 498–537).
† Bassin, 2005, p. 209.
‡ Lehmann's answer to Willy Müller, April 1, 1936; *Universitätsarchiv Tübingen*, UTA 126/373, 1.
§ The review was published in the *Kritische Vierteljahresschrift für Gesetzgebung und Rechtswissenschaft*, N.F., Vol. 29, pp. 363–240. *Universitätsarchiv Tübingen*, UTA 126/373, 10₂.
¶ Lehmann, "Totaler Krieg und Biologie," no date. *Universitätsarchiv Tübingen*, UTA 126/373, 7ₐ.
** Lehmann, Denkschrift des Deutschen Biologenverbandes, December 3, 1936; *Universitätsarchiv Tübingen*, UTA 126/373, 5.
†† Dr. W. Hoss an Universitätsrat, December 12, 1938; *Universitätsarchiv Tübingen*, UTA 126/373, 4.
‡‡ Geheime Staatspolizei, Staatspolizeistelle Stuttgart, an Herrn Regierungsrat Dr. Borst, Stuttgart, November 24, 1939; *Universitätsarchiv Tübingen*, UTA 126/373, 16.

questionable.* During the same procedures, the Dean of Natural Sciences, the paleontologist Edwin Hennig, appealed to the mandate for self-cleansing at German
Universities: "the main requirement" to remain a professor at a National Socialist
university "is, according to the words of the *Führer*, an unblemished character."†
Lehmann he considered a man who perhaps wants to be positive, but always ends up
creating tension and conflict.

In the campaign against Lehmann, the bigwigs of course sent graduate students
to the front line. The dispute they were to trigger centered primarily on scientific matters, but soon proliferated to a personal level and *ad personam* attacks. It did not help
that Lehmann was most difficult to get along with, as he constantly got entangled in
debates and fights with fellow faculty members, professional staff, as well as students.
One of those was Heinz Brücher, who for his PhD studied genetics in *Epilobium*
(willowherb) under Lehmann. In his own research, Lehmann defended the primary
role of the cell nucleus (*Genom*) in the transmission of hereditary traits, as opposed to
Fritz von Wettstein (1895–1945), director at the Kaiser-Wilhelm-Institute for Biology
in Berlin-Dahlem, and the Jena plant geneticist Otto Renner, who both ascribed to
the cytoplasm (*Plasmon*) an important role in the transmission of hereditary traits.‡
In his own research, Brücher not only found mistakes in a recent publication by
Lehmann and his deceased co-author Otto Schnitzler, but also obtained results that
he found irreconcilable with the stipulation of a monopoly of the cell nucleus in the
transmission of hereditary traits. Lehmann forced Brücher to remove the respective
parts from his thesis before submitting it to the faculty for consideration, with Walter
Zimmerman as second referee. As a result of Lehmann's request, Zimmermann
found the thesis a bit light, but waved it through given the amount of time Brücher
had already invested in it. Brücher, naturally incensed about Lehmann's infringement on academic freedom, complained about his advisor's behavior, while Lehman
made him an offer to stay on in his lab for another year on DFG funding, so he could
critically test his earlier—and for Lehmann controversial, indeed unacceptable—
findings. Fearing ongoing reprisals, Brücher left for the University of Jena, where he
became an independent researcher in Astel's institute, and published the missing part
of his thesis separately. Lehmann threatened to sue Brücher for scientific defamation,
but both Wettstein and Renner could not understand why Lehmann sought to take
these matters to court, rather than to respond to Brücher with a published rebuttal: "if Mr. Lehmann can refute Brücher's allegations on the basis of experimental
results, any journal would be at his disposal."§ Meanwhile, Astel informed Himmler

---

* Der Rektor (gez. Hoffmann) an Herrn Kultminister, Stuttgart, Februar 14, 1939; *Universitätsarchiv Tübingen*, UTA 126/373, 1.
† Zeugenvernehmung E. Hennig, April 26, 1940; *Universitätsarchiv Tübingen*, UTA 126/373,7.
‡ For an analysis of the broader debate see Harwood, 1993. For a detailed account of the Brücher—Lehmann dispute see Hoßfeld (1999b); Potthast and Hoßfeld (2010, pp. 452ff).
§ Wettstein to E. Hennig, Dean of Natural Sciences at the University of Tübingen, December 9, 1938; *Universitätsarchiv Tübingen*, UTA 126/373, 4. See also Renner to E. Hennig, December 12, 1938: "The language in Brücher's paper is unusually pointed, given that this is a discourse between student and teacher. But even more unusual is it that Lehmann would respond with legal action instead of a rebuttal"; *Universitätsarchiv Tübingen*, UTA 126/373, 6.

that Brücher had succeeded to thoroughly "refute Lehmann's theories," and hence to "mark the latter's publications as highly dubious."[*]

In 1932, the National Socialist Teacher's League (NSLB) spread rumors about Lehmann's non-Aryan ancestry,[†] which thwarted his attempt to become a party member in 1933, but which Lehmann could subsequently rebuke. However, during the de-nazification hearings for Lehmann, testimony was given of a conversation among members of the SS overheard in the wine bar *Lehmberger* in Tübingen to the effect that "after the attempt had failed to politically finish off Lehman as a Jew, Brücher will hopefully have managed to finish off Lehmann as a scientist."[‡] In his attempt to do so, Brücher—whom Astel found to be of "immaculate character."[§]—conspired with Werner Zündorf, another graduate student of Lehmann. Brücher and Zündorf, both plant geneticists and ardent Darwinians, were friends, fellow party members and both active in the NSDStB.[¶] Rumors had spread in the Botanical Institute in Tübingen that Lehmann was secretly opening staff mail, an unauthorized action that was later confirmed by Zimmermann in a hearing held on November 15, 1938.[**] To test Lehmann's actions, Brücher sent a specially marked letter from Jena to Zündorf at the Botanical Institute in Tübingen in September 1938.[††] As instructed by Brücher, Zündorf posted the letter back to its sender without opening it.[‡‡] Lehmann stepped into the trap and opened the letter.[§§] Brücher reported the incident to the State Prosecutor, who dropped the case, however, on October 14, 1938[¶¶]—"incomprehensibly" so for Brücher.[***] Zündorf, on the other hand, confronted Lehmann, who promptly requested an investigation of Zündorf's actions by the student leadership because of the damage his conspiracy had done to the reputation of the NSDStB. Zündorf was acquitted on November 27, 1938, with the verdict that "the behavior of the defendant is perhaps not entirely, but still to a certain degree justifiable."[†††] However, testimony given during that investigation brought to the fore a host of accusations against Lehmann, targeting him not only as scientist and administrator, but also as a dubitable, in fact intolerable character. The accusations were grave enough not only for Lehmann to seek an investigation against himself in an attempt to clear his name, but they also triggered an internal investigation of Lehmann by the Dean Edwin Hennig, following an order

---

[*] Letter from Astel to Himmler, April 28, 1939; Hendel et al. (2007) *Dok. Nr.* 128, p. 209.
[†] Kultusministerium Stuttgart, October 24, 1933; *Universitätsarchiv Tübingen*, UTA 126/373, 1.
[‡] Staatskommissariat für die politische Säuberung Tübingen-Lustnau; *Universitätsarchiv Tübingen*, UTA 126/373, 2.
[§] Astel to E. Hennig, Dean of Natural Sciences at the Univversity of Tübingen, December 20, 1938; *Universitätsarchiv Tübingen*, UTA 126/373, 19, #13.
[¶] *Universitätsarchiv Tübingen*, UTA 126/373, 19. #40.
[**] *Universitätsarchiv Tübingen*, UTA 126/373, 4.
[††] Brücher to E. Hennig, November 4, 1938; *Universitätsarchiv Tübingen*, UTA 126/373, 19, #1.
[‡‡] Zündorf, November 3, 1938, Beschleunigte Vernehmung; *Universitätsarchiv Tübingen*, UTA 126/373, 8.
[§§] In the proceedings against him, Lehmann would testify that the letter had ended up in his apartment and had accidentally been opened by his wife. Lehmann, Stellungnahme; *Universitätsarchiv Tübingen*, UTA 126/373, 14C.
[¶¶] Lehmann, "Abschliessende Bemerkungen"; *Universitätsarchiv Tübingen*, UTA 126/373, 18.
[***] Letter from Brücher to the Dean, E. Hennig, November 4, 1938; *Universitätsarchiv Tübingen*, UTA 126/373, 19, #1.
[†††] *Universitätsarchiv Tübingen*, UTA 126/373, 8.

issued by the University President Hoffmann on December 2, 1938.[*] Official disciplinary proceedings against Lehmann were initiated according to an injunction issued by the *Reichstatthalter* Württemberg Wilhelm Murr (1888–1945), dated November 1, 1939.[†] The procedure brought six charges against Lehmann: to have neglected his duty as a personality of public life to uncompromisingly support National Socialism; to have pursued selfish goals and interests in his research to the detriment of a person dependent on him; to have acted in a way that led to a loss of confidence in his staff; to have in the year 1925 sexually harassed a female member of the staff; to have inaccurately accounted for money he received from non-governmental sources; to have lied to an investigating member of the University Board, to the President of the University (Hoffmann), and to Vice-President Wetzel, who was also heading the National Socialist League of University Lecturers, and in this capacity was the most powerful person at the University of Tübingen.[‡] Obviously, the Brücher thesis figured large in those investigations, but so did the issue of sexual misconduct and political ethos. Anneliese Ehlers, a close associate of Brücher and Zündorf according to Lehmann,[§] testified that Lehmann's lectures were peppered with remarks that would on one hand "inflame those who align their thoughts with National Socialism, and on the other hand would drive blood into the cheeks of female students." "Female students," Heinrich Walter (1898–1989) from the Botanical Institute and Garden of the University of Stuttgart testified, "could under no circumstances work under Lehmann."[¶]

President Hoffmann was determined to get rid of Lehmann, in accord with Vice-President Wetzel, who did all he could do to drive Lehmann into voluntary retirement. As if stirring up all these troubles at the university were not enough, Zündorf went on to denounce Lehmann for subversive remarks about the *Führer* and his chief ideologue, Alfred Rosenberg, at the Gestapo on February 16, 1939.[**] Lehmann, Zündorf claimed, had remarked that "Rosenberg has two wives but no children." In his testimony from April 13, 1939, at the Gestapo office in Jena,[††] Zündorf further attested that Lehmann had said when talking about the *Führer*: "'Adolf Hitler, yes well, one hears so many things about him' … while making a thoroughly dubious gesture … I for one took this as an expression of hatred, not of matter-of-factness. I was personally all the more offended since his remarks referred to Adolf Hitler. This is why I still remember the whole scene so vividly. It appeared to me that Prof. Lehmann wanted to somehow express his view that Adolf Hitler for some reason either did not want to, or else could not beget children." In the same interview, Zündorf remarked that he was not alone in finding Lehmann's rubric on families of biologists with many children in *Der Biologe* tacky. The Gestapo concluded that

---

[*] Universitätsrat Knapp, an Rektor Hoffmann, January 18, 1939: *Universitätsarchiv Tübingen*, UTA 126/373, 13, Beilagen, Blatt 2.
[†] Der Reichsstatthalter in Würtenberg and Württ. Kultminister, Stuttgart, November 1, 1939, D7dK/680; *Universitätsarchiv Tübingen*, UTA 126/373, 13.
[‡] Uhsadel, 2011, p. 134.
[§] Vernehmung, December 6, 1938; *Universitätsarchiv Tübingen*, UTA 126/373, 4.
[¶] Testimony H. Walter, January 10, 1939; *Universitätsarchiv Tübingen*, UTA 126/373, 19, #25.
[**] *Universitätsarchiv Tübingen*, UTA 126/373, 16.
[††] *Universitätsarchiv Tübingen*, UTA 126/373, 16.

Lehmann was obviously a person who repeatedly dropped subversive remarks, yet always in a way that left insufficient grounds for prosecution: "All the same, Lehmann must be considered dangerous."*

Effective December 5, 1938, Lehmann was suspended from his official functions at the Botanical Institute, Zimmermann becoming his proxy. As a result, Zündorf was now supervised by Zimmermann, while Lehmann's insubordination resulted in an escalation of the disputes between him and Zimmermann. The official disciplinary proceedings against Lehmann were dropped on June 26, 1942.† Mentzel at the Reich Ministry for Education lifted the order of Lehmann's suspension from teaching, but maintained his ban from heading the Botanical Institute and Garden and also from PhD examinations by decree from April 13, 1943. Giving in to pressure from the University Tübingen, Mentzel withdrew his decree,‡ whereupon Lehmann pleaded by letter dated October 19, 1944: As National Socialist President of a National Socialist University "... I ask you to grant me again the unconstrained development of my research through assignment of the directorship of the Botanical Institute and Garden."§ On October 25, 1945, the French Military Government ordered the dismissal of Lehmann from tenured civil servant status (*Beamtenverhältnis*).¶ The de-nazification committee first classified Lehmann as "fellow traveller," later, in a reconsideration of his case at a meeting on October 27, 1949, as "exonerated," still noting his many "bows to the regime, which the person concerned does not deny."** In his self-defense, Lehmann famously claimed: "In the most degrading manner, my family and I were chased like villains from office to office, to the Gestapo and the SD, and all the way to the fence of the concentration camp." †† Lehmann was formally retired from the University of Tübingen in 1952.

PIONEERS OF PHYLOGENETIC SYSTEMATICS: THEIR BATTLE AGAINST IDEALISTIC MORPHOLOGY

The Heberer biographer and historian of biology Uwe Hoßfeld located the center of the fight against anti-Darwinian currents such as holism and Lamarckism at the University of Jena, but emphasized that this battle was not just scientifically, but also politically and ideologically motivated.

> It is the special merit of the botanist Werner Zündorf, during his years in Jena (1938–1943) and at Heberer's suggestion, to have confronted neo-Lamarckism in the 'classic country of idealistic morphology' in several publications.‡‡

After having earned his PhD in Tübingen on December 15, 1938, Zündorf was transferred to the University of Jena early in 1939 to join Heberer's Institute for General

---

* Staatspolizeistelle Stuttgart, November 24, 1939; *Universitätsarchiv Tübingen*, UTA 126/373, 16.
† For more details see Heiber, 1991, p. 438; also Potthast and Hoßfeld, 2010, p. 450.
‡ Heiber, 1991, p. 439.
§ *Universitätsarchiv Tübingen*, UTA 126/373, 1.
¶ *Universitätsarchiv Tübingen*, UTA 126/373, 1.
** *Universitätsarchiv Tübingen*, UTA 126/373, 2.
†† *Universitätsarchiv Tübingen*, UTA 126/373, 1. See also Heiber (1991, p. 440).
‡‡ Hoßfeld (2003, pp. 543–552).

Biology and Human Evolution, where he became scientific assistant. With the start of the war, Zündorf was conscripted to the army.[*] In 1943, Heberer published a splendidly produced multi-authored volume on *The Evolution of Organisms* (*Die Evolution der Organismen*), which was later hailed by historians of biology as the cornerstone of the German Synthesis of evolutionary theory.[†] In the preface, Heberer emphasized that this compendium had been written and compiled "amidst the European fight for freedom ... several contributors wrote their chapters as soldiers ... in that sense the book is also a gift from the fighting front."[‡] Heberer communicated to the publisher Astel's positive reaction to the fact that such a splendid contribution would come forth from the University of Jena, and proffered his suggestion to send a copy to Sauckel, who took such great interest in university affairs.[§] Uwe Hoßeld attributed the unusual quality of paper and printing to Heberer's good rapports with National Socialist leaders, especially the SS and its *Ahnenerbe*. The choice of contributors Heberer had made he found to be determined by scientific qualification on the one hand, personal friendships on the other.[¶] Since idealistic morphology was so heavily philosophy laden—especially in the writings of authors such as Naef, Troll, and Beurlen—Heberer did not want to leave his own initiative unguarded in that respect, and solicited an introductory chapter from Hugo Dingler, a politically like-minded friend. It was, after all, Dingler's *Der Zusammenbruch der Wissenschaft*[**] of 1926 where Heberer had first encountered a philosophical defense of the theory of descent.[††] Given the strong push for natural selection as a necessary and sufficient cause of evolutionary change propagated by the Astel phalanx in Jena, Dingler's contribution seems a bit out of place, however.[‡‡] Dingler's philosophy articulated a strong version of voluntarism rooted in the willing "I"; his understanding of the category of causality was anchored in the will of the subject—a position that Carnap of Vienna Circle fame dubbed "radical conventionalism."[§§] However, this dissonance would even escape one of the principal architects of the Modern Synthesis of evolutionary theory, Ernst Mayr, who found Dingler's essay "extremely well informed."[¶¶]

Heberer himself contributed a chapter in which he reduced macroevolutionary phenomena to a summation of microevolutionary events, attacking Troll, and especially Beurlen along the way. Given Dingler's introductory chapter, it is somewhat ironic that Heberer should dismiss Beurlen's appeal to the Will to Be—which Heberer equated with Blumenbach's (1781) *nisus formativus*—as metaphysically tainted and

---

[*] Junker, 2004, p. 251.
[†] "These authors really did succeed in achieving a synthesis": (Reif, 1983, p. 190); see also Junker and Hoßfeld (2002) and Hoßfeld and Junker (2003).
[‡] Heberer, 1943; see also Hoßfeld, 1999c, p. 199, n. 2; 2003, p. 541; Heberer penned the preface in the fall of 1942.
[§] Hoßfeld, 2005a, p. 271f.
[¶] Hoßfeld, 2005a, p. 270.
[**] Dingler, 1926.
[††] Heberer, 1956, p. 100.
[‡‡] Rieppel, 2012c.
[§§] Carnap, 1963 (1997, p. 15). For more on Dingler's background, and his contribution to Heberer's volume, see Rieppel (2012c).
[¶¶] Mayr, 1999, p. 22.

hence unworthy of any further discussion.* Zimmermann contributed to Heberer's volume a newly edited version of his classic paper from 1937[†] (on which more later), which laid the foundations for modern phylogenetic systematics.[‡] Zündorf—who went missing during the battle of Stalingrad[§]—used Heberer's project for his final assault on neo-Lamarckism and idealistic morphology, citing Zimmermann as a leading expert of phylogenetic systematics. Zimmermann, he argued, had turned phylogeny reconstruction on the basis of comparative morphology into an objective science, quite in contrast to the subjectivism, even mysticism, that in his view pervaded idealistic morphology: "for an explication as to how these [epistemological] problems relate to the soul of the race, see Clauβ (1936)," the latter the famous author of a classic book of the time on the psyche of the Nordic soul.[¶] Looking back on such orchestration of collaboration between Heberer, Zimmermann, and Zündorf, Lehmann commented after the war, in July 1945:

> How was it possible to transfer my duties [at the Botanical Institute and Garden] to a man [Zimmermann] who, during the final days of the National Socialist regime has published his major scientific papers in closest collaboration with an SS-*Führer* [i.e., Heberer], the latter someone who in the SS-*Ahnenerbe* was responsible for the ideologically so important rubric of phylogeny.[**]

In the attempt to whitewash himself, Lehmann branded Zimmermann's phylogenetic systematics as politically compromised; Heberer indeed signed off as co-editor on all entries on phylogenetic matters in *Der Biologe*. This is also the forum in which a polemic debate exploded, which Heberer hoped to once and for all bring to an end with his compendium, *Die Evolution der Organismen*, a book that in the coming decades was to go through two more editions. Although the debate eventually played out in *Der Biologe*, it was initially triggered by an entry penned by the Munich botanist and co-editor Ernst Bergdolt in the third volume of the *Zeitschrift für die gesamte Naturwissenschaft*,[††] the first one to be published under the Munich-based editorship. Writing in a journal for teachers in mathematics and natural sciences in 1934, the head of the division of racial hygiene at the Kaiser-Wilhelm-Institute for Anthropology, Human Genetics, and Eugenics.[‡‡] Fritz Lenz—a former student not only of Eugen Fisher but also of August Weismann, who wrote his *Habilitation* thesis on the inheritance of color patterns in butterfly hybrids[§§]—ascertained that Darwinian, that is, environmentally conditioned selection is the one and only way of race formation (see Chapter 6). Bergdolt in the ZGN criticized Lenz's examples as inconclusive and utilitarian, especially the anthropological ones, which Lenz had proffered in support of his conclusion with regards to the importance of environmentally

---

* Heberer, 1943a, p. 546.
† Zimmermann, 1937.
‡ Donoghue and Kadereit, 1992.
§ Heberer, 1943b, p. 253, n. 2; 1959, p. vii.
¶ Zündorf, 1943, p. 92.
** Lehmann, July 9, 1945; *Universitätsarchiv Tübingen*, UTA 126/373, 1.
†† Bergdolt, 1937/1938b.
‡‡ Lenz, 1927.
§§ Vita, Fritz Lenz; University Archive Göttingen (GWDG), Rektoratsakte, Fritz Lenz, Prof. Dr. med.

induced natural selection. Skin color of African people was not simply conditioned by the climate in which they live, argued Bergdolt: "The professor of anthropology [Lenz] who pursues his research under the African sun is unlikely to travel in a black frock."[*] Allying himself with the idealist morphological position that natural selection is not a sufficient explanation for all evolutionary change, Bergdolt appealed to the action of non-utilitarian creative forces in nature, the lawfulness of which Troll[†] had beautifully demonstrated through empirical research. Bergdolt's critique of Lenz's selectionist standpoint triggered a vitriolic retort by Zündorf, published in 1940 in *Der Biologe*.

Zündorf had already gained notoriety with a paper he had published the year before—in 1939, after he had moved from Tübingen to Jena—in the *Archiv für Rassen- und Gesellschaftsbiologie*, co-edited by Lenz and Heberer among others. His main target in that paper had been neo-Lamarckism, a doctrine he denounced as a "subversive movement" in the context of contemporary biology,[‡] as it could potentially undermine the efforts being made in racial hygiene based on eugenic legislation.[§] He portrayed Beurlen as a pure Lamarckian,[¶] who through his holism had provided new support to the Uexküllian environmentalism (*Milieutheorie*), which allowed for racial adaptation through the inheritance of acquired characteristics: "the force that in fact has overcome the environmentalism and with it the 'historical materialism' of Marxism, one that has been pursued since Mendel, is genetics and the racial theory built on it."[**] Holism quite generally he characterized as a "large-scale attack on scientific biology," launched primarily by Adolf Meyer (-Abich) and his allies.[††] Zimmermann,[‡‡] according to Zündorf, had elegantly shown that the inheritance of acquired characteristics is a purely genetic problem, and genetic research had after all conclusively refuted the Lamarckian doctrine. The propagation of Lamarckian ideas, as by the paleontologists Beurlen and Dacqué, can only serve to deprive—according to Zündorf—the *völkisch* state of its biological foundations, at a time when variation and selection have been proven to be the only relevant causes driving phylogeny.[§§]

With his 1940 entry in *Der Biologe*, Zündorf zeroed in on Troll, without missing the opportunity to expose the heresies spread by others such as Schindewolf, Dacqué, and Naef, all the way implying Bergdolt's co-conspiracy but without exposing the decorated party member to the same degree. Zündorf found idealistic morphology divorced from evolutionary theory; its reservations with respect to the causal efficacy of natural selection he claimed—following Heberer[¶¶]—to translate into a negation of the soil (*Wurzelboden*) in which racial theory is rooted. Idealistic morphologists project their minds into nature, instead of objectively learning from nature. Idealistic morphology

---

[*] Bergdolt, 1937/1938b, p. 109, n. 1.
[†] Troll, 1928.
[‡] Zündorf, 1939b, p. 299.
[§] Zündorf, 1939b, p. 301.
[¶] Zündorf, 1939b, p. 291.
[**] Zündorf, 1939b, p. 295.
[††] Zündorf, 1939b, p. 298.
[‡‡] Zimmermann, 1938.
[§§] Zündorf, 1939b, p. 301f.
[¶¶] Heberer, 1937c.

is consequently a subjective science, in contrast to phylogenetic systematics, which he claimed to be an objective science. In this context, Zündorf cited Zimmermann's 1937/1938 paper on the philosophical basis of biological sciences as the one that best reflects the scientific spirit of a man of the Nordic race. In that paper, Zimmermann had argued that a strict separation of object and subject is the prerequisite for all scientific inquiry, including comparative morphology and its use as a base for phylogeny reconstruction. Zimmermann had published this paper in the journal *Erkenntnis* that had been founded by members of the Vienna Circle, and Zimmermann duly cited not only the manifesto for a scientific world view issued in 1929 by the *Verein Ernst Mach*,[*] but also, and more specifically, the philosophers Carnap, Schlick, and Popper[†] among others: "Of course, my efforts to articulate a unified and unambiguous scientific method [in phylogenetic systematics] coincide completely with the endeavors of the Vienna Circle,"[‡] announced Zimmermann. Going from there, Zündorf censured Troll for his statement that "Darwinism was born in the stuffy air of English overpopulation," exclaiming in dispair: "will such frivolous attacks on Darwinism ever come to an end?"[§] He then established a link between idealistic morphology and Catholicism,[¶] an accusation that certainly must have irritated Troll who was—and continued to be—under pressure from Weigelt in that respect. In the fall of 1941, Heberer presented in Halle a public lecture on the theory of descent, which was attended by Troll. While accepting the fact of descent with modification, Troll in the discussion section commented critically on the selection theory. He later learnt that the Gestapo had been present on that occasion as well, and later inquired about his name. Troll subsequently complained to the university administration that the continued denunciation of him and his science as catholic, also enforced by the president of the university, Johannes Weigelt, might lead the Gestapo to believe that he was an outright opponent of evolutionary theory,[**] which is exactly how Zündorf wanted to portray him. Zündorf concluded his attack on Troll, and idealistic morphology more generally, with an appeal to pursue the unity of biology as embodied in Haeckel's monism.[††]

Zündorf's assault on Troll precipitated an immediate rebuttal by Bergdolt in *Der Biologe*, who rushed to Troll's support. Bergdolt continued to appeal to creative forces active in nature, and contrasted comparative (i.e., idealistic) with phylogenetic and experimental morphology. Phylogenetic morphology he characterized as a "fundamentally speculative comparison of [organismic] form enriched by the temporal dimension."[‡‡] The sufficiency of natural selection in driving evolutionary change he called into question by introducing examples of apparently functionless structures. Turning to issues of human race formation, Bergdolt rejected Fritz Lenz's example

---

[*] Neurath et al., 1929.

[†] Popper was not a member, but rather a satellite of the Vienna Circle; Stadler (1997).

[‡] Zimmermann, 1937/1938, p. 5.

[§] Zündorf, 1940, p. 19.

[¶] Zündorf, 1940, p. 20.

[**] Letter from Troll, November 18, 1941; archive of the Akademie der Wissenschaften Leopoldina (Wilhelm Troll, Matrikel Nr. MM 4178).

[††] Zündorf, 1940, p. 23.

[‡‡] Bergdolt, 1940, p. 398.

of Pygmies, which Lenz had argued remain restricted to relatively infertile relict areas because they were not able to compete with stronger neighboring races. "It is the struggle for existence which alone decides the fate of the races on Earth," Lenz is quoted as having said, to which Berdolt replied that dwarfism is widespread throughout the plant and animal kingdom without apparent or even necessary correlation with trophic conditions. Similar arguments he mounted against Lenz's example of the slit eyes that characterize Asian races, and that Lenz had linked to the intense light prevailing in Asian highlands.* In a footnote, he finally turned against Zündorf's critique of Troll's statement on Darwinism being born in the stuffy air of English overpopulation, capitalizing on the fact that Zündorf had not recognized that statement as a quote from Nietzsche: with his critique of Troll, "Zündorf has—even if unintentionally—accused Nietzsche of being frivolous."[†] Ernst Lehmann would later use Bergdolt, "who demonstrated so clearly the scientific ignorance and the flawed methods of Zündorf" in his defense.[‡] Recognizing such dangers early on, Heberer, the associate editor responsible for all phylogenetic content, was outraged that Bergdolt's paper was published in *Der Biologe* against his recommendation, and asked the managing editor Greite to release him from the editorial board: "It goes too far if polemics against the natural breeding of human races are launched from one's own camp"[§]—Bergdolt of course being an *Altnazi*. It took an intervention from Sievers to convince Heberer to withdraw his letter of resignation.[¶] The exchange and its ramifications furthermore enticed the animal psychologist Konrad Lorenz** to join the debate in a contribution to *Der Biologe*, where he again—like Heberer—accused the Catholic Church of being the main enemy of the theory of descent, which alone provides the scientific foundation for the "racial purification and improvement of the *Volk*."[††] With its hostility against scientific research, the Catholic Church acts "worse than the devil himself."[‡‡] Whether humanity will suffer the fate of the dinosaurs, or experience the further evolutionary perfection of the human brain to a presently unimaginable degree "is purely a question of biological potency and the Will to Life of our *Volk*."[§§] Domestication coupled with a denial of the efficacy of natural selection will undoubtedly have the consequences Oswald Spengler had so vividly sketched in his *The Decline of the West*: individuals with crippled instincts (*Instinktkrüppel*) unwilling or unable to procreate would be the expected outcome, according to Lorenz.[¶¶] "But exactly in this race to be or not to be, we Germans lead all other civilizations [*Kulturvölker*] by a thousand steps."**** When Lorenz wrote of "we Germans,"

---

* Bergdolt, 1940, p. 403.
† Bergdolt, 1940, p. 404, n. 8.
‡ Letter of Lehmann to the President of the University of Tübingen, August 3, 1942; *Universitätsarchiv Tübingen*, UTA 126/373, 7.
§ Heberer, cited from Deichmann, 1996, p. 274.
¶ Deichmann, 1996, p. 274.
** On Konrad Lorenz's involvement with the National Socialist regime see Bäumer (1990b), Deichmann (1996), and Burkhardt (2005).
†† Lorenz, 1940, p. 25.
‡‡ Lorenz, 1940, p. 25.
§§ Lorenz, 1940, p. 29.
¶¶ Lorenz, 1940, p. 31.
*** Lorenz, 1940, p. 29.

he meant the citizens of the *Nationalsozialistisches Grossdeutschland.*[*] Nothing can surpass the importance of teaching biology, and especially the theory of descent, in school, argued Lorenz: it provides the most reliable foundation for the new world view, motivating the students to pursue ideals and goals guided by a racially conditioned understanding of evolution.[†] Zimmermann raised a more moderate voice against Bergdolt, expressing his amazement that there still are authors out there who consider phylogenetics a mere "appendage" (*Anhängsel*)[‡] to morphology, a fact he related to the "strange historical development of phylogenetic problems": "The concept, that phylogenetics is also strictly a natural science with an independent complex of questions that need to be concisely dealt with, is still very young."[§]

Zündorf himself replied to Bergdolt as well, although delayed as he was serving at the western, later at the eastern front. He excused his slip with the Nietzsche quote as a misprint, and rehearsed his earlier remarks[¶] that "all cultures have not been created by indefinable and anonymous spirits of people and times, but solely by such racial powers as fashion an historical epoch."[**] He concluded his exposition by highlighting the alignment of phylogenetic systematics with racial ideology, a striking contrast to idealistic morphology. Starting with the identification of a type, and only subsequently working out its scope and variation, idealistic morphology pursues a top-to-bottom approach in systematics and classification: racial differentiation forms the trailing end of the inquiry. In contrast, phylogenetic systematics puts the question of race formation first, building the natural system from bottom up.[††]

Answering Zündorf's first attack on him in the journal he edited, Troll defended Nietzsche as "one of the most decisive critics of Darwinism"[‡‡] that the German *Volk* had brought forth. Without citing Zündorf in their joint paper on Goethe's morphology, Troll and Wolf nevertheless mentioned the title of Zündorf's 1940 paper, which in their judgment put forward an argument supporting the unification of comparative morphology with phylogenetics: "any attempt along these lines completely ignores the fact that the roots of the theory of descent are buried in idealistic morphology."[§§] Responding to Zimmermann's "groundless polemics," Troll judged him to articulate a "downright primitive understanding of the methods of comparative morphology."[¶¶] He attributed to Zimmermann a misguided understanding of the true nature of scientific theory construction, and rejected Zimmermann's reproach that he, Troll, was subject to "psychological inhibitions" and "emotional reactions" against phylogenetics.[***] Starting with his first publication, Troll asserted, he had always defended the fact of descent with modification. And anyway, if his

---

[*] Lorenz, 1940, p. 24.
[†] Lorenz, 1940, pp. 35–36.
[‡] Zimmermann, 1941, p. 49.
[§] Zimmermann, 1941, p. 47.
[¶] Zündorf, 1940, p. 19.
[**] Zündorf, 1942, p. 127.
[††] Zündorf, 1942, p. 129.
[‡‡] Troll, 1939, p. 634.
[§§] Troll and Wolf, 1940, p. 43.
[¶¶] Troll, 1943b, p. 72.
[***] Troll, 1943b, p. 73.

science were so anti-evolutionary, asked Troll,* how come that Zimmermann himself had adopted his own typological concept of *Rhynia* and interpreted it as the ancestral land plant (*Ur-Landpflanze*). Turning against Heberer, who had charged that through his elaborations on the phylogenetic origin of types Troll had turned idealistic morphology on its head,† Troll one more time defended the distinction between the fact of phylogeny, from the claim that natural selection is a sufficient cause for all evolutionary change—a distinction (going back to Tschulok's *Grundfrage* and *Faktorenfrage*) he found Heberer incapable of grasping. Troll again emphasized the a-causality of comparative morphology, which he characterized as the science that is called upon to deliver the phenomena that, once fully worked out, are to be subject to phylogenetic interpretation and explanation. As the editor of the *Botanisches Archiv*, Troll refused to accept a rebuttal of his defense by Heberer for publication. Heberer, who in defense of Zündorf legitimately challenged Nietzsche's competence in biology,‡ consequently placed his piece in *Der Biologe*.

The polemics between idealistic morphologists and phylogenetic systematists ultimately centered on the unity or disunity of biology: comparative (idealistic) versus phylogenetic morphology, macroevolution as distinct from microevolution, neo-Lamarckism as opposed to strict Darwinian, or more precisely Weismannian selectionism. Methodological issues of phylogeny reconstruction, if considered at all, receded into the background. The two people involved in the polemics against idealistic morphology who, in their early publications, made significant contributions to the methodological basis of phylogenetic systematics were Konrad Lorenz—at the time professor of psychology at the Immanuel Kant University in Königsberg, later Kaliningrad—and, especially, Walter Zimmermann. In 1941, Lorenz published in the prestigious *Journal für Ornithologie* a groundbreaking monograph, which showed that the concept of homology could be taken beyond comparative morphology and applied to innate behavioral patterns of ducks. Having thus captured phylogenetically informative behavioral characteristics, he proceeded to use them in the reconstruction of duck phylogeny, attempting to resolve "monophyletic"§ groups. It is in that context that Lorenz offered comments on the methodology of phylogenetic systematics, which he would again return to in his contribution to the Heberer volume of 1943.¶

Walter Zimmermann** was born on May 9, 1892, in Walldürn (Baden) into a catholic family, the father being a civil servant. He passed through Grade School and High School (*humanistisches Gymnasium*) in Karlsruhe, where the family had moved. Beginning with the winter semester 1910/1911, Zimmermann enrolled at the Karlsruhe Institute of Technology, later transferring to the universities of Freiburg i. Br., Berlin, and Munich, to study botany, zoology, geology, paleontology, mathematics, history, and literature. He was a student, among others, of Richard Hertwig and Otto Renner. Service in World War I delayed his studies, which he completed at

---

* Troll, 1943b, p. 438.
† Heberer, 1943a.
‡ Heberer, 1943c, p. 254.
§ Lorenz, 1941, p. 289.
¶ Lorenz, 1943.
** Junker, 2001a; *Universitätsarchiv Tübingen*, UTA 193/4679 (Personalakte des Rektoramtes, Walter Zimmermann).

the University of Freiburg i. Br., defending his PhD in March 1920. First scientific assistant at the Botanical Institute of the University of Freiburg (1920–1925), he moved to the University of Tübingen, again as an assistant (1925–1930), earning the *venia legendi* in botany in the summer of 1925 with the support of Ernst Lehmann. He was promoted to adjunct associate professor in 1929, to associate professor in 1930, and to full professor in 1960, shortly before his retirement. His long career in Tübingen was interrupted by military service in World War II, during which he was twice wounded. In 1948, Zimmermann passed on offers for employment in Karlsruhe (associate professor) and Greifswald (full professor). Very late into his career, the faculty of the University of Tübingen felt that Zimmermann should be awarded the title of full professor not only in recognition of his massive scientific achievements, but also to honor a deserving man. Zimmermann died in Tübingen on June 30, 1980 (Figure 7.2).

Zimmermann was not a member of the NSDAP, nor of any of its affiliated organizations, and only a collective member of the NSLB. He managed to keep a low political profile, as is well expressed in a report the Dean Edwin Hennig was required to send to Hoffmann, President of the University (dated October 14, 1938): "He has certain peculiarities that appear a bit contrived ... I believe he is vegetarian ... but everything he does is carried through with greatest responsibility ... he is through and through a clean-cut nature ... I cannot comment on his political inclinations more

FIGURE 7.2 Walter Zimmermann. (Universitätsbibliothek Tübingen, Bilddatenbank.)

specifically, but consider him a consummate patriot."* Lehmann, in 1939, praised the breadth and thoroughness of Zimmermann's expertise while commenting: "If erstwhile he stood on the ground of Southern German Democracy, he has always proven his true German persuasion in words and deeds. Today he is certainly positively inclined toward the new state."[†] Zimmermann's eugeneticist leanings became apparent in the last chapter of his book on *Inheritance of Acquired Characteristics and Selection* of 1938,[‡] where he explained laws of inheritance and issues of racial hygiene in evolutionary terms. The book was favorably reviewed by Zündorf in the *Zeitschrift für die gesamte Naturwissenschaft*: a "book that not just the Darwinist will pick up with pleasure."[§] Earlier, in 1936, Zimmermann promulgated a social-Darwinist interpretation of natural selection in a paper he published in the magazine *Rasse*, the outlet of the umbrella organization for the Nordic Movement, the *Nordischer Ring*: "The will to comprehend phylogeny as a process (i.e., dynamically, kinetically) must prevail, if we want to reach clarity in this question, one which is fundamental for the entire field of racial hygiene and *völkisch* renewal."[¶] On February 12, 1936, Zimmermann delivered a talk on *Selection and its Significance for our* Volk*, especially with regard to the Four Year Plan*: "Only the most severe selection of the strongest (*Lebenstüchtigsten*), one that presupposes a wealth of offspring, can prevent signs of decadence ... we have to make sure that the concept of selection and the commitment to selection, which in the end are intellectual problems, come to life in our *Volk*."[**] Apparently convinced of the virtues of eugenics in those days, Zimmermann emerged from the collapse of the Third Reich politically uncompromised.

In his classic 1937 paper on phylogenetic systematics, Zimmermann accused idealistic morphologists of not even trying to separate an objective from a subjective component in perception.[††] The strict separation of object and subject prevents natural science to lapse into metaphysics, and to get carried away by pseudo-problems, he argued.[‡‡] Zimmermann's embrace of logical positivism has already been mentioned. His plea for a strict separation of subject and object in science[§§]—not necessarily easy to achieve, but dependent on a willing intention[¶¶]—rests on the concept of the extramental world as an absolutely and objectively "Given," mediated in human cognition through sense data. The "sense datum" is not the object itself, nor is it a creation of the human mind; it was believed to be an independent, objective entity located somewhere in between the perceived object and the perceiving mind.[***] The material object is mentally (logically) reconstructed (sort of "patched together") from the sense data and their relation to one another, and hence reveals itself to the cognizing mind as a

---

* *Universitätsarchiv Tübingen*, UTA 193/4679 (Personalakte des Rektoramtes, Walter Zimmermann).
† Letter of reference with regards to an open position in botany at the University Königsberg, by E. Lehmann, dated January 9, 1939; *Universitätsarchiv Tübingen*, UTA 126/373, 10.
‡ Junker, 2001a; Potthast and Hoßfeld, 2010.
§ Zündorf, 1938/1939, p. 324.
¶ Zimmermann, 1936, p. 431.
** *Universitätsarchiv Tübingen*, UTA 193/4679 (Personalakte des Rektoramtes, Walter Zimmermann).
†† Zimmermann, 1937, p. 946.
‡‡ Zimmermann, 1937, p. 961.
§§ Zimmermann, 1937, p. 961; see also the discussion in Zimmermann, 1930, pp. v–vi.
¶¶ Zimmermann, 1937, p. 958.
*** Soames, 2003.

relational structure. For Zimmermann, this meant that "our picture of the world" is reduced to a "relational system," and it is this "relational system" that is the "object of our cognition."* All of this, of course, echoed contemporary developments in physics. In his classic *The Logic of Modern Physics*, Percy W. Bridgeman (1882–1961) had argued that theories of physics do not capture the "world-in-itself," but only its *relational structure*.† Similar ideas had previously been propagated by members of the Vienna Circle such as Schlick‡ (1917) and the young Carnap,§ or authors close to those such as Edgar Zilsel (1891–1944): "When we pursue natural sciences, mathematics, or logic, we try to explain existing structural relations through our theories, through *systems which are to be absolutely precise and completely determined*."¶ The elements of such systems are sense data, the system itself is determined by its relational structure, that is, by the way the sense data relate to one another. This quote from Zilsel nicely captures the spirit that determined the research program of Zimmermann.

Zimmermann's conception of objectivity in terms of sense datum theory is consistent with his deference to Schlick's definition of the "Given."** Schlick was well known for the foundationalism he defended in sessions of his circle, proposing to compare the description in the travel guide with the sight of the Vienna Cathedral. Zimmermann seems to have failed to recognize why Schlick felt the need to defend such foundationalism.†† The requirement resulted from Carnap's (and Neurath's) later turn to coherentism, where it is the coherence of statements, not their purportedly vain comparison with material objects, that justifies knowledge claims.‡‡ Carnap's turn to coherentism was motivated by Popper's fundamental insight that there cannot be any theory-free observation, that all observation statements are therefore fallible low-level hypotheses§§: "Each time when we take a reading from an instrument we rely on the hypotheses of geometrical optics, on the theory of solid bodies, on the correctness of Euclidean Geometry in small space, on the hypothesis of the existence of things, and innumerable other hypotheses."¶ Zimmermann, of course, felt he could brush aside such Popperian concerns as he sought not a philosophical, let alone psychologically motivated *theory* of cognition, but rather a practical *critique* of cognition, a philosophical stance he later aligned with empiriocriticism***—"a name introduced by Avenarius."††† The German-Swiss philosopher Richard Avenarius (1843–1896) introduced this name for a radical positivist doctrine

---

* Zimmermann, 1954; 1959, p. 49.
† Bridgeman, 1927.
‡ Schlick, 1917.
§ Carnap, 1922.
¶ Zilsel, 1916, cited from Stadler, 1997, p. 196; emphasis added.
** Zimmermann, 1937/1938, p. 6, n. 9.
†† "A statement the meaning of which depends on the properties instantiated by an object is a 'protocol sentence' in the sense of Carnap": Zimmermann, 1937/1938, p. 5.
‡‡ Creath, 1996, pp. 158ff.
§§ Carnap, 1963; 1997, p. 31f.
¶ Popper, 1979, p. 391. The book is a late publication of the parts that survived the chaos of war of the original manuscript on which Popper's *Logic of Scientificc Discovery* was based.
*** Zimmermann, 1954; 1959, p. 48f.
††† Zimmermann, 1953, p. 545.

influenced by Ernst Mach, according to which the major task of philosophy is a
natural understanding of the world based upon pure experience, and freed from all
metaphysics and materialism.

In the empirio-critical spirit, Zimmermann's prime goal was to liberate phylogenetic systematics from "mythical and intuitive pre-scientific concepts which
systematics, in contrast to astronomy and chemistry, had not yet fully overcome."*
Raging against idealistic morphology in his contribution to Heberer's 1943
compendium, and targeting Troll in particular, Zimmermann applauded the scientific
progress that is manifest in the change of astrology to become astronomy, or alchemy
morphing into chemistry. Only in phylogenetics, he complained, was such progress
notably lagging.[†] "Someone who merely follows his intuitions, or who recounts myths
or fairytales, is unwilling to separate the objective and the subjective components of
his experience."[‡] The move from nature mysticism to a rational science of phylogeny
reconstruction, he claimed, would primarily have to be a methodological one.[§]

> Indeed, it seems to me that all problems encountered in the discussion of phylogenetic
> problems have their root in the fact that we have not yet managed to adopt the vantage
> point of the 'new objectivity' (*neue Sachlichkeit*) which in that context is absolutely
> crucial.[¶]

Zimmermann, an adept of the *neue Sachlichkeit*: this put him, the logical positivist,
philosophically at the other end of the spectrum from Troll, the latter representing
almost paradigmatically the Mandarin tradition of the German professoriate.[**] *Neue
Sachlichkeit* was the term Neurath and Carnap used to characterize the philosophy
of the Vienna Circle, which entailed the rejection of metaphysics, theology, astrology,
and anthroposophy (nature mysticism).[††] What Carnap sought instead under that label
was a philosophy based on "mathematical logic, the most 'objective' and universal
discipline of all," hence a philosophy at least in principle amenable to "universal
agreement."[‡‡]

Given his forceful defense of objectivity in phylogenetic systematics, an objectivity
he located in "the scientific method," it seems odd that in his critique of idealistic
morphology Zimmermann would invoke the conventionalist voluntarism again
that was so characteristic of the time. Zündorf, in the Heberer volume, claimed
that the logical and epistemological independence of systematics from evolutionary
theory postulated by idealistic morphologists could not carry through, since every
systematist would at least *subconsciously* harbor evolutionary thoughts.[§§] The same
had previously been claimed by Zimmermann, who had stated that subconscious

---

* Junker, 2001a, p. 286.
† Zimmermann, 1943, p. 20.
‡ Zimmermann, 1943, p. 29.
§ Zimmermann, 1943, p. 20.
¶ Zimmermann, 1933, p. 346.
** Ringer, 1969 (1990).
†† Galison, 1990, p. 725; see also pp. 720–749. For a discussion of the *neue Sachlichkeit*, see also
Peuckert, 1987.
‡‡ Friedmann, 2000, p. 158.
§§ Zündorf, 1943, p. 95.

phylogenetic perspectives might certainly play a "central role" in idealistic morphology, for which reason a unity of comparative biology should be within easy reach.* Zimmermann consequently called for phylogenetics to become a *conscious* science,† and asked: "We have to group ... so, do we *want* to group [organisms] phylogenetically, or how else do we *want* to group [them]?"‡ Should classifications be intuitive, artificial, or phylogenetically motivated?§ In an unpublished draft manuscript, Zimmermann mused:

> What do we *call*, what do we *want* to call phylogenetics? A consensus in this issue is as important as a consensus regarding street names. Because if in a city everyone names streets differently, or if there is ambiguity about street names, then there is no surprise that we miss each other constantly and never come together. But we want to find each other, especially in issues concerning phylogenetics and morphology.¶

The phylogenetic system, for Zimmermann, was again a dichotomously structured hierarchy. Given the observation of cross-cutting character distribution, Zimmermann conceded that phylogenetic conclusions are always only probabilistic statements: the phylogenetic method contrasts mutually exclusive explanatory possibilities of different probability.** Nothing wrong with that, he contended, citing Henri Poincaré (1854–1912), who had argued that all of science ultimately moves in probabilistic terms, different degrees of probability differentiating hypotheses from theories, theories from laws—and those include historical laws.†† Zimmermann's first characterization of the natural *qua* phylogenetic as an enkaptic system— with reference to Tschulok—appears in 1953,‡‡ but was further elaborated in the second edition of his textbook published in 1959: "The natural system had been reconstructed long before the development of conscious phylogenetics ... since both work with the same principle," which is the subsumption of organisms into more or less inclusive groups.§§ Since the "natural" system was derived from idealistic morphology, however, Zimmermann could not assign to the concept of enkapsis any ontological grounding. The contrast between the concept of a complex whole subject to the part-whole relation, as opposed to a sum or an aggregate, he considered inherent categories of thought, ways of seeing rather than representations of natural entities.¶¶ Accordingly, Zimmermann would also not accept races, species, or higher taxa as complex wholes, or individuals. Taxonomic categories at all levels to him were abstract concepts; real was only the continuous flow of phylogeny through the past and present into the future. Evolutionary lineages do not form systematic categories such as "race" or "species" but rather are composed of a nexus of generations

---

* Zimmermann, 1930, p. 12; see also Zimmermann, 1959, p. 12.
† Zimmermann, 1930, p. 15; 1959, pp. 12–18.
‡ Zimmermann, 1937, p. 942f.
§ Zimmermann, 1937, p. 949f.
¶ *Universitätsarchiv Tübingen*, UTA 258/4 (Zimmermann Nachlass).
** Zimmermann, 1937, p. 953.
†† Zimmermann, 1937, p. 952, n. 1. See also Zimmermann, 1930, p. 370.
‡‡ Zimmermann, 1953, p. 489.
§§ Zimmermann, 1959, p. 6.
¶¶ Zimmermann, 1937, p. 966.

comprising individual organisms with new and different, heritable properties.[*] In a handwritten draft, Zimmermann mused:

> Certainly one could imagine definitive answers to the question 'what is a species'? This would obtain when that which we call a 'species' were a concrete individual thing, or at least would correspond to such a thing, one to which we could point a finger as to a house or to a human being whom we could identify either through personal acquaintance, or through reference to an address book. Then, and only then, would the question 'what is a species' have (at least provisionally) been answered with a simple, consistent and unambiguous indication.[†]

However, so he noted in the margin, "to a species always belong not just one individual, but many." To ascribe reality to a species would amount to Platonic realism (*Begriffsrealismus*), and hence imply an essentialism that is incompatible with evolution. For Zimmermann, the natural system and its categories are conceptual constructs, an "indispensable tool"[‡]; but what is of real interest to the phylogenetic systematist is the narration of the phylogenetic process that is reflected in phylogeny reconstruction. This, of course, lends an interesting twist to Zimmermann's views about eugenics and racial hygiene: races are abstract concepts, what is real is the process of inheritance, ultimately the process of descent with modification. Eugenics and racial hygiene are not means to improve this or that "race," the latter understood in any taxonomic sense, but rather to maintain and improve the *Volk*, the latter a historical collection of interbreeding individuals.

Turning to more practical aspects of phylogenetic systematics, Zimmermann introduced the concept *hologeny* (*Hologenie*),[§] which in recognition of the close correlation that necessarily must exist between ontogenetic transformation and phylogenetic transformation comprised both ontogeny and phylogeny. This naturally renders ontogeny an important tool in the search for homology, as well as for the discrimination of the primitive as opposed to the derived condition of form, a tool that can be complemented by paleontology. It is at this juncture that Zimmermann harked back on Othenio Abel's writings, which emphasized the necessity to distinguish between character transformation series (*Stufenreihen*), and ancestor–descendant series (*Ahnenreihen*). The distinction was necessitated, of course, by the ubiquitous cross-cutting character distribution, the *chevauchement des spécialisations* invoked by Abel's friend and mentor, Louis Dollo, and recognized by Zimmermann as well. In his textbook of 1930, Zimmermann cited the two book chapters contributed by Abel that provided the primary source of his thinking about these issues,[¶] and although only indirectly based on Abel, he went on to distinguish a *Sippenphylogenetik* (group-phylogenetics) from a *Merkmalsphylogenetik*

---

[*] Zimmermann, 1936, p. 429.

[†] *Universitätsarchiv Tübingen*, UTA 258/4 (Zimmermann Nachlass).

[‡] Zimmermann, 1936, p. 417.

[§] Zimmermann's excerpts from Herbert Spencer's *Principien der Biologie* (1876) indicate that he took the inspiration for the concept of hologeny from §160, p. 475, of that treatise. *Universitätsarchiv Tübingen*, UTA 258/4 (Zimmermann Nachlass).

[¶] Abel, 1911, 1914; see also Rieppel, 2013b.

(character-phylogenetics), stating emphatically in the preface: "In our book we deliberately accord priority to character-phylogenetics."* In the concluding chapter he expressed his hope that his book "has brought out with enough clarity *how much more safer character-phylogenetics is*" than group-phylogenetics.† In the second edition of his textbook, he called group-phylogenetics *Taxophylogenetik*, character-phylogenetics he referred to as *Semophylogenetik*,‡ recognizing a direct link between these concepts and Abel's work: "As *O. Abel* had already forcefully insisted upon, it is hardly possible to reconstruct true ancestor–descendant sequences (*Ahnenreihen*) from the Fossil Record, but rather only character transformation series (named *Stufenreihen* by *Abel* ...) ... following *Zimmermann*, a secure character-phylogeny is better than an unfounded group-phylogeny."§ The priority of character-phylogenetics over group-phylogenetics he anchored in the sound empirical base enjoyed by character analysis: the fundamental concerns that are frequently raised with respect to historical phylogenetics he found to relate to group-phylogenetics, and are consequently alleviated if a rigorous character-phylogenetics is pursued instead. And of course, like Abel, Zimmermann used seriability of character states as clue to phylogeny reconstruction,¶ introducing Tschulok's sketch of the evolution of the limb in horses as an example.** Character transformation series can only provide a clue to phylogeny if the primitive as opposed to the derived character state can be identified. However, it was most important to Zimmermann to recognize that while character transformation series can be so polarized using ontogeny or paleontology, it is the relative time of common ancestry that alone provides an accurate measure for degrees of phylogenetic relationships. The relative time of common ancestry in turn is indicated by the temporal sequence of phylogenetic character transformation.†† From this follows Zimmermann's fundamental, and influential definition of "phylogenetic relationship": "Those species or other taxa that share a more closely situated ancestor are more closely related to one another than those natural groups that go back to a more remotely situated ancestor."‡‡

It has been argued, and rightly so, that Zimmermann's classic paper of 1937, and its edited version in Heberer's compendium of 1943, have had an enormous impact on phylogenetic systematics, in particular in the hands of Willi Hennig[149], who called Zimmermann "one of the best recent theoreticians of systematic work."§§ As will be shown in Chapter 8, Hennig would address many core issues dealt with extensively by Zimmermann, not necessarily reaching the same conclusions, however. Those issues concerned the separation of a subjective from the objective component in observation; the nature of the enkaptic hierarchy; the reality of species; group-phylogenetics

---

* Zimmermann, 1930, p. vi.
† Zimmermann, 1930, p. 427; emphasis in the original.
‡ Zimmermann, 1959, p. 22; *Semophyletik* in Zimmermann 1930, p. 427, n. 1; see also Zimmermann, 1937, p. 984.
§ Zimmermann, 1953, p. 508.
¶ Zimmermann, 1937, p. 991.
** Zimmermann, 1937, p. 986; see also Tschulok, 1922, p. 169, Figure 40.
†† Zimmermann, 1937, p. 990f.
‡‡ Zimmermann, 1953, p. 10.
§§ Hennig, 1950, p. 14.

as opposed to character-phylogenetics; the polarization of character transformation series; and, perhaps most importantly, the use of the temporal dimension as the backbone of the phylogenetic system, rather than morphological similarity. Hennig would indeed, take up and continue to pursue the goal that Zimmermann had captured with the historian Leopold von Ranke's (1795–1886) famous line: to find out "how it really was."[*]

* Zimmermann, 1937, p. 981.

8 A New Beginning
From Speciation to Phylogenetics

THE STRESEMANN SCHOOL

Erwin Stresemann* was not one of many, but rather *the* towering figure in German ornithology during the first half of the twentieth century. Born in Dresden on November 22, 1889, his father a wealthy pharmacist, he enjoyed a carefree childhood and youth in a family cultivating arts, humanities, and natural sciences. "Money one didn't speak of, one had it; what one wanted to pursue was the only thing that was deemed important" for a career choice, a family member is reported to have said.[†] His interest in natural history kindled by his father, Erwin the child collected local beetles, lizards, and snakes, which he kept in jars or appropriate tanks. Later he turned to birds as pets. At the age of 17, he succeeded to interbreed two different bird species in captivity—the Redpoll Finch (*Birkenzeisig*) and the European Goldfinch (*Stieglitz*)—a rare success that earned him his first publication. Given the family tradition and concern for long-term economic security, Stresemann enrolled in Medical School at the University of Jena in 1908, at that time a common choice for those aspiring to a career in zoology. The encounter with Haeckel and his work motivated Stresemann to take two courses in marine biology in Bergen, Norway.[‡] In 1909, Stresemann transferred to the University of Munich, and from 1910 to 1912 participated—at his own expense—in a hugely productive expedition to the Maluku Islands, led by the Geologist Karl Deninger from the University of Freiburg i.Br., whose sister eventually became Stresemann's first wife (in 1916). The systematics of the rich collection of birds he brought back he worked out under the tutelage of the famous German ornithologist Ernst Hartert (1859–1933), who at the time worked in Lord Rothschild's private museum in Tring, Hertfordshire, England. It was most likely Hartert who recommended Stresemann, a rising star in ornithology, to Willy Kükenthal,[§] who was looking for an author to cover the birds as part of the Handbook of Zoology (*Handbuch der Zoologie*), which he co-edited. Stresemann was elated to be called upon for such a prestigious task, but World War I interrupted his research, as he was conscripted and sent first to the western frontline, later to northern Italy. After the war, Stresemann continued his studies in zoology under

---

* Biographical data on Stresemann were taken from Nöhring (1973); Wunderlich (1991); Rutschke (2001). Günther (1974, p. 42) lists additional obituaries. A comprehensive account of Stresemann's life and work was offered by Haffer et al. (2000).
† Wunderlich, 1991, p. 6.
‡ Nöhring, 1973, p. 456.
§ Nöhring, 1973, p. 458.

Richard Hertwig in Munich, earning his PhD *magna cum laude* in March 1920, with a thesis in ornithology. Kükenthal came back with his proposition that Stresemann should take on the volume on birds for the *Handbuch*, a project that would eventually take 15 years to complete. An early draft chapter that he sent to the editor struck such a mark, however, that Stresemann—then assistant at the Bavarian Zoological State Collections in Munich—was proposed for the succession of the Curator of the ornithological collections at the Zoological Museum in Berlin. Against the resistance of more senior staff, Kükenthal, the museum's Director, pushed through Stresemann's employment as research assistant, as of April 1, 1921. Stresemann was promoted to Curator in 1924, and named Professor at the Humboldt University in 1930. He would remain faithful to the Berlin Museum for the rest of his long and successful career, which ended with his mandatory retirement in 1961 (Figure 8.1).

From November 1935 to March 1936, Stresemann visited the United States, working at Yale University in New Haven, Connecticut. With the flags of fascism rising even higher in Germany, Stresemann was encouraged to stay in the United States, but declined to do so, as he felt too deeply entrenched not only in Germany—albeit one predating Hitler's dictatorial rule—but also in the European culture and science generally: "Although I could work in America, I could not live there,"* he confessed.

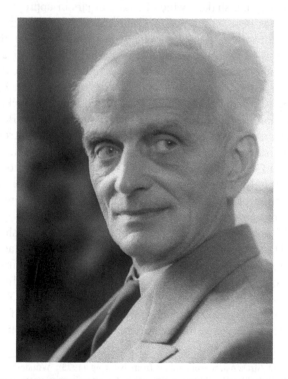

FIGURE 8.1   Erwin Stresemann. (From Museum für Naturkunde Berlin, Historische Bild-u. Schriftgutsammlungen [Sigl: MfN, HbSb]. Bestand: Zool. Mus., Orn. 225,4 [Stresemann].)

---

* Wunderlich, 1991, p. 299; Rutschke, 2001, p. 11.

Plowing through the rubble in Berlin after the collapse of the Third Reich, Stresemann was busy reorganizing the bird collections at the museum. Foreseeing the inevitable ultimate catastrophe, he had been instrumental in hiding the type collection in bomb-proof underground shelters. Living in the Western Sector of the city after the war, he commuted to work in the Eastern Sector, highly esteemed by the scientific establishment on both sides of the political divide. In 1954, he became a member of the prestigious German Academy of Sciences Leopoldina headquartered in Halle/Saale; a year later he was elected member of the German Academy of Sciences in Berlin. Given his stature as a pre-eminent ornithologist and honored biologist, Stresemann performed a most important function in uniting biology, especially ornithology, across the Iron Curtain that separated East and West during the cold war. Stresemann, like so many of his colleagues, was representative of the Mandarin tradition that characterized some of the German professoriate: deeply steeped in history, philosophy, and ancient languages, he was also a "subtle expert on Goethe,"* and himself published on languages of southeastern Asia he had encountered on the expedition he attended as a graduate student. Stresemann died in Berlin on November 20, 1972.

The great merit of Stresemann lies not only in his numerous publications, but also in the numerous graduate students he advised or tutored who went on to pursue an academic career.† Among the graduate students tutored by Stresemann was Ernst Mayr, later to become one of the main architects of the Modern Synthesis of evolutionary theory, together with the population geneticist Theodosius Dobzhansky (1900–1975), and the paleontologist George Gaylord Simpson (1902–1984). Another author commonly counted among the architects of the Modern Synthesis was Bernhard Rensch,‡ who like Mayr arrived—in his case PhD already in hand—at the Zoological Museum in Berlin in 1925. First a temporary research assistant for a few months, Rensch obtained regular permanent employment at the assistant level as of October 1, 1925. He had visited the Berlin Museum before, during the summer break in 1922, while still a graduate student in Halle. Then docent at the Zoological Institute in Halle, Friedrich Alverdes had recommended the gifted young zoologist to Stresemann, the latter an erstwhile fellow student of his. Rensch took the train to the capital to meet Stresemann, whom he described as "a slender, young man with sharp facial features, high forehead, a full head of brownish curled hair, and wearing a monocle most of the time, who in his adroitness appeared so different from a typical museum scientist."§ It would be Stresemann who, in 1925, reached out to Rensch again to hire him for the Berlin Museum. What impressed Rensch most about his mentor was how this deep-rooted systematist sought to integrate and enrich his science with biogeographic, ecological, and ethological perspectives.¶ The same was noted by the philosophically minded Klaus Günther (1907–1975),

---

* Nöhring, 1991, p. 458.
† Listed in Jahn et al. (1973); see also Steinbacher (1991). Glaubrecht (2007, pp. 71ff) speaks of a "Berlin School," with Stresemann, Mayr and Rensch as a central figures.
‡ Mayr, 1980, p. 1: Mayr "dedicated this paper to Bernhard Rensch, one of the architects of the evolutionary synthesis, on his eightieth birthday." Mayr later endowed the "Bernhard Rensch Prize" for the German Society of Biological Systematics.
§ Rensch, 1979, p. 40.
¶ Rensch, 1979, p. 49.

the renowned synthesizer of German efforts and achievements in systematics and phylogeny reconstruction during the period from 1939 to 1959.[*] Günther, first an entomologist at the Natural History Museum in Dresden, then a professor at the Free University of Berlin, recounted in his obituary for Stresemann how systematics had largely been dropped from German university curricula by the end of the nineteenth century, and thus had been reduced to a marginal science primarily pursued at museums only. He applauded Stresemann for his attempts, early on in his career already, to integrate ecology and biogeography with systematics, thus rendering the latter an interdisciplinary science.[†] Another relatively new field of research that Stresemann, and with him Mayr, would be heavily drawn to was genetics. Siding with the associated mutation theory, Stresemann took a stance against neo-Lamarckism, which at the time was widespread among ornithologists, including the young Mayr and Rensch.[‡] "Both Rensch and I," Mayr once wrote, "were confirmed anti-Darwinians in our earlier years before we undertook [*sic*] the inescapable consequences of modern genetics."[§]

Having studied under Haeckel in Jena, Stresemann the systematist left "phylogenetic or systematic morphology ... in the sense of Ernst Haeckel"[¶] behind as he turned to populational studies instead, that is, to the problem of species delimitation and speciation. It is interesting to note how much influence the essentialistic, indeed creationist species concept—the *Formenkreis*—propagated by Otto Kleinschmidt[**] exerted, especially in ornithology (see Chapter 6). The identity conditions of species under that concept were rooted in their Divine Creation, which rendered them immutable building blocks of nature. Against that background, however, Kleinschmidt emphasized the geographical heterogeneity and polymorphism of species, the latter hence composed of a variety of *Formen*. The *Formenkreislehre* thus entailed a number of elements that Stresemann could capitalize upon as he set about to "dynamize"[††] Kleinschmidt's essentially static species concept in an evolutionary context: reproduction and consequent inheritance, geographic variation, patterns of spatial distribution and correlated ecological parameters he came to recognize as major aspects of speciation. The evolution of the species concept in Stresemann's thought and work was sketched by his student and biographer Jürgen Haffer,[‡‡] who emphasized Ernst Mayr's assessment that Stresemann had formulated a biological conception of species as early as 1919: "Forms which have reached the species level have diverged physiologically [i.e., genetically] to the extent that, as proven in nature, they can come together again without interbreeding" (Figure 8.2).[§§]

After earning his PhD in June 1926 under Carl Zimmer (1873–1950), Professor for Systematics and Director of the Zoological Institute of the University of Berlin,

---

[*] Hennig, 1976.
[†] Günther, 1974; see also Glaubrecht, 2007, p. 73.
[‡] Rutschke, 2001, p. 304. On Rensch's early commitment to neo-Lamarckism see Rensch (1980, p. 294).
[§] Mayr, letter to Heinrich Frieling, March 28, 1947; Harvard University Archives, Ernst Mayr Correspondence, HUG(FP) 14.7, Box 4, Folder 186.
[¶] Haffer et al., 2000, p. 420; see also D. Williams, 2006, p. 4.
[**] Kleinschmidt, 1900.
[††] Nöhring, 1973, p. 462.
[‡‡] Haffer, 1991.
[§§] Stresemann, 1919, p. 64; the translation is from Mayr, 1942, p. 119.

FIGURE 8.2   Ernst Mayr. (From Museum für Naturkunde Berlin, Historische Bild- u. Schriftgutsammlungen [Sigl: MfN, HbSb]. Bestand: Zool. Mus., B I/2306 [Mayr].)

Mayr obtained a salaried position at the Berlin Natural History Museum. Stresemann arranged for him to join expeditions to New Guinea and the Solomon islands in 1928 and 1930, an experience which, in Mayr's own words, consolidated in his research on species delimitation and speciation, the recognition of the importance of vicariant geographic distribution, and variation in co-specific populations. "I owe to the Berlin Museum almost everything that I have learnt in the field of systematic zoology," Mayr would later explain.[*] In 1931, Mayr emigrated to the United States, where he worked as a Curator of Ornithology at the American Museum of Natural History from 1932 to 1953, before joining Harvard's Museum of Comparative Zoology, the director of which he would become in 1961. Just like his fatherly mentor Stresemann, with whom Mayr maintained a close, life-long friendship,[†] he saw a close connection between the study of speciation and the burgeoning population genetics,[‡] especially the kind developed by Theodosius Dobzhansky, and throughout his life would use these insights in his tireless campaign against typological thinking, idealistic morphology, and last but not least Willi Hennig's phylogenetic systematics that he would dub "cladism" or "the cladistic approach."[§] His aversion against all manifestations of typological thinking was certainly reinforced as he watched from afar the horrors of the Nazi regime unfold in his home country, based as they were on a typological

---

[*] Mayr, cited in Glaubrecht, 2007, p. 77; see also Mayr (1999, p. 23) "Virtually everything in Mayr's 1942 book was somewhat based on Stresemann's earlier publications."

[†] Rutschke, 2001, p. 315.

[‡] Compare Dobzhansky, 1937; Mayr, 1942.

[§] For the Ancient Greek term for "branch"; Mayr found Hennig's approach to target the "nearest common branching point" (Mayr, 1965, p. 78), or more generally the "recency of common descent" (Mayr, 1968, p. 547).

tradition in racial theory,[*] he himself a German citizen who was "considered a potential danger to the safety of the United States"[†] during World War II.

Rensch, in contrast, stayed at the Berlin Natural History Museum, although amidst the rapidly deteriorating economic conditions that plagued the Weimar Republic, he would repeatedly write to Mayr, inquiring about opportunities for employment in the United States. Already on November 12, 1931, Rensch inquired with Mayr:

> How long do you plan to stay over there. There is little reason for you to plan to return. The wages here are being successively reduced, two years of service have been crossed off, and another reduction is immanent, salary payments being made only every 10 days ... the financial crisis is expected to get even worse next year ... anyway, perhaps you could at some point recommend me for an assistantship at the museum in New York ... Although I fear the American resources will dry up as well.[‡]

When Rensch got into trouble with Nazi officials due to an exhibit he had developed at the Berlin Museum of Natural History in 1934 that mediated neo-Lamarckian tendencies, he wrote to Mayr again on October 8, 1935. Apologizing for not having written for so long, he explained this with "personal problems." "My situation here has been unresolved until recently, but now I have been extended for another year ... But I am also looking for other jobs. Should you hear about a possibility for me over there, I would greatly appreciate a tip."[§] Unfortunately, Mayr was unable to help.

Bernhard Rensch,[¶] just like Mayr himself, was a typical son of the *Bildungsbürgertum*.[**] Zoologist by profession, he was deeply interested in history, philosophy, psychology, arts, and Eastern cultures. Born on January 21, 1900, in Thale (Harz Mountains, Germany), he attended grammar school (*Reformgymnasium*) in Halle from 1912 to 1917. His whole class performed miserably in the final examinations (*Abitur*) in physics and mathematics. The reason was that the math teacher wanted as many students to fail the exam as possible, so that they would not be conscripted to the army for a war the teacher thought was already lost. Recognizing that strategy, which he considered misguided, the Principal ignored the exam results in math and physics, and promoted the students. In 1917/18, Rensch served in the army, at the end as lieutenant. After 2 years in French internment, he enrolled at the University of Halle to study zoology, botany, chemistry, and philosophy. During his student years, he was particularly impressed by the neurologist, psychiatrist, and philosopher Theodor Ziehen: "from the first hour I was spellbound by this preeminent personality."[††] Rensch found Ziehen's courses to explore the deepest problems of philosophy, and through his attendance in Ziehen's classes he became motivated to develop his own philosophical system that should complement his otherwise

---

[*] Weingart et al., 1992, p. 615.

[†] Junker, 1996, p. 35. These remarks are based on personal interaction between Thomas Junker and Ernst Mayr (Junker, 1996, p. 69).

[‡] Harvard University Archives, Ernst Mayr Correspondence, HUG(FP) 14.7, Box 1, Folder 14.

[§] Harvard University Archives, Ernst Mayr Correspondence, HUG(FP) 14.7, Box 1, Folder 7.

[¶] Dücker, 2003; see also Rensch, 1979.

[**] Harwood (1993, p. 362) characterized Rensch as a "comprehensive thinker." On Mayr see Junker (1996, p. 35).

[††] Rensch, 1979, p. 35.

scientific world-view. Rensch earned his PhD in 1922 under Valentin Haecker (1864–1927), with a study of cytological variation in different chicken breeds. After work as an assistant at the Institute for Agricultural Botany in Halle, Rensch was appointed at the Berlin Museum of Natural History in 1925, responsible for the mollusk collection. In 1927, he led an important zoological–anthropological expedition to the Lesser Sunda Islands, with Gerhard Heberer among the participants (Figure 8.3).

As was also the case for Mayr at the outset of his career, the focus of Rensch's early work was race (i.e., subspecies) formation and speciation in animals. Rensch conceptualized polymorphic species as groups of geographically vicariant races of common descent (*Rassenkreis*), where the neighboring races are able to freely interbreed.[*] Emphasizing geographical heterogeneity and geographic isolation among populations (races, i.e., subspecies) as important mechanisms of speciation, Rensch praised Kleinschmidt—whose *Formenkreise* he recognized as geographically vicariant species[†]—for having been the first to consequently employ the

FIGURE 8.3   Bernhard Rensch. (From Museum für Naturkunde Berlin, Historische Bild-u. Schriftgutsammlungen [Sigl: MfN, HbSb]. Bestand: Zool. Mus., Orn. 145,6 [Rensch].)

---

[*] Rensch, 1929, p. 13.
[†] Glaubrecht, 2007, p. 75.

"geographical principle" in microsystematic research.[*] Rensch's classic papers on speciation, published in 1929, 1933, and 1939, respectively, exerted a major influence on Ernst Mayr, and earned him a place in the pantheon of the architects of the Modern Synthesis of evolutionary theory.[†]

Rensch was a member of the National Socialist Teachers' Association, as also of the National Socialist German University Lecturers' Association, but otherwise eschewed political exposure in NS organizations. He is known to have bowed to ideological pressures of the time, but kept such commitments low profile and to an absolute minimum. With only two incriminating publications alluding to the race ideology of the Third Reich, Rensch only once referred to the slogan of "blood and soil," tied as it was to the *Lebensraum* doctrine, which he causally grounded in natural selection.[‡] Presenting a summary of his views on speciation in an anthropological journal, Rensch defended a neo-Lamarckian perspective with his claim that race formation in humans is not just the result of random mutation and natural selection, but also impacted by a direct influence of the environment.[§] The same neo-Lamarckian inclination—under severe attack by racial theorists of the time (see Chapter 7)[¶]—also transpired in an exhibit that Rensch had developed early in 1934, which got him into trouble with authorities that almost resulted in the termination of his museum employment. Rensch, in his autobiography, listed a number of additional motives for this reprimand, that is, his refusal to join the SA and NSDAP, his mocking the NS-camps that served the purpose to educate university docents to become politically mature persons, and his social as well as scientific interactions with German Jews.[**] He was eventually able to access the dossier that the authorities had compiled on him: in it Rensch was characterized as a "straight and honest person," although his "unbalanced focus on science" was taken to indicate a lack of enthusiasm for National Socialism. He was therefore judged to fail the qualification required for all university docents, namely to be "a scientific and a political person"[††] in the sense of Karl Escherich.

In February 1937, Rensch was appointed Director of the State Museum of Natural History of Westphalia in Münster, earning the *venia legendi* at the University of Münster in 1938, with the advice from Hermann Weber (see Chapter 5) to teach evolutionary theory, as well as comparative and functional anatomy. Conscripted to the army in 1940, Rensch developed health problems that led to his transfer to Aachen, where he was charged with training new recruits. He reported cunningly how he had managed to stay at the hotel, rather than in the barracks, where in the evenings he would read Ziehen's two volume treatise on epistemology,[‡‡] the second volume (second edition) of which had just appeared in print the year before. His condition (myocarditis) worsening, Rensch was dismissed from the army at the end

---

[*] Rensch, 1929, p. 7.
[†] Glaubrecht, 2007, p. 75.
[‡] Rensch, 1934, p. 704; see also Rensch, 1935; Junker and Hoβfeld, 2002, p. 238.
[§] Rensch, 1935, p. 333.
[¶] Holler, 1934a,b; Junker, 2001b, 2004.
[**] See also Junker, 2004, p. 171.
[††] Rensch, 1979, p. 78; A "political person" sensu Escherich (1934).
[‡‡] Rensch, 1979, p. 106.

of February 1942, to return to his teaching duties at the University of Münster as adjunct professor (*ausserplanmässiger Professor*). On January 12, 1944, Rensch was surprised by a telegram from the Reich Ministry of Education, ordering him to report as Chair of Zoology at the German University in Prague. With the Russians advancing from the East, the US Army approaching from the West, the German faculty retreated from Prague to Leipzig by train, Rensch continued on by foot to Jena via Apolda, and back to Pfiffelbach. The collapse of the Third Reich left him no way out of Thuringia for weeks, time which Rensch whiled away studying—among other authors—additional material published by Ziehen on the physiological basis of psychology.[*] Finally, however, Rensch made it back to Münster via Göttingen, where in 1947 he became Full Professor of Zoology and Director of the Zoological Institute and of the Natural History Museum. During times of distress and deprivation in the destroyed city of Münster, Rensch contacted Mayr in the United States again: "I don't know whether Stresemann has informed you about my personal fate." Rensch summarized his fate over the last few months and years, the war and the turmoil in Prague, the problems he had with Nazi officials because of his early neo-Lamarckism, but also stressed that he considered himself fortunate enough for having been immediately reinstated after the war as Director of the Natural History Museum in Münster, since politically he was "a white lamb," as he characterized himself. Offers for employment at the Natural History Museum in Dresden, or rather "what is left of it," as also at the *Phyletisches Museum* in Jena he deemed unattractive. "I have also completed a book manuscript on macroevolution for which I am now seeking a publisher in western Germany. A difficult feat, since Fischer, Bornträger and Springer are now all in the Eastern Zone."[†] Mayr was thrilled to hear from his old acquaintance: "Dear Rensch. What a pleasure to hear from you again and that you are alive and still able to work as best as possible in these horrible times."[‡] The cordial friendship between the two men deepened as they engaged in a lively correspondence that would carry on for years to come. Mayr quickly arranged for the shipment of recent American literature on evolutionary biology to Rensch, and informed him of the founding of a Society for the Study of Evolution by Dobzhansky, G.G. Simpson, and himself. Rensch was to be appointed to the editorial board of *Evolution*, the journal published by that society. Enlisting the help and support of Joseph J. Hickey[§] and Margaret Morse Nice,[¶] Mayr was leading the effort within the American Ornithologists' Union in sending C.A.R.E. packages to needy European ornithologists, and personally dispatched to Rensch and his wife food stuffs, coats, and shoes. "I can hardly emphasize enough," thanked Rensch, "what critical significance such aid for us has today, and how important it is for us in

---

[*] Rensch, 1979, p. 119.
[†] Rensch to Mayr, April 4, 1946; Harvard University Archives, Ernst Mayr Correspondence, HUG(FP) 14.7, Box 4, Folder 158.
[‡] Mayr to Rensch, June 5, 1946; Harvard University Archives, Ernst Mayr Correspondence, HUG(FP) 14.7, Box 4, Folder 158.
[§] Temple and Emlen, 1994.
[¶] Trautmann, 1977.

the spiritual sense as well."* Mayr asked Rensch for names and addresses of German ornithologists "who are in need. Also, what types of clothing and shoes they most need for themselves, their wives and children ... in order to avoid criticism it will be advisable not to include on this list anybody who was an out and out Nazi ... on the other hand I see no objections to people like Niethammer who were merely nominal Nazis."† Mayr's worries about the political past of German colleagues also became apparent when Heberer, upon his escape from Czech internment and consequent employment at the University of Göttingen (see previous chapter), requested from Mayr a copy of his 1942 book on *Systematics and the Origin of Species*. Mayr first turned to Rensch, inquiring: "... rumors have reached me that Heberer had Nazi sympathies."‡ Rensch and Heberer had been schoolmates,§ and of course Heberer had accompanied Rensch on his expedition to the Lesser Sunda Islands: "As far as my expedition companion Heberer is concerned," replied Rensch, "there is no way to conceal the fact that he was a *Sturmführer* in the SS, a rank, however, that was bestowed on him as an anthropologist. Since no other charges could be leveled against him, he has in the meantime been de-nazified, and is now teaching anthropology at the University of Göttingen. Since his school days, Heberer had an interest for Germanic prehistory: this explains why he was receptive of many of the ideas propagated by the Nazis. But he is certainly not to be counted amongst those who still today are inwardly convinced Nazis."¶ Mayr thereupon dispatched the book, not to Heberer personally, but to the library of Heberer's research station.

In hindsight, it is certainly Rensch's early studies on speciation that most influenced the Modern Synthesis of evolutionary theory, even though he adhered to neo-Lamarckian doctrines well into the early 1930s, to a point when he more fully learnt to understand the pleiotropic nature of most genes, and how pleiotropy relates to the workings of natural selection.** The second major contribution of Rensch to evolutionary theorizing was his book on macroevolution, *Problems of Trans-Specific Evolution*, published in Stuttgart in 1947.†† It was well received on the other side of the Atlantic: "Simpson is still reading your book and is deeply interested in it,"‡‡ Mayr related. Both Simpson and Dobzhansky supported the project of translating Rensch's book, albeit with some editing, "not just in the chapter on speciation." In his book, Rensch seconded Heberer's views that macroevolutionary phenomena could be reduced to a summation of microevolutionary events, and explained different

---

* Rensch to Mayr, December 12, 1946; Harvard University Archives, Ernst Mayr Correspondence, HUG(FP) 14.7, Box 4, Folder 176.

† Harvard University Archives, Ernst Mayr Correspondence, HUG(FP) 14.7, Box 4, Folder 176. Günther Niethammer (1908–1974; NSDAP, 1937; SS, 1940. 1940/41), was a graduate student and later collaborator of Stresemann; during the war he was a guard in the Auschwitz Concentration Camp, dispatched for special ornithological investigations in the surrounding area (Klee, 2003, p. 436).

‡ Mayr to Rensch, October 31, 1949; Harvard University Archives, Ernst Mayr Correspondence, HUG(FP) 14.7, Box 7, Folder 330.

§ Hoβfeld, 1999c, p. 195.

¶ Rensch to Mayr, December 2, 1949; Harvard University Archives, Ernst Mayr Correspondence, HUG(FP) 14.7, Box 7, Folder 330.

** Rensch, 1980, p. 296.

†† Rensch, 1947.

‡‡ Mayr to Rensch, February 26, 1948; Harvard University Archives, Ernst Mayr Correspondence, HUG(FP) 14.7, Box 5, Folder 253.

rates of evolutionary change through different levels of selection pressure. And yet, it should have been the all-pervading determinism propagated by Rensch that should have raised the eyebrows of the evolutionary biologists in the United States as they went about to integrate population genetics with ecology, behavioral studies, biogeography, and paleontology to achieve a Modern Synthesis of evolutionary theory.

It is the German (indeed Kantian) tradition to consider the goal of natural sciences, biology included, to be the discovery of laws of nature. Steeped in that tradition, Rensch described himself as a "consequent determinist (in the sense of the synergistic effects of all natural laws including the laws of logic and of probability)."* Statistical and probabilistic laws would seem to contradict an all-pervading determinism, however. It is therefore of special importance to correctly understand Rensch's concept of "chance," "probability," and "randomness,"† which he adopted from Ziehen's treatise on epistemology,‡ where randomness reflects not the fundamental structure of the universe, as it does in quantum mechanics. Instead, apparent randomness merely reflects ignorance on the part of the investigator, in cases where the complexity of natural processes defies the capacity of human cognition. It is therefore not surprising that Rensch spent a lifetime researching *laws* of evolution. To bring out the nomological nature of his evolutionary research both at the infra- and supra-specific level, he introduced the term *Bionomogenese*.§ He recognized, however, the apparent randomness of mutations, as well as apparent exceptions to the evolutionary trends he believed to be able to identify, and so allowed for probabilistic rules, yet under the same caveats that were articulated by Ziehen. But causal laws are not the only ones affecting evolutionary processes. Again following Ziehen, Rensch emphasized the significance of laws of logic in evolutionary theory as well: "how decisive the laws of logic are for evolution may easily be shown by the assumption that these laws would *not* be valid." Take a gene [A] that reduplicates, creating two identical genes [A, B]. Let one of these [B] reduplicate again: if the laws of logic (transitivity in this case) were not valid, "this third gene [C] would not be equal to the first [A]. Hence a continued identical reproduction of genes would not occur…"¶ What Rensch's argument here implies is Ziehen's postulate that the human capacity of logical thought is the result of an evolutionary adaptation to the structure of the extramental world. The significance of this postulate for phylogenetic systematics can hardly be overstated, as it allowed logic to be brought to the reconstruction of the Tree of Life. According to Rensch, "TH. ZIEHEN (1934, §22) was the first who pointed out that [human] thought has obviously adapted to the extramental logical and lawful structure of the world (*logische Weltgesetzlichkeit*) in the course of phylogeny."** In Rensch's rendition, the human capacity of logical thought developed phylogenetically as an adaptation to the universal laws of logic that reign over the universe. Mistaken thinking would be corrected by natural selection because of its failure to capture the logical

---

* Rensch, 1979, p. 170; see also Rensch, 1988.
† Rensch, 1968, p. 108.
‡ Ziehen, 1934/39.
§ Rensch, 1968, p. 108.
¶ Rensch, 1960, p. 98. Rensch's conclusion is based on the relation of transitivity: if A equals B, and B equals C, then A equals C.
** Rensch, 1968, p. 232; see also Ziehen, 1934, p. 87.

structure of either the material, or of the physical world. "Hence logical laws as well as causal laws were also valid before man existed and before there were any organisms on earth."[*]

THE BERLIN SCHOOL BRANCHES OUT TO DRESDEN

Wilhelm Meise[†] (1901–2002) was an early graduate student of Erwin Stresemann, a resident at the Berlin Natural History Museum almost simultaneously with Mayr. He earned his PhD in 1928 with a thesis on the hybrid zone between the western Carrion Crow and the eastern Hooded Crow that stretches across Europe.[‡] His became a textbook example for hybridization along a zone of secondary contact of previously geographically separated populations that had acquired incipient species status. Meise treated the two species again as a *Formenkreis*, which differently from Kleinschmidt he defined dynamically as "geographically vicariant forms that are interconnected by transitional or hybrid zones."[§] In 1929, Meise was appointed Curator at the Natural History Museum in Dresden, responsible for all collections except insects. Interned for 3 years in Siberia after World War II, Meise returned to his native country in 1948, where Stresemann arranged for his temporary appointment in his division at the museum in Berlin. In 1951, Meise transferred to the position of Curator of Ornithology at the Natural History Museum in Hamburg, where he also became an adjunct professor at the university.

Meise was one of the leading German ornithologists of the twentieth century, but while in Dresden, and being responsible for zoological collections other than birds as well, he was also interested to work on other flying vertebrates such as snakes, lizards, and frogs. These were the times when a grammar school student (*Gymnasiast*) with an obvious knack for systematic research spent a significant part of his leisure time in the collections of the Dresden Natural History Museum. To be delivered on May 4, 1931, Willi Hennig's homework required the composition of an essay, for which he chose the title *The Position of Systematics in Zoology*.[¶] The essay was motivated by Hennig's concern for a lack of interest and support for systematic research in German biology. It is the same concern that by then had been articulated by Austrian and German systematists for over 50 years already. The essay is an amazing document, as it reveals Hennig's deep interest in systematics at that early age. An introductory historical sketch includes zoological as well as botanical, neontological as well as paleontological perspectives. Darwin's theory is praised for having turned the natural system into a phylogenetic tree. The importance of von Baer's insights for character analysis is illustrated with an example drawing on fish and frog larvae, and tadpoles, respectively. Oscar Heinroth (1871–1945) is (correctly) cited as the first author to have used behavioral characteristics in systematics. The issue of "geographical subspecies" is touched upon, in which context Hennig cited

---

[*] Rensch, 1960, p. 99.
[†] Haffer, 2003a,b; Hoerschelmann and Neumann, 2003.
[‡] Meise, 1928.
[§] Cited after Haffer, 2003a, p. 119.
[¶] The essay was posthumously published by Dieter Schlee, Hennig's assistant and collaborator at the Staatliches Museum für Naturkunde in Stuttgart, in *Entomologica Germanica*: Hennig (1978/79).

the "Jordan's Rule" in support of the claim that "closely related forms do not live in the same area, but in adjacent areas that are separated by some kind of barrier." In a footnote he added: "Unfortunately I could not secure literature on this *Formenkreis* theory before due-date."[*] Additional similar rules he invoked in support of the claim that systematists are not simple-minded "species manufacturers" (*Speziesmacher*),[†] a perspective he did not get from Othenio Abel but from the hugely popular encyclopedia *Brehms Thierleben* instead, and he continued to insist that systematics is a true generalizing science. The essay ends with an outlook as to how systematics has illuminated neighboring sciences, not just biogeography but paleobiogeography as well.

At the Dresden museum, Hennig had attracted the attention of Meise, as well as of the entomologist Fritz van Emden (1898–1958), who both introduced him to their systematic work. Meise suggested to the 18-year-old volunteer to pursue a taxonomic revision of the snake genus *Dendrophis*, a "flying snake" from Australia, New Guinea and Asia today known as *Dendrelaphis*. The result was a joint publication in the *Zoologischer Anzeiger* in 1932, with Willi Hennig—who had just graduated from grammar school—as second author.[‡] It was his second publication; his first one dated from the year before, reporting his findings of an entomological—faunistic survey he had conducted in the Dresden school district.[§] Three years later, Meise and Hennig published further results from continued collaborative research, now expanded to include the "flying snake" genus *Chrysopelea* from Southeast Asia.[¶] While a student at the University of Leipzig, Hennig conducted independent studies on the systematics of the flying lizard genus *Draco*, acknowledging Meise for help in obtaining comparative material, as well as general support and much appreciated advice. After the war, Hennig met his former tutor Meise in Berlin again, most probably enthusiastically explaining to him the thoughts and methods he had in the meantime developed in order to render phylogenetic systematics a proper science. Meise was attracted to the stringency of Hennig's method, subjecting his book that was published in 1950, as well as later contributions of 1953 and 1957, to attentive scrutiny. While Hennig himself had long left work on reptiles behind, and specialized exclusively in entomology, Meise became the ambassador for Hennig's phylogenetic systematics in German ornithology.[**]

In Hennig's early and independently researched papers on the systematics of the agamid lizard genus *Draco*,[††] some of the old and widespread motivations of German biologists come to the fore again, in particular a view of systematics as a science in search of natural laws. His exposure to Meise must have sharpened in Hennig the appreciation of the species as the fundamental building block of biodiversity, and also triggered in him insights into the intricacies of speciation and species delimitation.

---

[*] Hennig, 1978/79, p. 196.
[†] Hennig, 1978/79, p. 193.
[‡] Meise and Hennig, 1932.
[§] Hennig, 1931.
[¶] Meise and Hennig, 1935.
[**] The account of Hennig's interactions with Meise follows Haffer (2003a, pp. 124–127). But see also Schmitt (2001, 2013).
[††] Hennig, 1936a,b.

In his work on *Draco*, Hennig deployed Rensch's concept of *Rassenkreise*. He recognized 14 species (or, better, *Rassenkreise*) in the genus *Draco*, and found as a rule that "the most derived coloration of the patagonium is found in those subspecies that have departed the furthest from the center of origin of their *Rassenkreis*."[*] Albeit "not quite as lawful," the same he claimed to hold for the relative length of the hind limb, which in all *Rassenkreisen* (i.e., polymorphic species) increases from West to East in those cases where subspecies differed from one another in that character.[†] And again, yet this time with respect to the gular fold: given an expansion from East to West, the youngest races furthest to the West have the most derived gular fold structure.[‡] Similar lawfulness he found to hold in the insect world, licensing the use of geographical distribution for the polarization of character transformation series in support of species delimitation and phylogeny reconstruction.[§] Significantly, in his attempt to delineate the *Rassenkreise* from one another, he recognized at the level of races (subspecies) the same phenomenon that had previously been recognized at the level of species comparison, that is, the ubiquity of the crossing of specializations (*Spezialisationskreuzungen*, i.e., Dollo's *chevauchement des spécialisations*).[¶] These early studies foreshadow some issues that would figure prominently in Hennig's later phylogenetic systematics: problems of the species category and issues of speciation, the aspiration to discover lawfulness in variation, the polarization of primitive and derived characteristics in relation to the evolutionary origin of a species or subspecies, and—perhaps most importantly—an all-pervading crossing of specializations.

The papers on *Draco* reported Hennig's last hands-on systematic investigation of reptiles. He had, under the influence of Fritz van Emden, found his love for entomology, with dipterans in particular becoming the focus of his entire career. With their complex lifecycle involving larval, pupal, and imaginal stages, insects pose a far more intricate problem to the microsystematist than lizards or snakes. When discussing "laws of geographical distribution" in dipterans, Hennig drew a distinction between classifications that are built on the subjective choice of so-called "index features," as opposed to a system that is strictly phylogenetically structured, that is, based on phylogenetic relationships. Hennig conceded that the latter are by no means easily deciphered, but, he argued, "that if one believes in the possibility for progress in phylogenetic systematics at all, then one would expect the phylogenetic system to gradually approach a stable final structure."[**] How, then, to deal with larval, pupal, and imaginal stages of insects in phylogenetic systematics? Here, Hennig drew a fundamental distinction between classifications based on degrees of morphological similarity or dissimilarity, versus a classification strictly based on degrees of phylogenetic relationships. While he admitted the possibility to base classifications on morphology for practical purposes, he

---

[*] Hennig, 1936b, p. 551.
[†] Hennig, 1936b, p. 555.
[‡] Hennig, 1936a, p. 171.
[§] Hennig, 1936c, p. 161.
[¶] Hennig, 1936b, p. 557.
[**] Hennig, 1936c, p. 172.

insisted that such classifications were artificial, and could not be used as a basis for the inference of scientific conclusions or laws of phylogeny.[*]

But many systematists understood microsystematics to be primarily about species description, identification, and re-identification, not necessarily about phylogeny reconstruction. A great theoretical mind in ecology, August Thienemann (see Chapter 6) was more of a practical man when it came to systematics, that is, the identification and classification of species. As an ecologist focusing on limnology, Thienemann took an interest in natural as well as artificial lakes, and their colonization patterns by aquatic organisms. Among artificial lakes ranked water reservoirs that "arrived in Germany about the same time that limnology, the study of lakes, emerged as a discipline,"[†] with Thienemann as one of its leading pioneers. The management of water reserves would surely involve not only limnologists and fisheries biologists, but also water hygienists, who would see themselves confronted not with the imaginal stages of chironomids, but with their larvae and pupae instead, as Thienemann emphasized. Non-specialists must be able to readily identify larvae and pupae without having to breed them first to obtain the imaginal stage, for which reason Thienemann was not the only one to argue in support of different nomenclatorial systems that would apply to different stages in the lifecycle of insects. That way, "short names replace lengthy descriptions"[‡] in species identification at any developmental stage. Besides, Thienemann contended, systematics and classification are always built on morphological similarity relations. Without a Fossil Record, phylogenetic interpretations are a theoretical add-on: "It is impossible to simply ignore these questions!"[§]

Hennig retorted only 6 years later, in 1943, accusing Thienemann and his co-author of identifying comparative morphology with systematics, and of insinuating that morphology provides the foundation for phylogenetics. Even with the availability of potential ancestors (*Stammformen*) in the Fossil Record, "I explicitly deny that comparative morphology should be the sole foundation of 'phylogenetic systematics'!... The law according to which 'the more similar two forms or groups of forms are to each other, the more closely they are phylogenetically related, and vice versa' is fundamentally flawed."[¶] This is because the putative law calls for overall similarity rather than careful character analysis as an indicator of phylogenetic relationships. Indeed, it is precisely the discordance between a systematics of larvae, pupae, and imagos, each based on morphological similarity relations, which highlights the need for a unifying, phylogenetically grounded systematics.[**] To a unique process of descent with modification must correspond a unique classification, or system, of biodiversity. Although Hennig conceded that morphological similarity ultimately provides a clue

---

[*] Hennig, 1936c, p. 174f.
[†] Blackbourne, 2006, p. 232.
[‡] Thienemann and Krüger, 1937, p. 266.
[§] Thienemann and Krüger, 1937, p. 258. In his letter to the Chair of the search committee for the Zschokke succession at the University of Basel, dated February 3, 1931, Thienemann highlighted Tschulok's (1910, p. 197) distinction of "seven equally important yet completely incommensurate material perspectives in biological research," which rendered systematics and classification independent from phylogenetics (State Archive, Basel, StABS ED-REG 1a 2 1402).
[¶] Hennig, 1943, p. 139.
[**] Hennig, 1943, p. 140.

to phylogenetic relationships, he emphasized that it is not mere measurements, nor the sheer number of shared characteristics that is relevant, but that shared morphological characteristics must carefully be weighted with regards to their significance for phylogeny reconstruction, and in that process must be enriched with insights from biogeography, ecology, and paleontology (Figure 8.4).

To arrive at a truly phylogenetic system, Hennig argued, careful character analysis and weighting is thus required across the "(onto-) genetic"* dimension, and where the ontogenetic dimension includes stages of disparate morphologies, these still needed to be accommodated by a unique phylogenetic system. Hennig later introduced the concept of the *semaphoront*, the "character-bearer," which replaced the organism as empirical basis of comparison—a concept that will require a more detailed discussion. Most readers and commentators took and take Hennig's *Semaphoront* as a conceptual devise to accommodate different stages of ontogeny, particularly in holometabolous insects such as those he was himself working on. True enough—but it was more. It was a philosophical move to accomplish what Walter Zimmermann, as well as Theodor Ziehen, had requested for natural sciences: to separate the objective from the subjective component in perception, and to do so on a putatively more secure basis than Zimmermann's voluntarism (see Chapter 7).

FIGURE 8.4   Willi Hennig, around 1973. (From original M. Linke; Willi Hennig Archive, Senckenberg Museum Görlitz [Director: Prof. Dr. W.E.R. Xylander].)

* Hennig, 1943, p. 140f.

WILLI HENNIG

Willi Hennig's life and work was chronicled in detail on the occasion of the centenary of Hennig's birthday by the Hennig biographer Michael Schmitt,[*] on whose work the following biographical sketch is based. The eldest of three sons, Willi Hennig was born on April 20, 1913, in Dürrhennersdorf near Löbau in Saxony. His father was a railroad worker, whose job required repeated relocations of the family during Willi's childhood. His mother zealously promoted optimal school education for her children on his father's modest salary. From 1927 to 1932, Willi attended the *Reformrealgymnasium*, a boarding school in the Klotzsche District of Dresden. It was his teacher in natural history, Maximilian Rost, with whose family Willi lived, who first took Hennig to the Dresden Natural History Museum and introduced him to Wilhelm Meise. Hennig graduated from the *Gymnasium* in February 1932, after which he enrolled at the University of Leipzig to study zoology, botany, and geology, starting with the summer semester of 1932. His wife being Jewish, the entomologist at the Dresden Natural History Museum, Fritz van Emden, was laid off in September 1933 following the enactment of the Law for the Restoration of the Career Civil Service in the spring of that year. His successor was Klaus Günther (1907–1975),[†] whom Hennig met on the occasion of his frequent visits to the Dresden museum. The two men would engage in a deep, lifelong friendship based both on mutual professional interests and personal affinity. In Michael Schmitt's estimation, Günther exerted a most important influence on Hennig's intellectual development. Hennig graduated from the University of Leipzig with a PhD on April 15, 1936; his doctoral thesis treated the genital apparatus of a group of dipterans and its significance for systematics. A stipend from the *Deutsche Forschungsgemeinschaft* allowed him to pursue systematic studies at the DEI (*Deutsches Entomologisches Institut* of the *Kaiser-Wilhelm Gesellschaft*) in Berlin-Dahlem as of January 1, 1937. By January 1, 1939, Hennig had secured employment at the DEI in a permanent position as scientific assistant. During World War II, Hennig served as infantryman in the German army in western and eastern deployments, until he was wounded by shrapnel at the eastern frontline in 1942. After his convalescence, he served the German army in his capacity as entomologist in a unit tasked with disease control, deployed in northeastern Italy near the end of the war. It was in Lignano, situated on the Gulf of Trieste, that Hennig and his comrades were captured by British forces and taken prisoners of war. Hennig was released in the fall of 1945, on which occasion he returned to the University of Leipzig to stand in for the missing Professor of Zoology and Director of the Zoological Institute, Paul Buchner (1886–1978), his former PhD advisor.[‡]

His heart was set on returning to Berlin, however, where he resumed his position at the DEI in April 1947. As of November 1, 1949, Hennig was promoted to Vice Director and Head of the Section for Systematic Entomology at the DEI, which after the war was relocated to Berlin-Lichterfelde in the Eastern Sector of the town. Hennig and his family lived in the Western Sector, in Berlin-Steglitz, which made for a hefty commute. Hennig earned the *Habilitation* in 1950 at the University of Potsdam,

---

[*] Schmitt, 2013; see also Schlee, 1978; Dupuis, 1990; Schmitt, 2001.
[†] Hennig, 1976.
[‡] Senglaub, 2000.

where he started to teach courses in zoology in 1951. During the 1950s, Hennig kept visiting the Natural History Museum in Berlin, where he promoted his phylogenetic systematics in seminars, and engaged in discussions primarily with Klaus Günther, but also with the herpetologist Heinz Wermuth (1918–2002).[*] A talk presented by Hennig at the museum on October 30, 1953, was followed by a heated discussion that lasted for over three hours. In the course of the debate, Hennig insisted that his method could successfully decipher phylogenetic relationships even in the absence of fossils. The paleontologist Walter R. Gross countered that fossils alone can deliver direct proof of descent with modification. Hennig replied "your fossils are of no interest to me," whereupon Gross jumped to his feet, exclaiming "in that case your theories are of no interest to me either," banging the door as he stormed out of the room.[†] On August 13, 1963, the government of the German Democratic Republic (GDR) started to erect the Berlin Wall, which would completely cut off East Berlin from West Berlin. Hennig was offered continuation of his employment at the DEI only on the condition that he moved his family to East Berlin and filed for citizenship of the GDR. Hennig protested against such "coercion,"[‡] and preferred to stay in West Berlin, where he found immediate temporary employment at the Berlin Institute of Technology. He rejected offers for employment from overseas as he did not want to leave behind the European historic and cultural background, and also feared possible disadvantages for the education of his sons in a foreign country. In April 1963, Hennig moved his family to Ludwigsburg near Stuttgart, where—like Heinz Wermuth—he had been offered a position at the State Museum of Natural History of Baden-Württemberg. Hennig was appointed head of the section for phylogenetic systematics, the *Abteilung für stammesgeschichtliche Forschung* that had been set up specifically for him. Hennig had forfeited the opportunity to become Director of the DEI when he decided against moving to East Berlin. Had he ever harbored hope to become Director of the State Museum of Natural History of Baden-Württemberg, he must have been disappointed by the appointment, in 1969, of Bernhard Ziegler (1929–2013) to that post. Ziegler, a well-connected native of Stuttgart, studied geology and paleontology in Tübingen, and earned his PhD in paleontology under Helmut Hölder (1915–2014) in 1955. In 1957, he transferred to the University of Zürich, where he obtained his *Habilitation* under Emil Kuhn-Schnyder (1905–1994) in 1962. With a solid background in stratigraphy and traditional invertebrate paleontology, it is hardly surprising that Ziegler was not exactly a supporter of Hennig's phylogenetic systematics. Tensions between the two men were fueled by Hennig's growing national and global academic recognition. In 1969, Hennig was presented with the certificate of an honorary doctorate from the Free University of Berlin, an honor that had been instigated by his friend Klaus Günther. Students managed to have him appointed as an adjunct professor (*Honorarprofessor*) at the University of Tübingen in 1970. Among many other accolades may be mentioned his membership in the German Academy of Sciences Leopoldina, and the award of the Linnean Medal (Gold Medal) of the Linnean Society of London, as well as of the Gold Medal

---

[*] Peters, 1995, p. 5.
[†] Schmitt, 2001, p. 327; see also Peters, 1995, p. 5.
[‡] Peters, 1995, p. 9.

of the American Museum of Natural History in New York. Hennig died in his home of a heart attack on November 5, 1976.

A PROSPECTUS FOR PHYLOGENETIC SYSTEMATICS

When introducing his *Grundzüge* of 1950 to his entomological peer community, Hennig started out by saying that "biological 'systematics' lacks a sufficient theoretical foundation," a situation he said he attempted to remedy with his book.[*] Although the reception of Hennig's work has mainly focused on methodological and terminological aspects, his own intention in addition had been to improve the conceptual foundations of biosystematics. Exploring the multidimensionality of biodiversity, amenable to a multitude of different systems along different dimensions (morphological, physiological, ecological, biogeographical, etc.), Hennig called for a foundational system to which all other possible systems could be related. That foundational system would have to be the phylogenetic system which, given the fact that species originate exclusively through the division (*Teilung*) of ancestral species, could be represented by an accurately determined schema, one which corresponds precisely to an enkaptic hierarchy.[†]

Hennig had completed the first draft of the manuscript for the *Grundzüge* in 1945. In 1947, Hennig expressed in a footnote his chagrin that due to a shortage of paper, his (by that time refined) extensive investigation of the conceptual foundations of phylogenetic systematics could not appear in print.[‡] He consequently published what he considered to be the most salient points in two essays, which appeared in *Forschungen und Fortschritte—Nachrichtenblatt der deutschen Wissenschaft und Technik*, in 1947 and 1949, respectively. Given the dense and meandering flow of thought that is manifest in the *Grundzüge*, these two essays serve as a welcome source for the identification of those issues that Hennig saw as the most important ones in need of clarification in contemporary systematics.

Hennig opened the first essay of 1947 with a rejection of the concept of theory-free observation. Going against Zimmermann's strict separation of object and subject, without citing him though, Hennig invoked an irreducible active relation (*aktionale Struktur*)[§] between the object and the cognizing subject. Systematics cannot be theory-free—an important argument Hennig would eventually deploy against the tyranny of overall-similarity. Since raw observation cannot yield phylogenetically informative characters, there is therefore a need to develop a theory of character analysis. Like Beurlen and Naef, Hennig talked about *Gestalt*, but took such *Gestalt* to be a multidimensional organization. Idealistic morphology, he claimed, orders biodiversity according to degrees of morphological similarity, and—citing Naef—he conceded that it is often true that results of comparative morphology can be translated into a phylogenetic system. But morphology is only one dimension of the many that constitute an organismic *Gestalt*. Other dimensions are ontogeny, physiology, ecology, biogeography, and the more a truly phylogenetic system is in accordance with

---

[*] Hennig, 1952, p. 329.
[†] Hennig, 1952, p. 330.
[‡] Hennig, 1947, footnote on p. 276.
[§] Hennig, 1947, p. 276.

data gleaned from all these dimensions, the more can it be trusted to represent the real phylogenetic process. Such argumentation, Hennig insisted, is in no way circular, but instead is based on the principle of reciprocal illumination that had so well been articulated in humanities by Wilhelm Dilthey (1833–1911).

So what is it that the phylogenetic systematist is working with in his search for consilience of evidence within the multidimensional multiplicity of organismic diversity? Here Hennig introduced the concept of the "character-bearer"—later to be named the *Semaphoront*—which in this early writing he characterized as a time-slice through an individual organism thin enough such that, for the duration of that time-slice, the organism neither changes itself nor changes its relations to other organisms.* The very invocation of a metaphysics of time-slices reveals Hennig's adoption of a process philosophy, an important component of logical positivism (logical empiricism).† The organism is a developing, ever-changing entity, most drastically so in holometabolous insects or other organisms with complex lifecycles. For the practical purposes of the systematist, the organism must be chopped up into time-slices that yield the characters required for the reconstruction of its phylogenetic relationships. But the characters so obtained are only pointers; reality resides with the organism and its phylogenetic history as part of a species. Whereas the "character-bearer" provides the empirical base for the working systematist, it is the species that provides the building block for the phylogenetic system. Without citing Oscar Hertwig's seminal paper of 1917, nor Naef's elaborations on it, Hennig rejected reticulating *phylogenetic* relationships, restricting reticulating (later called *tokogenetic*) relations to bisexually reproducing species. The organismic diversity is organized into reproductive communities, that is, species, Hennig argued, and *phylogenetic* relationships obtain from the division of species, from which results the hierarchical structure of the phylogenetic system:

> The system corresponds to a division hierarchy, with the species as the fundamental divisional unit, in the same way as the hierarchy of the histosystems in the singular organism is based on the cell as the basic divisional unit.‡

There is no evidence that Hennig read original publications of Heidenhain, but in the first volume of his *Theoretische Biologie*, Bertalanffy (with 27 citations the most cited reference in Hennig's *Grundzüge*) presented a succinct summary of Heidenhain's views (see Chapter 6), introducing the enkaptic hierarchy based on histosystems that form divisional entities at increasing levels of inclusiveness and complexity.§ The consequence, correctly recognized by Hennig, is the reality and individuality not only of organisms, but also of species and more inclusive phylogenetic entities.

---

* Hennig, 1947, p. 276.
† Logical positivism and logical empiricism are often treated as synonyms. Philip Kitcher (1993, p. 4f, n. 4) introduced a useful distinction however, characterizing logical positivism as focused on demarcation criteria of "cognitive significance" for science. Logical empiricism he characterized as focused on the study of scientific confirmation/refutation, problems of scientific explanation, and the structure of scientific theories—the demarcation issue no longer being central.
‡ Hennig, 1947, p. 279.
§ Bertalanffy, 1932, pp. 261ff.

Treating the species as the fundamental divisional unit of an enkaptic hierarchy led Hennig to compare a species lineage-splitting event to the asexual reproduction in protozoans. And just as the parent protozoan continues to live in all and only its offspring, so, Hennig claimed, would the ancestral species live on in all and only is descendants. All the species included in a taxonomic unit of supraspecific rank would thus embody the phylogenetic potential of their common stem-species.[*]

These conclusions go to the very heart of Hennig's later analysis of monophyly: both species, as well as more inclusive taxa, confront their environment as historically conditioned, causally efficacious entities (*Wirkungseinheiten*), that is, as individuals of higher order. Causal relations are existence implying: it is hard to imagine things that do not exist to enter into causal relations. Cause and effect are linked through time and located in space: individuality obtains. The species for Hennig was a dynamic entity, a genealogical lineage that splits and splits again, thus creating more inclusive entities of higher complexity, yet themselves again dynamic and causally efficacious. Hennig's phylogenetic system is a dynamic, indeed a processual system. The goal of phylogenetic systematics, so Hennig's conclusion, is to picture that process of repeated species lineage splitting in a "structure image" (*Strukturbild*), which is a dichotomously structured phylogenetic tree.

The structure image that pictures the phylogenetic relations between species Hennig derived from what was known about the origin of new species. New species, he contended, originate through the division of an ancestral species lineage into two or more daughter species.[†] The figure Hennig inserted at this point shows a strictly dichotomously structured graph leading from one stem species through three speciation events to four daughter species. Species, like cells, or Heidenhain's histomeres more generally, multiply by fission. As they diversify through geographical dispersal and adaptation, species become different from one another in many different ways, each one of which Hennig called—following the empiricist philosopher Theodor Ziehen—a "dimension." Each species thus represents a multidimensional *Gestalt*. There consequently exists for a group of phylogenetically related species, derived from a common stem-species, a multidimensional "property space" (*Eigenschaftsraum*). Hennig emphasized that such abstract, or theoretical, use of the concept of "space" is consistent with Carnap's analysis of that concept in his doctoral thesis *Der Raum* of 1922. Given those theoretical premises, the task of the phylogenetic systematist then must be to measure, or calculate, the location of each species in that "property space," not in the sense of an absolute or individual position, but rather in terms of their position relative to one another. The phylogenetic system is after all a relational system, based on phylogenetic relations. However, to do so would require that those properties, which collectively determine the multidimensional *Gestalt* of a species, could be rendered in numerical values. To do so Hennig judged theoretically possible, but practically—at least at the current time—impossible.[‡] What could be done, instead, is to represent the phylogenetic relations between species in only one dimension, namely along the axis of time, depicting the successive species lineage splitting

---

[*] Hennig, 1947, p. 279.
[†] Hennig, 1949, p. 136.
[‡] Hennig, 1949, p. 136.

events: the result in Hennig's view is an enkaptic system.[*] Going from there, Hennig moved on to introduce a number of new terms that would eventually catch on, such as *apomorph* and *plesiomorph*. The concept of apomorphy, Hennig explained, would designate the position of a phylogenetic group in the morphological property space relative to the stem-form (*Stammform*). The concept of plesiomorphy, in contrast, would designate the position of a group near the point where the stem-species was located.[†] Herein lies the root of Hennig's theory of character analysis, which claims that only apomorphic, never plesiomorphic characters can successfully indicate relative degrees of phylogenetic relationships. Introducing many additional neologisms, of which more later, Hennig finally raised the question whether these are indeed necessary, especially since phylogenetic systematics had been able to proceed without them in the past. Phylogenetic systematics, Hennig argued, must be cast as a true science in its own right, and as is true of all sciences, the goal for phylogenetic systematics should likewise be a mathematization, a "*mathesis universalis*." But to achieve such a mathematization, phylogentic systematics must aspire, as do all other sciences, to a clarification of its fundamental concepts, that is, to a precision language (*Präzisionssprache*), where the terms acquire their precise meaning through a decoupling from natural language.[‡]

Indeed, Hennig concluded his essay with the proposition that the maturity of a special science can be judged by the degree to which a special vocabulary has been created that would unambiguously and with ultimate precision refer to the "facts" researched by that science. Later, Hennig would argue that the terminology he introduced was specifically "calculated" to fulfill the needs of phylogenetic systematics. He talked about the need to "calculate" monophyletic groupings, and insisted that once the apomorphic and plesiomorphic characteristics have been identified, the reconstruction of the phylogenetic system is governed by methods that correspond to a "calculation rule."[§] This "calculation rule" would later become known as "Hennig's argumentation scheme." Hennig concluded his introduction to his *Grundzüge* with a warning: the book, he said, was not intended to offer easy reading and a guide to effortless understanding. Instead, it seeks the development of a sound conceptual foundation for a science, which he felt had fallen into disrepute. The task to complete would be to develop a system that treats organisms, and groups of organisms, not just as bearers of characteristics that indicate degrees of similarity, and hence degrees of relatedness. Instead, the system should take into account the multidimensional *Gestalt* (*Totalgestalt*) of organisms, and order them in the exhaustive space (*Gesamtraum*) of their many dimensions.[¶]

Not an easy feat, indeed!

---

[*] Hennig, 1949, p. 136.
[†] Hennig, 1949, p. 138.
[‡] Hennig, 1949, p. 138.
[§] Hennig, 1955, pp. 21, 22, 25.
[¶] Hennig, 1952, p. 331.

9 Grundzüge
The Conceptual Foundations of Phylogenetic Systematics

SYSTEMATICS—"*SPEZIESMACHEREI*" OR A TRUE SCIENCE?

Hennig's ultimate ambition was to portray systematics as a science on par with the so-called hard or exact sciences such as physics or chemistry. Accordingly, Hennig opened his *Grundzüge* by identifying systematics as an essential component of all true sciences. Science, after all, is not just about the recording and describing of individual episodes or events, but about generalization instead. Sciences seek to systematize, rationalize, or order phenomena or events in a way that would result not just in a catalog or a classification, but in a way that would reveal the laws of nature.[*] The old distinction drawn by Wilhelm Windelband (1848–1915) and Heinrich Rickert (1863–1936), one that ultimately dates back to Wilhelm Dilthey (1833–1911), thus surfaced again right at the beginning of Hennig's 1950 book.[†] Nomothetic sciences deal with universal laws of nature; they play out in a deterministic world subject to necessity rooted in the lawful connection of cause to effect. Idiographic sciences deal, instead, with the individuality of historical events, the uniqueness of the *hic et nunc*, of the here and now. Hennig harbored important reservations *vis à vis* Windelband and Rickert's distinction of nomothetic versus idiographic sciences, however. That distinction, he conceded, might hold on a conceptual level, as had been argued by Tschulok,[‡] but it fails with respect to practical research. If valid, such classification would threaten the unity of science,[§] reason enough for Hennig to reject it,[¶] as he had come to value the unity of science through his study of Theodor Ziehen's epistemology. All sciences, Hennig asserted, incorporate nomothetic as well as idiographic components in their research strategies, and biology especially so—as had already been recognized by Windelband himself. To consider an organism from a strictly physiological point of view would be a nomothetic research program. To consider an individual organism, or a species-individual, as a historically conditioned part of the phylogenetic system would result in an idiographic research program. The question Hennig set out to investigate was the degree to which phylogenetic systematics should or could incorporate a nomothetic component as part of its theoretical foundation.

---

[*] Hennig, 1950, p. 4.
[†] See the commentary by Oakes in Rickert, 1986.
[‡] Tschulok, 1910, cited by Hennig, 1950, p. 293.
[§] Richardson, 1998, p. 37.
[¶] Hennig, 1950, p. 2.

If the Kantian ideal is upheld that all natural science should strive to discover natural laws, preferably laws that can be expressed in the language of mathematics, how must science then proceed in order to discover those very laws, asked Hennig? What for Hennig seemed to be required in every special science concerned with the discovery of natural laws—and that for him included phylogenetic systematics—was the *classification* of events, each one experienced in the *hic et nunc* of space and time, as either similar, or dissimilar, or more strongly put, as the same or different. Such systematization, or classification puts things of the same kind together, and sets those apart from things of a different kind. Things of the same kind tend to submit to the same causal relations, or laws, that range over the kind. Conversely, different kinds of things are set apart from one another through the different causal relations, or laws, that govern them. It is, therefore, such *systematization* of experience, which imparts a nomothetic component on systematics, and it is in that sense that systematics must itself be considered to be a foundational component of all special sciences, not just of biology. It is in that sense, Hennig argued, that science and systematics, as an integral part of it, can straightaway be defined as the systematic, that is, rational orientation of humans in their world.[*] To relegate systematics as mere *Speziesmacherei* to the backburner of biology was, Hennig argued, a fundamentally flawed perspective.

Rensch had provided numerous powerful examples for the interplay of idiographic and nomothetic research strategies in biology that Hennig had argued for. Starting with the idiographic analysis of numerous particular examples of adaptation of birds and mammals to local climatic and geographic conditions here and there, he ascended to a nomothetic level when invoking rules, or laws, that reign over and thus unify the particular examples as exemplars of the same kind of evolutionary change.[†] But the relevant paper by Rensch that Hennig cited in this context was as much about evolutionary rules and laws, as it was about speciation. The origin of a new species, the splitting of a species lineage into two daughter lineages, is a particular event that plays out at a particular location in space, and in the course of a particular duration in time. Species, as Hennig had recognized, are individuals. Species individuality obtains from the fact that species are spatio-temporally located, causally (tokogenetically, i.e., reproductively) integrated systems. But events, such as the splitting of a stem-species lineage into two daughter species, are just as much spatio-temporally extended, and causally determined. Events that link cause and effect naturally stretch through time and hence are individuals, too. The phylogenetic system for Hennig is thus a concatenation of historical (individual) events. From a philosophical perspective, deterministic, that is, necessitating laws of nature impart identity on the objects or events over which they range. Gold, for example, is characterized by the atomic number 79; every nugget of gold, no matter where it comes from, is an exemplar of its kind, characterized by the identical atomic number and hence sharing identical properties and causal propensities with other nuggets of gold. It is this microphysical

---

[*] Hennig, 1950, p. 4.
[†] A summary of Rensch's research, with many of those rules confirmed (Rensch, 1939), was cited by Hennig (1950, in addition to other papers by Rensch).

structure in virtue of which gold obeys deterministic laws of nature.* Gold in that sense is a paradigmatic example of a *natural kind*. The relation that delimits such natural kinds is—on the classical, strong reading—the identity relation: tokens of natural kinds submit to the identity relation in virtue of their shared (i.e., identical) causally efficacious properties.[†] The question then becomes: Can there be different kinds, or types, of speciation governed by different laws, as the title of Rensch's 1939 publication—*Types of Speciation*—seems to insinuate? If there were to be types, or kinds, of speciation, all speciation events of the same kind, or type, would have to submit to the identity relation, as they would be governed by the same law(s).

The identity relation is traditionally analyzed in terms of Leibniz's law, which has two parts to it. The first, uncontroversial principle is the "Indiscernability of Identicals," which states that numerically identical entities share exactly the same properties. In other words, it is impossible for two (or more) numerically identical individuals ("things" or "substances"[‡]) to exist. The converse, controversial principle is the "Identity of Indiscernibles": it states that no two (i.e., numerically different) objects can have exactly the same properties. Obviously, the first, uncontroversial identity relation cannot range over successive speciation events, as it only applies to numerically identical, that is, self-identical individuals. But the second, controversial identity relation could potentially apply. Speciation events are historical, each in itself unique, playing out at different times in different places and starting out from different, indeed historically unique initial conditions. Such historicity renders any speciation event inherently different from any other one: on that account, there cannot be recurrent, identical speciation events through phylogenetic history. On the strong, indeed classical reading, there can therefore be no kinds, or types of speciation, hence also no lawfulness ranging over multiple speciation events. But should that be the case, there would be no regularities to speciation, nor rules or laws that range over multiple speciation events. This is the juncture at which Hennig[§] sided with Ziehen,[¶] who had adopted Kant's rejection of Leibniz's *Principium identitatis indiscernibilium*, that is, of the "Identity of Indiscernibles." Hennig therefore cautioned to accept an absolute uniqueness and consequent unrepeatability of individuals and events in which they engage, that is, an absolute historicity of all living beings.[**] The reason is that the doctrine of absolute uniqueness and unrepeatability of organic beings and events would not allow for any lawfulness in historical biology, as Heinrich Rickert[††] had cogently argued. To render systematics a true science as he understood it, Hennig sought to transcend such an intransigent idiographic perspective on historical sciences.

---

* See discussion in Kirkam (2001); for a discussion of examples that relate to biosystematics, and further references, see Rieppel (2008).

[†] For an easily accessible recent account on the strong, essentialistic conception of natural kinds, see Ellis (2002). The literature on natural kinds, conceptualizing those both in a strong and a weak sense, is vast. Rieppel (2009a, 2013c) may provide a starting point in discussions that relate to biosystematics.

[‡] Hennig, 1950, p. 23.

[§] Hennig, 1950, p. 23.

[¶] Ziehen, 1934, p. 58.

[**] Hennig, 1950, p. 23.

[††] Rickert, 1902.

Hennig was cognizant of the fact that successive speciation events could not be subject to universal, necessitating laws, but he was also not ready to surrender to utter, unconstrained historical contingency, as Darwin had done with his twin principles of random variation and natural selection. He chose to chart middle ground. He could have done so by weakening the essentialistic concept of natural kinds as sketched above, but no such arguments developed by philosophers were available to him at the time of his writing.* So he chose to tackle the second horn of the dilemma by weakening the notion of lawfulness. The fundamental laws of physics, such as the Newtonian laws of mechanics, had classically been given universal scope, as they were believed to range over the fundamental structure of the universe (a belief that faltered with the advent of quantum mechanics). Hennig recognized this to be a characterization of natural laws that is far too strong for biology, especially historical biology such as phylogenetics. He insisted, however, that any particular, singular (individual) event—an event-token—is interesting for science, including phylogenetics, only if it is a token (a "model," as he called it) of a type event, that is, an exemplar of some certain kind of events.† Were such not the case, there would be no potential to discover any regularities such as those that underlie Rensch's rules of speciation, and of adaptational evolutionary change in general. Since Hennig—in contrast to von Bertalanffy's aspirations—was fully aware of the fact that a historical science such as phylogenetics cannot be hypothetico-deductively structured,‡ any evolutionary rules, or laws, would have to be inductively inferred from observed regularity, as Rensch had so aptly demonstrated. When Rensch talked of evolutionary rules, he implicitly allowed for exceptions, however. The incomplete comprehension of the complexity of biological processes rendered prediction chancy, and made a hypothetico-deductively structured argument downright impossible. There was therefore, Hennig recognized, an important difference between the understanding of the classical laws of physics (e.g., Newtonian mechanics), and biological laws: the first were typically accorded universal scope, something that was blatantly impossible for biological, or at least for evolutionary laws. Gravity is a classical physical law of universal scope, even if a particular body falling here and now through air rather than through a vacuum only approximates—if very closely—its predictions. But even if Rensch had claimed that geographic separation and subsequent genetic isolation is a *sine qua non* for speciation to occur, this did not necessarily rule out other modes of speciation (of which three are known: allopatric, parapatric, and sympatric). Biological rules, or laws, have in general less predictive force than laws in physics, but this for Hennig§ was a difference of degree only, not a difference of kind: it was for him a difference of scope. Gravity has a broader scope than rules, or laws, of speciation. Everything that partakes in speciation is subject to gravity, but

---

* Brian Ellis defends an essentialistic conception of natural kinds in chemistry, but notes that "according to the strict criteria of modern essentialism, biological species are not natural kinds" (Ellis, 2002, p. 154). He consequently invokes "variable kinds" (p. 28), or talks of species as "cluster concepts" (pp. 12, 156). Richard Boyd's (1991, 1999) "homeostatic property cluster natural kind" has indeed proven very fruitful in discussions of biological species (see, e.g., MacLeod and Reydon, 2013, and references therein).
† Hennig, 1950, p. 291.
‡ Hennig, 1950, p. 300.
§ Hennig, 1950, p. 306.

not everything that is subject to gravity also partakes in speciation. The way Hennig saw to increase the precision, and with it the predictive power, of biological rules or laws was via scope restriction: "One can increase the precision of the predictions of a law by construing the type-event as similar as possible to a concrete event, and then restrict the genus (class) of events that fall under the law's predictions as much as possible."* Hennig chose scope restriction as a way of "mopping up the exceptions"† to the laws and rules of phylogeny. The ideal physical law would have universal scope, the worst "biological rule" would have a very restricted scope, but could still be considered a causal law as long as it made better than chance predictions.‡

THE EMPIRICAL BASE OF PHYLOGENETIC SYSTEMATICS

Hennig's interest and insistence of lawfulness in phylogeny reconstruction must in part have been motivated by his desire to separate objective from subjective components of perception, a problem that had been raised by Zimmermann, and which Hennig strove to solve along the somewhat contorted and highly counterintuitive lines sketched by Theodor Ziehen. Unlike Rensch, Hennig had no personal exposure to Ziehen as academic teacher, but he read Ziehen's two-volume compendium on epistemology, and used it—in conjunction with other sources—in the development of the conceptual foundations of phylogenetic systematics. As outlined in Chapter 8, the neurologist, psychiatrist, and philosopher Ziehen rejected an indeterministic world-view, and based his epistemology on universal lawfulness that governs not only the physical world (Ziehen's causal laws) but also the mental world (Ziehen's so-called parallel laws, running parallel to the causal laws of the extramental world). To understand Ziehen's system, it is necessary, however, to understand that he based his epistemology on process philosophy and on Ernst Mach's neutral monism.

Everyday common-sense intuition pictures the world in terms of substantial objects and their properties, scattered in space and time. But, argued Mach, modern physics had spoilt this seemingly easy access to the world, as it replaced "matter" with fields of energy. Space, he explained, can no longer be captured in terms of the absolute position of particular objects, but only by the totality of the spatial relations between the so-called elements, such that "natural laws should be formulated as functional (i.e., mathematical) relations between the elements."§ Given that modern physics had dissolved matter into energy, these "elements" cannot therefore be the perceived objects themselves,¶ but can only correspond to sensations of those objects. Mach held these sensations as directly, immediately, and indubitably given in perceptual experience, which is how he secured an objective epistemic access to the world. Stripped of all metaphysical baggage, an "element" for Mach becomes a "neutral entity." The observation of a red apple, for example, results in the sensation of a "neutral red patch." The repeated examination of a red apple, perhaps manipulating

---

* Hennig, 1950, p. 306.
† See the discussion of scope restriction in Kitcher, 1993, p. 121, n. 34.
‡ See Griffiths (1999) for a more in-depth modern discussion of this issue.
§ Joergensen, 1970, p. 854.
¶ "As a boy already I recognized the [Kantian] 'thing-in-itself' as a useless metaphysical invention, an idle metaphysical illusion" (Mach, quoted from Stadler, 1997, p. 138).

it and inspecting it from all sides, yields a multiplicity of such perceptional "neutral elements," which then combine in the observer's mind to form a complex whole. This is how the apple, as an object of interest and investigation, is phenomenologically reconstructed by the observer's perceptional apparatus. The central point of Mach's phenomenological epistemology is that the red apple becomes simultaneously comprehensible as an object of the physical world, subject to causal laws of nature, as well as, and at the same time, as an object of the cognizing mind, subject to psychological laws of cognition. This is Mach's neutral monism, which was adopted and further elaborated by Ziehen, as it held a great promise: the unity of physics and psychology, ultimately the unity of science. The reason for Ziehen to adopt Mach's neutral monism was that he recognized in this doctrine a solid epistemic foundation on which to build an objective science. Similarly, the reason for Hennig to follow Ziehen's lead in that direction was to fulfill Walter Zimmermann's requirement that in order to count as an objective science cleansed of metaphysics, phylogenetic systematists must strive to separate the subjective from the objective component of perception—but not merely on a voluntary basis as Zimmermann had argued.

In Ziehen's epistemology,* the objects of the extramental world that are represented in perception became neutral "somethings," *Etwasse* in German. The perceptional sensation of a neutral "something" Ziehen called *Gignomen* (plural: *Gignomene*). As taught by modern physics, the "somethings" that cause the *Gignomene* for Ziehen were not material objects, but instead fields of condensation of energy. *Gignomene*, Ziehen went on to argue, comprise two components or parts, both equally involved in the cognition of the material world: the first derives from the physical cause of the perception, the second component comprises the mental (i.e., psychologically conditioned) idea that this sensation invokes about its physical cause. Ziehen called the first component of the *Gignomen* its "reduction part" (*Reduktionsteil* or *Redukt*), the second, psychological component he called "parallel part" (*Parallelteil*). The "reduction part," also abbreviated as *R*-part, represents the extramental world, not as a collection of material objects and their properties, but in terms of the lawful causal relations that structure the extramental world. The "parallel part" would be subject to parallel laws (running parallel to the physical laws), which reign over the psychological realm. The quintessence of that epistemology for Ziehen was that through the *reduction* of the *Gignomen* to its *R*-part, all subjective (mental) components of perception (and consequent cognition) could be eliminated. The *R*-part of the *Gignomen* reveals the causal laws of nature, and these in turn determine the relations that prevail in nature—and it is the relational structure of the world that in process philosophy is deemed relevant, not material (substantial) objects *per se*. This, then, is the philosophical machinery Hennig borrowed from Ziehen. Remember that Hennig's phylogenetic system is a relational system, as it is determined by phylogenetic relations, not by properties instantiated by objects— although he took such properties as pointers to phylogenetic relations. Small wonder that Ernst Mayr would judge his writing "turgid," even "unintelligible" in some places.

---

* Ziehen, 1934, 1939.

Mayr's remarks[*] were primarily aimed at Hennig's *Phylogenetic Systematics* of 1966, but might as well have targeted his *Grundzüge* of 1950, in which he first developed his adoption of Ziehen's epistemology. The translator of the manuscript of Hennig's 1966 book, Rainer Zangerl, likewise noted that he encountered serious linguistic challenges as he proceeded to translate Hennig's manuscript.[†] Consider Hennig's definition of order, which closely followed Ziehen's: *order* is

> the totality of progressively graduated vicinal similarities of more or less determined positional relationships of several or many, even an infinite number, maximally all, 'somethings' within a finite or infinite whole, whereby *positional relationship* is meant to be a more or less defined relation of a simple or complex something to other somethings that belong to a unified whole in regard to quality, intensity, locality, temporality, or number.[‡]

For Hennig, the ultimate task of systematics was to locate the relative position of each species in a multidimensional property space (see Chapter 8). The concept of a multidimensional property space underlies Ziehen's, and hence Hennig's definition of order: it is within this multidimensional property space that the position of perceptional "somethings" relative to one another needs to be determined on the basis of "progressively graduated vicinal similarities." Ziehen, who adopted such relationalism from the neo-Kantian philosopher Ernst Cassirer[§] (1874–1945), considered the "ordinability,"[¶] or "seriability"[**] of phenomena to reveal the lawful relations prevailing between them. Accordingly, it is, for example, the ordinability or seriability of morphological features that reveals the relevant relations between them—be those ideal, or real (causal). As outlined in previous chapters, Carl Gegenbaur, Othenio Abel, Adolf Naef, Walter Zimmermann, as well as Hennig turned to the seriability of character states in terms of a character transformation series in order to identify relations that would determine the natural system. Gegenbaur in particular is the first and paradigmatic example of someone who subjected a morphological character transformation series to an evolutionary (phylogenetic) explanation, thus theoretically replacing ideal with causal relations. Hennig would do the same, but in so doing asked the theoretically highly relevant question—one already articulated by Abel—as to how to polarize such character transformation series, so that the process of phylogenetic transformation would correctly be revealed. However, to talk about character transformation series presupposes the

---

[*] Mayr, 1974, p. 99; 1982, p. 226.

[†] Zangerl, Preface, in Hennig, 1966.

[‡] Hennig (1950, p. 6; 1966, p. 3f); Hennig adopted this definition from Ziehen (1939, p. 10): "the totality of progressively graduated vicinal similarities of more or less determined positional relationships of several or many, even an infinite number, maximally all, 'somethings' within a finite or infinite whole," whereby *position* (according to Ziehen, 1939, p. 9) is meant to be "a more or less defined relation of a simple or complex something to other somethings that belong to a unified whole in regard to quality, intensity, locality, temporality, or number."

[§] Ziehen (1934, p. 72) acknowledged that his "relationalism" was based on Cassirer's (1923 [1953]) insight that "the concept of relations takes priority over the concept of objects in the theory of knowledge."

[¶] Ziehen, 1934, pp. 93–98.

[**] Ziehen, 1934, pp. 93–98, 123–142.

antecedent identification of characters useful for phylogenetic inference in the first place. And for phylogenetic systematics to be a true science, such identification of phylogenetically informative characters must be objective, at least in Zimmermann's and Hennig's view.

The objective description and ordering (classification) of the biological furniture of the world would have to be, according to Bertalanffy, the first step in the development of a *Theoretical Biology* (1932). Hennig held up against such a program his concerns about the theory-ladenness of observation. He found Bertalanffy's postulate to be premised on the mistaken idea that any science, biology, or biological systematics included, could be pursued without any underlying assumptions. Hennig recognized an active relation between an observer and the observed object, which alone he found to render theory-free classification already impossible. Ordering, classification, for him always proceeded against a certain background, that is, in the light of a specific perspective adopted by the practicing scientist.*

But what, then, is the empirical base for ordering, according to degrees of phylogenetic relationships, in systematics? What is the empirical base for the identification of characters that would provide the clue to such order? In the living world, that is, in biology, the "somethings" invoked by Ziehen become "animated natural things" that form a "multidimensional multiplicity"†—a view which Hennig fully endorsed. When characterizing these "animated natural things," Hennig followed Nicolai Hartmann‡ in calling organisms systems that are defined not by substantial parts that instantiate properties, but by the causal interactions prevalent between the parts that integrate the system and hold it together.§—a view he found also articulated by Bertalanffy, who wrote

> The old contrast between [substantive] 'form' and [relational] 'function' is to be reduced to the relative speed of processes within the organism. Structures are extended, slow processes, functions are transitory, rapid processes. The organic form must be comprehended as a cross-section through a spatio-temporal flow of events.¶

There is therefore, Hennig concluded, no principled distinction any more to be drawn between a "thing" and an "event."** Organic form at any stage of development, or of a life cycle, its thus a cross section through a flow of events. On a strong reading of process philosophy, "there is no holding nature still and looking at it."†† But that is exactly what the classifying biologists need to achieve, as Adolf Naef had so clearly recognized. To do so, Hennig invoked his "character bearer," the *semaphoront*, a concept he might have borrowed from Carnap's thesis *Der Raum*.‡‡

---

* Hennig, 1966, pp. 11–12; emphasis added.
† Ziehen, 1934, p. 20; approvingly cited by Hennig, 1950, p. 6.
‡ Hartmann, 1912, p. 17; see also the discussion in Tremblay (2013). Nicolai Hartmann likewise adopted an "event ontology": "the concept of absolute matter is gone" (Hartmann, 1912, p. 23).
§ Hennig, 1950, pp. 5–23.
¶ Bertalanffy (1941, p. 251; approvingly cited by Hennig, 1950, p. 5).
** Hennig, 1950, p. 5.
†† Whitehead, 1920, p. 14.
‡‡ Carnap (1922, p. 17) spoke of the "bearer" of characters such as colors.

He characterized the *"character-bearing semaphoront"*[*] as "the ultimate element of the biological system," defined as "the organism or the individual ... during a certain, theoretically infinitely small, period of its life."[†] For Hennig, the organism as a whole constitutes a multidimensional multiplicity of causal relations, located in the four-dimensional continuum of space and time that Hennig claimed to be familiar to the physicist and mathematician, but not necessarily to biologists.[‡] Hennig[§] approvingly cited Bertalanffy, who had characterized an organism as a complex whole composed of a multitude of parts,[¶] that is, as an enkaptic hierarchy. But, the essence of life, or of an enkaptic hierarchy that constitutes an organism, is not any "substance," but rather its "organization,"[**] Bertalanffy insisted. A "living substance," according to Bertalanffy, is not a material object located in space and time, but rather an immensely complex system of causal (chemical and physiological) relations.[††] It is not the substantial basis of the organism, and its parts that are of interest to the life scientist, but the relations between them that enable their participation in the process of life[‡‡]: the living organism constitutes a fluid organizational system of relations that hold lawfully between physicochemical events. In that sense, Bertalanffy drew parallels between the biological concept of an organism, and the concept of energy as deployed in physics.[§§]

Hennig introduced the semaphoront to pull the organism so understood, that is, a multidimensional system of causal relations, into the *Anschaulichkeit* of the three-dimensional space. The semaphoront is the tool Hennig used to slice through the process that constitutes an organism in order to obtain an invariant part of it from which characters can be lifted. In order to ensure the objectivity of such character analysis, Hennig presented the semaphoront as the *R*-part of the *Gignomen* that results from the observation of an organism. Consider an entomologist studying a holometabolous insect. Following the creature through its successive developmental stages yields in perception a succession of semaphoronts. To the degree that these semaphoronts can be ordered, or simply fall into an unbroken series of form stages, they disclose the causal laws of development that structure the developing organism. Conversely, if only an incomplete series of stages is available to the systematist, it is possible to infer the existence of the missing (i.e., unobserved) stages on the basis of these laws of development, as Ziehen had argued.[¶¶] The whole organism thus obtains not in perception, but through phenomenological reconstruction from "ordinable" or "seriable" semaphoronts in conjunction with the causal laws that are disclosed

---

[*] Hennig, 1966, p. 6; emphasis in the original.
[†] Hennig, 1950, p. 9; 1966, p. 6.
[‡] "Torrey (1939) is correct when he complains that the biologist still works far too little with those concepts familiar to the physicist and mathematician, of the four-dimensional continuum of space and time": Hennig (1966, p. 6); see also Hennig (1950, p. 8).
[§] Hennig, 1950, p. 5.
[¶] Bertalanffy, 1932, p. 83.
[**] Bertalanffy, 1932, p. 82.
[††] Bertalanffy, 1932, p. 49.
[‡‡] Bertalanffy, 1941.
[§§] Bertalanffy, 1932, p. 86.
[¶¶] "Intermediate stages need not always to be observed, but can be inferred from causal laws": Ziehen (1934, p. 60).

through that series of semaphoronts. Here is how Hennig put Ziehen's epistemological machinery to work, using Ziehen's example of an oak tree that develops from an acorn: Are they the same, or different at different times? Hennig wrote

> 'We ascribe substantial sameness to two successive R's, R_1 and R_2, if the difference between R_1 and R_2 is causally understandable through a continuous sequence of intermediate states' (Ziehen, I)[*]. Only on the basis of this criterion can we recognize, for example, the acorn and the oak tree that develops from it as one and the same individual: 'thus two phases of the same thing may be completely different in their R-components [in this sense, peculiarities or characters][†], and yet on the basis of the continuous causal connection we speak of a single thing' (Ziehen, I).[‡]

From this, Hennig concluded to the identity of the acorn with the oak tree: They both represent the same yet changing individual through time, a system of causal relations that is located in time and space and hence has both individuality and reality.[§] But it is the acorn and the oak tree that have reality as one and the same individual, not the R_1 and R_2 they cause in perception.[¶]

And yet, Hennig spoke about the "spatial, three-dimensional body"[**] of the semaphoront, and cautioned that in practice, the duration during which a semaphoront persists as a constant, that is, unchanging entity amenable to systematic analysis is not invariant across organisms, but rather depends on the rate at which characters undergo ontogenetic change.[††] This is a realist account of a semaphoront: a material (substantial) body extending in the three spatial dimensions and with a variable duration in time, in short, an individual.[‡‡] Such a realist account of the semaphoront is possible because the R-part of a *Gignomen* reveals the causal laws that structure the object of perception, which implies the real existence of the object of perception, even if it is—on Ziehen's account—revealed to the observer as a *Gignomen* only. Ziehen's *Gignomen* is a phenomenological concept, but if it reveals the causal structure of the world, and if an organism is conceived of as a system of causal relations, rather than a complex of substantial parts instantiating properties, then the *Gignomen* will capture the reality of the semaphoront.

PHYLOGENETIC HIERARCHY

In his *Theoretical Biology*, a book frequently cited by Hennig, Bertalanffy defined phylogeny as a process that plays out in the generational nexus, the latter revealing the transformation-relations between organic form conditions. If so defined, he contended, talk about phylogeny can only make sense in terms of events, and

---

[*] Hennig (1966, p. 81); the quote is from Ziehen (1934, p. 60); R's are the reduction parts, in this case of two consecutive *Gignomene*.
[†] Hennig's insertion into his citation from Ziehen (1934).
[‡] Hennig, 1966, p. 81; see also Hennig, 1950, p. 114. The quote is from Ziehen, 1934, p. 63.
[§] Hennig, 1966, p. 81.
[¶] Rieppel, 2006.
[**] Hennig, 1950, p. 9; 1966, p. 7.
[††] Hennig, 1966, p. 6.
[‡‡] Tremblay, 2013.

relations between events (of transformation).* Hennig's phylogenetic system is such a relational system, indeed a processual one; the events on which it is based is the succession of speciation events through evolutionary time, the successive splitting of species lineages that gives rise to the phylogenetic system. In his prospectus of 1947 (see Chapter 8), Hennig stated very clearly that the phylogenetic system that results from the repeated dichotomous splitting of species lineages forms an enkaptic hierarchy, a nested hierarchy of complex wholes or supraorganismal individuals. As the stem-species lineage splits, it gives rise to two daughter species. Relative to one another, these daughter species represent sister-species; together, the sister-species form a complex whole, or individual, of higher rank that cannot simply be reduced to the sum of its parts due to emergent properties that become manifest at this higher level of complexity. Nicolai Hartmann (1882–1950), another philosopher who influenced Hennig,[†] had argued that hierarchical levels characterized by emergent properties each represent a *categorical novum*. For Hennig, a complex whole of higher rank was just such a categorical novum[‡]: Laws that govern a certain level of the enkaptic hierarchy cannot simply be presumed to also be valid at other levels of inclusiveness, even if any level of inclusiveness of the hierarchy is constituted by parts that look similar to, or are even identical with, parts that constitute different levels of inclusiveness.[§]

Sexually reproducing species—like the ones Hennig was most concerned about—form a reticulating pattern of relationships, one that Hennig dubbed *tokogenetic*. Speciation disrupts these tokogenetic relationships and replaces those with *phylogenetic* relationships, that is, the relation of descent between the stem-species and its two daughter species. On Hennig's account, if a stem-species splits to give rise to two daughter species, the stem-species itself ceases to exist as a separate species; its two daughter species are united in a phylogenetic relationship of first degree. If each of the daughter species splits again, the four descendant species are united in a phylogenetic relationship of second degree.[¶] The work the tokogenetic relations do for the species, the phylogenetic relations do for supraspecific units in the enkaptic hierarchy, that is, the phylogenetic system: they mark them out as individuals.

To consider species to be complex wholes, that is, individuals, has a long tradition in German biology, one that can be traced back all the way to the nature romanticism of Friedrich Wilhelm Schelling.[**] Nicolai Hartmann had argued that reality presupposes individuality, the latter anchored in temporal duration; Hennig found those criteria of individuation to perfectly apply to species.[††] Furthermore, Nicolai Hartmann had characterized objects of nature again not as substantial bodies, but as effects of causal relations.[‡‡] Taking species as such Hartmannian objects of nature allowed Hennig to adopt the dynamic conception of species offered by Karl Friederichs in his

---

* Bertalanffy, 1932, p. 20.
† For a detailed discussion, see Tremblay, 2013.
‡ See Rieppel, 2009b, for more details.
§ Hennig, 1950, p. 116.
¶ Hennig, 1950, p. 102.
** See Rieppel, 2011a, for a full discussion.
†† Hennig, 1950, p. 115; see also Tremblay, 2013.
‡‡ *"Naturgegenstände sind ... nichts anderes als Wirkungen"*: Hennig (1950, p. 5).

classic 1927 paper on community ecology. The species, according to Friederichs, is a causally efficacious entity, a *Wirkungseinheit*, Hennig agreed,[*] and applied to species Friederichs' concept of an "organization," that is, a complex whole (*Ganzheit*) that endures through self-regulation.[†] This is an eminently processual conception of species,[‡] in which context Hennig approvingly cited Haase-Besell's widely read theoretical treatise on evolutionary theory of 1941, which attempted to reduce the species problem to the ontology of modern physics.[§] Gertraud Haase-Besell (1876–?) is a relatively obscure figure. She was a plant geneticist living in Dresden. Without institutional affiliation, she pursued research on foxgloves (*Digitalis*) in her own private garden.[¶] The constant "brewing and bubbling"[**] of the genetic background of species, according to her, renders a static conception of species impossible: "the species is never in a state of static arrest, but is always in dynamic motion,"[††] she wrote in the booklet that attracted Hennig's attention.

As building blocks of the enkaptic hierarchy that is the phylogenetic system, species become "speciation entities" (*Speziationseinheit*).[‡‡] The sexually reproducing species itself forms a reticulated system. Its fission results in phylogenetic relations that tie the stem-species to its two daughter species. Together, the two daughter species form a complex whole of higher rank, or inclusiveness. Hennig called such complex wholes of increasing inclusiveness, which include all and only the descendants of an ancestral stem-species, monophyletic taxa. Monophyly is thus a property that emerges from the splitting of an ancestral species lineage, uniting sister-taxa in a group of unique evolutionary origin at all levels of inclusiveness in the properly reconstructed phylogenetic system.[§§]

Monophyly has widely been recognized as *the* centerpiece of Hennig's phylogenetic systematics. But the concepts of "monophyly" and "paraphyly" as used in the past have remained notoriously ambiguous, Hennig rightly lamented as he opened the discussion of these topics in his *Grundzüge*. The question cannot be solved through arbitrary decisions, Hennig insisted; a disambiguation of these concepts can only be achieved on the basis of our understanding of the actual course of evolution. Bisexually reproducing species are, on his account, individuals, real entities located in space and time, and he thought that new species can originate solely through the fission of already existing species. There follows from this account the insight that the definition of the concept of "phylogenetic relationship" must somehow tie descendant species to their common stem-species. And it is that tie of the stem-species to its descendants that must ultimately determine the meaning of monophyly. For Hennig, only monophyletic taxa are admissible in phylogenetic systematics. Hennig considered all supraspecific taxa to be groups of species, and they are monophyletic

---

[*] Hennig, 1950, p. 118.
[†] Hennig, 1950, p. 117f.
[‡] See also the discussion in Rieppel, 2009c.
[§] Hennig, 1950, p. 119.
[¶] Junker, 2004, p. 146; see also Haffer, 1999, p. 137.
[**] Haase-Besell, 1941a, p. 247.
[††] Haase-Besell, 1941b, p. 71.
[‡‡] Hennig, 1950, p. 83.
[§§] See Rieppel, 2009b, for a more detailed discussion.

if all the species included in a taxon can be traced back to a common stem-species. Or, in other words, a monophyletic taxon must include all, and only, the descendants of a common stem-species.[*]

In support of the reality and individuality not only of species but also of monophyletic taxa of higher rank, Hennig cited from Möbius's 1886 paper on species concepts the views of the botanist Carl Wilhelm von Nägeli, who in 1865 already had ascertained that not only species but also "the genera and higher concepts are not abstractions, but concrete things, complexes of forms that belong together, that have a common origin"[†] (see Chapter 2). While historically interesting, that argument alone he considered not sufficient, however. Defining "phylogenetic relationships" in his *Grundzüge*, Hennig specified that a stem-species that splits into two daughter species through the interruption of tokogenetic relations itself ceases to exist as a separate species.[‡] But in his prospectus of 1947, Hennig clarified that the stem-species, upon its division, continues to live on in its two daughter species.[§] Hennig introduced this somewhat opaque statement on the continued life of the stem-species in its daughter species to more forcefully justify the reality, and consequent individuality, of supraspecific monophyletic taxa, but how is it to be understood? Hennig cited Ziehen, who had commented on Nicolai Hartmann's discourse on ontology and realism: things for Hartmann are real in their becoming and passing, in their location and duration in time.[¶] Monophyletic taxa, as constituted by two sister species that share a common stem-species, have a discrete location of origin in time and hence are real, according to Hartmann's criteria. Hennig cited Hartmann to the effect that the status of reality is not tied to matter or space, but rather to time and individuality, which in turn bestows singularity and uniqueness on objects.[**] Supraspecific monophyletic taxa are thus not only real but also individuals. "In the phylogenetic system, the name of groups of animals at all levels of inclusiveness … are *proper names*," Hennig would later claim.[††] Hennig explained his idea using asexually reproducing protozoans as an analogy. Consider a protozoan individual that splits into two daughter individuals; on Hennig's account, the mother individual can be considered identical with both of its daughter individuals, because it continues to live on in its offspring. Such conclusion is licensed by the fact that there persists an uninterrupted causal nexus between the condition when there existed only one individual, and the condition where there exist two daughter individuals. As outlined above, Ziehen[‡‡] had argued that the acorn and the oak tree are identical, hence the same individual, because the two conditions of form are causally connected through an uninterrupted series of causally interdependent intermediate stages. Same for protozoans, same for species: the two daughter species that originate from a common stem-species can together be considered identical with their stem-species, as the latter continues to live on in both

---

[*] This, and the quotes in the preceding section concerning monophyly, are from Hennig (1950, p. 307f).
[†] Nägeli, cited from Hennig, 1966, p. 77; see also Nägeli, 1865.
[‡] Hennig, 1950, p. 102.
[§] Hennig, 1947, p. 279; see also the discussion in Chapter 8.
[¶] Hennig, 1950, p. 115; Ziehen, 1939, p. 146.
[**] Hennig, 1950, p. 115; quote from Hartmann, 1942; see also Tremblay, 2013, for further discussion.
[††] Hennig, 1953, p. 3.
[‡‡] Hennig, 1950, p. 114; Ziehen, 1934, p. 63.

of them together. This is true because of an unbroken causal connection between the stem-species and its two daughter species. The upshot is that supraspecific taxa are individuals, and hence have real existence.[*]

Hennig conceded, however, that this argument was based on two presuppositions. The first is the existence of true genealogical, that is, blood relationships between a species and all of its descendant species, the second is the fact that the stem-species continues to exist in both its daughter species together. It is important at this juncture to remember Hennig's conception of the species as a multidimensional entity, the totality of its dimensions defining the property space of that species: morphology, physiology, behavior, ecology, geography, and so on. Hennig wanted the phylogenetic system to be based exclusively on one single property: phylogenetic relationships. Now, assume the existence of a stem-species A at an earlier time horizon, and its two daughter species B and C at a later time horizon. Assume it further to be the case that A and one of its descendants, B, are identical in their morphological, physiological, behavioral, and ecological properties, whereas C has deviated from the ancestral condition A. On the basis of morphology and ecology, A could be claimed to be identical with B but not with C. But phylogenetically, the very same relationship of descent that prevails between A and B also prevails between A and C. On phylogenetic grounds, it therefore cannot be claimed that A is identical with B but not with C. Instead, the conclusion must be that A is identical with both B and C together.[†]

The two daughter species that originate from a species lineages splitting event are each individuals in their own right, each the starting point for an individual evolutionary trajectory that extends into the future. But as sister species, they are united in an identity of higher order, one that is mediated by the stem-species that continues to live on in them, infusing them with its persistent evolutionary potential. That is the emergent property that Hennig called "monophyly."

THE CLADOGRAM

When confronted with the task to graphically represent the phylogenetic hierarchy, Hennig emphasized the impossibility to picture the multidimensional property space into which species have to be sorted in its totality.[‡] What can be depicted, though, is the succession of species lineage splitting events that give rise to the phylogenetic hierarchy. The result is a dichotomously branching diagram arranged along the axis of time. Such a diagram specifies a nested hierarchy of sister-group relationships as they obtain from the species-lineage splitting event to which the respective sister-groups can be traced back. Hennig called such a branching diagram a *Strukturbild*,[§] that is, the picture of a structure (or system), where the structure is defined by the relations that obtain between its constituent parts. The template for such a representation of a relational structure, or system, he may well have found in Carnap's

---

[*] Hennig, 1950, p. 115.
[†] Hennig in Schlee, 1971, p. 28. Hennig's argument runs like this: if $A \rightarrow B$ (Paul is the father of Sam); and if $A \rightarrow C$ (Paul is the father of Mark); then also $A \rightarrow (B \cup C)$ (Paul is the father of Sam and Mark together). See also Rieppel (2005).
[‡] Hennig, 1950, p. 279.
[§] Hennig, 1947, p. 279.

thesis of 1922, *Der Raum*, discussed as it was in Bernhard Bavink's book on nature philosophy that was also cited by Hennig. In his doctoral thesis, Carnap had distinguished a purely "formal space," an "intuitive space," and the "physical space." It was Carnap's treatment of the "formal space" that caught Hennig's attention,[*] one he found characterized by Bavink as "the purely formal space of the mathematician, which represents nothing but a basic ordering scheme for multiplicity."[†] In Carnap's words,

> We call a 'general system of order' a system of relations between un-interpreted relational entities ... for which the most diverse things can be substituted (such as numbers, colors, *degrees of relationships*, circles, judgments, humans) in so far as there exist between them relations which satisfy certain *formal requirements*.[‡]

According to Carnap, "Cassirer has shown that a science that has the goal of determining the individual through lawful connections without losing its individuality must use not the class (species) concept, but relational concepts, since these can lead to the formation of series and, thus, the erection of systems of order."[§] An ordering system that forms a system of relations is just what Hennig wanted his phylogenetic system to be. For Carnap, the relations of physical entities relative to one another are matters of fact, the metrics used to express these relations are conventionally determined.[¶] For Hennig, a multitude of metrics could potentially be used to place species into their relative positions in the multidimensional property space they represent, such as those delivered by morphology, physiology, ecology, biogeography, or behavioral studies. The dimension[**] along which Hennig chose to express the phylogenetic relationships of species was time, which would yield a phylogenetic system to which all other dimensions and their systems could be related. The semaphoront is the representation of the living organism during some—theoretically infinitely small— duration of its existence. As such, it delivers the characters on the basis of which to reconstruct phylogenetic relationships. But the characters—at the time of Hennig's writing mostly morphological, but also serological—he considered mere pointers,[††] relating to the phylogenetic system like symptoms relate to an illness. The branching diagram that came to be known as the cladogram[‡‡] of phylogenetic systematics is for Hennig a Carnapian *Strukturbild*, the graphic representation of a relational system, that is, of the phylogenetic system, one that results from species lineages splitting and splitting again. Hennig started from the components that will build up the system, and these are species. He then called for a structure description that would

---

[*] Hennig, 1950, p. 153.

[†] Bavink (1933, p. 132). "... *den rein formalen Raum des Mathematikers, der weiter nichts als eine Mannigfaltikeitsordung überhaupt ... darstellt.*" See also Friedmann (2000, p. 64); Carus (2007, p. 102).

[‡] Carnap (1922, p. 5f; emphasis added).

[§] Carnap, cited in Richardson, 1998, p. 38; Friedman, 1999, p. 125.

[¶] Carus, 2007, p. 134.

[**] The *Principium divisionis*: Hennig (1953, p. 4).

[††] *Steckbriefmerkmale* in Hennig (1957, p. 52).

[‡‡] The technical term for a "branching diagram" is cladogram, *clados* being the ancient Greek word for "branch."

capture the relevant relations that would exist between these components of the system, which are phylogenetic relations. Since on his account, the relevant phylogenetic relations are those between a stem-species and its two daughter species, the appropriate structure description for the entire (phylogenetic) system is a dichotomously branching diagram—the cladogram.[*]

Hennig's early outlook was organicist in nature, as is evidenced by his adoption of the concept of the enkaptic hierarchy. The species for him was a causally efficacious entity, a *Wirkungseinheit* as Friederichs had called them in his seminal paper of 1927.[†] The early Hennig tried to anchor the strictly bifurcating structure of the phylogenetic system in the nature of speciation. In his *Grundzüge,* he confronted *"the problem of the dichotomy of the phylogenetic tree"* head-on, and asked the questions whether a dichotomously structured cladogram does, in fact, represent the true phylogenetic process, or whether it is just an artifact generated by the systematist's desire for symmetry?[‡] The end result of Hennig's elaborate musings[§] was a justification of the bifurcating nature of the phylogenetic system derived from biological theories about speciation. In his later writings, Hennig would increasingly distance himself from such biological speculations, and emphasize the methodological aspects of phylogeny reconstruction instead. He must have recognized that his early theorizing about patterns and rates of speciation entails an unwelcome orthogenetic element, as also the fact that ongoing research on population biology and genetics would not support his elaborate models of rates of speciation that are supposed to vary in correlation with the phylogenetic age of genealogical lineages. In his critical review of the phylogenetic system of insects, published in 1953, Hennig announced the virtues of set theory instead, identified as a branch of mathematical logic, which he declared to be the required tool with which to classify the multidimensional multiplicity that is manifest in organismic diversity.[¶] In order to get a classificatory grip on nature, the phylogenetic systematist in the same paper is called upon to resolve a *"Regel-de-tri*-problem."[**] *"Regel-de-tri"* is the German rendition for *"Regula de tribus termini,"* "the rule of three." That rule is captured by Hennig's revised definition of "phylogenetic relationship": species A is more closely related to species B than to species $C,$ if A and B share an ancestral species that is not also ancestral to C.[††] The *principium divisionis* of Hennig's phylogenetic system is thus revealed to be the temporal dimension, that is, time as reflected in the sequence of successive lineage splitting (bifurcating) events. The essence of phylogeny reconstruction along this line of reasoning is consequently not the search for ancestors and descendants in the Fossil Record, but the search for *sister groups* of species, or supraspecific taxa quite generally, be they represented by fossils or extant organisms. On the basis of Hennig's methodology, fossils cannot claim any ontological privilege

---

[*] Hennig, 1957, pp. 55–57.
[†] Hennig, 1950, p. 118.
[‡] Hennig, 1950, p. 332; emphasis in the original.
[§] See Rieppel, 2011g, for a more detailed discussion.
[¶] Hennig, 1953, p. 6.
[**] Hennig, 1953, p. 14.
[††] Hennig, 1965, p. 97; same in Hennig, 1953, p. 7.

in phylogeny reconstruction—contrary to what Beurlen had claimed.[*] Hennig's methodology, mediating the search for sister species, or sister groups at higher levels of inclusiveness, on the basis of the "rule of three" thus implies a dichotomous structure of the phylogenetic system at a most fundamental level: it simply cannot admit trichotomies, or multitomies, as it also cannot accommodate speciation through evolution within a lineage between two branching events. That speciation should always be dichotomous, and that the ancestral species should itself always go extinct (i.e., cease to exist in its own right as an individual) when it splits into two daughter species: these are perhaps the most controversial principles inherent in Hennig's phylogenetic systematics that provoked critique from his fellow systematists in Germany, paleontologists in particular. From a biological point of view, Ernst Mayr criticized the two principles as "strictly arbitrary" and "unrealistic,"[†] but for Hennig they were indispensible in his quest for a disambiguation of phylogenetic relationships. Still grappling with the underlying necessity of dichotomous speciation, Hennig emphasized that modern population genetics as pursued by Dobzhansky has thrown new light on the problem of speciation, recognizing the major importance of geographic isolation.[‡] The necessity of geographic isolation (allopatric speciation) Hennig then took as the basis for his claim that the phylogenetic system must *"necessarily"*[§] correspond to a fully dichotomously structured diagram. If correctly reconstructed, the phylogenetic system would thus reveal the exact sequence of all speciation events that brought forth the currently recognized biodiversity.[¶]

In his 1956 review of German literature in systematics and phylogenetics, Hennig's friend and discussion partner Klaus Günther praised the "significance and fertility" of Hennig's "description or *definition* of speciation as a strictly dichotomous process of lineage splitting."[**] The same he repeated in his follow-up review published in 1962, now adding in support of such praise a quote from the philosopher Kurt Bloch: "the dichotomy is the most adequate and logically best founded form of classification ... it stands at the beginning of analysis, and hence at the beginning of systematics."[††] Going to the original source reveals that Bloch continued this statement with the phrase: "All forms of classification except for the dichotomous one are initially suspect, because they do not imply necessity."[‡‡] The necessity here appealed to is grounded in logic, as it follows from the Law of Excluded Middle. In a strictly dichotomous system, every species, or taxon, must be placed on one or the other side of the fork; there is no third place to go. Hennig concurred, citing Beurlen to the effect that the common stem-form of crossopterygian fishes and stegocephalians must necessarily have been either a fish, or a tetrapod[§§]—there is nothing in between.

---

[*] Hennig, 1950, p. 19; citing Beurlen, 1936 [should be 1930], p. 534.
[†] Mayr, 1982, p. 229.
[‡] Hennig, 1957, p. 58; see also Hennig, 1950, p. 326.
[§] Hennig, 1957, p. 58, emphasis added.
[¶] Hennig, 1957, p. 62.
[**] Günther, 1956, p. 45, emphasis added.
[††] Günther, 1962, p. 279.
[‡‡] Bloch, 1956, p. 71.
[§§] Hennig, 1950, p. 144f.

However, a certain uneasiness must have persisted in Hennig's mind with respect to the question how theories of speciation relate to a strictly dichotomously branching diagram. The claim that speciation is always dichotomous was relativized by Hennig in his *Grundzüge* already as one issued for "simplicity's sake."[*] By 1966, Hennig declared the principle of dichotomy to be primarily a methodological tool.[†] There is, nevertheless, an important shift in the evolution of Hennig's thought. In his *Grundzüge*, Hennig endorsed the holistic–organicist concept of the enkaptic hierarchy, subject to the part—whole relation and located in space and time, thus tied to individuality, causality, ultimately to reality. Metaphysically, this is a far cry from his switch to set theory, the "rule of three," and the consequent enforcement of the "cladistic principle of dichotomy" as a mere methodological tool.[‡]

HETEROBATHMY OF CHARACTERS

As detailed in previous sections of this book, Carl Gegenbaur already recognized heterochrony in the evolutionary transformation of organs or organ systems among organisms (see Chapter 1). This means that no organism is characterized by either exclusively primitive or exclusively derived traits. Based on the same insight, Louis Dollo proposed an evolutionary law, which he called the *chevauchement des spécialisations* (the crossing of specializations). On the basis of this law, Othenio Abel proposed a major conceptual shift, one that would take phylogeny reconstruction from a search for ancestors and their descendants, to a search of common evolutionary origins instead (see Chapter 3). In his *Grundzüge*, Hennig used the German term *Spezialisationskreuzugen* for such crossing of specializations, and on the same basis declared the search for ancestors and their descendants in the Fossil Record a futile enterprise. Instead, phylogeny reconstruction must be based, he argued, on a search for sister-group relationships. Later, Hennig switched to Armen Takhtajan's term "heterobathmy of characters"[§] to refer to such crossing of specializations, which he declared "a precondition for the establishment of the phylogenetic relationships."[¶]

Hennig realized that most ancestor—descendant sequences reconstructed by paleontologists correspond not to *Ahnenreihen*, but to *Stufenreihen sensu* Othenio Abel,[**] that is, to character transformation series, not to series of ancestors and descendants. This for him had two major consequences. Phylogeny reconstruction had to rely not on a search for ancestors and descendants, but on Walter Zimmermann's *Merkmalsphylogenetik* (character-phylogenetics), which the latter had proposed in a classic paper published in the Heberer compendium of 1943 that Hennig had read.[††] It is the careful analysis of character transformation series that reveals the heterobathmy of characters. The latter, in turn, explains why the study of fossils provides no

---

[*] Hennig, 1950, p. 352.
[†] Hennig (1966, p. 210); see also Hull (1979, p. 425), who interestingly claimed that "no cladist has ever maintained that the 'principle of dichotomy' is an empirical claim about the process of speciation."
[‡] Hennig, 1974, p. 292.
[§] Takhtajan, 1959, p. 13; Hennig, 1965, p. 107.
[¶] Hennig, 1965, p. 107.
[**] Hennig, 1950, p. 134.
[††] Hennig, 1950, p. 343.

direct access to phylogeny.[*] The Russian zoologist Alexej N. Sewertzoff[†] (1866–1936), in his landmark study of the laws of morphological evolution published in 1931, as well as Adolf Naef, Walter Zimmermann, and Abel's student Kurt Ehrenberg (1896–1979)—they all had warned of exaggerated expectations with respect to the paleontological method, Hennig insisted. The same was true of Wilhelm Kühnelt (1905–1988), whose paper on the principles of systematics Hennig likewise cited. In his discussion of "what has become known as *Spezialisationskreuzungen*," Kühnelt noted the "frequently encountered difficulties of placing fossils on the phylogenetic tree." The reason is that they may exhibit a mixture of characteristics, which render the direct derivation of surviving descendants from extinct ancestors impossible. For this reason, fossils will often have to be placed "not on branching points of the phylogenetic tree, but at the tip of small side branches" that sprout at "deeper levels" of the tree.[‡] The conclusion Hennig drew from all this was that there cannot exist groups that are either completely primitive or alternatively completely derived in all of their characteristics. A monophyletic group must be characterized by at least one derived characteristic in order to be recognized as such, which implies that the primitive state of the same character has to be present in its sister group.[§]

This is how the heterobathmy of characters becomes a precondition for the reconstruction of sister-group relationships. Hennig thought that it is precisely this mixture of primitive and derived characters across extant species that allows the reliable inference of the sequence of speciation events through which extant species have derived from ancestral species through time.[¶] Given the absence of exclusively primitive, and exclusively derived groups, ancestor–descendant series cannot be reconstructed. But if a mixture of primitive and derived conditions of form characterizes each group under consideration, then it becomes possible to recognize sister-group relationships on the basis of shared derived characters. Shared derived characters imply common ancestry, because the implication of phylogeny reconstruction is that shared derived characters are inherited from the most recent common ancestors of those organisms that share the derived character condition. Shared derived characters are homologues of those organisms that share them, and homology points to common ancestry.

The paleontologist Otto Schindewolf had already highlighted, in a theoretical paper of 1937, that the "crossing of specializations" does away with the overall primitive, in no way specialized "root form."[**] Hennig disputed Schindewolf's consequent claim, however, that the "crossing of specializations" renders the reconstruction of the morphology of hypothetical ancestors (stem-forms, *Stammformen*) impossible. In Hennig's view, *Spezialisationskreuzungen*, the heterobathmy of characters, means only that every species, in a comparison with others, related species, will show a mixture of relatively more primitive (plesiomorphic) and more derived (apomorphic) characters. The careful analysis of character evolution across a number of related

---

[*] Hennig, 1950, p. 134.
[†] Levit et al., 2004.
[‡] Kühnelt, 1942, p. 13.
[§] Hennig, 1965, p. 107.
[¶] Hennig, 1965, p. 107.
[**] Schindewolf, 1937, p. 207.

species allows in Hennig's view the reconstruction of the morphology of the stem-species (of the hypothetical ancestor at the node of the bifurcation), as it will simply comprise the plesiomorphic characters present in the two daughter species.[*] The method of phylogenetic systematics thus does allow the inference of the ancestral morphology of a monophyletic taxon, but it does not allow the discovery of the ancestor of a monophyletic taxon—precisely because of the heterobathmy of characters.

Tschulok, in 1922 already, had argued that taking the cross-cutting distribution of characters at face value would result in the reconstruction of reticulated relationships. The philosopher Bernhard Bavink, whom Hennig had read, had likewise commented on *Spezialisationskreuzungen*, which he took to indicate that the systematist has to deal with a network of characters subject to Mendelian patterns of inheritance.[†] The paleontologist Edwin Hennig had abandoned the phylogenetic tree in favor of a phylogenetic sheaf, or bush.[‡] All of these difficulties disappear if one heeds Tschulok's advice of 1922: "the *conditio sine qua non* for the reconstruction of phylogenetic trees is the distinction of the primitive and derived condition of form."[§] If phylogeny reconstruction is *Merkmalsphylogenese* as argued by Zimmermann and Hennig, and if the recognition of sister-group relationships is to be based on shared derived characters, then the correct polarization of character transformation series is, indeed, a *sine qua non* for phylogeny reconstruction. Birds are not recognized as a monophyletic group on the basis of the presence of forelimbs, because forelimbs are a character that marks out a more inclusive group, the tetrapods. Birds are recognized as a monophyletic taxon on the basis of the presence of bird wings (of different structure than bat wings), an evolutionary transformation of forelimbs and hence a shared derived character of birds. And as argued by Hennig, the primitive condition of the same character, that is, the forelimb, must be present in the nearest relatives of birds, which are theropod dinosaurs. Tetrapods in turn are recognized as a monophyletic group because of the presence of tetrapod limbs, an evolutionary transformation of ancestral fins, and hence a shared derived character of tetrapods.

If monophyly is the centerpiece of Hennig's metaphysics, then the distinction of the primitive (plesiomorphic) from the derived (apomorphic) condition of form in character analysis is the centerpiece of his epistemology, which claims that phylogenetic relationships are exclusively revealed by shared derived characters (synapomorphies). To classify organisms based on shared primitive characters (symplesiomorphies) runs the risk of creating non-monophyletic groups, which in turn results in an artificial system.

---

[*] Hennig, 1950, pp. 144ff.
[†] Bavink, 1933, p. 42.
[‡] Hennig, 1950, p. 336.
[§] Tschulok, 1922, p. 197.

Epilogue

Hennig was not a particularly outgoing man. He was the archetypal collections-based museum researcher, who shied away from public exposure and far-flung professional travel.[*] Except for a debate with Ernst Mayr in 1974,[†] he did not engage in print in scientific disputes concerning his phylogenetic systematics. The reception of his ideas was slow and restrained in Germany, less so in the East than in West Germany. A philosophical analysis of systematics by the Marxist philosopher Rolf Löther[‡] (b. 1933) is credited with having introduced Hennig's phylogenetic systematics in East German University curricula.[§] Günther Peters (b. 1932), herpetologist at the Natural History Museum in Berlin and professor for Zoology at the Humboldt University, was successful in establishing Hennig's systematics in the East German High School textbook for biology.[¶] Günter Osche (1926–2009) published an early West German explication of Hennig's methodology, without achieving any broader impact, however.[**] The history of annual West German meetings on systematics and phylogeny, the *Phylogenetisches Symposium*, initiated in 1956 by Adolf Remane, Wolf Herre and Gerhard Heberer, among others and continuing today, shows that some prominent zoologists such as Bernhard Rensch, Willi Hennig, and Klaus Günther were never invited to present.[††] During Hennig's lifetime, only one such meeting—held in 1963 in Marburg—was devoted to monophyly and polyphyly, concepts dear to Hennig's heart. Although Hennig attended the meeting,[‡‡] he was not asked to give a formal presentation. The former Remane student Peter Ax (1927–2013) was the only speaker who defended a rigorous conception of monophyly in the sense of Hennig,[§§] the others still adhering to cross-cutting conceptions of monophyly such as the one promoted by George Gaylord Simpson.[¶¶] It was Ax's textbook on phylogenetic systematics, published in 1984, which eventually established a broad acceptance of Hennig's argumentation pattern among West German zoologists.

The impulse for a widespread, indeed international transformation of biosystematics, and therewith ironically a transformation of Hennig's own phylogenetic systematics as well, did not originate in Germany, however, but in the Natural History Museums in London and New York.[***] The year 1966 was not only the year of publication of the English translation of Hennig's *Phylogenetic Systematics* but also the year of publication of a monograph on Transantarctic chironomid midges

---

[*] Schmitt, 2010, 2013.
[†] Mayr, 1974; Hennig, 1974.
[‡] Löther, 1972; on Löther, see his autobiographical sketch in Löther (2010).
[§] Schmitt, 2001, p. 341.
[¶] Peters, 1995, p. 9.
[**] Osche, 1963.
[††] Kraus, 1984; Kraus and Hoßfeld, 1998.
[‡‡] Hennig, 1964.
[§§] Ax, 1964.
[¶¶] Simpson, 1961.
[***] Hull, 1988; see also Rieppel, 2014.

by the Swedish entomologist Lars Brundin,[*] working at the Natural History Museum of Stockholm. The Stockholm Museum was at the time a world center for the study of fossil fishes, which motivated a young graduate from the University of Hawaii to spend some time there as a postdoctoral fellow, again in 1966. After arriving in Stockholm, the ichthyologist Gareth Nelson (b. 1937) soon became aware of Brundin's monograph just published. Brundin had prefaced his monograph with a 50-page account of Hennig's method of phylogenetic systematics, an exposition that galvanized Nelson's interest. According to Nelson, Brundin's "critique (1966: 11–64) began the [cladistic] revolution."[†] At the second Annual Willi Hennig Society Meeting, held in Ann Arbor, MI, in October 1981, Colin Patterson (1933–1998) from the Natural History Museum London reminisced how he first came into contact with Hennig's systematics: "Then, one day early in 1967, Gary Nelson, who was spending six months in the BM [British Museum (Natural History)], told me that something had just appeared in the library that I might find interesting … it was Brundin's monograph on chironomids just arrived. I was bowled over by it – it was like discovering logic for the first time."[‡] Upon his return to New York, Nelson set like-minded colleagues at the American Museum of Natural History on fire. Brundin was urged to explicate his understanding of Hennig's principles of phylogeny reconstruction at the 1967 Nobel Symposium on "Current Problems of Lower Vertebrate Phylogeny,"[§] a presentation that both Nelson and Patterson attended.[¶] The impact of Brundin's presentation was succinctly summed up by Patterson: "the heart of Brundin's paper was one message: 'phylogenetics is the search for the sister group.' That message eventually got through…."[**]

A crucial point in this account is Patterson's emotion upon reading Brundin, which was that of having discovered "logic for the first time," to which Nelson added "my reaction was much the same as Colin Patterson's a few months later."[††] On that basis, Patterson went on to develop his "test of congruence,"[‡‡] which takes a set-theoretical approach to the resolution of conflicting character distribution. Shared derived characters are grouped in a way so they form a hierarchy of sets that maximize the relation of inclusion and exclusion, but minimize the relation of overlap. This would seem to be in accord with Hennig's claim that the multidimensional multiplicity of biodiversity should be classified using the tools of set theory.[§§] As was pointed out in the Introduction, Hennig aimed with his *Grundzüge* of 1950 not just to deliver a new method of phylogeny reconstruction, but beyond that to lay the conceptual foundation for a phylogenetic systematics that would synthesize much of the current historical, that is, evolutionary biology. In pursuit of this project, Hennig

---

[*] Brundin, 1966.
[†] Nelson, 2004, p. 133. See also Nelson (2014).
[‡] Patterson's talk, delivered on October 3, 1981, was transcribed and made available by David M. Williams, Department of Life Sciences, The Natural History Museum, London, United Kingdom. See also Nelson (2014, p. 141).
[§] Brundin, 1968.
[¶] Hull, 1988, p. 144.
[**] Patterson, 1989, p. 472.
[††] Nelson, 2014, p. 141.
[‡‡] Patterson, 1982.
[§§] Hennig, 1953, p. 6.

found himself exposed to two incommensurable styles of thought that characterized German biology in the first half of the twentieth century. As sketched in this book, the holistic-organicist camp of comparative biologists, some of which drifting off into *romantisch-völkisch* nature mysticism, was confronted by the *neue Sachlichkeit* placated by the neo-positivist, or empiricist camp of German biologists. Not surprisingly, Hennig remained rather unsuccessful in his attempt to resolve that tension, as the same divide can be located in his own synthesis.

With the enkaptic hierarchy, Hennig endorsed a core concept of German holism-organicism. The concept was put to work in anatomy (Heidenhain, Benninghoff), developmental biology (Alveres, Dürken), ecology (Friederichs, Thienemann), phylogeny (Beurlen), and Nazi "biopolitics."[*] Hennig's choice to present the phylogenetic system as an enkaptic hierarchy entails important ontological commitments. From this perspective, the phylogenetic system is a nested hierarchy of complex wholes, or individuals, subject to the part-whole relation; the system is built up through the successive splitting of species lineages, the species being the system's fundamental divisible building block; supraspecific levels of complexity are characterized by emergent properties (monophyly). The whole phylogenetic system is thus grounded in space and time, in individuality and causality, ultimately in reality.

Such metaphysical commitments clash with Hennig's epistemology, that is, his appeal to logic generally, and set-theory in particular, his employment of the "rule of three" in the resolution of phylogenetic relationships, and his enforcement of the principle of dichotomy as a methodological tool in phylogeny reconstruction. There are, therefore, two hierarchies at work in Hennig's phylogenetic systematics: the hierarchy of species-lineages splitting and splitting again, and the hierarchy of boxes within boxes, or sets within sets.[†] Whereas both hierarchies suggest the same classification, and hence afford the same inferences about the organisms classified, metaphysically the two hierarchies are profoundly different.[‡]

Looking beyond 1966 and into the twenty-first century, it can be said that the second, logistical perspective prevailed. Hennig already called for the principle of reciprocal illumination, as the phylogenetic system is compared to other dimensions of the property space of species. He emphasized in particular the strong phylogenetic signal that obtains when morphology, ontogeny, paleontology, and biogeography deliver a congruent picture of relationships. Modern, computer-based phylogeny reconstruction is completely coherentist in its approach, in that it seeks maximal congruence of phylogenetically informative characters relative to a hierarchy under some optimality criterion (maximum parsimony, maximum likelihood, or posterior Bayesian probability). A coherentist epistemology has traditionally been associated with an instrumentalist philosophy of science, and it can fairly be said that modern systematics took an instrumentalist turn that led away from Hennig's early ontological commitments.[§]

---

[*] NS "biopolitics" combined biologism with the concept of *Lebensraum* in a vision of a racially pure national community seeking spatial expansion: Weingart et al. (1992, p. 370); see also Caplan (2008, p. 19).
[†] P.A. Williams, 1992.
[‡] Hull, 1988, p. 399.
[§] Rieppel, 2007b,c; 2014.

Literature Cited

ARCHIVAL SOURCES

Archive, Academy of Sciences Leopoldina, Halle / Saale
Archive of the Max Planck Society, Berlin-Dahlem
Archives of the American Museum of Natural History, Henry Fairfield
  Osborn Papers
Bundesarchiv Berlin (Berlin Documentation Center)
Harvard University Archives, Ernst Mayr Correspondence
State-Archive Kanton Basel-Stadt
State-Archive Kanton Zürich
University Archive Göttingen
University Archive Greifswald
University Archive Halle
University Archive Tübingen
University Archive Zürich

PUBLISHED SOURCES

Part of *Chapter 1* is derived from Rieppel, Olivier. 2011. The Gegenbaur
  Transformation: A paradigm change in comparative biology, *Systematics and
  Biodiversity*, 9: 177–190; © The Natural History Museum, with permission of
  Taylor & Francis Group on behalf of The Natural History Museum, available
  online: http://www.tandfonline.com/ [DOI: 10.1080/14772000.2011.602754].
Part of *Chapter 2* is derived from Rieppel, Olivier. 2011. Ernst Haeckel
  (1834–1919) and the monophyly of life. *Journal for Zoological Systematics
  and Evolutionary Research*, 49: 1–5; © Blackwell Verlag GmbH, published
  by Wiley-Blackwell, available online: http://www.onlinelibrary.wiley.com/
  [DOI: 10.1111/j.1439-0469.2010.00580.x].
Part of *Chapter 3* is derived from Rieppel, Olivier. 2012. Adolf Naef (1883–1949),
  systematic morphology and phylogenetics. *Journal for Zoological Systematics
  and Evolutionary Research*, 50: 2–13; © Blackwell Verlag GmbH, published by
  Wiley-Blackwell, available online: http://www.onlinelibrary.wiley.com/ [DOI:
  10.1111/j.1439-0469.2011.00635.x]. Rieppel, Olivier, David M. Williams, and
  Malte C. Ebach. 2013. Adolf Naef (1883–1949): On foundational concepts
  and principles of systematic morphology. *Journal of the History of Biology*,
  46: 445–510; © Springer Science + Business Media, available online: http://
  www.link.springer.com/journal/10739/ [DOI: 10.1007/s10739-012-9338-4].
  Rieppel, Olivier. 2013. Othenio Abel (1875–1946) and "the phylogeny of the
  parts." *Cladistics*, 29: 328–335; © The Willi Hennig Society 2012, published
  by John Wiley & Sons, available online: http://www.onlinelibrary.wiley.com/
  [DOI: 10.1111/j.1096-0031.2012.00428.x].

Part of *Chapter 5* is derived from Rieppel, Olivier. 2012. Wilhelm Troll (1897–1978): Idealistic morphology, physics, and phylogenetics. *History and Philosophy of Life Sciences*, 33: 321–342; © 2011 Stazione Zoologica Anton Dohrn. Rieppel, Olivier. 2012. Karl Beurlen (1901–1985), Nature mysticism, and Aryan paleontology. *Journal of the History of Biology*, 45: 253–299; © Springer Science + Business Media, available online: http://www.link.springer.com/journal/10739/ [DOI: 10.1007/s10739-011-9283-7].

Part of *Chapter 6* is derived from Rieppel, Olivier. 2012. Karl Beurlen (1901–1985), Nature mysticism, and Aryan paleontology. *Journal of the History of Biology*, 45: 253–299; © Springer Science + Business Media, available online: http://www.link.springer.com/journal/10739/ [DOI: 10.1007/s10739-011-9283-7].

And: Rieppel, Olivier. 2013. Othenio Abel (1875–1946): the rise and decline of paleobiology in German paleontology. Historical Biology, 25: 77–97; © Taylor & Francis Group, available online: http://www.tandfonline.com/ [DOI: 10.1080 / 08912963.2012.697899].

And: Rieppel, Olivier. 2013. Styles of scientific reasoning: Adolf Remane (1898–1976) and the German evolutionary synthesis. *Journal of Zoological Systematics and Evolutionary Research*, 51: 1–12; © Blackwell Verlag GmbH, published by Wiley-Blackwell, available online: http://www.onlinelibrary.wiley.com/ [DOI: 10.1111/jzs.12003].

Part of *Chapter 7* is derived from Rieppel, Olivier. 2012. Wilhelm Troll (1897–1978): Idealistic morphology, physics, and phylogenetics. *History and Philosophy of Life Sciences*, 33: 321–342; © 2011 Stazione Zoologica Anton Dohrn.

And: Rieppel, Olivier. 2012. Karl Beurlen (1901–1985), Nature mysticism, and Aryan paleontology. *Journal of the History of Biology*, 45: 253–299; © Springer Science + Business Media, available online: http://www.link.springer.com/journal/10739/ [DOI: 10.1007/s10739-011-9283-7].

REFERENCES

Abel, Othenio. 1900. *Untersuchungen über die fossilen Platanistiden des Wiener Beckens.* Denkschriften der Kaiserlichen Akademie der Wissenschaften, Mathematisch-Naturwissenschaftliche Classe, 68, Wien, Germany, pp. 839–874.

Abel, Othenio. 1907. Die Aufgaben und Ziele der Paläozoologie. *Verhandlungen der Zoologisch-Botanischen Gesellschaft in Wien*, 57: 67–78.

Abel, Othenio (Ed.). 1909. Was verstehen wir unter monophyletischer und polyphyletischer Abstammung? *Verhandlungen der Zoologisch-Botanischen Gesellschaft in Wien*, 59: 243–256.

Abel, Othenio. 1911. Die Bedeutung der fossilen Wirbeltiere für die Abstammungslehre, pp. 198–250. In: Hertwig, R. (Ed.), *Die Abstammungslehre.* Zwölf gemeinverständliche Vorträge über die Deszendenztheorie im Licht der neueren Forschung. Jena, Germany: Gustav Fischer.

Abel, Othenio. 1912. *Grundzüge der Palaeobiologie der Wirbeltiere.* Stuttgart, Germany: E. Schweizerbart.

Abel, Othenio. 1913. Neuere Wege phylogenetischer Forschung. *Verhandlungen der Gesellschaft Deutscher Naturforscher und Aerzte*, 85: 116–124.

Abel, Othenio. 1914. Paläntologie und Paläozoologie, pp. 303–395. In: Hertwig, R. and R.v. Wettstein (Eds.), *Die Kultur der Gegenwart, Dritter Teil, Vierte Abteilung, Vierter Band. Abstammungslehre. Systematik. Paläontologie. Biogeographie.* Leipzig, Germany: B.G. Teubner.

Abel, Othenio. 1919. *Die Stämme der Wirbeltiere.* Berlin, Germany: Walter de Gruyter & Co.

Abel, Othenio. 1920. *Lehrbuch der Paläozoologie.* Jena, Germany: Gustav Fischer.

Abel, Othenio. 1922. Diskussionsbeitrag an der Frankfurter Tagung, 8–10 August 1921. *Palaeontologische Zeitschrift*, 4: 112.

Abel, Othenio. 1923. Vererbungswissenschaft und Morphologie. *Verhandlungen der Zoologisch-Botanischen Gesellschaft in Wien*, 73: 199–209.

Abel, Othenio. 1927. Veranstaltungen. *Verhandlungen der Zoologisch-Botanischen Gesellschaft in Wien*, 77: 243–249.

Abel, Othenio. 1928a. Louis Dollo. Zur Vollendung seines siebzigsten Lebensjahres. *Palaeobiologica*, 1: 7–12.

Abel, Othenio. 1928b. Die Festgabe der "Palaeobiologica." *Palaeobiologica*, 1: 1–6.

Abel, Othenio. 1928c. Das biologische Trägheitsgesetz. *Biologia Generalis*, 4: 1–102.

Abel, Othenio. 1929a. *Paläobiologie und Stammesgeschichte.* Jena, Germany: Gustav Fischer.

Abel, Othenio. 1929b. Otto Jaekel (21. Februar 1863–6. März 1929). *Palaeobiologica*, 2: 143–186.

Abel, Othenio. 1929c. Das biologische Trägheitsgesetz. *Palaeontologische Zeitschrift*, 11: 7–17.

Adam, Uwe D. 1977. *Hochschule und Nationalsozialismus: Die Universität Tübingen im Dritten Reich.* Tübingen, Germany: J.C.B. Mohr (Paul Siebeck).

Adatte, Thierry and Iwan Stössel-Sittig. 2004. Jürgen Remane. (1934–2004). *Eclogae Geologicae Helvetiae*, 97: I–II.

Alfert, Max. 1972. Martin Heidenhain, pp. 223–224. In: Gillispie, C.C. (Ed.), *Dictionary of Scientific Biography*, vol. VI. New York: Charles Scribner's Sons.

Alverdes, Friedrich. 1932. Die Ganzheitsbetrachtung in der Biologie. *Sitzungsberichte der Gesellschaft zur Beförderung der gesamten Naturwissenschaften zu Marburg*, 67: 89–118.

Alverdes, Friedrich. 1935a. *Die Totalität des Lebendigen.* Leipzig, Germany: Barth.

Alverdes, Friedrich. 1935b. Kausalität, Finalität und Ganzheit. *Acta Biotheoretica*, 3: 167–180.

Alverdes, Friedrich. 1936a. Der Beriff des 'Ganzen' in der Biologie. *Zeitschrift für Rassenkunde und ihre Nachbargebiete*, 4: 1–9.

Alverdes, Friedrich. 1936b. Organizismus und Holismus. Neuere theoretische Strömungen in der Biologie. *Der Biologe*, 5: 121–128.

Alverdes, Friedrich. 1939. Biologische Ganzheitsbetrachtung. *Zeitschrift für die gesamte Naturwissenschaft*, 5: 58–67.

Alverdes, Friedrich and Ernst Krieck. 1937. Zwiegespräch über völkisch-politische Anthropologie und biologische Grenzbetrachtung. *Der Biologe*, 6: 49–55.

Amidon, Kevin S. 2009. Adolf Meyer-Abich, holism, and the negotiation of theoretical biology. *Biological Theory*, 3: 357–370.

Anonymous. 1931/32. Biologie als Staatliche Notwendigkeit. *Der Biologe*, 1: 96–98.

Anonymous. 1932/33. Mitteilungen. *Der Biologe*, 2: 128.

Anonymous. 1933. Verein zur Förderung des mathematischen und naturwissenschaftlichen Unterrichts. *Unterrichtsblätter für Mathematik und Naturwissenschaften*, 39: 113–115.

Anonymous. 1936. Personen-Nachrichten. *Zeitschrift der Deutschen Geologischen Gesellschaft*, 89: 668.

Anonymous. 1937a. Zeitschriftenwesen. *Zeitschrift der Deutschen Geologischen Gesellschaft*, 89: 603.

Anonymous. 1937b. Kleine Mitteilungen. *Zeitschrift der Deutschen Geologischen Gesellschaft* 89: 292–296.

Anonymous. 1937c. Bericht über die Hauptversammlung in Aachen vom 21. bis 29. August 1937. *Zeitschrift der Deutschen Geologischen Gesellschaft*, 89: 553–566.

Anonymous. 1938. Bericht über die Hauptversammlung in München vom 14. bis 21. Juli 1938. *Zeitschrift der Deutschen Geologischen Gesellschaft*, 90: 531–548.

Anonymous. 1939a. Bericht über die Jahresversammlung der Palaeontologischen Gesellschaft in Bayreuth (11–14 Juli 1938). *Palaeontologische Zeitschrift*, 21: 1–5.

Anonymous. 1939b. Wissenschaftliche Sitzung der Gesellschaft. *Zeitschrift der Deutschen Geologischen Gesellschaft*, 91: 253–254.

Appel, Toby A. 1987. *The Cuvier-Geoffroy Debate. French Biology in the Decades before Darwin*. Oxford: Oxford University Press.

Arthaber, Gustav von, Karl Diener. 6. January 1928. *Neue Freie Presse*, Vienna, Austria, No. 2741, January 8, 1928, p. 13. http://anno.onb.ac.at/cgi-content/anno?apm=0&aid=nfp&datum=19280108&seite=13 (accessed July 25, 2012).

Ash, G. Mitchell. 1995. *Gestalt Psychology in German Culture, 1890–1967. Holism and the Quest for Objectivity*. Cambridge, UK: Cambridge University Press.

Asher, Leon. 1922. Theoretische Biologie und biologisches Weltbild. *Die Naturwissenschaften*, 10: 474–477.

Aumüller, Gerhard and Kornelia Grundmann. 2002. Anatomy during the Third Reich. The Institute of Anatomy at the University of Marburg, as an example. *Annals of Anatomy*, 184: 295–303.

Ax, Peter. 1964. Der Begriff Polyphylie ist aus der Terminologie der natürlichen, phylogenetischen Systematik zu eliminieren. *Zoologischer Anzeiger*, 173: 52–56.

Ax, Peter. 1984. *Das Phylogenetische System*. Stuttgart, Germany: Gustav Fischer.

Axhausen, Georg. 1912. Über Implantation und Transplantation, pp. 301–320. In: Abderhalden, E. (Ed.), *Fortschritte der Naturwissenschaftlichen Forschung*. Fünfter Band. Berlin, Germany: Urban und Schwarzenberg.

Ayer, Alfred J. 1936. *Language, Truth & Logic*. London: Gollancz.

Baer, Carl Ernst von. 1828. *Ueber Entwickelungsgeschichte der Thiere. Beobachtung und Reflexion, Theil I*. Königsberg, Germany: Gebr. Bornträger.

Baer, Carl Ernst von. 1876a. Ueber Zielstrebigkeit in den organischen Körpern insbesondere, pp. 170–234. In: *Reden gehalten in wissenschaftlichen Versammlungen und kleinere Aufsätze vermischten Inhalts. Zweiter Theil. Studien aus dem Gebiete der Naturwissenschaften*. St. Petersburg, Russia: H. Schmitzdorff.

Baer, Carl Ernst von. 1876b. Ueber Darwin's Lehre, pp. 235–480. In: *Reden gehalten in wissenschaftlichen Versammlungen und kleinere Aufsätze vermischten Inhalts. Zweiter Theil. Studien aus dem Gebiete der Naturwissenschaften*. St. Petersburg, Russia: H. Schmitzdorff.

Baer, Carl Ernst von. 1876c. Ueber den Zweck in den Vorgängen der Natur.—Erste Hälfte. Ueber Zweckmässigkeit oder Zielstrebigkeit überhaupt, pp. 49–105. In: *Reden gehalten in wissenschaftlichen Versammlungen und kleinere Aufsätze vermischten Inhalts. Zweiter Theil. Studien aus dem Gebiete der Naturwissenschaften*. St. Petersburg, Russia: H. Schmitzdorff.

Barfurth, Dietrich. 1920. Wilhelm Roux zum siebzigsten Gedburtstage. *Die Naturwissenschaften*, 8: 431–435.

Bassin, Mark. 2005. Blood or Soil? The *völkisch* movement, the Nazis, and the legacy of *Geopolitik*, pp. 204–242. In: Brüggemeier, F.-J., M. Cioc, and T. Zeller (Eds.), *How Green Were the Nazis? Nature, Environment, and Nation in the Third Reich*. Athens, OH: Ohio University Press.

Bather, Francis A. 1930. Otto Jaekel. *Quarterly Journal of the Geological Society of London*, 86: lviii–kix.

Bäumer, Änne. 1989. Die Politisierung der Biologie zur Zeit des Nationalsozialismus. *Biologie in unserer Zeit*, 19: 76–80.

Bäumer, Änne. 1990a. Die Zeitschrift 'Der Biologe' als Organ der NS-Biologie. *Biologie in unserer Zeit*, 20: 42–47.

Bäumer, Änne. 1990b. *NS-Biologie*. Stuttgart, Germany: S. Hirzel.

Bavink, Bernhard. 1933. *Ergebnisse und Probleme der Naturwissenschaften. Eine Einführung in die heutige Naturphilosophie.* Fünfte, neu bearbeitete und erweiterte Auflage. Leipzig, Germany: S. Hirzel.

Bavink, Bernhard. 1941. *Ergebnisse und Probleme der Naturwissenschaften. Eine Einführung in die heutige Naturphilosophie.* 7. Auflage. Leipzig, Germany: S. Hirzel.

Beiser, Frederick C. 2002. *German Idealism: The Struggle against Subjectivism, 1781–1801.* Cambridge, MA: Harvard University Press.

Benninghoff, Alfred. 1935/36. Form und Funktion. 1 Teil. *Zeitschrift für die gesamte Naturwissenschaft,* 1: 149–160.

Benninghoff, Alfred. 1938. Über Einheiten und Systembildungen im Organismus. *Deutsche Medizinische Wochenschrift,* 64: 1377–1382.

Bergdolt, Ernst. 1937/38a. Troll, W. Vergleichende Morphologie der höherern Pflanzen, I. Band. I. Teil, 2. Lieferung. Berlin, Bornträger, 1937. *Zeitschrift für die gesamte Naturwissenschaft,* 3: 181–182.

Bergdolt, Ernst. 1937/38b. Zur Frage der Rassenbildung beim Menschen. *Zeitschrift für die gesamte Naturwissenschaft,* 3: 109–113.

Bergdolt, Ernst. 1940. Über Formwandlungen—zugleich eine Kritik von Artbildungstheorien. *Der Biologe,* 9: 398–407.

Bertalanffy, Ludwig von. 1927. Über die Bedeutung der Umwälzungen in der Physik für die Biologie. *Biologisches Zentralblatt,* 47: 653–622.

Bertalanffy, Ludwig von. 1932. *Theoretische Biologie, Band I.* Berlin, Germany: Gebr. Bornträger.

Bertalanffy, Ludwig von. 1941. Die organische Auffassung und ihre Auswirkungen. *Der Biologe,* 10: 247–258, 337–345.

Bethe, Albrecht. 1940. Erinnerungen an die Zoologische Station in Neapel. *Die Naturwissenschaften,* 28: 820–822.

Beurlen, Karl. 1930. Vergleichende Stammesgeschichte. Grundlagen, Methoden, Probleme unter besonderer Berücksichtigung der höheren Krebse. *Fortschritte der Geologie und Paläntologie,* 8: 317–586.

Beurlen, Karl. 1932. Funktion und Form in der organischen Entwicklung. *Die Naturwissenschaften,* 20: 73–80.

Beurlen, Karl. 1935. Der Aktualismus in der Geologie. *Zentralblatt für Mineralogie, Geologie und Paläontologie, B* 1935: 520–525.

Beurlen, Karl. 1935/36. Das Gestaltproblem in der Natur. *Zeitschrift für die gesamte Naturwissenschaft,* 1: 445–457.

Beurlen, Karl. 1936. Der Zeitbegriff in der modernen Wissenschaft und das Kausalitätsprinzip. *Kant-Studien,* 41: 16–37.

Beurlen, Karl. 1936/37. Überblick über die Werke von Edgar Dacqué. *Zeitschrift für die gesamte Naturwissenschaft,* 1: 445–457.

Beurlen, Karl. 1937a. *Die stammmesgschichtlichen Grundlagen der Abstammungslehre.* Jena, Germany: G. Fischer.

Beurlen, Karl. 1937b. Bedeutung und Aufgabe der Deutschen Geologischen Gesellschaft. *Zeitschrift der Deutschen Geologischen Gesellschaft,* 89: 52–58.

Beurlen, Karl. 1937c. Mitteilung über die Fachgliederung Bodenkunde (Geologie, Mineralogie, Geophysik) im Reichsforschungsrat. *Zeitschrift der Deutschen Geologischen Gesellschaft,* 89: 360.

Beurlen, Karl. 1942. Erdgeschichte und Naturgesetz. *Zeitschrift der Deutschen Geologischen Gesellschaft,* 94: 192–199.

Bieler, Rüdiger. 1992. Gastropod phylogeny and systematics. *Annual Review of Ecology and Systematics,* 23: 311–338.

Blackbourne, David. 2006. *The Conquest of Nature Water, Landscape, and the Making of Modern Germany.* London: W.W. Norton.

Blandford, Walter F.H. 1897. Fritz Müller. *Nature*, 56: 546–548.

Bloch, Kurt. 1952. Zur Frage der Realität der sog. Typen in der biologischen Systematik. *Acta Biotheoretica*, 10: 1–10.

Bloch, Kurt. 1956. *Zur Theorie der naturwissenschaftlichen Systematik, unter besonderer Berücksichtigung der Biologie.* Acta Biotheoretica, Supplementum Primum, 9 (i.e., Bibliotheca Biotheoretica, 7). Leiden, the Netherlands: E.J. Brill.

Boelsche, Wilhelm. 1934. Haeckel als Erlebnis. *Der Biologe*, 3: 34–38.

Böker, Hans. 1924. Begründung einer biologischen Morphologie. *Zeitschrift für Morphologie und Anthropologie*, 24: 1–22.

Boletzky, Sigurd von. 1999. Systematische Morphologie und Phylogenetik—zur Bedeutung des Werkes von Adolf Naef (1883–1949). *Vierteljahrsschrift der Naturforschenden Gesellschaft in Zürich*, 144: 73–82.

Boletzky, Sigurd von. 2000. Adolf Naef (1883–1949). A biographical note, pp. ix–xiii. In: Naef, A. 1928. *Cephalopoda. Embryology. Fauna and Flora of the Bay of Naples. Translated from German.* Washington, DC: Smithsonian Institution Libraries.

Bouquet, Mary. 1996. Family trees and their affinities: The visual imperative of the genealogical diagram. *The Journal of the Royal Anthropological Institute*, 2: 43–66.

Bowler, Peter J. 1975. The changing meaning of evolution. *Journal of the History of Ideas*, 36: 95–114.

Bowler, Peter J. 2007. Fins and limbs and fins into limbs: The historical context, 1840–1940, pp. 7–14. In: Hall, B.K. (Ed.), *Fins into Limbs. Evolution, Development, and Transformation.* Chicago, IL: University of Chicago Press.

Boyd, Richard. 1991. Realism, anti-fondationalism and the enthusiasm for natural kinds. *Philosophical Studies*, 61: 127–148.

Boyd, Richard. 1999. Homeostasis, species, and higher taxa, pp. 141–185. In: Wilson, R.A. (Ed.), *Species. New Interdisciplinary Essays.* Cambridge, MA: MIT Press.

Brauer, Friedrich. 1885. Systematisch-zoologische Studien. I. System und Stammbaum. *Sitzungsberichte der Akademie der Wissenschaften in Wien* (mathematisch-naturwissenschaftliche Klasse), 91: 237–413.

Brauer, Friedrich. 1887. Beziehungen der Deszendenzlehre zur Systematik. *Schriften des Vereins zur Verbreitung Naturwissenschaftlicher Kenntnisse*, 27: 579–614.

Braun, Alexander. 1853. *Das Individuum der Pflanze in seinem Verhältnis zur Spezies. Generationsfolge, Generationswechsel und Generationstheilung der Pflanze.* Berlin, Germany: Königliche Akademie der Wissenschaften.

Braun, Alexander. 1876. Die Frage nach der Gymnospermie der Cycadeen erläutert durch die Stellung dieser Familie im Stufengang des Gewächsreichs. *Monatsberichte der Königlich. Preussischen Akademie der Wissenschaften zu Berlin aus dem Jahre* 1875: 241–267.

Braus, Hermann. 1920. Wilhelm Roux und die Anatomie. *Die Naturwissenschaften*, 8: 435–442.

Breidbach, Olaf. 2003. Post-Haeckelian comparative biology—Adolf Naef's idealistic morphology. *Theory in Biosciences*, 122: 174–193.

Bridgeman, Percy W. 1927. *The Logic of Modern Physics.* New York: Macmillan.

Brigandt, Ingo. 2002. Homology and the origin of correspondence. *Biology and Philosophy*, 17: 389–407.

Brigandt, Ingo. 2009. Natural kinds in evolution and systematics: Metaphysical and epistemological considerations. *Acta Biotheoretica*, 57: 77–97.

Brohmer, Paul. 1927. Die pädagogischen Akademien und der biologische Unterricht der höheren Schulen. *Unterrichtsblätter für Mathematik und Naturwissenschaften*, 33: 337–339.

Bronn, Heinrich G. 1853. Allgemeine Einleitung in die Naturgeschichte, pp. 1–138. In: Agassiz, L., H.G. Bronn, A.C. v. Leonhard, M. Perth, E.A. Guitzmann, and M. Seubert (Eds.), *Volks-Naturgeschichte der drei Reiche für Schule und Haus.* Stuttgart, Germany: J. B. Müller.

Bronn, Heinrich G. 1858a. *Morphologische Studien über die Gestaltungs-Gesetze der Naturkörper überhaupt und der organischen insbesondere.* Leipzig, Germany: G.F. Winter.

Bronn, Heinrich G. 1858b. *Untersuchungen über die Entwickelungs-Gesetze der organischen Welt während der Bildungs-Zeit unserer Erd-Oberfläche: Eine von der französischen Akademie im Jahre 1857 preisgekrönte Preisschrift.* Stuttgart, Germany: E. Schweizerbart.

Bronn, Heinrich G. 1859. On the laws of evolution of the organic world during the formation of the crust of the earth. *Annals and Magazine of Natural History,* 4 (33): 81–90.

Bronn, Heinrich G. 1860. CH. DARWIN: On the Origin of Species by means of Natural Selection, or the preservation of favoured races in the struggle for life (502 pp. 8°, London 1859). *Neues Jahrbuch für Mineralogie, Geognosie, Geologie und Petrefakten-Kunde,* 1860: 112–116.

Brooks, W.K. 1883. The Metamorphosis of Penaeus. *Annals and Magazine of Natural History,* 11(5): 147–149.

Brooks, W.K. 1887. Lucifer: A study in morphology. *Memoirs from the Biological Laboratory of the Johns Hopkins University,* 1: 57–137.

Brücher, Heinz. 1936. *Ernst Haeckel's Blut- und Geistes-Erbe. Eine kulturbiologische Monographie.* Munich, Germany: J.F. Lehmann.

Brücher, Heinz. 1941. Okkultismus in der Naturforschung. *Der Biologe,* 10: 265–266.

Brücke, Ernst W. 1861. Die Elementarorganismen. *Sitzungsberichte der Kaiserlichen Academie der Wissenschaften in Wien, Mathematisch-Naturwissenschaftliche Classe,* Abtheilung 3, 44: 381–406.

Brundin, Lars. 1966. Transantarctic relationships and their significance, as evidenced by chironomid midges. *Kungl. Svenska Vetenskaps Academiens Handlingar,* 4(11): 1–472.

Brundin, Lars. 1968. Application of phylogenetic principles in systematics and evolutionary theory, pp. 473–495. In: Orvig, T. (Ed.), *Current Problems of Lower Vertebrate Phylogeny: Proceedings of the 4th Nobel Symposium held in June 1967 at the Swedish Museum of Natural History in Stockholm.* Stockholm, Sweden: Almqvist and Wiskell.

Brunner von Wattenwyl, Karl. 1873. Georg Ritter von Frauenfeld. Ein Nachruf. *Verhandlungen der Zoologisch-Botanischen Gesellschaft in Wien,* 23: 535–538.

Brunner von Wattenwyl, Karl. 1901. Geschichte der K. K. Zoologisch-Botanischen Gesellschaft, pp. 3–16. In: Handlirsch, A. and R.v. Wettstein (Eds.), *Botanik und Zoologie in Österreich in den Jahren 1850 bis 1900. Festschrift.* Wien, Austria: Alfred Hölder.

Burkamp, Wilhelm. 1929. *Die Struktur der Ganzheiten.* Berlin, Germany: Junker und Dünnhaupt.

Burkhardt, Richard W. 2005. *Patterns of Behavior: Konrad Lorenz, Niko Tinbergen, and the Founding of Ethology.* Chicago, IL: University of Chicago Press.

Cannon, W.F. 1961. The impact of uniformitarianism. Two letters from John Herschel to Charles Lyell. 1836–1837. *Proceedings of the American Philosophical Society,* 105: 301–314.

Caplan, Jane. 2008. Introduction, pp. 1–25. In: Caplan, J. (Ed.), *Nazi Germany.* Oxford: Oxford University Press.

Carnap, Rudolf 1922. Der Raum. Ein Beitrag zur Wissenschaftslehre, pp. 1–87. In: Vaihinger, H., Frischeisen-Köhler, and M., Liebert, A. (Eds.), *Kant Studien.* Ergänzungshefte im Auftrag der Kant-Gesellschaft, Nr. 56. Berlin, Germany: Reuther & Reichard.

Carnap, Rudolf. 1963/1997. Intellectual autobiography, pp. 1–84. In: Schilpp, P.A. (Ed.), *The Philosophy of Rudolf Carnap.* La Salle, IL: Open Court.

Carus, André W. 2007. *Carnap and Twentieth-Century Thought.* Cambridge, UK: Cambridge University Press.

Cassirer, Ernst. 1923/1953. *Substance and Function, and Einstein's Theory of Relativity.* Translated by W.C. Swabey and M.C. Swabey. New York: Dover.

Cassirer, Ernst. 1950/1978. *The Problem of Knowledge. Philosophy, Science, & History since Hegel.* New Haven, CT: Yale University Press.

Cernajsek, Tillfried and Johannes Seidl. 2007. Zwischen Wissenschaft, Politik und Praxis. 100 Jahre Österreichische Geologische Gesellschaft (vormals Geologische Gesellschaft in Wien). *Austrian Journal of Earth Sciences*, 100: 252–274.

Chambers, Robert. 1844. *Vestiges of the Natural History of Creation.* London: John Churchill.

Chickering, R.A. 1973. A voice of moderation in Imperial Germany: The 'Verband für Internationale Verständigung' 1911–1914. *Journal for Contemporary History*, 88: 147–164.

Churchill, Fredrick B. 1975. Roux, Wilhelm, pp. 570–575. In: Gillispie, C.C. (Ed.), *Dictionary of Scientific Biography*, vol. 11. New York: Charles Scribner's Sons.

Churchill, Fredrick B. 2008. Roux, Wilhelm. Complete Dictionary of Scientific Biography. 2008. *Encyclopedia.com*, http://www.encyclopedia.com/topic/Wilhelm_Roux.aspx#1-1G2:2830903754-full (accessed August 23, 2012).

Clauβ, Ludwig F. 1936. *Die nordische Seele. Eine Einführung in die Rassenseelenkunde*, 5. Aufl. Munich, Germany: Lehmann Verlag.

Coleman, William. 1964. *Georges Cuvier, Zoologist.* Cambridge, MA: Harvard University Press.

Coleman, William. 1976. Morphology between type concept and descent theory. *Journal of the History of Medicine*, 31: 149–175.

Cornwell, John. 2003. *Hitler's Scientists.* London: Penguin Books.

Craw, R. 1992. Margins of cladistics: Identity, difference and place in the emergence of phylogenetic systematics, 1864–1975, pp. 65–107. In: Griffiths, P. (Ed.), *Trees of Life: Essays in Philosophy of Biology.* Dordrecht, the Netherlands: Kluwer.

Creath, Richard. 1996. The unity of science: Carnap, Neurath, and beyond, pp. 158–169. In: Galison, P. and D.J. Stump (Eds.), *The Disunity of Science. Boundaries, Contexts, and Power.* Stanford, CA: Stanford University Press.

Dabelow, Adolf. 1953/54. Alfred Benninghoff †. *Anatomischer Anzeiger*, 100: 157–165.

Dacqué, Edgar. 1915. *Grundlagen und Methoden der Paläogeographie.* Jena, Germany: G. Fischer.

Dacqué, Edgar. 1921. *Vergleichende biologische Formenkunde der fossilen niederen Tiere.* Berlin, Germany: Gebr. Bornträger.

Dacqué, Edgar. 1924. *Urwelt, Sage und Menschheit.* Munich, Germany: R. Oldenbourg.

Darwin, Charles. 1838. On the formation of mould. *Proceedings of the Geological Society of London, November 1833 to June 1838*, 2: 574–576.

Darwin, Charles. 1859. *On the Origin of Species.* London: John Murray.

Darwin, Charles. 1881. *The Formation of Vegetable Mould through the Action of Worms, with Observations on Their Habits.* London: John Murray.

Darwin, Charles. 1888. *On the Origin of Species.* 6th Edition, with Additions and Corrections. London: John Murray.

Darwin, Francis. 1887. *The Life and Letters of Charles Darwin*, Including an Autobiographical Chapter. 3rd Edition, in Three Volumes. London: John Murray.

Darwin, Francis. 1959. *The Life and Letters of Charles Darwin*, vol. II. New York: Basic Books.

Davidoff, Michael von. 1879. Beiträge zur vergleichenden Anatomie der hinteren Gliedmasse der Fische. *Morphologisches Jahrbuch*, 5: 450–520.

Dayrat, Benoît. 2003. The roots of phylogeny: How did Haeckel build his trees? *Systematic Biology*, 52: 515–527.

De Bont, Raf. 2010. Organisms in their milieu. Alfred Giard, his pupils, and early ethology, 1870–1930. *Isis*, 101: 1–29.

de Queiroz, Kevin. 1985. The ontogenetic method for determining character polarity and its relevance to phylogenetic systematics. *Systematic Zoology*, 34: 280–299.

Deichmann, Ute. 1996. *Biologists under Hitler*. Translated by Thomas Dunlap. Cambridge, MA: Harvard University Press.

Deichmann, Ute and Benno Müller-Hill. 1994. Biological research at universities and Kaiser Wilhelm Institutes in Nazi Germany, pp. 160–183. In: Rennberg, M. and M. Walker (Eds.), *Science, Technology, and National Socialism*. Cambridge, UK: Cambridge University Press.

Desmond, Adrian. 1982. *Archetypes and Ancestors*. Chicago, IL: University of Chicago Press.

Di Gregorio, Mario A. 2005. *From Here to Eternity. Ernst Haeckel and Scientific Faith*. Göttingen, Germany: Vandenhoeck & Ruprecht.

Diemberger, A. 1941. Lebensgemeinschaften und Unterricht. *Der Biologe*, 10: 218–219.

Dietrich, Michael R. 1996. On the mutability of genes and geneticists: The 'Americanization' of Richard Goldschmidt and Victor Jollos. *Perspectives on Science*, 4: 321–345.

Dingler, Hugo. 1926. *Der Zusammenbruch der Wissenschaft und der Primat der Philosophie*. Munich, Germany: Ernst Reinhardt.

Dingler, Hugo. 1943. Über den Kern einer fruchtbaren Diskussion über die moderne theoretische Physik. *Zeitschrift für die gesamte Naturwissenschaft*, 99: 212–221.

Dingus, Lowell and Timothy Rowe. 1998. *The Mistaken Extinction. Dinosaur Evolution and the Origin of Birds*. New York: W.H. Freeman & Co.

Dobzhansky, Theodosius. 1937. *Genetics and the Origin of Species*. New York: Columbia University Press.

Dolezal, Helmut. 1969. Hayek, August Gustav Joseph Endler v., pp. 151–152. In: Wagner, F. (Ed.), *Neue Deutsche Biographie*, vol. 8. Berlin, Germany: Duncker und Humblot.

Dollo, Louis. 1895. Sur la phylogénie des dipneustes. *Bulletin de la Société Belge de Géologie, de Paléontologie, et d'Hydrologie*, 9: 79–128.

Dollo, Louis. 1909. La paléontologie éthologique. *Bulletin de la Société Belge de Géologie, de Paléontologie, et d'Hydrologie*, 23: 377–421.

Donoghue, Michael J. and Joachim W. Kadereit. 1992. Walter Zimmermann and the growth of phylogenetic theory. *Systematic Biology*, 41: 74–85.

Drevermann, Fritz. 1929. Otto Jaekel †. *Palaeontologische Zeitschrift*, 11: 183–184.

Driesch, Hans. 1890a. Tektonische Studien an Hydroidpolypen. I. Die Campanulariden und Sertulariden. *Jenaische Zeitschrift für Naturwissenschaft*, 24 (N.F. 17), 189–225.

Driesch, Hans. 1890b. Tektonische Studien an Hydroidpolypen II. Plumularia und Aglaophenia. Die Tubulariden. *Jenaische Zeitschrift für Naturwissenschaft*, 24 (N.F. 17), 657–688.

Driesch, Hans. 1892. Entwicklungsmechanische Studien. I. Der Werth der beiden ersten Furchungszellen. Experimentelle Erzeugung von Theil- und Doppelbildungen. *Zeitschrift für wissenschaftliche Zoologie*, 53: 160–178.

Driesch, Hans. 1893a. Entwicklungsmechanische Studien. VI. Über einige allgemeine Fragen der Theoretischen Morphologie. *Zeitschrift für wissenschaftliche Zoologie*, 55: 34–62.

Driesch, Hans. 1893b. Entwicklungsmechanische Studien. IV. Experimentelle Veränderung des Typus der Furchung und ihre Folgen (Wirkungen von Wärmezufuhr und von Druck). *Zeitschrift für wissenschaftliche Zoologie*, 55: 10–29.

Driesch, Hans. 1894. Analytische Theorie der organischen Entwicklung. Leipzig, Germany: Wilhelm Engelmann.

Driesch, Hans. 1899a. Von der Methode der Morphologie. Kritische Erörterungen. *Biologisches Centralblatt*, 19: 33–58.

Driesch, Hans. 1899b. Die Lokalisation morphogenetischer Vorgänge. *Ein Beweis vitalistischen Geschehens*. Leipzig, Germany: Wilhelm Engelmann.

Driesch, Hans. 1910. *Zwei Vorträge zur Naturphilosophie. I. Die logische Rechtfertigung der Lehre von der Eigengesetzlichkeit des Belebten. II. Über Aufgabe und Begriff der Naturphilosophie*. Leipzig, Germany: Wilhelm Engelmann.

Driesch, Hans. 1911. *Die Biologie als selsbtändige Grundwissenschaft und das System der Biololgie*. Zweite Auflage. Leipzig, Germany: Wilhelm Engelmann.

Driesch, Hans. 1914. *The Problem of Individuality*. London: Macmillan.

Driesch, Hans. 1920. Wilhelm Roux als Theoretiker. *Die Naturwissenschaften*, 8: 446–450.

Driesch, Hans. 1921. *Philosophie des Organischen. Gifford-Vorlesungen gehalten an der Universität Aberdeen in den Jahren 1907–1908*. 2. Auflage. Leipzig, Germany: Wilhelm Engelmann.

Driesch, Hans. 1935. Zur Kritik des Holismus. *Acta Biotheoretica*, 1: 185–202.

Driesch, Hans. 1951. *Lebenserinnerungen. Aufzeichnungen eines Forschers und Denkers in entscheidender Zeit*. Basel, Switzerland: Ernst Reinhardt.

Driesch, Margarete. 1951. Das Leben von Hans Driesch, pp. 7–10. In: Wenzl, A. (Ed.), *Hans Driesch. Persönlichkeit und Bedeutung für Biologie und Philosophie von Heute*. Basel, Switzerland: Ernst Reinhardt.

Du Bois-Reymond, Emil. 1872. *Über die Grenzen des Naturerkennens*. Leipzig, Germany: Veit & Co.

Du Bois-Reymond, Emil. 1881. *Über die Übung. Rede gehalten zur Feier des Stiftungstages der militärärztlichen Bildungs-Anstalten am 2. August 1881*. Berlin, Germany: Gustav Lange (Paul Lange).

Dücker, Gerti. 2003. Rensch, Bernhard Karl Emanuel, *Neue Deutsche Biographie*, 21: 437–438 [Onlinefassung]; http://www.deutsche-biographie.de/pnd118599771.html (accessed August 5, 2013).

Duméril, Constant. 1805. *Zoologie analytique, ou method naturelle de classification des animaux, rendue plus facile à l'aide de tableaux synoptiques*. Paris, France: Allais.

Dupuis, Claude. 1984. Willi Hennigs impact on taxonomic thought. *Annual Review of Ecology and Systematics*, 15: 1–24.

Dupuis, Claude. 1990. Hennig, Emil Hans Willi, pp. 407–410. In: Holmes, F.L. (Ed.), *Dictionary of Scientific Biography*, vol. 17, Supplement II. New York: Charles Scribner's Sons.

Dürken, Bernhard. 1935. Das Verhältnis der Teile zum Ganzen im Organismus. *Zeitschrift für mathematischen und naturwissenschaftlichen Unterricht aller Schulgattungen*, 66: 57–65.

Dürken, Bernhard. 1936. *Entwicklungsbiologie und Ganzheit. Ein Beitrag zur Neugestaltung des Weltbildes*. Leipzig, Germany: B.G. Teubner.

Dürken, Bernhard and Hans Salfeld. 1921. *Die Phylogenese; Fragestellungen zu ihrer exakten Erforschung*. Berlin, Germany: Gebr. Bornträger.

Ehrenberg, Kurt. 1975. *Othenio Abels Lebensweg. Unter Benützung autobiographischer Aufzeichnungen*. Wien, Austria: Im Selbstverlag.

Ehrenberg, Rudolf. 1929. Über das Problem einer "theoretischen Biologie." *Die Naturwissenschaften*, 17: 777–781.

Ehrenfels, Christian von. 1890. Über Gestaltqualitäten. *Vierteljahrsschrift für wissenschaftliche Philosophie*, 14: 242–292.

Eickstedt, Egon von. 1935. Einführung. *Zeitschrift für Rassenkunde und ihre Nachbargebiete*, 1: 1–2.

Eisig, Hugo. 1916. Arnold Lang und die Zoologische Station in Neapel, 1878–1885, pp. 56–126. In: Haeckel, E., K. Hescheler, and H. Eisig. *Aus dem Leben und Wirken von Arnold Lang. Dem Andenken des Freundes und Lehrers gewidmet*. Jena, Germany: Gustav Fischer.

Ellis, Brian D. 2002. *The Philosophy of Nature. A Guide to the New Essentialism*. Montreal, Québec, Canada: McGill-Queen's University Press.

Elz, Wolfgang. 2009. Foreign policy, pp. 50–77. In: McElligott, A. (Ed.), *The Short Oxford History of Germany: Weimar Germany*, Oxford: Oxford University Press.

Ernst, Paul. 1926. Das morphologische Bedürfnis. *Die Naturwissenschaften*, 14: 1075–1080.

Escherich, Karl L. 1934. *Termitenwahn. Eine Münchner Rektoratsrede über die Erziehung zum politischen Menschen*. Munich, Germany: G. Müller.

Federley, Harry. 1929. Weshalb lehnt die Genetik die Annahme einer Vererbung erworbener Eigenschaften ab? *Palaeontologische Zeitschrift*, 11: 287–310.

Feuerborn, Heinrich J. 1935. Das Kernstück der deutschen Volksbildung: die Biologie. *Der Biologe*, 4: 99–105.

Flachowsky, Sören. 2010. "Werkzeuge der deutschen Kriegsführung." Die Forschungspolitik der Deutschen Forschungsgemeinschaft und des Reichsforschungsrates zwischen 1920 und 1945, pp. 53–69. In: Orth, K. and W. Oberkrome (Eds.), *Die Deutsche Forschungsgemeinschaft 1920–1970. Forschungsförderung im Spannungsfeld von Wissenschaft und Politik*. Stuttgart, Germany: Franz Steiner.

Forman, Paul. 1971. Weimar Culture, causality, and quantum theory, 1918–1927: Adaptation by German physicists and mathematicians to a hostile intellectual environment. *Historical Studies in the Physical Sciences*, 3: 1–115.

Fowler, Henry H. 1945. *German Scientific Research and Engineering, from the Standpoint of International Security. T.I.D.C. Project 3*. Washington, DC: The National Academy of Sciences.

Friederichs, Karl. 1927. Grundsätzliches über die Lebenseinheiten höherer Ordnung und den ökologischen Einheitsfaktor. *Die Naturwissenschaften*, 15: 153–157; 182–186.

Friederichs, Karl. 1937. *Ökologie als Wissenschaft von der Natur*. Leipzig, Germany: J.A. Barth.

Friedman, Michael. 1999. *Reconsidering Logical Positivism*. Cambridge, UK: Cambridge University Press.

Friedmann, Michael. 2000. *A Parting of the Ways. Carnap, Cassirer, and Heidegger*. Chicago, IL: Open Court.

Froese, H. 1937/38. Neuere Anschauungen in der Biologie. Arbeitsgemeinschaft der Fachgruppe Naturwissenschaft der Studentenschaft der Universität München. *Zeitschrift für die gesamte Naturwissenschaft*, 3: 304–306.

Fürbringer, Max. 1903a. Carl Gegenbaur, pp. 389–454. In: *Heidelberger Professoren aus dem 19. Jahrhundert*, vol. 2. Heidelberg, Germany: Carl Winter.

Fürbringer, Max. 1903b. Carl Gegenbaur. *Anatomischer Anzeiger* 23: 589–608.

Galison, Peter. 1990. Aufbau/Bauhaus: Logical positivism and architectural modernism. *Critical Inquiry*, 16: 709–752.

Gardner, Sebastian. 1999/2008. *Kant, and the Critique of Pure Reason*. London: Routledge.

Garstang, Walter. 1922. The theory of recapitulation: A critical restatement of the Biogenetic Law. *Proceedings of the Linnean Society of London*, 35: 81–101.

Gasman, Daniel. 1998. *Haeckel's Monism and the Birth of Fascist Ideology*. New York: Peter Lang.

Gasman, Daniel. 2002. Haeckel's scientific monism as a theory of history. *Theory in Biosciences*, 121: 260–279.

Gasman, Daniel. 2004. *The Scientific Origins of National Socialism*. New Brunswick, NJ: Transaction Publishers.

Gegenbaur, Carl. 1859. *Grundzüge der vergleichenden Anatomie*. Leipzig, Germany: Wilhelm Engelmann.

Gegenbaur, Carl. 1865. Schultergürtel der Wirbelthiere, pp. 1–135. In: Gegenbaur, C. (Ed.), *Untersuchungen zur vergleichenden Anatomie der Wirbelthiere, zweites Heft*. Leipzig, Germany: Wilhelm Engelmann.

Gegenbaur, Carl. 1870a. *Grundzüge der vergleichenden Anatomie*. Zweite, umgearbeitete Auflage. Leipzig, Germany: Wilhelm Engelmann.

Gegenbaur, Carl. 1870b. Ueber das Skelet der Gliedmassen der Wirbelthiere im Allgemeinen und der Hintergliedmassen der Selachier insbesondere. *Jenaische Zeitschrift für Medizin und Naturwissenschaft*, 5: 397–447. Reprinted in Fürbringer, M. and H. Bluntschli. 1912. *Gesammelte Abhandlungen von Carl Gegenbaur, Band II*. Leipzig, Germany: Wilhelm Engelmann.

Gegenbaur, Carl. 1872. Ueber das Archipterygium. *Jenaische Zeitschrift für Medizin und Naturwissenschaft*, 7: 131–141. Reprinted in Fürbringer, M. and H. Bluntschli. 1912. *Gesammelte Abhandlungen von Carl Gegenbaur, Band II*. Leipzig, Germany: Wilhelm Engelmann.

Gegenbaur, Carl. 1874. *Grundriss der vergleichenden Anatomie*. Leipzig, Germany: Wilhelm Engelmann.

Gegenbaur, Carl. 1876. Die Stellung und Bedeutung der Morphologie. *Morphologisches Jahrbuch. Eine Zeitschrift für Anatomie und Entwickelungsgeschichte*, 1: 1–19.

Gegenbaur, Carl. 1878. *Grundriss der vergleichenden Anatomie*. Zweite verbesserte Auflage. Leipzig, Germany: Wilhelm Engelmann.

Gegenbaur, Carl. 1879. Zur Gliedmassenfrage. An die Untersuchungen v. Davidoff's angeknüpfte Bemerkungen. *Morphologisches Jahrbuch*, 5: 521–515.

Gegenbaur, Carl. 1889. Ontogenie und Anatomie, in ihren Wechselwirkungen betrachetet. *Morphologisches Jahrbuch*, 15: 1–9.

Gegenbaur, Carl. 1898. *Vergleichende Anatomie der Wirbelthiere, mit Berücksichtigung der Wirbellosen. Erster Band. Einleitung, Integument, Skeletsystem, Muskelsystem, Nervensystem und Sinnesorgane*. Leipzig, Germany: Wilhelm Engelmann.

Gegenbaur, Carl. 1901. *Erlebtes und Erstrebtes*. Leipzig, Germany: Wilhelm Engelmann.

Gerhard, Gesine. 2005. Breeding pigs and people for the Third Reich. Richard Walther Darré's agrarian ideology, pp. 129–146. In: Brüggemeier, F.-J., M. Cioc, and T. Zeller (Eds.), *How Green Were the Nazis? Nature, Environment, and Nation in the Third Reich*. Athens, OH: Ohio University Press.

Gerstengarbe, Sybille, Heidrun Hallmann, and Wieland Berg. 1995. Die Leopoldina im Dritten Reich, pp. 168–204. In: Scriba, C.J. (Ed.), *Die Elite der Nation im Dritten Reich. Das Verhältnis von Akademien und ihrem wissenschaftlichen Umfeld zum Nationalsozialismus*. Halle (Saale), Germany: Deutsche Akademie der Naturforscher Leopoldina.

Geus, Armin. 2010. *Natura infinita est*—Artbegriff und Artwandel bei Anton Friedrich Spring (1814–1872), pp. 17–34. In: Jahn, I. and A. Wessel (Eds.), *For a Philosophy of Biology*. Berliner Studien zur Wissenschaftsphilosophie und Humanontogenetik, 26. Munich, Germany: Kleine Verlag.

Ghiselin, Michael. 1974. A radical solution to the species problem. *Systematic Zoology*, 23, 536–544.

Ghiselin, Michael. 1997. *Metaphysics and the Origin of Species*. Albany, NY: SUNY Press.

Gicklhorn, Josef. 1957. Claus, Carl Friedrich, Zoologe, pp. 268–269. In: Stollberg-Wernigerode, O. (Ed.), *Neue Deutsche Biographie*, vol. 3. Berlin, Germany: Duncker und Humblot.

Gillis, J. Andrew, Randall D. Dahn, and Neil H. Shubin. 2009. Shared developmental mechanisms pattern the vertebrate gill arch and paired fin skeletons. *Proceedings of the National Academy of Sciences*, 106: 5720–5724.

Gilmour, John S.L. 1940. Taxonomy and philosophy, pp. 461–474. In: Huxley, J. (Ed.), *The New Systematics*. Oxford: Oxford University Press.

Glaubrecht, Matthias. 2007. Die Ordnung des Lebendigen. Zur Geschichte und Zukunft der Zoosystematik in Deutschland, pp. 59–110. In: Wägele, J.W. (Ed.), *Höhepunkte der zoologischen Forschung im deutschen Sprachraum*. Marburg, Germany: Basilisken Presse.

Gliboff, Sander. 2008. *H.G. Bronn, Ernst Haeckel, and the Origins of German Darwinism. A Study in Translations and Transformation*. Cambridge, MA: MIT Press.

Goebel, Karl von. 1884. Vergleichende Entwicklungsgeschichte der Pflanzenorgane, pp. 99–430. In: Schenk, A. (Ed.), *Handbuch der Botanik. Dritter Band, Erste Hälfte*. Breslau, Poland: Eduard Trewendt.

Goebel, Karl von. 1905. Die Grundprobleme der heutigen Pflanzenmorphologie. *Biologisches Centralblatt*, 25: 66–83.

Goette, Alexander. 1875. *Die Entwickelungsgeschichte der Unke (Bombinator ignaeus), als Grundlage einer vergleichenden Morphologie der Wirbelthiere.* Leipzig, Germany: Leopold Voss.

Goodrich, Edwin S. 1930. *Studies on the Structure and Development of Vertebrates.* London: Macmillan & Co.

Gould, Steven J. 1977. *Ontogeny and Phylogeny.* Cambridge, MA: Belknap Press at Harvard University Press.

Gould, Steven J. 1989. *Wonderful Life: The Burgess Shale and the Nature of History.* New York: W.W. Norton.

Greite, Walter. 1939a. Zum Geleit. *Der Biologe*, 8: 1–2.

Greite, Walter. 1939b. Aufbau und Aufgaben des Reichsbundes für Biologie. *Der Biologe*, 9: 1.

Greite, Walter. 1940. Zum Geleit. *Der Biologe*, 8: 233–241.

Griffiths, Paul E. 1999. Squaring the circle: Natural kinds with historical essences, pp. 209–228. In: Wilson, R.A. (Ed.), *Species. New Interdisciplinary Essays.* Cambridge, MA: MIT Press.

Gross, Walter R. 1943. Paläontologische Hypothesen zur Faktorenfrage der Deszendenzlehre. Über die Typen- und Phasenlehren von Schindewolf und Beurlen. *Die Naturwissenschaften*, 31: 237–245.

Grün, Bernd. 2010. Die Medizinische Fakultät Tübingen im Nationalsozialismus, pp. 239–277. In: Wiesing, U., K.-R. Brintzinger, B. Grün, H. Junginger, and S. Michl (Eds.), *Die Universität Tübingen im Nationalsozialismus.* Stuttgart, Germany: Franz Steiner Verlag.

Grüttner, Michael. 2004. *Biographisches Lexikon zur nationalsozialistischen Wissenschaftspolitik.* Heidelberg, Germany: Synchron Publishers.

Günther, Klaus. 1956. Systematik und Stammesgeschichte der Tiere, 1939–1953. *Fortschritte der Zoologie*, N.F. 10: 33–278.

Günther, Klaus. 1962. Systematik und Stammesgeschichte der Tiere, 1954–1959. *Fortschritte der Zoologie*, N.F. 14: 268–547.

Günther, Klaus. 1974. Erwin Stresemann (22. XI. 1889–20. XI. 1972). *Sizungsberichte der Gesellschaft Naturforschender Freunde zu Berlin*, N.F. 14: 37–42.

Gütt, Arthur. 1937. Bevölkerungspolitik und Biologie. *Der Biologe*, 6: 375–379.

Haase-Bessell, Gertraud. 1941a. Evolution. *Der Biologe*, 10: 233–247.

Haase-Bessell, Gertraud. 1941b. *Der Evolutionsgedanke in seiner heutigen Fassung.* Jena, Germany: Gustav Fischer.

Haeckel, Ernst. 1862. *Die Radiolarien (Rhizopoda Radiolaria). Eine Monographie*, vol. 1. Berlin, Germany: Georg Reimer.

Haeckel, Ernst. 1864. Über die Entwickelungstheorie Darwins. *Amtlicher Bericht über die Versammlung Deutscher Naturforscher und Ärzte*, 38: 17–30.

Haeckel, Ernst. 1866. *Generelle Morphologie der Organismen*; 2 vols. Berlin, Germany: Georg Reimer.

Haeckel, Ernst. 1868. *Natürliche Schöpfungsgeschichte.* Berlin, Germany: Georg Reimer.

Haeckel, Ernst. 1869. *Zur Entwicklungsgeschichte der Siphonophoren.* Utrecht, the Netherlands: C. van der Post Jr.

Haeckel, Ernst. 1870a. *Natürliche Schöpfungsgeschichte.* Zweite, verbesserte und vermehrte Auflage. Berlin, Germany: Georg Reimer.

Haeckel, Ernst. 1870b. Ueber Entwicklungsgang und Aufgabe der Zoologie. Rede gehalten beim Eintritt in die philosophische Facultät zu Jena am 12. Januar 1869. *Jenaische Zeitschrift für Medicin und Naturwissenschaft*, 5: 353–370.

Haeckel, Ernst. 1872a. *Natürliche Schöpfungsgeschichte.* Dritte verbesserte Auflage. Berlin, Germany: Georg Reimer.

Haeckel, Ernst. 1872b. *Die Kalkschwämme. Eine Monographie.* Erster Band (Genereller Theil). Berlin, Germany: Georg Reimer.

Haeckel, Ernst. 1873. *Natürliche Schöpfungsgeschichte.* Vierte verbesserte Auflage. Berlin, Germany: Georg Reimer.

Haeckel, Ernst. 1874a. Die Gastraea-Theorie, die phylogenetische Classification des Thierreichs und die Homologie der Keimblätter. *Jenaische Zeitschrift für Medizin und Naturwissenschaft*, N.F. 8: 1–55.

Haeckel, Ernst. 1874b. *Anthropogenie. Keimes- und Stammesgeschichte des Menschen.* Leipzig, Germany: Wilhelm Engelmann.

Haeckel, Ernst. 1875a. *Natürliche Schöpfungsgeschichte.* Sechste verbesserte Auflage. Berlin, Germany: Georg Reimer.

Haeckel, Ernst. 1875b. *Ziele und Wege der heutigen Entwickelungsgeschichte.* Jena, Germany: Hermann Dufft.

Haeckel, Ernst. 1875c. Die Gastrula und die Eifurchung der Thiere. *Jenaische Zeitschrift für Medizin und Naturwissenschaft*, 9: 402–508.

Haeckel, Ernst. 1877a. Ueber die heutige Entwickelungslehre im Verhältnisse zur Gesamtwissenschaft. *Verhandlungen der Gesellschaft Deutscher Naturforscher und Ärzte*, 50: 14–22.

Haeckel, Ernst. 1877b. *Anthropogenie oder Entwickelungsgeschichte des Menschen.* Dritte umgearbeitete Auflage. Leipzig, Germany: Wilhelm Engelmann.

Haeckel, Ernst. 1878a. *Freie Wissenschaft und freie Lehre. Eine Entgegnung auf Rudolf Virchow's Münchener Rede über 'Die Freiheit der Wissenschaft im modernen Staat'.* Stuttgart, Germany: E. Schweizerbart.

Haeckel, Ernst. 1878b. Zellseelen und Seelenzellen, pp. 143–181. In: Haeckel, E. (Ed.), *Gesammelte populäre Vorträge aus dem Gebiete der Entwicklungslehre, erstes Heft.* Bonn, Germany: Emil Strauss.

Haeckel, Ernst. 1879a. *Freedom in Science and Teaching.* New York: D. Appleton & Co.

Haeckel, Ernst. 1879b. Ueber die Wellenzeugung der Lebenstheilchen oder die Perigenesis der Plastidule, pp. 24–79. In: Haeckel, E. (Ed.), *Gesammelte populäre Vorträge aus dem Gebiete der Entwickelungslehre, zweites Heft.* Bonn, Germany: Emil Strauss.

Haeckel, Ernst. 1887. *Die Radiolarien (Rhizopoda Radiaria). Eine Monographie. Zweiter Theil.* Berlin, Germany: Georg Reimer.

Haeckel, Ernst. 1888. *Die Radiolarien (Rhizopoda Radiaria). Eine Monographie. Dritter Theil.* Berlin, Germany: Georg Reimer.

Haeckel, Ernst. 1894. *Systematische Phylogenie. Entwurf eines Natürlichen Systems der Organismen auf Grund ihrer Stammesgeschichte. Erster Theil: Systematische Phylogenie der Protisten und Pflanzen.* Berlin, Germany: Georg Reimer.

Haeckel, Ernst. 1895. *Systematische Phylogenie. Entwurf eines Natürlichen Systems der Organismen auf Grund ihrer Stammesgeschichte. Dritter Theil: Systematische Phylogenie der Wirbelthiere.* Berlin, Germany: Georg Reimer.

Haeckel, Ernst. 1896. *Systematische Phylogenie. Entwurf eines Natürlichen Systems der Organismen auf Grund ihrer Stammesgeschichte. Zweiter Theil: Systematische Phylogenie der wirbellosen Thiere.* Berlin, Germany: Georg Reimer.

Haeckel, Ernst. 1916. Arnold Lang, pp. 1–21. In: Haeckel, E., K. Hescheler, and H. Eisig. *Aus dem Leben und Wirken von Arnold Lang. Dem Andenken des Freundes und Lehrers gewidmet.* Jena, Germany: Gustav Fischer.

Haeckel, Werner. 1934. Zum Gleit! Ernst Haeckel und die Gegenwart. *Der Biologe*, 3: 33–34.

Haering, Theodor. 1935. Philosophie und Biologie. *Der Biologe*, 4: 393–397.

Haffer, Jürgen. 1991. Artbegriff und Artbegrenzung im Werk des Ornithologen Erwin Stresemann (1889–1972). *Mitteilungen aus dem Zoologischen Museum in Berlin*, 67 (Supplementheft, Annalen für Ornithologie, 15): 77–91.

Haffer, Jürgen. 1999. Beiträge zoologischer Systematiker und einiger Genetiker zur Evolutionären Synthese in Deutschland (1937–1950), pp. 121–150. In: Junker, T. and E.M. Engels. *Die Entstehung der Synthetischen Theorie. Beiträge zur Geschichte der Evolutionsbiologie in Deutschland. 1930–1950.* Berlin, Germany: VWB—Verlag für Wissenschaft und Bildung.

Haffer, Jürgen. 2003a. WILHELM MEISE (1901–2002), ein führender Ornithologe Deutschlands im 20. Jahrhundert. *Verhandlungen des Naturwissenschaftlichen Vereins in Hamburg* (N.F.) 40: 117–140.

Haffer, Jürgen. 2003b. In memoriam: Wilhelm Meise (1901–2002). *The Auk*, 120: 540.

Haffer, Jürgen, Erich Rutschke, and Klaus Wunderlich. 2000. Erwin Stresemann 1889–1972— Leben und Werk eines Pioniers der wissenschaftlichen Ornithologie. *Acta Historica Leopoldina*, 34: 1–465.

Haldane, John S. 1931. *The Philosophical Basis of Biology*. New York: Doubleday, Doran & Co.

Hall, Brian K. 2000. Balfour, Garstang and de Beer: The first century of evolutionary embryology. *American Zoologist*, 40: 718–728.

Hammarsten, Olof D., und John Runnström. 1926. Ein Beitrag zur Diskussion über die Verwandtschaftsbeziehungen der Mollusken. *Acta Zoologica*, 7: 1–67.

Handlirsch, Anton. 1901. Morphologisch-systematische Richtung mit Einschluss der Biologie und Thiergeographie, pp. 249–251. In: Handlirsch, A. and R.v. Wettstein (Eds.), *Botanik und Zoologie in Österreich in den Jahren 1850 bis 1900. Festschrift*. Wien, Austria: Alfred Hölder.

Handlirsch, Anton. 1903. Zur Phylogenie der Hexapoden (vorläufige Mitteilung). *Sitzungsberichte der kaiserlichen Akademie der Wissenschaften, mathematisch-naturwissenschaftliche Klasse*, 112: 716–738.

Handlirsch, Anton. 1905. Friedrich Moritz Brauer. *Verhandlungen der Zoologisch-Botanischen Gesellschaft in Wien*, 55: 129–166.

Handlirsch, Anton. 1908. *Die fossilen Insekten und die Phylogenie der rezenten Formen. Textband*. Leipzig, Germany: Wilhelm Engelmann.

Handlirsch, Anton. 1925a. Aus der Geschichte der Entomologie, pp. 1–21. In: Schröder, C. (Ed.), *Handbuch der Entomologie*, Band III. Jena, Germany: Gustav Fischer.

Handlirsch, Anton. 1925b. Die systematischen Grundbegriffe, pp. 61–78. In: Schröder, C. (Ed.), *Handbuch der Entomologie*, Band III. Jena, Germany: Gustav Fischer.

Harrington, Anne. 1996. *Reenchanted Science. Holism in German Culture from Wilhelm II to Hitler*. Princeton, NJ: Princeton University Press.

Harten, Hans-Christian, Uwe Neirich, and Matthias Schwerendt. 2006. *Rassenhygiene als Erziehungsideologie des Dritten Reichs: biographisches Handbuch*. Berlin, Germany: Akademie Verlag.

Hartmann, Max. 1950. Deutsche philosophisch-biologische Veröffentlichungen der Jahre 1939–1945. Philosophia Naturalis. *Archiv für Naturphilosophie und die philosophischen Grenzgebiete der exacten Wissenschaften und Wissenschaftsgeschichte*, 1: 132–139.

Hartmann, Nicolai. 1912. *Philosophische Grundfragen der Biologie*. Stuttgart, Germany: W. Kohlhammer.

Hartmann, Nicolai. 1942. *Systematische Philosophie*. Göttingen, Germany: Vandenhoeck & Ruprecht.

Harwood, Jonathan. 1993. *Styles of Scientific Thought. The German Genetics Community. 1900–1933*. Chicago, IL: University of Chicago Press.

Harwood, Jonathan. 1996. Weimar Culture and biological theory: A study of Roichard Woltereck (1877–1944). *History of Science*, 34: 347–377.

Hashagen, Ulf. 2010. Ein ausländischer Mathematiker im NS-Staat: Constantin Carathéodory als Professor an der Univerität München. *Deutsches Museum Preprints*, www.deutsches-museum.de/fileadmin/Content/.../Caratheodory15.9.10.pdf.

Hatschek, Berthold. 1888. *Lehrbuch der Zoologie. Eine morphologische Übersicht des Thierreiches zur Einführung in das Studium dieser Wissenschaft*. Jena, Germany: Gustav Fischer.

Heads, Michael. 2005. The history and philosophy of panbiogeography, pp. 67–123. In: Llorente, J. and J.J. Morrone (Eds.), *Regionalización Biogeográfica en Iberoamérica y Tópicos Afines*. Mexico City, Mexico: Universidad Nacional Autónoma de México.

Heberer, Gerhard. 1935. Michaelsen, W.: Das Wesen der Systematik, dem jungen Kollegen an dem Beispiel des modernen Oligochätensystems erläutert. Zool. Anz. CIX, 1–19, 1935. *Zeitschrift für Rassenkunde*, 1: 321.

Heberer, Gerhard. 1937a. Dürken, B.: Entwicklungsbiologie und Ganzheit. Ein Beitrag zur Neugestaltung desWeltbildes 1936. Leipzig, Teubner. 207 S, 56 Abb. Preis geh. RM 5.80, geb. RM 6.80. *Volk und Rasse*, 12: 271–272.

Heberer, Gerhard. 1937b. Brücher, Heinz: Ernst Haeckel's Bluts- und Geisteserbe. Eine kulturbiologische Monographie. Mit einem Geleitwort von Präs. Prof. Dr. C. Astel. Verl. Lehmann, München. 188 S, 15 Abb. U. 2 Sippschaftstafeln. Brosch. RM. 8.80, geb. RM. 10.- *Der Biologe*, 6: 65.

Heberer, Gerhard. 1937c, Neue Funde zur Urgeschichte des Menschen und ihre Bedeutung für Rassenkunde und Weltanschauung. *Volk und Rasse*, 12: 422–427, 435–444.

Heberer, Gerhard. 1939. Die gegenwärtigen Vorstellungen über den Stammbaum der Tiere und die "Systematische Phylogenie" Ernst Haeckel's. *Der Biologe*, 8: 264–273.

Heberer, Gerhard. 1942. Thienemann, A. Leben und Umwelt. Bios 12. 122 Seiten. J.H. Barth, Leipzig 1941. *Der Biologe*, 11: 168.

Heberer, Gerhard. 1943a. Das Typenproblem in der Stammesgechichte, pp. 545–585. In: Heberer, G. (Ed.), *Die Evolution der Organismen. Ergebnisse und Probleme der Stammesgeschichte*. Jena, Germany: Gustav Fischer.

Heberer, Gerhard. 1943b. Vorwort des Herausgebers, pp. iii–v. In: Heberer, G. (Ed.), *Die Evolution der Organismen. Ergebnisse und Probleme der Stammesgeschichte*. Jena, Germany: Gustav Fischer.

Heberer, Gerhard. 1943c. Über die Eindeutigkeit in der Darstellung wissenschaftlicher Auffassungen (Antwort an Herrn W. Troll). *Der Biologe*, 12: 253–255.

Heberer, Gerhard. 1956. Die Stellung Hugo Dingler's zur Evolutionstheorie, pp. 99–110. In: Krampf, W. (Ed.), *Hugo Dingler—Gedenkbuch zum 75. Geburtstag*. Munich, Germany: Eidos Verlag.

Heberer, Gerhard. 1959. Vorwort des Herausgebers, pp. vii–viii. In: Heberer, G. (Ed.), *Die Evolution der Organismen. Ergebnisse und Probleme der Abstammungslehre,* zweite Auflage. Stuttgart, Germany: Gustav Fischer.

Hecht, Gerhard. 1937/38. Biologie und Nationalsozialismus. *Zeitschrift für die gesamte Naturwissenschaft*, 3: 280–290.

Heiber, Helmut. 1991. *Universität unterm Hakenkreuz*. Teil I. Der Professor im Dritten Reich. Bilder aus der akademischen Provinz. Munich, Germany: K.G. Saur.

Heiber, Helmut. 1994. *Universität unterm Hakenkreuz*. Teil II. Kapitulation der Hohen Schulen. Das Jahr 1933 und seine Themen. Band 2. Munich, Germany: K.G. Saur.

Heidenhain, Martin. 1907. *Plasma und Zelle. Erste Abteilung. Allgemeine Anatomie der lebendigen Masse*. Jena, Germany: Gustav Fischer.

Heidenhain, Martin. 1920. Neue Grundlegungen zur Morphologie der Speicheldrüsen. *Anatomischer Anzeiger*, 52: 305–331.

Heidenhain, Martin. 1921. Über die teilungsfähigen Drüseneinheiten oder Adenomeren. Sowie über die Grundbegriffe der morphologischen Systemlehre. *Wilhelm Roux' Archiv für Entwicklungsmechanik*, 49: 1–178.

Heidenhain, Martin. 1923. *Formen und Kräfte in der lebendigen Natur*. Berlin, Germany: Springer.

Heidenhain, Martin. 1932. *Die Spaltungsgesetze der Blätter. Eine Untersuchung über Teilung und Synthese der Anlagen, Organisation und Formbildung sowie über die Theorie korrelativer Systeme. Beitrag XVI zur synthetischen Morphologie*. Jena, Germany: Gustav Fischer.

Heidenhain, Martin. 1937. *Synthetische Morphologie der Niere des Menschen. Bau und Entwicklung dargestellt auf neuer Grundlage*. Leiden, the Netherlands: E.J. Brill.

Heider, Karl. 1919. Ernst Haeckel. *Die Naturwissenschaften*, 7: 945–946.

Heisenberg, Werner. 1943. Die Bewertung der modernen theoretischen Physik. *Zeitschrift für die gesamte Naturwissenschaft*, 9: 201–212.

Hendel, Joachim, Uwe Hoßfeld, Jürgen John, Oliver Lehmuth, and Rüdiger Stutz. 2007. *Wege der Wissenschaft im Nationalsozialismus. Dokumente zur Universität Jena, 1933–1945.* Stuttgart, Germany: Franz Steiner.

Henig, Ruth. 1998. *The Weimar Republic, 1919–1933.* Abingdon, UK: Routledge.

Hennig, Edwin. 1916. Paläntologie und Entwicklungslehre. *Die Naturwissenschaften*, 4: 514–518.

Hennig, Edwin. 1932. *Wesen und Wege der Paläontologie. Eine Einführung in die Versteinerungslehre als Wissenschaft.* Berlin, Germany: Gebr. Bornträger.

Hennig, Edwin. 1937. Die Paläontologie in Deutschland. *Der Biologe*, 6: 1–6.

Hennig, Edwin. 1938. Eine neue Schau der Perm-Trias Landsaurier aus 4 Erdteilen. *Der Biologe*, 7: 300–303.

Hennig, Edwin. 1944. Organisches Werden, paläontologisch gesehen. *Paläontologische Zeitschrift*, 23: 281–316.

Hennig, Willi. 1931. Einiges über die Insketen des Landesschulgebietes. *Mitteilungen aus der Landesschule Dresden*, 8: 1–6.

Hennig, Willi. 1936a. Revision der Gattung *Draco* (Agamidae). Temminckia, 1: 153–220.

Hennig, Willi. 1936b. Über einige Gesetzmässigkeiten der geographischen Variation in der Reptiliengattung *Draco* L.: "parallele", "konvergente" Rassenbildung. *Biologisches Zentralblatt*, 56: 549–559.

Hennig, Willi. 1936c. Beziehungen zwischen geographischer Verbreitung und systematischer Gliederung bei einigen Dipterenfamilien: ein Beitrag zum Problem der Gliederung systematischer Kategorien höherer Ordnung. *Zoologischer Anzeiger*, 166: 161–175.

Hennig, Willi. 1943. Ein Beitrag zum Problem der "Beziehungen zwischen Larven- und Imaginalsystematik." *Arbeiten über morphologische und taxonomische Entomologie aus Berlin-Dahlem*, 10: 38–144.

Hennig, Willi. 1947. Probleme der biologischen Systematik. *Forschungen und Fortschritte. Nachrichten der deutschen Wissenschaft und Technik*, 21/23: 276–279.

Hennig, Willi. 1949. Zur Klärung einiger Begriffe der phylogenetischen Systematik. *Forschungen und Fortschritte. Nachrichten der deutschen Wissenschaft und Technik*, 25: 136–138.

Hennig, Willi. 1950. *Grundzüge einer Theorie der Phylogenetischen Systematik.* Berlin, Germany: Deutscher Zentralverlag.

Hennig, Willi. 1952. Autorreferat. *Beiträge zur Entomologie*, 2: 329–331.

Hennig, Willi. 1953. Kritische Bemerkungen zum phylogenetischen System der Insekten. *Beiträge zur Entomologie*, 3 (Beilageband): 1–85.

Hennig, Willi. 1955. Meinungsverschiedenheiten über das System der niederen Insekten. *Zoologischer Anzeiger*, 155: 21–30.

Hennig, Willi. 1957. Systematik und Phylogenese, pp. 50–71. In: Hannemann, H.-J. (Ed.), *Bericht über die Hundertjahrfeier der Deutschen Entomologischen Gesellschaft Berlin.* Berlin, Germany: Akademie Verlag.

Hennig, Willi. 1964. Diskussionsbemerkung. *Zoologischer Anzeiger*, 173: 63.

Hennig, Willi. 1965. Phylogenetic Systematics. *Annual Review of Entomology*, 10, 97–116.

Hennig, Willi. 1966. *Phylogenetic Systematics.* Urbana, IL: University of Illinois Press.

Hennig, Willi. 1974. Kritische Bemerkungen zur Frage "Cladistic analysis or cladistic classification?" *Zeitschrift für zoologische Systematik und Evolutionsforschung*, 12: 279–294.

Hennig, Willi. 1976. Klaus Günther. 7. 10. 1907 bis 1. 8. 1975. *Verhandlungen der Deutschen Zoologischen Gesellschaft*, 1976: 297–298.

Hennig, Willi. 1978/79. Die Stellung der Systematik in der Zoologie. *Entomologica Germanica*, 4: 193–199.

Henschel, Klaus. 1993. Bernhard Bavink (1879–1947): der Weg eines Naturphilosophen vom deutschnationalen Sympathisanten der NS-Bewegung bis zum unbequemen Non-Konformisten. *Sudhoffs Archiv*, 77: 1–32.

Herbst, Curt. 1924. Jacques LOEB. Ein kurzer Überblick über sein Lebenswerk. *Die Naturwissenschaften*, 12: 397–406.

Herbst, Curt. 1941. Hans Driesch (geb. 28. 10. 1867, gest. 16l 4. 1941). Als experimenteller und theoretischer Biologe. *Roux's Archiv für Entwicklungsmechanik der Organismen*, 141: 111–153.

Hermann, Armin. 1984. Naturwissenschaft und Technik im Dienste der Kriegswirtschaft, pp. 157–167. In: Tröger, J. (Ed.), *Hochschule und Wissenschaft im Dritten Reich*. Frankfurt, Germany: Campus Verlag.

Herre, Wolf. 1936. Ueber Rasse und Artbildung. Studien an Salamandriden. *Abhandlungen und Berichte aus dem Museum für Naturkunde und Vorgeschichte und dem naturwissenschaftlichen Verein zu Magdeburg*, 6: 193–221.

Herter, Konrad and Reinhard Bickerich. 1973. Die Mitglieder der Gesellschaft Naturforschender Freunde zu Berlin in den ersten 200 Jahren des Bestehens der Gesellschaft, 1773–1972. *Sitzungsberichte der Gesellschaft Naturforschender Freunde zu Berlin*, N.F. 13: 59–156.

Hertler, Chistine and Michael Weingarten. 2001. Ernst Haeckel (1834–1919), pp. 51–78. In: Jahn, J. and M. Schmitt (Eds.), *Darwin & Co. Eine Geschichte der Biologie in Portraits*, vol. 1. Munich, Germany: C.H. Beck.

Hertwig, Oscar. 1879. Die Geschichte der Zellentheorie. *Deutsche Rundschau*, 20: 417–429.

Hertwig, Oscar. 1892. *Lehrbuch der Zoologie*. Jena, Germany: Gustav Fischer.

Hertwig, Oscar. 1893. *Die Zelle und die Gewebe. Grundzüge der allgemeinen Anatomie und Physiologie*. Jena, Germany: Gustav Fischer.

Hertwig, Oscar. 1894. *Zeit- und Streitfragen der Biologie*. Heft 1. Präformation oder Epigenetik? Gründzüge einer Entwicklungstheorie der Organismen. Jena, Germany: Gustav Fischer.

Hertwig, Oscar. 1897. *Zeit- und Streitfragen der Biologie*. Heft 2. Mechanik und Biologie. Jena, Germany: Gustav Fischer.

Hertwig, Oscar. 1898. *Die Zelle und die Gewebe. Grundzüge der allgemeinen Anatomie und Physiologie*. Zweites Buch. Allgemeine Anatomie und Physiologie der Gewebe. Jena, Germany: Gustav Fischer.

Hertwig, Oscar. 1899. *Die Lehre vom Organismus und seine Beziehung zur Sozialwissenschaft*. Jena, Germany: Gustav Fischer.

Hertwig, Oscar. 1906a. Einleitung und allgemeine Literaturübersicht, pp. 1–85. In: Hertwig, O. (Ed.), *Handbuch der vergleichenden und experimentellen Entwicklungslehre der Wirbeltiere*, Band 1, Teil 1. Jena, Germany: Gustav Fischer.

Hertwig, Oscar. 1906b. *Allgemeine Biologie*. Zweite Auflage des Lehrbuchs "Die Zelle und die Gewebe." Jena, Germany: Gustav Fischer.

Hertwig, Oscar. 1909. Darwins Einfluss auf die deutsche Biologie. *Internationale Wochenschrift für Wissenschaft, Kunst und Technik*, 3(31): 953–958.

Hertwig, Oscar. 1917. Das genealogische Netzwerk und seine Bedeutung für die Frage der monophyletischen oder der polyphyletischen Abstammungshypothese. *Archiv für Mikroskopische Anatomie*, 89: 227–242.

Hertwig, Oscar. 1918a. *Das Werden der Organismen. Zur Widerlegung von Darwins Zufallstheorie durch das Gesetz in der Entwicklung*, 2. Auflage. Jena, Germany: Gustav Fischer.

Hertwig, Oscar. 1918b. *Zur Abwehr des ethischen, des sozialen, des politischen Darwinismus*. Jena, Germany: Gustav Fischer.

Hertwig, Oscar. 1922. *Der Staat als Organismus. Gedanken zur Entwicklung der Menschheit*. Jena, Germany: Gustav Fischer.

Hertwig, Richard. 1919. Haeckels Verdienste um die Zoologie. *Die Naturwissenschaften*, 7: 951–958.

Hescheler, Karl. 1916. Arnold Lang in Zürich, 1889–1914, pp. 127–266. In: Haeckel, E., K. Hescheler, and H. Eisig. *Aus dem Leben und Wirken von Arnold Lang. Dem Andenken des Freundes und Lehrers gewidmet.* Jena, Germany: Gustav Fischer.

Heß, W. 1906. Müller, Johann Friedrich Theodor, pp. 516–518. In: *Allgemeine Deutsche Biographie,* vol. 52. Munich, Germany: Historische Kommission der Bayerischen Akademie der Wissenschaften. http://de.wikisource.org/w/index.php?title=ADB:M%C3%BCller,_Fritz&oldid=1688506.

Heydemann, Berndt. 1977. Zum Tode von Professor Dr. Dr. h.c. Adolf Remane. *Faunistisch-Ökologische Mitteilungen,* 5: 85–91.

Hicks, Stephen R.C. 2010. *Nietzsche and the Nazis. A Personal View.* Roscoe, IL: Ockham's Razor Publishing.

Hildebrandt, Kurt. 1935/36. Positivismus und Natur. *Zeitschrift für die gesamte Naturwissenschaft,* 1: 1–22.

Hildebrandt, Kurt. 1937/38. Die Bedeutung der Abstammungslehre für die Weltanschauung. *Zeitschrift für die gesamte Naturwissenschaft,* 3: 15–34.

Hitler, Adolf. 1941. *Mein Kampf. Complete and Unabriged, Fully Annotated.* New York: Reynal & Hitchcock.

Hoerschelmann, Heinrich and J. Neumann. 2003. Prof. Dr. Wilhelm Meise 12. 9. 1901–24. 8. 2002. *Journal für Ornithologie,* 144: 110–111.

Höffe, Otfried. 1981. Immanuel Kant (1724–1804), pp. 7–39. In: Höffe, O. (Ed.), *Klassiker der Philosophie,* vol. II. Munich, Germany: C.H. Beck.

Hoffmann, Friedrich. 1959. Fabricius, Johann Christian, pp. 736–737. In: Stollberg-Wernigerode, O. (Ed.), *Neue Deutsche Biographie,* vol. 4. Berlin, Germany: Duncker und Humblot.

Höflechner, Walter. 2007. Schulze, Franz Eilgard, Zoologe, Mediziner, pp. 723–724. In: Hockerts, H.G. (Ed.), *Neue Deutsche Biographie,* vol. 23. Berlin, Germany: Duncker und Humblot.

Hofmann, Christoph. 1941. Kampf um eine angewandte Wissenschaft. *Der Biologe,* 10: 414–416.

Hölder, Helmut. 1976. Ein halbes Hundert Bände. *Paläontologische Zeitschrift,* 50: 6–8.

Holler, Kurt. 1934a. Übersicht über die Nordische Bewegung im letzten Jahre. *Rasse. Monatsschrift der Nordischen Bewegung,* 1: 31–37.

Holler, Kurt. 1934b. Nationalsozialistisch getarnte Umweltlehre. *Rasse. Monatsschrift der Nordischen Bewegung,* 1: 37–38.

Hopster, Norbert. 1985. Ausbildung und politische Funktion der Deutschlehrer im Nationalsozialismus, pp. 113–139. In: Lundgreen, P. (Ed.), *Wissenschaft im Dritten Reich.* Frankfurt, Germany: Suhrkamp.

Hopwood, Nick. 2015. *Haeckel's Embryos. Images, Evolution, and Fraud.* Chicago, IL: The University of Chicago Press.

Horn, Wolfgang. 1968. Ein unbekannter Aufsatz Hitlers aus dem Frühjahr 1924. *Vierteljahreshefte für Zeitgeschichte,* 16: 280–294.

Hoßfeld, Uwe. 1997. *Gerhard Heberer (1901–1973). Sein Beitrag zur Biologie im 20. Jahrhundert.* Berlin, Germany: VWB—Verlag für Wissenschaft und Bildung.

Hoßfeld, Uwe. 1999a. Zoologie und Synthetische Theorie: Interview mit Wolf Herre, p. 241–257. In: Junker, T. and E.M. Engels. *Die Entstehung der Synthetischen Theorie. Beiträge zur Geschichte der Evolutionsbiologie in Deutschland. 1930–1950.* Berlin, Germany: VWB—Verlag für Wissenschaft und Bildung.

Hoßfeld, Uwe. 1999b. Die *Epilobium*-Kontroverse zwischen den Botanikern Heinz Brücher und Ernst Lehmann. *NTM International Journal of History & Ethics of Natural Sciences, Technology & Medicine,* 7: 140–160.

Hoßfeld, Uwe. 1999c. Die Moderne Synthese und *Die Evolution der Organismen,* pp. 189–225. In: Junker, T. and E.M. Engels. *Die Entstehung der Synthetischen Theorie. Beiträge zur Geschichte der Evolutionsbiologie in Deutschland. 1930–1950.* Berlin, Germany: VWB—Verlag für Wissenschaft und Bildung.

Hoβfeld, Uwe. 2000. Staatsbiologie, Rassenkunde und Moderne Synthese in Deutschland während der NS Zeit, pp. 249–305. In: Brömer, R. (Ed.), *Evolutionsbiologie von Darwin bis heute*. Berlin, Germany: VWB—Verlag für Wissenschaft und Bildung.

Hoβfeld, Uwe. 2003. Von der Rassenkunde, Rassenhygiene und biologischen Erbstatistik zur synthetischen Theorie der Evolution: eine Skizze der Biowissenschaften, pp. 519–574. In: Hoβfeld, U., J. John, O. Lemuth, and R. Stutz (Eds.), *"Kämpferische Wissenschaft." Studien zur Universität Jena im Nationalsozialismus*. Weimar, Germany: Böhlau Verlag.

Hoβfeld, Uwe. 2005a. *Geschichte der biologischen Anthropologie in Deutschland. Von den Anfängen bis in die Nachkriegszeit*. Stuttgart, Germany: Franz Steiner Verlag.

Hoβfeld, Uwe. 2005b. Nationalsozialistische Wissenschaftsinstrumentalisierung. Die Rolle von Karl Astel und Lothar Stengel von Rutkowski bei der Genese des Buches "Ernst Haeckel's Blut- und Geistes-Erbe" (1936), pp. 171–194. In: Krauβe, E. (Ed.), *Der Brief als wissenschaftshistorische Quelle*. Berlin, Germany: VWB—Verlag für Wissenschaft und Bildung.

Hoβfeld, Uwe. 2007. Haeckel als NS-Philosoph?, pp. 445–463. In: John, J. and J.H. Ulbricht (Eds.), *Jena, ein nationaler Erinnerungsort?* Weimar, Germany: Böhlau Verlag.

Hoβfeld, Uwe and Olaf Breidbach. 2005. Haeckels Politisierung der Biologie. *Thüringer Blätter zur Landeskunde*, 54.

Hoβfeld, Uwe and Thomas Junker. 2003. Anthropologie und synthetischer Darwinismus im Dritten Reich: "Die Evolution der Organismen." *Anthropologischer Anzeiger*, 61: 85–114.

Hoβfeld, Uwe and Lennart Olsson. 2003. The history of comparative anatomy in Jena—An overview. *Theory in Biosciences*, 122: 109–126.

Hoβfeld, Uwe and Lennart Olsson. 2006. Freedom of the mind got *Nature* banned by the Nazis. *Nature*, 443: 271.

Hoβfeld, Uwe, Lennart Olsson, and Olaf Breidbach. 2003. Editorial: Carl Gegenbaur (1826–1903) and his influence on the development of evolutionary morphology. *Theory in Biosciences*, 122: 105–108.

Hueck, Werner. 1926. Die Synthesiologie von Martin Heidenhain als Versuch einer allgemeinen Theorie der Organisation. *Die Naturwissenschaften*, 14: 149–158.

Hull, David L. 1973. *Darwin and His Critics. The Reception of Darwin's Theory of Evolution by the Scientific Community*. Cambridge, MA: Harvard University Press.

Hull, David L. 1976. Are species really individuals. *Systematic Zoology* 25: 174–191.

Hull, David L. 1979. The limits of cladism. *Systematic Zoology*, 28: 416–440.

Hull, David L. 1988. *Science as a Process. An Evolutionary Account of the Social and Conceptual Development of Science*. Chicago, IL: University of Chicago Press.

Hull, David L. 1989. *The Metaphysics of Evolution*. Albany, NY: State University of New York Press.

Hull, David L. 1999. On the plurality of species: Questioning the party line, pp. 23–48. In: Wilson, R.A. (Ed.), *Species. New Interdisciplinary Essays*. Cambridge, MA: MIT Press.

Huser-Bugmann, Karin. 1998. *Schtetl an der Sihl. Einwanderung, Leben und Alltag der Ostjuden in Zürich 1880–1939*. Zürich, Switzerland: Chronos.

Huser, Karin. 2005. Vom ersten Weltkrieg bis in die heutige Zeit, pp. 283–426. In: Bär, U. and M.R. Siegel (Eds.), *Geschichte der Juden im Kanton Zürich. Von den Anfängen bis in die heutige Zeit*. Zürich, Switzerland: Orell Füssli.

Hutton, Christopher. 2005. *Race and the Third Reich*. Cambridge, UK: Polity Press.

Huxley, Thomas H. 1879. Prefatory Note, pp. iii–xx. In: Haeckel, E. *Freedom in Science and Teaching*. New York: D. Appleton & Co.

Illies, Joachim. 1976. *Das Geheimnis des Lebendigen. Leben und Werk des Biologen Adolf Portmann*. Munich, Germany: Kindler.

Imort, Michael. 2005. Eternal Forest—Eternal Volk. The Rhetoric and Reality of National Socialist Forest Policy, pp. 43–72. In: Brüggemeier, F.-J., M. Cioc, and T. Zeller (Eds.), *How Green Were the Nazis? Nature, Environment, and Nation in the Third Reich*. Athens, OH: Ohio University Press.

Jacobj, Walther. 1952/53. Martin Heidenhain. *Anatomischer Anzeiger*, 99: 80–94.

Jacobshagen, Eduard. 1927. *Zur Reform der allgemeinen vergleichenden Formenlehre*. Jena, Germany: Gustav Fischer.

Jaekel, Otto. 1914a. Bericht über die Gründung und erste Jahresversammlung der Palaeontologischen Gesellschaft. *Palaeontologische Zeitschrift*, 1: 58–60.

Jaekel, Otto. 1914b. Wege und Ziele der Palaeontologie. *Palaeontologische Zeitschrift*, 1: 1–58.

Jaekel, Otto. 1916. *Die natürlichen Grundlagen staatlicher Organisation*. Kriegsausgabe—Im Selbstverlag des Verfassers (Bezug durch Georg Stilke—Berlin und Brüssel).

Jaekel, Otto. 1922. Funktion und Form in der organischen Entwicklung. *Palaeontologische Zeitschrift*, 4: 147–166.

Jahn, Ilse. 1966. Handlirsch, Anton, p. 608. In: Stollberg-Wernigerode, O. (Ed.), *Neue Deutsche Biographie*, vol. 7. Berlin, Germany: Duncker und Humblot.

Jahn, Ilse. 2001. Matthias Jacob Schleiden (1804–1881), pp. 139–156. In: Jahn, J. and M. Schmitt (Eds.), *Darwin & Co. Eine Geschichte der Biologie in Portraits*, vol. 1. Munich, Germany: C.H. Beck.

Jahn, Ilse, Wilhelm Meise, and Rolf Nöhring. 1973. Bibliographie der Publikationen von Erwin Stresemann und der von ihm angeregten Dissertationen, redigierten Zeitschriften und der Festschriften, die ihm gewidmet sind. *Journal für Ornithologie*, 114: 482–500.

Janavay, Christopher. 2002. *Schopenhauer. A Very Short Introduction*. Oxford: Oxford University Press.

Janik, Allan and Stephen E. Toulmin. 1973. *Wittgenstein's Vienna*. New York: Simon & Schuster.

Jatta, Guiseppe. 1896. 1. Cefalopodi viventi nel Golfo di Napoli (Sistematica). *Fauna und Flora des Golfes von Neapel und der angrenzenden Meeres-Abschnitte*, 23: 1–268.

Jax, Kurt. 1998. Holocoen and ecosystem—On the origin and historical consequences of two concepts. *Journal of the History of Biology* 31: 113–142.

Joergensen, Joergen. 1970. The development of logical empiricism, pp. 847–932. In: Neurath, O., R. Carnap, and C. Morris (Eds.), *Foundations of the Unity of Science*. Chicago, IL: University of Chicago Press.

Johannsen, Wilhelm. 1909. *Elemente der exackten Erblichkeitslehre: Deutsche wesentlich erweiterte Ausgabe in fünfundzwanzig Vorlesungen*. Jena, Germany: Gustav Fischer.

Junge, F. 1885. *Der Dorfteich als Lebensgemeinschaft*. Kiel, Germany: Lipsius & Tischer.

Junker, Thomas. 1996. Factors shaping Ernst Mayr's concepts in the history of biology. *Journal of the History of Biology*, 29: 29–77.

Junker, Thomas. 2000. Adolf Remane und die Synthetische Theorie, pp. 131–157. In: Höxtermann, E., J. Kaasch, M. Kaasch, and R.K. Kinzelbach (Eds.), *Berichte zur Geschichte der Hydro- und Meeresbiologie und weitere Beiträge zur 8. Jahrestagung der DGGTB in Rostock 1999*. Berlin, Germany: VWB—Verlag für Wissenschaft und Bildung.

Junker, Thomas. 2001a. Walter Zimmermann (1892–1980), pp. 275–295. In: Jahn, I. and M. Schmitt (Eds.), *Darwin & Co. Eine Geschichte der Biologie in Porträts*, vol. 2. Munich, Germany: C.H. Beck.

Junker, Thomas. 2001b. Wandte sich Bernhard Rensch in den Jahren 1934–38 aus politischen Gründen vom Lamarckismus ab?, pp. 287–311. In: Hoßfeld, U. and R. Brömer (Eds), *Darwinismus und/als Ideologie*. Berlin, Germany: Verlag für Wissenschaft und Bildung, VWB.

Junker, Thomas. 2004. *Die zweite Darwinsche Revolution. Geschichte des Synthetischen Darwinismus in Deutschland 1924 bis 1950*. Marburg, Germany: Basilisken Presse.

Junker, Thomas and Uwe Hoßfeld. 2002. The architects of the evolutionary synthesis in national socialist Germany: Science and politics. *Biology and Philosophy*, 17: 223–249.

Junker, Thomas and Hannelore Landsberg. 1994. Die zwei Tode eines Naturforschers. Der Weg Julius Schusters (1886–1949). *Medizinhistorisches Journal*, 29: 149–170.

Kaasch, Joachim and Michael Kaasch. 2003. Hallesche Naturwissenschaftler (Emil Abderhalden und Johannes Weigelt) in der Zeit des Nationalsozialismus. Eine Fallstudie, pp. 1027–1064. In: Hoßfeld U., J.J.O. Lemuth, and R. Stutz (Eds.), "*Kämpferische Wissenschaft.*" *Studien zur Universität Jena im Nationalsozialismus.* Köln, Germany: Böhlau Verlag.

Kaiser, Friedhelm. 1939. *Germanenkunde als politische Wissenschaft. Bericht über die Forschungs- und Lehrgemeinschaft 'das Ahnenerbe' 1939 zu Kiel.* Neumünster, Germany: Karl Wachholz.

Kangro, Hans. 2008. Müller, Johann Heinrich Jacob. Complete Dictionary of Scientific Biography. 2008. *Encyclopedia.com,* http://www.encyclopedia.com/doc/1G2-2830903080.html (accessed August 23, 2012).

Kant, Immanuel. 1786. *Metaphysische Anfangsgründe der Wissenschaft.* Riga, Latvia: J.F. Hartknoch.

Kant, Immanuel. 1919. *Kritik der Reinen Vernunft. Neu herausgegeben von Theodor Valentiner.* Hamburg, Germany: Felix Meiner.

Kater, Michael H. 1974. *Das Ahnenerbe der SS, 1935–1945. Ein Beitrag zur Kulturpolitik des Dritten Reiches.* Stuttgart, Germany: Deutsche Verlags-Anstalt.

Kelly, Alfred. 1981. *The Descent of Darwin. The Popularization of Darwinism in Germany, 1860–1914.* Chapel Hill, NC: The University of North Carolina Press.

Kirchner, Walter. 1984. Ursprünge und Konsequenzen rassistischer Biologie, pp. 77–91. In: Tröger, J. (Ed.), *Hochschule und Wissenschaft im Dritten Reich.* Frankfurt, Germany: Campus Verlag.

Kirkam, Richard L. 2001. *Theories of Truth. A Critical Introduction.* Cambridge, MA: MIT Press.

Kitcher, Philip. 1993. *The Advancement of Science. Science without Legend, Objectivity without Illusions.* Oxford: Oxford University Press.

Klee, Ernst. 2003. *Das Personenlexikon zum Dritten Reich. Wer war was vor und nach 1945.* Frankfurt, Germany: S. Fischer.

Klein, Gustav. 1932. Richard Wettstein. Ein Charakterbild. *Österreichische Botanische Zeitschrift,* 81: 1–4.

Kleinschmidt, Otto. 1900. Arten oder Formenkreise? *Journal für Ornithologie,* 48: 134–139.

Kleinschmidt, Otto. 1926a. *Die Formenkreislehre und das Weltwerden des Lebens.* Halle, Germany: Schwetschke.

Kleinschmidt, Otto. 1926b. Der weitere Ausbau der Formenkreislehre. *Journal für Ornithologie,* 74: 405–408.

Kleßmann, Christoph. 1985. Osteuropaforschung und Lebensraumpolitik im Dritten Reich, pp. 350–383. In: Lundgreen, P. (Ed.), *Wissenschaft im Dritten Reich.* Frankfurt, Germany: Suhrkamp.

Koechlin, F. 2004. Don Quijote der Laboratorien. Die Wochenzeitung (Basel), June 10, 2004.

Kowalewsky, Wladimir. 1873. Herrn CHARLES DARWIN in tiefster Verehrung gewidmet. *Palaeontographica,* 22: i–iv.

Kowalewsky, Wladimir. 1873/74. Monographie der Gattung *Anthracotherium. Palaeontographica,* 22: 131–210, 211–190, 291–346.

Kraus, Otto. 1984. Die Veranstaltung "Phylogenetisches Symposion": Rückblick auf 25 Tagungen (1955–1982). *Verhandlungen des Naturwissenschaftlichen Vereins in Hamburg* (N.F.) 27: 277–289.

Kraus, Otto and Uwe Hoßfeld. 1998. 40 Jahre "Phylogenetisches Symposium" (1956–1997): eine Übersicht—Anfänge, Entwicklung, Dokumentation und Wirkung. *Jahrbuch für Geschichte und Theorie der Biologie,* 5: 157–186.

Krumbach, Thilo. 1919. Ernst Haeckel's Person und Werk im Urteil der Zeitgenossen. *Die Naturwissenschaften,* 7: 966–971.

Kühn, Alfred. 1935. Karl Heider und ein Entwicklungsabschnitt der Zoologie. *Die Naturwissenschaften,* 23: 791–796.

Kuhn, Dorothea and Rike Wankmüller (Eds.). 1955. *Goethe. Naturwissenschaftliche Schriften.* Hamburg, Germany: Christian Wegner.

Kuhn, Thomas S. 1962. *The Structure of Scientific Revolutions.* Chicago, IL: The University of Chicago Press.

Kühne, Walter G. 1978/79. Willi Hennig 1913–1976: Die Schaffung einer Wissenschaftstheorie. *Entomologica Germanica*, 4: 374–376.

Kühnelt, Wilhelm. 1942. Prinzipien der Systematik, pp. 1–16. In: Bertalanffy, L. von (Ed.), *Handbuch der Biologie*, Band VI, Heft 1. Potsdam, Germany: Akademische Verlagsgesellschaft Athenaion.

Kühnl, Reinhard. 1984. Reichsdeutsche Geschichtswissenschaft, pp. 92–104. In: Tröger, J. (Ed.), *Hochschule und Wissenschaft im Dritten Reich.* Frankfurt, Germany: Campus Verlag.

Kuhn-Schnyder, Emil. 1976. Karl Ernst von Baer. Begründer der modernen Embryologie (1792–1876). *Acta Teilhardiana*, 13: 3–32.

Kuhn-Schnyder, Emil. 1982. Lang, Arnold, pp. 529–530. In: *Neue Deutsche Biographie.* Munich, Germany: Historische Kommission bei der Bayerischen Akademie der Wissenschaften. http://www.deutsche-biographie.de/pnd119435861.html.

Küppers, Horst. 2007. *Die Geschichte der Mineralogie in Kiel.* www.ifg.uni-kiel.de/AGs/Depmeier/MinKiel-2007.pdf (accessed December 4, 2013).

Kutschera, Ulrich. 2007. Paleobiology: The origin and evolution of a scientific discipline. *Trends in Ecology and Evolution*, 22: 172–173.

Lam, Herman J. 1936. Phylogenetic symbols, past and present. *Acta Biotheoretica*, 2: 153–194.

Lang, Arnold. 1887. *Mittel und Wege phylogenetischer Erkenntnisse.* Jena, Germany: Gustav Fischer.

Larson, Erik. 2011. *In the Garden of Beasts.* New York: Broadway Paperbacks.

Laubichler, Manfred D. 2003. Integrating comparative anatomy and embryology. *Journal of Experimental Zoology (Molecular and Developmental Evolution)*, 300B: 23–31.

Laves, Fritz. 1953. Paul Niggli, 26. Juni 1888 bis 13. Januar 1953. *Experientia*, 9: 197–198.

Lehmann, Ernst. 1931/32. Zur Einführung und Begründung. *Der Biologe*, 1: 1–5.

Lehmann, Ernst. 1934a. Uexküll, J. von. Staatsbiologie, zweite Auflage. Hamburg: Hanseatische Verlagsanstalt. *Der Biologe*, 3: 25.

Lehmann, Ernst. 1934b. Rektoratsreden zweier 'Biologen—Rektoren'. *Der Biologe*, 3: 59–60.

Lehmann, Ernst. 1934c. Zum 70. Geburtstage des Verlegers J. F. Lehmann. *Der Biologe*, 3: 305–307.

Lehmann, Ernst. 1934d. Schmidt, Heinrich. Ernst Haeckel, Denkmal eines grossen Lebens. Fromannsche Buchhandlung (Walter Biedermann), Jena 1934. Kart. RM. 2.80, geb. RM. 3.80. *Der Biologe*, 3: 132.

Lehmann, Ernst. 1935a. Nachruf auf Verleger Dr. Julius Friedrich Lehmann. *Der Biologe*, 4: 143.

Lehmann, Ernst. 1935b. Die Biologie an der Zeitwende. *Der Biologe*, 4: 375–381.

Lehmann, Ernst. 1935c. Nachruf auf Hans Schemm. *Der Biologe*, 4: 98.

Lehmann, Ernst. 1936a. Die Ecke des kinderreichen Biologen. *Der Biologe*, 5: 416.

Lehmann, Ernst. 1936b. Biologie und Bolschewismus. *Der Biologe*, 5: 160.

Lehmann, Ernst. 1936c. Die Aufgaben einer Deutschen Biologie. Kulturpolitik und Unterhaltung. Tägliches Beiblatt zum Völkischen Beobachter. 262. Ausgabe vom 18. September 1936.

Lehmann, Ernst. 1937. Dürken, Bernhard: Entwicklungsbiologie und Ganzheit. Ein Beitrag zur Neugestaltung des Weltbildes. Verl. G.B. Teubner, Leipzig u. Berlin, 1936. VI, 207 S. u. 56 Abb. Geh. RM.. 5.80, beg. RM. 6.80. *Der Biologe*, 6: 396–400.

Lengerken, 1934. Ernst Lehmann. 1934. Biologischer Wille. Wege und Ziele biologischer Arbeit im neuen Reich. J.F. Lehmanns Verlag, München. *Der Biologe*, 3: 244.

Lenoir, Timothy. 1982. *The Strategy of Life. Teleology and Mechanics in Nineteenth Century German Biology.* Chicago, IL: University of Chicago Press.

Lenz, Friedrich. 1943. Vorwort. *Archiv für Hydrobiologie*, 40 (August Thienemann Festband): v–viii.

Lenz, Fritz. 1921. Oskar Hertwigs Angriff gegen den 'Darwinismus' und die Rassenhygiene. *Archiv für Rassen- und Gesellschaftsbiologie*, 13: 194–203.

Lenz, Fritz. 1927. Ein deutsches Forschungsinstitut für Anthropologie, menschliche Erblehre und Eugenik. *Archiv für Rassen- und Gesellschaftsbiologie*, 19: 457–458.

Lenz, Fritz. 1934. Über Rassen und Rassenbildung. *Unterrichtsblätter für Mathematik und Naturwissenschaften*, 40: 177–189.

Leuckart, Rudolf. 1851. *Ueber den Polymorphismus der Individuen oder die Erscheinungen der Arbeitstheilung in der Natur*. Giessen, Germany: J. Ricker.

Levit, Gregory, Uwe Hoβfeld, and Lennart Olsson. 2004. The integration of Darwinism and evolutionay morphology: Alexej Nikolajevich Sewertzoff (1866–1936) and the developmental basis of evolutionary change. *Journal of Experimental Zoology (Molecular and Developmental Evolution)*, 302B: 343–345.

Levit, Gregory S. and Lennart Olsson. 2006. "Evolution on rails": Mechanisms and levels of orthogenesis. *Annals for the History and Philosophy of Biology*, 11: 97–136.

Lipsius, Friedrich. 1934. Ernst Haeckel als Naturphilosoph. *Der Biologe*, 3: 43–46.

Longerich, Peter. 2012. *Heinrich Himmler*; translated by Jeremy Noakes and Lesley Sharpe. Oxford: Oxford Univrsity Press.

Lorenz, Konrad. 1940. Nochmals: Systematik und Entwicklungsgedanke im Unterricht. *Der Biologe*, 9: 24–36.

Lorenz, Konrad. 1941. Vergleichende Bewegungsstudien an Anatinen. *Journal für Ornithologie*, 89 (Suppl.): 194–293.

Lorenz, Konrad. 1943. Psychologie und Stammesgeschichte, pp. 105–127. In: Heberer, G. (Ed.), *Die Evolution der Organismen. Ergebnisse und Probleme der Stammesgeschichte*. Jena, Germany: Gustav Fischer.

Löther, Rolf. 1972. *Die Beherrschung der Mannigfaltigkeit. Philosophische Grundlagen der Taxonomie*. Jena, Germany: VEB Gustav Fischer.

Löther, Rolf. 2010. Erinnerungen an meinen wissenschaftlichen Werdegang, pp. 81–95. In: Jahn, I. and A. Wessel (Eds.), *For a Philosophy of Biology*. Berliner Studien zur Wissenschaftsphilosophie und Humanontogenetik, 26. Munich, Germany: Kleine Verlag.

Lovejoy, Arthur. O 1936. *The Great Chain of Being*. Cambridge, MA: Harvard University Press.

Løvtrup, Soren 1978. On von Baerian and Haeckelian recapitulation. *Systematic Zoology*, 27: 348–352.

Lubosch, Wilhelm. 1926. Kritische Bemerkungen über den Begriff der Biologischen Morphologie. *Gegenbaurs Morphologisches Jahrbuch*, 55: 655–666.

Lubosch, Wilhelm. 1931. Geschichte der vergleichenden Anatomie, pp. 3–76. In: Bolk, L., E. Göppert, E. Kallius, and W. Lubosch (Eds.), *Handbuch der vergleichenden Anatomie der Wirbeltiere, Erster Band*. Berlin, Germany: Urban und Schwarzenberg.

Lundgreen, Peter. 1985. Hochschulpolitik und Wissenschaft im Dritten Reich, pp. 9–30. In: Lundgren, P. (Ed.), Wissenschaft im Dritten Reich. Frankfurt, Germany: Suhrkamp.

MacLeod, Miles and Thomas A.C. Reydon. 2013. Natural kinds in philosophy and in the life sciences: Scholastic twilight or new dawn? *Biological Theory*, 7: 89–99.

Mägdefrau, Karl. 1973. *Geschichte der Botanik. Leben und Leistung grosser Forscher*. Stuttgart, Germany: Gustav Fischer.

Maier, Wolfgang. 2008/09. Zur morphologischen und phylogenetischen Methodologie von Hermann Weber. *Entomologia Generalis*, 31: 113–117.

Mainx, F. 1971. Foundations of biology, pp. 568–654. In: Neurath, O., R. Carnap, and C. Morris (Eds.), *Foundations of the Unity of Science*, vol. 1. Chicago, IL: University of Chicago Press.

May, Eduard. 1937/38. K. Friederichs. Ökologie als Wissenschaft von der Natur oder biologische Raumforschung 1937. *Zeitschrift für die gesamte Naturwissenschaft*, 3: 486–487.

May, Walter. 1904. *Goethe, Humboldt, Darwin, Haeckel. Vier Vorträge*. Berlin, Germany: Enno Quehl.

Mayr, Ernst. 1942. *Systematics and the Origin of Species*. New York: Columbia University Press.

Mayr, Ernst. 1965. Numerical phenetics and taxonomic theory. *Systematic Zoology*, 14: 73–97.

Mayr, Ernst. 1968. Theory of biological classification. *Nature*, 220: 545–548.

Mayr, Ernst. 1974. Cladistic analysis or cladistic classification. *Zeitschrift für zoologische Systematik und Evolutionsforschung*, 12: 94–128.

Mayr, Ernst. 1980. Prologue: Some thoughts on the history of the evolutionary synthesis, pp. 1–48. In: Mayr, E. and Provine, W.B. (Eds.), *The Evolutionary Synthesis. Perspectives on the Unification of Biology*. Cambridge, MA: Harvard University Press.

Mayr, Ernst. 1982. *The Growth of Biological Thought. Diversity, Evolution, and Inheritance*. Cambridge, MA: The Belknap Press of Harvard University Press.

Mayr, Ernst. 1999. Thoughts on the evolutionary synthesis, pp. 19–29. In: Junker, T. and E.M. Engels. *Die Entstehung der Synthetischen Theorie. Beiträge zur Geschichte der Evolutionsbiologie in Deutschland. 1930–1950*. Berlin, Germany: VWB—Verlag für Wissenschaft und Bildung.

McIntosh, Robert P. 2011. The history of early British and US-American ecology to 1950, pp. 277–285. In: Schwarz, A. and K. Jax (Eds.), *Ecology Revisited: Reflecting on Concepts, Advancing Science*. Heidelberg, Germany: Springer.

McKinney, H. L. 1974. Müller, Fritz (Johann Friedrich Theodor), pp. 559–561. In: Gillispie, C.C. (Ed.), *Dictionary of Scientific Biography*, vol. 9. New York: Charles Scribner's Sons.

Meise, Wilhelm. 1928. Die Verbreitung der Aaskrähe (Formenkreis *Corvus corona* L.). *Journal für Ornithologie*, 766: 1–203.

Meise, Wilhelm and Willi Hennig. 1932. Die Schlangengattung *Dendrophis. Zoologischer Anzeiger*, 99: 273–297.

Meise, Wilhelm and Willi Hennig. 1935. Zur Kenntnis von *Dendrophis* und *Chrysopelea*— Ein Beitrag zur systematischen Bewertung der Opisthoglypha. *Zoologischer Anzeiger*, 109: 138–150.

Meister, Kay. 2005a. Wilhelm Troll (1897–1978). The tradition of idealistic morphology in the German botanical sciences of the 20th century. *History and Philosophy of Life Sciences*, 27: 221–247.

Meister, Kay. 2005b. Metaphysiscche Konsequenz—Die idealistische Morphologie Edgar Dacqués. *Neues Jahrbuch für Geologie und Paläntologie, Abhandlungen*, 235: 197–233.

Mertens, Lothar. 2004. *'Nur politisch Würdige'. Die DFG-Forschungsförderung im Dritten Reich 1933–1937*. Berlin, Germany: Akademie Verlag.

Mertens, Robert. 1953. Böker, p. 397. In: *Neue Deutsche Biologie. Zweiter Band*. Berlin, Germany: Duncker und Humblot.

Meyer (-Abich), Adolf. 1929. Das Wesen der idealistischen Biologie und ihre Beziehung zur modernen Biologie. *Archiv für Geschichte der Mathematik, der Naturwissenschaften und der Technik*, 11: 149–178.

Meyer (-Abich), Adolf. 1934. Die Axiome der Biologie. *Nova Acta Leopoldina*, N.F. 1: 474–551.

Michaelsen, Wilhelm. 1935. Das Wesen der Systematik, dem jungen Kollegen an dem Beispiel des modernen Oligochätensystems erläutert. *Zoologischer Anzeiger*, 109: 1–19.

Mildenberger, Florian. 2007. *Umwelt als Vision: Leben und Werk Jakob von Uexküll (1864–1944)*. Stuttgart, Germany: Franz Steiner (Sudhoffs Archiv, Beihefte; Heft 56).

Mildenberger, Florian. 2010. Die Geburt der Umwelt. Werk und Wirkung Jakob v. Uexküll (1864–1944), pp. 1–25. In: Hermann, B. (Ed.), *Beiträge zum Göttinger Umwelthistorischen Kolloquium 2009–2010*. Göttingen, Germany: Universiätsverlag.

Milne-Edwards, Henri. 1851. *Introduction à la zoologie générale*. Paris, France: V. Masson.

Möbius, Karl A. 1877. *Die Auster und die Austernwirtschaft*. Berlin, Germany: Wiegand, Hempel & Parey.

Möbius, Karl A. 1886. Die Bildung, Geltung und Bezeichnung der Artbegriffe und ihr Verhältnis zur Abstammungslehre. Zoologische Jahrbücher. *Zeitschrift für die Systematik, Geographie und Biologie der Thiere*, 1: 241–274.

Mocek. Reinhard. 2001. Wilhelm Roux (1850–1924), pp. 456–476. In: Jahn, J. and M. Schmitt (Eds.), *Darwin & Co. Eine Geschichte der Biologie in Portraits*. Vol. 1. Munich, Germany: C.H. Beck.

Möller, Alfred. 1915/1920/1921. *Fritz Müller, Werke, Briefe und Leben*, 3 vols. in 4. Jena, Germany: Gustav Fischer.

Mörike, Klaus D. 1988. *Geschichte der Tübinger Anatomie*. Tübingen, Germany: J.C.B. Mohr (Paul Siebeck).

Morrone, Juan J. 2000. Entre el escarnio y el encomio: Léon Croizat y la Panbiogeografía. *Interciencia*, 25: 41–47.

Moss, Brian, Penny Johnes, and Geoffroy Phillips. 1994. August Thienemann and Loch Lomond—An approach to the design of a system for monitoring the state of north-temperate waters. *Hydrobiologia*, 290: 1–12.

Mosse, George L. 1966. *Nazi Culture. Intellectual, Cultural and Social Life in the Third Reich*. Madison, WI: University of Wisconsin Press.

Mosse, George L. 1998. *The Crisis of German Ideology. Intellectual Origins of the Third Reich*. New York: Howard Fertig.

Müller, Fritz. 1863. Die Verwandlung der Garnelen. Erster Beitrag. *Archiv für Naturgeschichte*, 29: 7–23.

Müller, Fritz. 1864. *Für Darwin*. Leipzig, Germany: Wilhelm Engelmann.

Müller, Johannes. 1837. *Handbuch der Physiologie des Menschen*, vol. 1. Koblenz, Germany: J. Hölscher.

Müller, Johannes. 1840. *Handbuch der Physiologie des Menschen*, vol. 2. Koblenz, Germany: J. Hölscher.

Müller, Johannes. 1852a. *Über Synapta digitata und über die Erzeugung von Schnecken in Holothurien*. Berlin, Germany: Georg Reimer.

Müller, Johannes. 1852b. Upon the development of mollusks in Holothuriae. *Annals and Magazine of Natural History*, 9: 22–37.

Mulligan, William. 2009. The Reichswehr and the Weimar Republic, pp. 78–101. In: McElligott, A. (Ed.), *Weimar Germany*, Oxford: Oxford University Press.

Naef, Adolf. 1909. Die Organogenese des Cölomsystems und der zentralen Blutgefässe von Loligo. *Jenaische Zeitschrift für Naturwissenschaft* 45, N.F. 38: 221–266.

Naef, Adolf. 1911. Studien zur Generellen Morphologie der Mollusken. 1. Teil: Über Torsion und Asymmetrie der Gastropoden. *Ergebnisse und Fortschritte der Zoologie*, 3: 73–164.

Naef, Adolf. 1913. Studien zur generellen Morphologie der Mollusken. 2. Teil: Das Cölomsystem in seinen topographischen Beziehungen. *Ergebnisse und Fortschritte der Zoologie*, 3: 329–462.

Naef, Adolf. 1917. *Die individuelle Entwicklung organischer Formen als Urkunde ihrer Stammesgeschichte (Kritische Bemerkungen über das sogenannte "biogenetische Grundgesetz")*. Jena, Germany: G. Fischer.

Naef, Adolf. 1919. *Idealistische Morphologie und Phylogenetik (zur Methodik der systematischen Morphologie)*. Jena, Germany: Gustav Fischer.

Naef, Adolf. 1921. *Die Cephalopoden. Fauna und Flora des Golfes von Neapel und der angrenzenden Meeres-Abschnitte*. Berlin, Germany: R. Friedländer & Sohn.

Naef, Adolf. 1922. *Die fossilen Tintenfische. Eine paläozoologische Monographie*. Jena, Germany: Gustav Fischer.

Naef, Adolf. 1923. Kritische Biologie und ihre Gliederung. *Vierteljahrsschrift der Naturforschenden Gesellschaft in Zürich*, 68: 329–334.

Naef, Adolf. 1925. Über Morphologie und Stammesgeschichte. *Vierteljahrsschrift der Naturforschenden Gesellschaft in Zürich*, 70: 234–240.

Naef, Adolf. 1926. Zur Diskussion des Homologiebegriffes und seiner Anwendung in der Morphologie. *Biologisches Zentralblatt*, 46: 405–427.

Naef, Adolf. 1927. Die Definition des Homologiebegriffes. *Biologisches Zentralblatt*, 47: 187–190.

Naef, Adolf. 1931a. *Phylogenie der Tiere* (Handbuch der Vererbungswissenschaft, vol. 3). Berlin, Germany: Gebr. Bornträger.

Naef, Adolf. 1931b. Allgemeine Morphologie. I. Die Gestalt als Begriff und Idee. (Diagnostik und Typologie der organischen Formen), pp. 77–118. In: Bolk, L., E. Göppert, E. Kallius, and W. Lubosch (Eds.), *Handbuch der vergleichenden Anatomie der Wirbeltiere,* erster Band. Berlin, Germany: Urban und Schwarzenberg.

Naef, Adolf. 1932. Morphologie der Tiere, pp. 3–17. In: Dittler, R., G. Joos, E. Korschelt, G. Linck, F. Oltmanns, F., and K. Schaum, K. (Eds.), *Handwörterbuch der Naturwissenschaften*, 2nd edition, vol. 7. Jena, Germany: Gustav Fischer.

Naef, Adolf. 1933. *Die Vorstufen der Menschwerdung. Eine anschauliche Darstellung der menschlichen Stammesgeschichte und eine kritische Betrachtung ihrer allgemeinen Voraussetzungen.* Jena, Germany: Gustav Fischer.

Nagel, Anne C. 2012. *Hitlers Bildungsreformer. Das Reichsministerium für Wissenschaft, Erziehung und Volksbildung 1934–1945*. Frankfurt, Germany: S. Fischer.

Nägeli, Carl W. 1856. Die Individualität in der Natur. Mit besonderer Berücksichtigung des Pflanzenreiches. *Monatsschrift des wissenschaftlichen Vereins Zürich*, 1: 171–212.

Nägeli, Carl W. 1865. *Entstehung und Begriff der naturhistorischen Art*. Munich, Germany: Verlag der königlichen Akademie der Wissenschaften.

Nägeli, Carl W. 1884. *Mechanisch-physiologische Theorie der Abstammungslehre*. Munich, Germany: R. Oldenbourg.

Nelson, Gareth. 2004. Cladistics: Its arrested development, pp. 127–147. In: Williams, D.M. and P.L. Forey (Eds.), *Milestones in Systematics*. Berkeley, CA: The University of California Press.

Nelson, Gareth. 2014. Cladistics at an earlier time, pp. 139–149. In: Hamilton, A. (Ed.), *The Evolution of Phylogenetic Systematics*. Boca Raton, FL: CRC Press.

Neubauer, H. 1932/33. Die biologisch-dynamische Wirtschaftsweise in Landwirtschaft und Gartenbau. *Der Biologe*, 2: 217–219.

Neurath, Otto, Hans Hahn, and Rudolf Carnap. 1929. *Wissenschaftliche Weltauffassung. Der Wiener Kreis*. Wien, Austria: Artur Wolf.

Nickel, Gisela. 1996. *Wilhelm Troll (1897–1978). Eine Biographie*. Halle (Saale), Germany: Deutsche Akademie der Naturforscher Leopoldina.

Nietzsche, Friedrich. 1901/1968. *The Will to Power*. Translated by Walter Kaufmann and R.J. Hollingdale. Edited by Walter Kaufmann. New York: Vintage Books.

Noakes, Jeremy. 2008. Hitler and the Nazi state: Leadership, hierarchy, and power, pp. 73–98. In: Caplan, J. (Ed.), *Nazi Germany*. Oxford: Oxford University Press.

Nöhring, Rolf. 1973. Erwin Stresemann. 22. 11. 1889–20. 11. 1973. *Journal für Ornithologie*, 114: 455–471.

Nyhart, Lynn K. 1987. The disciplinary breakdown of German morphology, 1870–1900. *Isis*, 78: 365–389.

Nyhart, Lynn K. 1995. *Biology Takes Form. Animal Morphology and the German Universities, 1800–1900*. Chicago, IL: University of Chicago Press.

Nyhart, Lynn K. 2002. Learning from history: Morphology's challenges in Germany ca. 1900. *Journal of Morphology*, 252: 2–14.

Nyhart, Lynn K. 2003. The importance of the "Gegenbaur School" for German morphology. *Theory in Biosciences*, 122: 162–173.

Nyhart, Lynn K. 2009a. *Modern Nature. The Rise of the Biological Perspective in Germany.* Chicago, IL: University of Chicago Press.

Nyhart, Lynn K. 2009b. Embryology and morphology, pp. 194–214. In: Ruse, R. and R.J. Richards (Eds.), *The Cambridge Companion to the "Origin of Species."* Cambridge, UK: Cambridge University Press.

Nyhart, Lynn K. and Scott Lidgard. 2011. Indviduals at the center of biology: Rudolf Leuckart's *Polymorphismus der Individuen* and the ongoing narrative of parts and wholes. With an annotated translation. *Journal of the History of Biology*, 44: 373–443.

Oberdan, Thomas. 1993. *Protocols, Truth, and Convention.* Amsterdam, the Netherlands: Rodopi.

Olby, R. 1974. Naegeli, Carl Wilhelm von, pp. 600–602. In: Gillispie, C.C. (Ed.), *Dictionary of Scientific Biography*, vol. IX. New York: Charles Scribners's Sons.

Oppenheimer, Jane M. 1971. Driesch, Hans Adolf Eduard, pp. 186–189. In: Gillispie, C.C. (Ed.), *Dictionary of Scientific Biography*, vol. IV. New York: Charles Scribner's Sons.

Oppenheimer, Jane M. 1991. Curt Herbst's contributions to the concept of embryonic induction, pp. 83–90. In: Gilbert. S.F. (Ed.), *A Conceptual History of Modern Embryology.* New York: Plenum Press.

Osborn, Henry F. 1889. The paleontological evidence for the transmission of acquired characters. *American Naturalist*, 23: 561–566.

Osborn, Henry F. 1891. Evolution and heredity, pp.130–141. In: *Biological Lectures Delivered at the Marine Laboratory of Woods Hole in the Summer Session of 1890.* Boston, MA: Ginn & Co.

Osborn, Henry F. 1892. The difficulties in the heredity theory. *American Naturalist*, 26: 537–567.

Osborn, Henry F. 1905. The present problems of paleontology. *Popular Science Monthly*, 66: 226–242.

Osborn, Henry F. 1920. Renewal of our relations with the scientific men of Europe. *Science*, 51: 567–568.

Osche, Günther. 1963. Grundzüge der allgemeinen Phylogenetik, pp. 817–906. In: Gessner, F. (Ed.), *Handbuch der Biologie*, vol. III/2. Frankfurt, Germany: Akademische Verlagsgesellschaft Athenaion.

Ospovat, Dov. 1981. *The Development of Darwin's Theory: Natural History, Natural Theology, and Natural Selection, 1838–1859.* Cambridge, UK: Cambridge University Press.

Otten, Paul. 1943. Wie behandle ich die Lebensgemeinschaft 'Der Wald' in meinem lebenskundlichen Unterricht. *Der Biologe*, 12: 73–77.

Owen, Richard. 1843. Lectures on the Comparative Anatomy and Physiology of the Invertebrate Animals, delivered at the Royal College of Surgeons, from Notes taken by William White Cooper, M.R.C.S., and Revised by Professor Owen. London, UK: Longman, Brown, Green, and Longmans.

Owen, Richard. 1866. *On the Anatomy of Vertebrates*, Vol. I. Fishes and Reptiles. London: Longmans, Green, and Co.

Passing, H. 1937/38. Der Vierjahresplan. *Zeitschrift für die gesamte Naturwissenschaft*, 3: 162–163.

Patterson, Colin. 1982. Morphological characters and homology, pp. 21–74. In: Joysey K.A. and Friday A.E. (Eds.), *Problems of Phylogenetic Reconstruction*. London: Academic Press.

Patterson, Colin. 1989. Phylogenetic relationships of major groups: Conclusions and prospects, pp. 471–488. In: Fernholm B., K. Bremer, and H. Jörnvall (Eds.), *The Hierarchy of Life: Molecules and Morphology*. Amsterdam, the Netherlands: Elsevier.

Pauley, Bruce F. 1992. *From Prejudice to Persecution. A History of Austrian Anti-Semitism.* Chapel Hill, NC: The University of North Carolina Press.

Peters, Günther. 1995. Über Willi Hennig als Forscherpersönlichkeit. *Sitzungsberichte der Gesellschaft Naturforschender Freunde zu Berlin* (N.F.) 34: 3–10.

Petzold, J. 1922. Zur Krisis des Kausalitätsbegriffs. *Die Naturwissenschaften*, 10: 693–695.

Peuckert, Detlev J.K. 1987. *The Weimar Republic*, New York: Hill and Wang.

Peyer, Bernhard. 1942. Jean Strohl (1886–1942). *Vierteljahrsschrift der Naturforschenden Gesellschaft in Zürich*, 87: 533–539.

Philiptschenko, Jurij. 1927. *Variabilität und Variation*. Berlin, Germany: Gebr. Borntäger.

Pine, Lisa. 2007. *Hitler's "National Community." Society and Culture in Nazi Germany*. London: Hodder Arnold.

Pine, Lisa. 2010. *Education in Nazi Germany*. Oxford: Berg.

Plate, Ludwig. 1914. Prinzipien der Systematik mit besonderer Berücksichtigung des Systems der Tiere, pp. 119–159. In: Hertwig, R. and Wettstein, R.v. (Eds.), *Kultur der Gegenwart, Dritter Teil, Vierte Abteilung, Vierter Band. Abstammungslehre. Systematik. Paläontologie. Biogeographie*. Leipzig, Germany: B.G. Teubner.

Platnick, Norman I. and Gareth Nelson. 1988. Spanning-tree biogeography: Shortcut, detour, or dead-end? *Systematic Zoology*, 37: 410–419.

Poggi, Stefano. 1989. Positivistische Philosophie und naturwissenschaftliches Denken, pp. 11–151. In: Poggi, S. and W. Röd. (Eds.), *Geschichte der Philosophie*, Band X. Die Philosophie der Neuzeit 4. Positivismus, Sozialismus und Spiritualismus im 19. Jahrhundert. Munich, Germany: C.H. Beck.

Popper, Karl R. 1972. *Objective Knnowledge. An Evolutionary Approach*. Oxford: University of Oxford Press.

Popper, Karl R. 1979. *Die beiden Grundprobleme der Erkenntnistheorie*. Tübingen, Germany: J.C.B. Mohr (Paul Siebeck).

Porsch, O. 1931. Richard Wettstein. *Berichte der Deutschen Botanischen Gesellschaft*, 49: 180–199.

Portmann, Adolf. 1944. *Biologische Fragmente zu einer Lehre vom Menschen*. Basel, Switzerland: Schwabe.

Portmann, Adolf. [1944?]. *Vom Ursprung des Menschen. Ein Querschnitt durch die Forschungsergebnisse. Fünftes und sechstes Tausend*. Basel, Switzerland: Friedrich Reinhardt.

Portmann, Adolf. 1948. *Einführung in die vergleichende Morphologie der Wirbeltiere*. Basel, Switzerland: Schwabe.

Portmann, Adolf. 1960. *Neue Wege der Biologie*. Munich, Germany: R. Piper.

Portmann, Adolf. 1964. *Don Quijote und Sancho Pansa. Vom gegenwärtigen Stand der Typenlehre*. Basel, Switzerland: Friedrich Reinhardt.

Potthast, Thomas. 1999. Theorien, Organismen, Synthesen: Evolutionsbiologie und Ökologie im angloamerikanischen und deutschsprachigen Raum von 1920–1960, pp. 259–292. In: Junker, T. and E.-M. Engels (Eds.), *Die Entstehung der Synthetischen Theorie. Beiträge zur Geschichte der Evolutionsbiologie in Deutschland. 1930–1950*. Berlin, Germany: VWB—Verlag für Wissenschaft und Bildung.

Potthast, Thomas and Uwe Hoßfeld. 2010. Vererbungs- und Entwicklungslehren in Zoologie, Botanik und Rassenkunde/Rassenbiologie: zentrale Forschungsfelder der Biologie an der Universität Tübingen im Nationalsozialismus, pp. 435–482. In: Wiesing, U., K.-R. Brintzinger, H. Grün, and S. Michl (Eds.), *Die Universität Tübingen im Nationalsozialismus*. Stuttgart, Germany: Franz Steiner Verlag.

Preuß, Ulrich K. 1984. Die Perversion des Rechtsgedankens, pp. 116–128. In: Tröger, J. (Ed.), *Hochschule und Wissenschaft im Dritten Reich*. Frankfurt a.M., Germany: Campus Verlag.

Prinz, Wolfgang. 1985. Ganzheits- und Gestaltpsychologie und Nationalsozialismus, pp. 55–81. In: Lundgreen, P. (Ed.), *Wissenschaft im Dritten Reich*. Frankfurt, Germany: Suhrkamp.

Quenstedt, Wener and Manfred Schröter. 1957. Dacqué Edgar Viktor August. *Neue Deustche Biographie* 3: 465–467. http://www.deutsche-biographie.de/pnd116011564.html.

Rabes, Otto. 1934. Biologieunterricht und völkische Erziehung. 84 Seiten, Verlag Diesterweg. *Unterrichtsblätter für Mathematik und Naturwissenschaften*, 40: 112.

Rammstedt, Otthein. 1985. Theorie und Empirie des Volksfeindes. Zur Entwicklung einer "deutschen Soziologie," pp. 253–313. In: Lundgreen, P. (Ed.), *Wissenschaft im Dritten Reich*. Frankfurt, Germany: Suhrkamp.

Range, Paul and Gerhard Mempel. 1936. Geschäftliche Sitzung der Hauptversammlung in Kassel. *Zeitschrift der Deutschen Geologischen Gesellschaft*, 88: 585–586.

Rasch, Manfred. 1994. Mentzel, Rudolf, pp. 96–98. In: Hockerts, H.G. (Ed.), *Neue Deutsche Biographie*, vol. 17. Berlin, Germany: Duncker und Humblot. http://www.deutsche-biographie.de/pnd116885947.html.

Ratzel, Friedrich. 1897. Ueber den Lebensraum. Eine biogeographische Skizze. *Die Umschau. Übersicht über die Fortschritte und Bewegungen auf dem Gesamtgebiet der Wissenschaft, Technik, Literatur und Kunst*, 1: 363–367.

Ratzel, Friedrich. 1901. Der Lebensraum. Eine biogeographische Studie, pp. 101–189. In: Bücher, R., K.V. Fricker , F.X. Funk, G.v. Mandry, G.v. Mayr, and F. Ratzel (Eds.), *Festgaben für Albert Schäffle zur siebenzigsten Wiederkehr seines Geburtstages am 24. Januar 1901*. Tübingen, Germany: Laupp.

Rauh, Werner. 1979. Wilhelm Troll (1897–1978). *Akademie der Wissenschaften und der Literatur Mainz, Jahrbuch* 1979: 88–91.

Rehm, Sigmund. 1992. Heinz Brücher †. *Angewandte Botanik*, 66: 1.

Reif, Wolf-Ernst. 1983. Evolutionary theory in German paleontology, pp. 173–203. In: Grene, M. (Ed.), *Dimensions of Darwinism*. Cambridge, UK: Cambridge University Press.

Reif, Wolf-Ernst. 1986. The search for a macroevolutionary theory in German paleontology. *Journal of the History of Biology*, 19: 79–130.

Reif, Wolf-Ernst. 1998. Adolf Naef's idealistische Morphologie und das Paradigma typologischer Makroevolutionstheorien, pp. 411–424. In: Engels, E.-M., Th. Junker, and M. Weingarten (Eds.), *Ethik der Biowissenschaften. Geschichte und Theorie*. Berlin, Germany: VWB—Verlag für Wissenschaft und Bildung.

Reif, Wolf-Ernst. 1999. Deutschsprachige Paläontologie im Spannungsfeld zwischen Makroevolution und Neo-Darwinismus (1920–1950). *Verhandlungen zur Geschichte und Theorie der Biologie*, 2: 152–188.

Reimann, Bruno W. 1984. Die "Selbst-Gleichschaltung" der Universitäten, 1933, pp. 38–52. In: Tröger, J. (Ed.), *Hochschule und Wissenschaft im Dritten Reich*. Frankfurt, Germany: Campus Verlag.

Reisch, George A. 2007. From 'The Life of the Present' to the "Icy Slopes of Logic." Logical Empiricism, the Unity of Science Movement, and the Cold War, pp. 58–87. In: Richardson, A. and T. Uebel (Eds.), *The Cambridge Companion to Logical Empiricism*. Cambridge, UK: Cambridge University Press.

Reiser, Frederick, C. 2002. *German Idealism. The Struggle against Subjectivism, 1781–1801*. Cambridge, MA: Harvard University Press.

Reiβ, Christian. 2007. No evolution, no heredity, just development—Julius Schaxel and the end of the Evo-Devo agenda in Jena, 1906–1933: A case study. *Theory in Biosciences*, 126: 155–164.

Reiβ, Christian, Susan Springer, Uwe Hoβfeld, Lennart Olsson, and Georgy S. Levit. 2007. Introduction to the autobiography of Julius Schaxel. *Theory in Biosciences*, 126: 165–175.

Remane, Adolf. 1927. Art und Rasse. *Verhandlungen der Gesellschaft für Physische Anthropologie*, 2: 2–33.

Remane, Adolf. 1935. Aufgaben zoologischer Heimatforschung in Schleswig-Holstein. *Die Heimat*, 45: 22–23.

Remane, Adolf. 1939a. Der Geltungsbereich der Mutationstheorie. *Verhandlungen der Deutschen Zoologischen Gesellschaft*, 41 (Zoologischer Anzeiger, Suppl., 12): 206–220.

Remane, Adolf. 1939b. Die Gemeinschaft als Lebensform in der Natur. *Kieler Blätter*, 1939: 43–61.

Remane, Adolf. 1940. Artbild und Vererbung. *Jahresbände der Wissenschaftlichen Akademien des NSD-Dozentenbundes*, 1: 117–126.

Remane, Adolf. 1941. Die Abstammungslehre im gegenwärtigen Meinungskampf. *Archiv für Rassen- und Gesellschaftsbiologie*, 35: 89–122.

Remane, Adolf. 1942. Das biologische Weltbild im Wandel der Zeiten. Kieler Blätter, 1942: 73–85.

Remane, Adolf. 1948. Die Theorie sprunghafter Typenbildung und das Spezialisationsgesetz. *Die Naturwissenschaften*, 35: 257–261.

Remane, Adolf. 1951. Das Problem des Typus in der morphologischen Biologie. *Studium Generale*, 4: 390–399.

Remane, Adolf. 1952. *Die Grundlagen des Natürlichen Systems, der Vergeichenden Anatomie, und der Phylogenetik. Theoretische Morphologie und Systematik I*. Leipzig, Germany: Akademische Verlagsgesellschaft.

Remane, Adolf. 1964. Begrüssungsansprache des Ersten Vorsitzenden. *Zoologischer Anzeiger*, Supplementband 27: 35–36.

Remane, Jürgen. 2003. Remane, Robert Gustav Adolf. *Neue Deutsche Biographie*, 21: 412–413. http://www.deutsche-biographie.de/pnd11874447X.html.

Renner, Otto. 1938/39. Troll Wilhelm, Vergleichende Morphologie der höherern Pflanzen, Erster Band: Vegetationsorgane. 3. Lieferung. S. I-XII u. 509–955. Mit 368 Abb. Berlin 1937 (Gebr. Bornträger). *Zeitschrift für Botanik*, 33: 526–528.

Rensch, Bernhard. 1929. *Das Prinzip der Rassenkreise und das Problem der Artbildung*. Berlin, Germany: Gebr. Bornträger.

Rensch, Bernhard. 1933. Zoologische Systematik und Artbildungsproblem. *Verhandlungen der Deutschen Zoologischen Gesellschaft*, Suppl. 6: 19–83.

Rensch, Bernhard. 1934. Über einige Beziehungen von Rasse und Klima bei Säugetieren. *Die Medizinische Welt (Erblehre und Rassenpflege)*, 8: 703–704.

Rensch, Bernhard. 1935. Umwelt und Rassenbildung. *Archiv für Anthropologie*, N.F. 23: 326–333.

Rensch, Bernhard. 1939. Typen der Artbildung. *Biological Reviews*, 14: 180–222.

Rensch, Bernhard. 1947. *Neuere Probleme der Abstammungslehre: Die Transspezifische Evolution*. Stuttgart, Germany: Ferdinand Enke.

Rensch, Bernhard. 1960. The laws of evolution, pp. 95–115. In: Tax, S. (Ed.), *Evolution after Darwin*, vol. 1. Chicago, IL: The University of Chicago Press.

Rensch, Bernhard. 1968. *Biophilosophie auf erkenntnistheoretischer Grundlage (Panpsychistischer Identismus)*. Stuttgart, Germany: Gustav Fischer.

Rensch, Bernhard. 1979. Lebensweg eines Biologen in einem turbulenten Jahrhundert. Stuttgart, Germany: Gustav Fischer.

Rensch, Bernhard. 1980. Historical development of the present synthetic neo-Darwinism in Germany, pp. 284–303. In: Mayr, E. and W.B. Provine (Eds.), *The Evolutionary Synthesis. Perspectives on the Unification of Biology*. Cambridge, MA: Harvard University Press.

Rensch, Bernhard. 1988. *Probleme genereller Determiniertheit allen Geschehens*. Berlin, Germany: Paul Parey.

Reynolds, Andrew. 2007. The theory of the cell state and the question of cell autonomy in nineteenth and early twentieth-century biology. *Science in Context*, 20: 771–795.

Reynolds, Andrew. 2008. Ernst Haeckel and the theory of the cell state: Remarks on the history of a bio-political metaphor. *History of Science*, 46: 123–152.

Richards, Robert J. 1992. *The Meaning of Evolution. The Morphological Construction and Ideological Reconstruction of Darwin's Theory*. Chicago, IL: University of Chicago Press.

Richards, Robert J. 2002. *The Romantic Conception of Life. Science and Philosophy in the Age of Goethe*. Chicago, IL: University of Chicago Press.

Richards, Robert J. 2007. Ernst Haeckel's alleged anti-Semitism and contributions to Nazi biology. *Biological Theory*, 2: 97–103.

Richards, Robert J. 2008. *The Tragic Sense of Life. Ernst Haeckel and the Struggle over Evolutionary Thought*. Chicago, IL: University of Chicago Press.

Richardson, Alan W. 1998. *Carnap's Construction of the World. The Aufbau and the Emergence of Logical Empiricism*. Cambridge, UK: Cambridge University Press.

Richter, Stefan and Rudolf Meier. 1994. The development of phylogenetic concepts in Hennig's early theoretical publications. *Systematic Biology*, 43: 212–221.

Rickert, Heinrich. 1902. *Die Grenzen der naturwissenschaftlichen Begriffsbildung. Eine logische Einleitung in die historischen Wissenschaften*. Tübingen, Germany: J.C.B. Mohr (Paul Siebeck).

Rickert, Heinrich. 1986. *The Limits of Concept Formation in Natural Science*. Translated by G. Oakes. Cambridge, UK: Cambridge University Press.

Rieppel, Olivier. 2005. Les raciness de positivism logique dans la philosophie empiriocritique de Willi Hennig. *Biosystema*, 24 (*Philosophie de la Systematique*): 9–22.

Rieppel, Olivier. 2006. On concept formation in systematics. *Cladistics*, 22: 474–492.

Rieppel, Olivier. 2007a. The metaphysics of Hennig's phylogenetic systematics: Substance, events and laws of nature. *Systematics and Biodiversity*, 5: 345–360.

Rieppel, Olivier. 2007b. Parsimony, likelihood, and instrumentalism in systematics. *Biology & Philosophy*, 22: 141–144.

Rieppel, Olivier. 2007c. The nature of parsimony and instrumentalism in systematics. *Journal for Zoological Systematics and Evolutionary Research*, 45: 177–183.

Rieppel, Olivier. 2008. Total evidence in phylogenetic systematics. *Biology & Philosophy*, 24: 607–622.

Rieppel, Olivier. 2009a. Origins, taxa, names and meanings. *Cladistics*, 24: 598–610.

Rieppel, Olivier. 2009b. Hennig's enkaptic system. *Cladistics*, 25: 311–317.

Rieppel, Olivier. 2009c. Species as a process. *Acta Biotheoretica*, 57: 33–49.

Rieppel, Olivier. 2010. Sinai Tschulok (1875–1945)—A pioneer of cladistics. *Cladistics*, 26: 103–111.

Rieppel, Olivier. 2011a. Species are individuals—The German tradition. *Cladistics*, 27: 629–645.

Rieppel, Olivier. 2011b. The Gegenbaur Transformation: A paradigm change in comparative biology. *Systematics and Biodiversity*, 9: 177–190.

Rieppel, Olivier. 2011c. *Evolutionary Theory and the Creation Controversy*. Heidelberg, Germany: Springer.

Rieppel, Olivier. 2011d. Ernst Haeckel and the monophyly of life. *Journal of Zoological Systematics and Evolutionary Research*, 49: 1–5.

Rieppel, Olivier. 2011e. Wilhelm Troll (1897–1978): Idealistic morphology, physics, and phylogenetics. *History and Philosophy of Life Sciences*, 33: 321–342.

Rieppel, Olivier. 2011f. Karl Beurlen (1901–1985), Nature mysticism, and Aryan paleontology. *Journal of the History of Biology*, 45: 253–299.

Rieppel, Olivier. 2011g. Willi Hennig's dichotomization of nature. *Cladistics*, 27: 103–112.

Rieppel, Olivier. 2012a. Othenio Abel (1875–1946): The rise and decline of paleobiology in German paleontoloy. *Historical Biology: An International Journal of Paleobiology*, 25: 77–97.

Rieppel, Olivier. 2012b. Adolf Naef (1883–1949), systematic morphology and phylogenetics. *Journal for Zoological Systematics and Evolutionary Research*, 50: 2–13.

Rieppel, Olivier. 2012c. Hugo Dingler (1881–1954) and the philosophical foundation of the German Evolutionary Synthesis. *Biological Theory*, 6: 162–168.

Rieppel, Olivier. 2013a. Styles of scientific reasoning: Adolf Remane (1898–1976) and the German evolutionary synthesis. *Journal of Zoological Systematics and Evolutionary Research*, 51: 1–12.

Rieppel, Olivier. 2013b. Othenio Abel (1875–1946) and the "phylogeny of the parts." *Cladistics*, 29: 328–335.

Rieppel, Olivier. 2013c. Biological individuals and natural kinds. *Biological Theory*, 7: 162–169.

Rieppel, Olivier. 2014. The early cladogenesis of cladistics, pp. 117–137. In: Hamilton, A. (Ed.), *The Evolution of Phylogenetic Systematics*. Berkeley, CA: University of California Press.

Rieppel, Olivier, David M. Williams, and Malte C. Ebach. 2013. Adolf Naef (1883–1949): On foundational concepts and principles of systematic morphology. *Journal of the History of Biology*, 46: 445–510.

Ringer, Fritz K. 1969/1990. *The Decline of the German Mandarins. The German Academic Community, 1890–1933*. Hanover, Germany: University Press of New England.

Ritter, Markus. 2000. Die Biologie Adolf Portmanns in zeitgeschichtlichem Kontext. *Basler Zeitschrift für Geschichte und Altertumskunde*, 100: 207–254.

Roger, Jacques. 1971. *Les Sciences de la Vie dans la Pensée Française du XVIIIe Siècle*, 2nd edition. Paris, France: Armand Collin.

Rogers, Ben. 1999. *A.J. Ayer. A Life*. New York: Grove Press.

Rohkrämer, Thomas. 2005. Martin Heidegger, national socialism, and environmentalism, pp. 171–203. In: Brüggemeier, F.-J., M. Cioc, and T. Zeller (Eds.), *How Green Were the Nazis? Nature, Environment, and Nation in the Third Reich*. Athens, OH: Ohio University Press.

Romanes, George J. 1881. The struggle of the parts in the organism. *Nature*, 24: 505–506.

Roper, Clyde F. 1990. Cephalopodologie. *Science*, 248: 898–899.

Rosa, D. 1899. *La riduzione progressiva della variabilità e i suoi rapporti coll'estinzione e coll'origine delle specie*. Torino, Italy: C. Clausen.

Roux, Wilhelm. 1878. Ueber die Verzweigungen der Blutgefässe des Menschen. Eine morphologische Studie. *Jenaische Zeitschrift für Naturwissenschaft*, 12: 205–266.

Roux, Wilhelm. 1879. Ueber die Bedeutung der Ablenkung des Arterienstammes bei der Astabgabe. *Jenaische Zeitschrift für Naturwissenschaft*, 13 (N.F. 6): 321–337.

Roux, Wilhelm 1880/1895. Ueber die Leistungsfähigkeit der Principien der Descendenzlehre zur Erklärung der Zweckmässigkeiten des thierischen Organismus, pp. 102–133. In: Roux, W. (Ed.), *Gesammelte Abhandlungen über Entwicklungsmechanik der Organismen*, erster Band. Leipzig, Germany: Wilhelm Engelmann.

Roux, Wilhelm. 1881a. *Der Kampf der Theile im Organismus. Ein Beitrag zur Vervollständigung der mechanischen Zweckmässigkeitslehre*. Leipzig, Germany: Wilhelm Engelmann.

Roux, Wilhelm. 1881b. Der Kampf der Theile im Organismus. *Biologisches Centralblatt*, 1: 241–251.

Roux, Wilhelm (Ed.). 1883/1895. Referat, pp. 21–23. In: *Gesammelte Abhandlungen über Entwicklungsmechanik der Organismen*, erster Band. Leipzig, Germany: Wilhelm Engelmann.

Roux, Wilhelm (Ed.). 1885/1895. Einleitung zu den 'Beiträgen zur Entwickelungsmechanik des Embryo', pp. 1–23. In: *Gesammelte Abhandlungen über Entwicklungsmechanik der Organismen*, zweiter Band. Leipzig, Germany: Wilhelm Engelmann.

Roux, Wilhelm (Ed.). 1888/1895. Beiträge zur Entwicklungsmechanik des Embryo. Nr. V. Ueber die künstliche Hervorbringung 'halber' Embryonen durch Zerstörung einer der beiden ersten Furchungszellen, sowie über die Nachentwicklung (*Postgeneration*) der fehlenden Körperhäfte, pp. 419–521. In: *Gesammelte Abhandlungen über Entwicklungsmechanik der Organismen*, zweiter Band. Leipzig, Germany: Wilhelm Engelmann.

Roux, Wilhelm (Ed.). 1889/1895. Die Entwicklungsmechanik der Organismen, eine anatomische Wissenschaft der Zukunft, pp. 24–54. In: *Gesammelte Abhandlungen über Entwicklungsmechanik der Organismen*, zweiter Band. Leipzig, Germany: Wilhelm Engelmann.

Roux, Wilhelm (Ed.). 1892/1895. Ueber das entwicklungsmechanische Vermögen jeder der beiden ersten Furchungszellen des Eies, pp. 766–817. In: *Gesammelte Abhandlungen über Entwicklungsmechanik der Organismen*, zweiter Band. Leipzig, Germany: Wilhelm Engelmann.

Roux, Wilhelm (Ed.). 1895. Vorwort zum zweiten Abdrucke, pp. 139–149. In: *Gesammelte Abhandlungen über Entwicklungsmechanik der Organismen*, zweiter Band. Leipzig, Germany: Wilhelm Engelmann.

Roux, Wilhelm. 1923. Wilhelm Roux in Halle a. S., pp. 1–66. In: Grote, L.R. (Ed.), *Die Medizin der Gegenwart in Selbstdarstellungen*, vol. I. Leipzig, Germany: Felix Meiner.

Rübel, Eduard. 1947. *Geschichte der Naturforschenden Gesellschaft in Zürich. Neujahrsblatt der Naturforschenden Gesellschaft in Zürich auf das Jahr 1947*. Zürich, Switzerland: Gebr. Fretz.

Rügemer, Hans. 1937/38. Die "Nature" eine Greuelzeitschrift. *Zeitschrift für die gesamte Naturwissenschaft*, 3: 475–479.

Rupke, Nicolas A. 1994. *Richard Owen. Victorian Naturalist*. New Haven, CT: Yale University Press.

Rürup, Reinhard and Michael Schüring. 2008. *Schicksale und Karrieren. Gedenkbuch für die von den Nationalsozialisten aus der Kaiser-Wilhelm-Gesellschaft vertriebenen Forscherinnen und Forscher*. Göttingen, Germany: Wallenstein.

Russell, Edward S. 1916/1982. *Form and Function: A Contribution to the History of Animal Morphology. With a New Introduction by George V. Lauder*. Chicago, IL: University of Chicago Press.

Rüting, Torsten. 2004. History and significance of Jacob von Uexküll and of his institute in Hamburg. *Sign Systems Studies* 32: 35–72.

Rutschke, Erich. 2001. Erwin Stresemann (1889–1972), pp. 296–315. In: Jahn, J. and M. Schmitt (Eds.), *Darwin & Co. Eine Geschichte der Biologie in Portraits*, vol. 2. Munich, Germany: C.H. Beck.

Ryle, Gilbert. 1949. *The Concept of Mind*. London, UK: Hutchinson.

Salfeld, Hans. 1921. Kiel- und Furchenbildung auf der Schalenaussenseite der Ammonoideen in ihrer Bedeutung für die Systematik und Festlegung von Biozonen. *Zentralblatt für Mineralogie, Geologie und Paläontologie*, 1921: 343–347.

Salfeld, Hans. 1924. *Die Bedeutung der Konservativstämme für die Stammesentwicklung der Ammoniten*. Leipzig, Germany: M. Wey.

Salvini-Plawen, Luitfried and Maria Mizzaro. 1999. 150 Jahre Zoologie an der Universität Wien. *Verhandlungen der Zoologisch-Botanischen Gesellschaft in Wien*, 136: 1–76.

Sapper, Karl. 1938. Zur Kritik der Ganzheitsbiologie. *Acta Biotheoretica*, 4: 111–118.

Sauter, Johann. 1934/35. Der VIII Internationale Philosophenkongress 1934 (vom 2.-7. September in Prag). *Blätter für Deutsche Philosophie*, 8: 437–448.

Scharfe, Martin. 1984. Einschwörung auf den völkisch-germanischen Kulturbegriff, pp. 105–115. In: Tröger, J. (Ed.), *Hochschule und Wissenschaft im Dritten Reich*. Frankfurt, Germany: Campus Verlag.

Schaxel, Julius. 1919. *Grundzüge der Theoriebildung in der Biologie*. Jena, Germany: Gustav Fischer.

Schindewolf, Otto H. 1936. *Paläntologie, Entwicklungslehre, und Genetik. Kritik und Synthese*. Berlin, Germany: Gebr. Bornträger.

Schindewolf, Otto H. 1937. Beobachtungen und Gedanken zur Deszendenzlehre. *Acta Biotheoretica*, 3: 195–212.

Schlee, Dieter. 1971. Die Rekonstruktion der Phylogenese mit HENNIG's Prinzip. Frankfurt, Germany: Verlag Waldemar Kramer.

Schlee, Dieter. 1978. In Memoriam Willi Hennig 1913–1976. Eine biographische Skizze. *Entomologica Germanica*, 4: 377–391.

Schleicher, August. 1860. *Die Deutsche Sprache*. Stuttgart, Germany: Cotta

Schleicher, August. 1863. *Die Darwin'sche Theorie und die Sprachwissenschaft.* Weimar, Germany: Hermann Böhlau.

Schleiden, Matthias J. 1838. Beiträge zur Phytogenesis. *Archiv für Anatomie, Physiologie, und wissenschaftliche Medicin*, 1838: 137–176.

Schleiden, Matthias J. 1842. *Grundzüge der wissenschaftlichen Botanik, nebst einer methodologischen Einleitung als Anleitung zum Studium der Pflanze. Erster Theil.* Leipzig, Germany: Wilhelm Engelmann.

Schleiermacher, Sabine. 2005. Rassenhygiene und Rassenanthropologie an der Universität Berlin, pp. 89–98. In: Jahr, C. and R. Schaarschmidt (Eds.) *Die Berliner Universität in der NS Zeit.* Band I. Strukturen und Personen. Wiesbaden, Germany: Franz Steiner Varlag.

Schlick, Moritz. 1917. *Raum und Zeit in der gegenwärtigen Physik.* Berlin, Germany: Springer.

Schmidt, Heinrich. 1934a. Ernst Haeckel—Denkmal eines grossen Lebens. Jena, Germany; Fromann.

Schmidt, Heinrich. 1934b. Ein lebendiges Denkmal Ernst Haeckels. *Der Biologe*, 3: 49–51.

Schmidt, Jutta. 1975. Jakob von Uexküll und Houston Stewart Chamberlain. Ein Briefwechsel in Auszügen. *Medizinhistorisches Journal*, 10: 121–129.

Schmitt, Michael. 2001. Willi Hennig (1913–1976), pp. 157–175. In: Jahn, J. and M. Schmitt (Eds.), *Darwin & Co. Eine Geschichte der Biologie in Portraits*, vol. 2. Munich, Germany: C.H. Beck.

Schmitt, Michael. 2010. WILLI HENNIG, the cautious revolutioniser. *Paleodiversity*, 3 (Supplement): 3–9.

Schmitt, Michael. 2013. *From Taxonomy to Phylogenetics—Life and Work of Willi Hennig.* Leiden, the Netherlands: Brill Academic Publishers.

Schottky, Walter. 1921. Das Kausalproblem der Quantentheorie als eine Grundfrage der modernen Naturforschung überhaupt. *Die Naturwissenschaften*, 9: 492–496; 506–511.

Schreiner, Klaus. 1985. Führertum, Rasse, Reich. Wissenschaft von der Geschichte nach der nationalsozialistischen Machtergreifung, pp. 163–252. In: Lundgreen, P. (Ed.), *Wissenschaft im Dritten Reich.* Frankfurt, Germany: Suhrkamp.

Schultze, Walter. 1939/1966. The nature of academic freedom, pp. 314–316. In: Mosse, G.L. (Ed.), *Nazi Culture. Intellectual, Cultural and Social Life in the Third Reich.* Madison, WI: University of Wisconsin Press.

Schuster, Julius. 1929. Idealistische Morphologie als Gegenwartsproblem. *Sitzungsberichte der Gesellschaft Naturforschender Freunde zu Berlin, Jahrgang* 1928: 189–212.

Schwabe, G.H. and Lars Brundin. 1961. August Thienemann in Memoriam. *Oikos*, 12: 310–319.

Schwarz, Astrid and Kurt Jax. 2011. Early ecology in the German-speaking world through WWII, pp. 231–275. In: Schwarz, A. and K. Jax (Eds.), *Ecology Revisited: Reflecting on Concepts, Advancing Science.* Heidelberg, Germany: Springer.

Secord, James A. (Ed.).1994. Introduction, pp. vii–xlv. In: *Vestiges of the Natural History of Creation, and Other Evolutionary Writings. Robert Chambers.* Chicago, IL: University of Chicago Press.

Secord, James. A. 2000. *Victorian Sensation. The Extraordinary Publication, Reception, and Secret of Vestiges of the Natural History of Creation.* Chicago, IL: University of Chicago Press.

Seising, Rudolf. 2005. *Die Fuzzifizierung der Systeme. Die Entstehung der Fuzzy Set Theory und ihrer erster Anwendungen—Ihre Entwicklung bis in die 70er Jahre des 20. Jahrhunderts.* Stuttgart, Germany: Franz Steiner Verlag.

Senglaub, Konrad. 2000. Erinnerungen eines ehemaligen Zoologie-Studenten an seinen akademischen Lehrer Willi Hennig im Winter-Semester 1946/47 in Leipzig. *Sitzungsberichte der Gesellschaft Naturforschender Freunde zu Berlin* (N.F.) 39: 165–168.

Siewing, Rolf. 1977. A. Remane. 10. 8. 1898 bis 22. 12. 1976. *Verhandlungen der Deutschen Zoologischen Gesellschaft*, 1977: 342–343.

Simon, Christian. 2010. Naturwissenschaften in Basel im 19. Und 20. Jahrhundert. Die Philosophisch-Naturwissenschaftliche Fakultät der Universität. www.unigeschichte. unibas.ch

Simon, Gerd. 2008. Der Krieg als Krönung der Wissenschaft. homepages.uni-tuebingen.de/ gerd.simon/krieg1.htm (accessed April 18, 2013).

Simpson, George G. 1961. *Principles of Animal Taxonomy*. New York: Columbia University Press.

Sloan, Phillip R. 2002. Reflections on the species problem: What Marjorie Grene can teach us about a perennial issue, pp. 225–255. In: Auxier, R.E. and L.E. Hahn (Eds.), *The Philosophy of Marjorie Grene*. La Salle, IL: Open Court.

Sloan, Phillip R. 2009. Originating species. Darwin on the species problem, pp. 67–86. In: Ruse, M. and R.J. Richards (Eds.), *The Cambridge Companion to the "Origin of Species."* Cambridge, UK: Cambridge University Press.

Soames, Scott. 2003. *Philosophical Analysis in the Twentieth Century*, vol. 1. The Dawn of Analysis. Princeton, NJ: Princeton University Press.

Sommerfeld, Arnold. 1921. *Atombau und Spektrallinien*, zweite Auflage. Braunschweig, Germany: Friedr. Vieweg & Sohn.

Sommerfeld, Arnold. 1924. Grundlagen der Quantentheorie und des Bohrschen Atommodells. *Die Naturwissenschaften*, 12: 1047–1049.

Soury, Jules. 1879. Préface du Traducteur, pp. i–xxxvi. In: Haeckel, E. (Ed.), *Les Preuves du Transformisme. Réponse à Virchow*. Paris, France: Librairie Germer Baillière et Cie.

Soyfer, Valery N. 2003. Tragic history of the VII international congress of genetics. *Genetics*, 165: 1–9.

Spencer, Herbert. 1876. *Die Principien der Biologie*. Autorisierte deutsche Ausgabe nach der zweiten englischen Ausgabe übersetzt von B. Vetter. I. Band. Stuttgart, Germany: E. Schweizerbart.

Spengler, Oswald. 2006. *The Decline of the West*. An Abridged Edition by Helmut Werner. New York: Vintage Books.

Spring, Anton Fr. 1838. *Ueber die naturhistorischen Begriffe von Gattung, Art und Abart, und die Ursachen der Abartungen in den organischen Reichen. Eine Preisschrift*. Leipzig, Germany: Friedrich Fleischer.

Stadler, Friedrich. 1997. *Studien zum Wiener Kreis. Ursprung, Entwicklung und Wirkung des logischen Empirismus im Kontext*, Frankfurt, Germany: Suhrkamp.

Stamm, Alfred and Pio Fioroni. 1984. Adolf Portmann: ein Rückblick auf seine Forschungen. *Verhandlungen der Naturforschenden Gesellschaft in Basel*, 94: 87–120.

Steinbacher, Joachim. 1991. Über Stresemann und seine Schüler. *Mitteilungen aus dem Zoologischen Museum in Berlin*, 67 (Supplementheft, Annalen für Ornithologie, 15): 31–36.

Stella, Marco and Karel Kleisner. 2010. Uexküllian *Umwelt* as science and as ideology: The light and the dark side of a concept. *Theory in Biosciences*, 129: 39–51.

Stengel-von Rutkowski, Lothar. 1937. Lehmann, E.: Wege und Ziele einer deutschen Biologie. J.F. Lehmannns Verlag, München 1936. Preis: RM. 1.80. *Der Biologe*, 6: 33.

Stephenson, Jill. 2008. Inclusion: Building the national community in propaganda and practice, pp. 99–121. In: Caplan, J. (Ed.), *Nazi Germany*. Oxford: Oxford University Press.

Sterelnikov, Ivan and Roman Hecker. 1968. Wladimir Kowalewsky's sources of ideas and their importance for his work and for Russian evolutionary paleontology. *Lethaia*, 1: 219–229.

Stevens, Peter F. 1994. *The Development of Biological Systematics, Antoine-Laurent de Jussieu, Nature, and the Natural System*. New York: Columbia University Press.

Storch, Otto. 1949. Berthold Hatschek. *Österreichische Akademie der Wissenschaften, Almanach*, 99: 284–296.

Strasburger, Eduard A. 1872. *Die Coniferen und die Gnetaceen*. Jena, Germany: Hermann Dabis.

Strasburger, Eduard A. 1874. Ueber die Bedeutung phylogenetischer Methoden für die Erforschung lebender Wesen. Rede gehalten beim Eintritt in die philosophische Facultät der Universität Jena am 2. August 1873. *Jenaische Zeitschrift für Medizin und Naturwissenschaft*, 8 (N.F. 1): 56–80.

Stresemann, Erwin. 1919. Über die europäischen Baumläufer. *Verhandlungen der Ornithologischen Gesellschaft Bayern*, 14: 39–74.

Stroemer, Monika. 1995. Die Bayerische Akademie der Wisssenschaften im Dritten Reich, pp. 89–111. In: Scriba, C.J. (Ed.), *Die Elite der Nation im Dritten Reich. Das Verhältnis von Akademien und ihrem wissenschaftlichen Umfeld zum Nationalsozialismus*. Halle (Saale), Germany: Deutsche Akademie der Naturforscher Leopoldina.

Strohl, Jean. 1940. Karl Hescheler, 1868–1940. *Verhandlungen der Schweizerischen Naturforschenden Gesellschaft*, 120: 445–450.

Strübel, Gustav. 1984. 1945—Neuanfang oder versäumte Gelegenheit?, pp. 168–179. In: Tröger, J. (Ed.), *Hochschule und Wissenschaft im Dritten Reich*. Frankfurt, Germany: Campus Verlag.

Stubbe, J. 1942. Bruno Thüring: Albert Einstein's Umsturzversuch der Pysik und seine inneren Möglichkeiten und Ursachen. Berlin: Dr. Georg Lüttke Verlag. 1941. *Zeitschrift für die gesamte Naturwissenschaft*, 8: 306–307.

Takhtajan, Armen. 1959. *Die Evolution der Angiospermen*. Jena, Germany: Gustav Fischer.

Tanner, Michael. 2000. *Nietzsche. A very short Introduction*. Oxford: Oxford University Press.

Taschwer, Kurt. 2012. Kämpfer gegen die "Verjudung" der Universität. Der Standard, 10. Oktober 2012, p. 15.

Temkin, O. 1959. The idea of descent in post-Romantic German biology: 1848–1858, pp. 323–355. In: Glass, B., O. Temkin, and W.L. Stauss Jr. (Eds.) *Forerunners of Darwin, 1745–1859*. Baltimore, MD: Johns Hopkins University Press.

Temple, Stanley A. and John T. Emlen. 1994. In Memoriam: Joseph J. Hickey, 1907–1993. *Auk* 111: 450–452.

Thacher, James, K. 1877a. Median and paired fins, a contribution to the history of vertebrate limbs. *Transactions of the Connecticut Academy*, 3: 281–310.

Thacher, James, K. 1877b. Ventral fins of ganoids. *Transactions of the Connecticut Academy*, 4: 233–242.

Thienemann, August. 1918. Lebensgemeinschaft und Lebensraum. *Naturwissenschaftliche Wochenschrift* N.F. 17, #20: 281–290; #21: 297–303.

Thienemann, August. 1925. Der See als Lebensgemeinschaft. *Die Naturwissenschaften*, 13: 598–600.

Thienemann, August. 1935a. Lebensgemeinschaft und Lebensraum. *Unterrichtsblätter für Mathematik und Naturwissenschaften*, 41: 337–350.

Thienemann, August. 1935b. *Die Bedeutung der Limnologie für die Kultur der Gegenwart*. Stuttgart, Germany: Schweizerbart.

Thienemann, August. 1939. Grundzüge einer allgemeinen Ökologie. *Archiv für Hydrobiologie*, 35: 267–285.

Thienemann, August. 1940. Unser Bild der lebenden Natur. 90. u. 91. *Jahresbericht der Naturhistorischen Gesellschaft zu Hannover*, 1940: 27–51.

Thienemann, August. 1941. *Leben und Umwelt*. Leipzig, Germany: Johann Ambrosius Barth.

Thienemann, August. 1959. *Erinnerungen und Tagebuchblätter eines Biologen. Ein Leben im Dienste der Limnologie*. Stuttgart, Germany: Schweizerbart.

Thienemann, August and Friedrich Krüger. 1937. "*Orthocladius*" *abiskoensis* Edwards und *rubicundus* (Mg.), zwei "Puppenspezies" der Chironomiden (Chironomiden aus Lappland. II.). *Zoologischer Anzeiger*, 117: 257–267.

Thienemann, August and Jean-Jacques Kieffer. 1916. Schwedische Chironomiden. *Archiv für Hydrobiologie*, Suppl. 2: 483–554.

Thomas, H.H. 1932/33. The old morphology and the new. *Proceedings of the Linnean Society of London*, 145th Session, London, UK: 17–46.

Thüring, Bruno. 1941. *Albert Einstein's Umsturzversuch der Pysik und seine inneren Möglichkeiten und Ursachen*. Berlin, Germany: G. Lüttke.

Tirala, Lothar G. 1934a. Ist der Untergang der Kulturvölker eine biologische Notwendigkeit? *Der Biologe*, 3: 52–53.

Tirala, Lothar G. 1934b. Dialog über die biologische Weltanschauung. *Der Biologe*, 3: 273–284.

Tollmann, Alexander. 1986. Karl Beurlen. 17. April 1901–27. Dezember 1985. *Mitteilungen der österreichischen geologischen Gesellschaft*, 79 (Umweltgeologie-Band): 373–374.

Tönnies, Ferdinand. 1917. Jaekel, Dr. Otto, ord. Professor an der Universität in Greifswald, die naürlichen Grundlagen staatlicher Organisation. Kriegsausgabe. Erstes bis drittes Tausend. 1916. Im Selbstverlage des Verfassers. (Bezug durch Georg Stilke—Berlin und Brüssel.) 196 S. M.2.–*Weltwirtschaftliches Archiv*, 11: 135–137.

Tönnnies, Ferdinand. 1923. Hertwig, Oscar. Der Staat als Organismus. Gedanken zur Entwicklung der Menscheit. Jena 1922. Gustav Fischer, vi u. 264 S. 30 Mk. *Niemeyers Zeitschrift für Internationales Recht*, 30: 303–304.

Torrey, Theodore W. 1939. Organisms in time. *Quarterly Review of Biology*, 14: 275–288.

Trautmann, Milton B. 1977. In Memoriam: Margaret Morse Nice. *Auk* 94: 430–441.

Tremblay, Frederic. 2013. Nicolai Hartmann and the metaphysical foundation of phylogenetic systematics. *Biological Theory*, 7: 56–68.

Trembley Abraham. 1744. *Mémoires pour servir a l'Histoire d'un Genre de Polypes d'Eau Douce*. Paris, France: Durand.

Trembley, Abraham. 1943. *Correspondence inédite entre Réaumur et Abraham Trembley. Introduction by Emile Guyénot*. Geneva, Switzerland: Georg & Cie.

Tröger, Jörg (Ed.). 1984. Einleitung des Herausgebers, pp. 7–10. In: *Hochschule und Wissenschaft im Dritten Reich*. Frankfurt, Germany: Campus Verlag.

Troll, Wilhelm. 1925. Gestalt und Gesetz. Versuch einer geistesgeschichtlichen Grundlegung der morphologischen und physiologischen Forschung. *Flora*, N.F. 118 & 119: 536–565.

Troll, Wilhelm. 1926. *Goethe's Morphologische Schriften*. Jena, Germany: Eugen Diederichs.

Troll, Wilhelm. 1928. *Organisation und Gestalt im Bereich der Blüte*, Berlin, Germany: Springer.

Troll, Wilhelm. 1929. Grundprobleme der Pflanzenmorphologie und der Biologie überhaupt. *Biologisches Zentralblatt*, 49: 43–60.

Troll, Wilhelm. 1932. Morphologie der Pflanzen, pp. 1–3. In: Dittler, R., G. Joos, E. Korschelt, G. Linck, F. Oltmanns, and K. Schaum (Eds.), *Handwörterbuch der Naturwissenschaften*, 2nd edition, vol. 7. Jena, Germany: Gustav Fischer.

Troll, Wilhelm. 1935/36. Die Wiedergeburt der Morphologie aus dem Geiste deutscher Wissenschaft. *Zeitschrift für die gesamte Naturwissenschaft*, 1: 349–356.

Troll, Wilhelm. 1937. *Vergleichende Morphologie der höheren Pflanzen*. Erster Band. Vegetationsorgane. Berlin, Germany: Gebr. Bornträger.

Troll, Wilhelm. 1939. Phylogenetische oder idealistische Morphologie. *Botanisches Archiv*, 40: 634–635.

Troll, Wilhelm. 1943a. Um die Objektivität in der Wiedergabe wissenschaftlicher Auffassungen. Eine Auseinandersetzung mit Herrn G. HEBERER. *Botanisches Archiv. Zeitschrift für die gesamte Botanik und ihre Grenzgebiete*, 44: 431–438.

Troll, Wilhelm. 1943b. Zur Morphologie und Phylogenie des Ophioglossaceenblattes. Eine Erwiderung an Herrn W. ZIMMERMANN. *Berichte der Deutschen Botanischen Gesellschaft*, 61: 70–74.

Troll, Wilhelm. 1944. Urbild und Ursache in der Biologie. *Botanisches Archiv. Zeitschrift für die gesamte Botanik und ihre Grenzgebiete*, 45: 396–416.

Troll, Wilhelm. 1949. Die Urbildlichkeit der organischen Gestaltung und Goethe's Prinzip der "Variablen Proportionen." *Experientia*, 5: 491–495.

Troll, Wilhelm. 1951a. Das Analogieproblem in seiner Bedeutung für die Naturerkenntnis. *Experientia*, 7: 436–440.

Troll, Wilhelm. 1951b. Biomorphologie und Biosystematik als typologische Wissenschaften. *Studium Generale* 4: 376–389.

Troll, Wilhelm. 1952. Über die Grundlagen des Naturverständnisses. *Scientia*, 87: 11–18.

Troll, Wilhelm and A. Meister. 1951. Wesen und Aufgabe der Biosystematik in ontologischer Beleuchtung. *Philosophisches Jahrbuch*, 61: 105–131.

Troll, Wilhelm and Karl L. Wolf. 1940. Goethes morphologischer Auftrag. Zum 150. Jahr des Erscheinens von Goethes Versuch über die Metamorphose der Pflanzen. *Botanisches Archiv. Zeitschrift für die gesamte Botanik und ihre Grenzgebiete*, 41: 1–71.

Tschulok, Sinai. 1908. Zur Methodologie und Geschichte der Deszendenztheorie. *Biologisches Centralblatt*, 1908: 4–18, 33–51, 73–96, 97–117.

Tschulok, Sinai. 1910. *Das System der Biologie in Forschung und Lehre. Eine historisch-kritische Studie.* Jena, Germany: Gustav Fischer.

Tschulok, Sinai. 1912. Logisches und Methodisches. Die Stellung der Morphologie im System der Wissenschaften und ihre Beziehung zur Entwicklungslehre, pp. 1–49. In: Lang, A. (Ed.), *Handbuch der Morphologie der wirbellosen Tiere, zweiter Band, erste Lieferung.* Jena, Germany: Gustav Fischer.

Tschulok, Sinai. 1922. *Deszendenzlehre (Entwicklungslehre). Ein Lehrbuch auf historisch-kritischer Grundlage.* Jena, Germany: Gustav Fischer.

Uexküll, Jakob V. 1920. *Staatsbiologie (Anatomie-Physiologie-Pathologie des Staates).* Berlin, Germany: Gebr. Paetel.

Uexküll, Jakob V. 1920/1926. *Theoretical Biology*; translated by D.L. Mackinnon. New York: Hartcourt, Brace & Co.

Uexküll, Jakob V. 1931. Die Rolle des Subjekts in der Biologie. *Die Naturwissenschaften*, 19: 386–391.

Uexküll, Jakob V. 1935/36. Die Bedeutung der Umweltforschung für die Erkenntnis des Lebens. *Zeitschrift für die gesamte Naturwissenschaft*, 1: 257–272.

Uexküll, Thure. 1953. Der Mensch und die Natur. *Grundzüge einer Naturphilosophie*. Berne, Switzerland: Francke.

Uhsadel, Christoph. 2011. Die Universität Tübingen im NS-Staat. http://www.tribuene-verlag.de/T196_Uhsadel.pdf (acccessed July 15, 2013).

Ungerer, Emil. 1922. *Die Teleologie Kants und ihre Bedeutung für die Logik der Biologie.* Berlin, Germany: Gebr. Bornträger.

Ungerer, Emil. 1936/37. Wilhelm Troll: Vergleichende Morphologie der höherern Pflanzen, I. Band: Vegetationsorgane. 1. Lieferung. Berlin, Gebr. Bornträger, 1935. 172 S. Mit 104 Abbildungen. *Zeitschrift für die gesamte Naturwissenschaft*, 2: 86–88.

Ungerer, Emil. 1941. Hans Driesch. Der Naturforscher und Naturphilosoph (1867–1941). *Die Naturwissenschaften*, 29: 457–462.

Uschmann, Georg H. 1969. Hertwig, Oscar Wilhelm August, pp. 706–707. In: Wagner, F. (Ed.), *Neue Deutsche Biographie*, vol. 8. Berlin, Germany: Duncker und Humblot.

Uzzell, Thomas, Rainer Günther, and Leszek Berger. 1976. *Rana ridibunda* and *Rana esculenta*: A leaky hybridogenetic system. *Proceedings of the Academy of Natural Sciences of Philadelphia*, 128: 147–171.

Vareschi, Volkmar. 1942. A. Thienemann: Leben und Umwelt. Barth, Leipzig. 1941. *Zeitschrift für die gesamte Naturwissenschaft*, 8: 261.

Virchow, Rudolf. 1855. Cellular-Pathologie. *Virchows Archiv für pathologische Anatomie und Physiologie und für klinische Medizin*, 8: 3–39.

Virchow, Rudolf. 1858. *Die Cellularpathologie in ihrer Begründung auf physiologische und pathologische Gewebelehre*. Berlin, Germany: August Hirschwald.

Virchow, Rudolf (Ed.). 1862. Atome und Individuen. Vortrag gehalten im wissenschaftlichen Vereine der Singakademie zu Berlin am 12. Februar 1859, pp. 35–76. In: *Vier Reden über Leben und Kranksein*. Berlin, Germany: Georg Reimer.

Virchow, Rudolf. 1877. Hochverehrte Anwesende! *Verhandlungen der Gesellschaft Deutscher Naturforscher und Ärzte*, 50: 65–78.

Vogt, Carl. 1854. *Recherches sur les animaux inférieures de la Méditerranée. Premier Mémoir. Sur les siphonophores de la mer de Nice.* Geneva, Switzerland: Section des Sciences Naturelles et Mathématiques de l'Institut Genevois.

Volkmann, R.V. 1935. Martin Heidenhain und die mikroskopische Technik (Zu seinem 70. Geburtstag). *Zeitschrift für Mikroskopie*, 51: 309–315.

Voss, Wilhelm. 1934. *Die lebensgesetzlichen Grundlagen des Nationalsozialismus*. Frankfurt, Germany: Moritz Diesterweg.

Waaser, Friedrich. 1940a. L. Wolf und W. Troll: Goethe's morphologischer Auftrag. Leipzig, Akad. Verlagsgesellschaft, 1940. *Zeitschrift für die gesamte Naturwissenschaft*, 6: 329–331.

Waaser, Friedrich. 1940b. Goethe's Typus-Idee und das Problem der organischen Form. *Zeitschrift für die gesamte Naturwissenschaft*, 6: 6–16.

Walker, Mark. 2003. Nazi Science? Natural science in national socialism, pp. 993–1012. In: Hoßfeld, U., J. John, O. Lemuth, and R. Stutz (Eds.), *"Kämpferische Wissenschaft." Studien zur Universität Jena im Nationalsozialismus*. Weimar, Germany: Böhlau Verlag.

Walther, Johannes. 1919. Ernst Haeckel als Mensch und Lehrer. *Die Naturwissenschaften*, 7: 946–951.

Weber, Hermann. 1935/36. Lage und Aufgabe der deutschen Biologie in der deutschen Gegenwart. *Zeitschrift für die gesamte Naturwissenschaft*, 1: 95–106.

Weber, Hermann. 1937. DÜRKEN, B. Entwicklungsbiologie und Ganzheit. Ein Beitrag zur Neugestaltung unseres Weltbildes. Leipzig und Berlin: B. G. Teubner 1936. VI, 207 S. und 56 Abbild. 15 cm × 23 cm. Preis geh. RM 5.80, geb. RM 6.80. *Die Naturwissenschaften*, 27: 590–591.

Weber, Hermann. 1942. Organismus und Umwelt. *Der Biologe*, 11: 57–68.

Weber, Hermann. 1958. Konstruktionsmorphologie. *Zoologische Jahrbücher, Abteilung für allgemeine Zoologie und Physiologie der Tiere*, 68: 1–112.

Weberling Focko. 1981. Wilhelm Troll, 1897–1978. *Berichte der Deutschen Botanischen Gesellschaft*, 94: 311–413.

Weberling Focko. 1999. Wilhelm Troll, his work and influence. *Systematics and Geography of Plants*, 68: 9–24.

Weidenreich, Franz. 1921. Das Evolutionsproblem und der individuelle Gestaltungsanteil am Entwicklungsgeschehen. *Vorträge und Aufsätze über Entwicklungsmechanik der Organismen*, 27: 1–115.

Weigelt, Johann. 1939. Geiseltalforschung und Phylogenie. *Der Biologe*, 8: 35–38.

Weigmann, Gerd. 1973. Verzeichnis der wissenschaftlichen Schriften von Prof. Dr. Dr. h.c. Adolf Remane. *Faunistisch-Ökologische Mitteilungen*, 4: 275–281.

Weikart, Richard. 2004. *From Darwin to Hitler. Evolutionary Ethics, Eugenics, and Racism in Germany*. New York: Palgrave Macmillan.

Weindling, Paul J. 1981. Theories of the cell state in Imperial Germany, pp. 99–155. In: Webster, C. (Ed.), *Biology, Medicine and Society 1840–1940*. Cambridge, UK: Cambridge University Press.

Weindling, Paul J. 1985. Darwinism in Germany, pp. 685–698. In: Kohn, D. (Ed.), *The Darwinian Heritage*. Princeton, NJ: Princeton University Press.

Weindling, Paul J. 1989a. Ernst Haeckel, Darwinismus, and the secularization of nature, pp. 311–327. In: Moore, J.E. (Ed.), *History, Humanity and Evolution. Essays for John C. Greene*. Cambridge, UK: Cambridge University Press.

Weindling, Paul J. 1989b. *Health, Race, and German Politics between National Unification and Nazism, 1870–1945*. Cambridge, UK: Cambridge University Press.

Weindling, Paul J. 1991. *Darwinism and Social Darwinism in Imperial Germany: The Contribution of the Cell Biologist Oscar Hertwig (1849–1922)*. Stuttgart, Germany: Gustav Fischer.

Weingart, Peter. 1985. Eugenik—Eine angewandte Wissenschaft. Utopien der Menschenzüchtung zwischen Wissenschaftsentwicklung und Politik, pp. 314–349. In: Lundgreen, P. (Ed.), *Wissenschaft im Dritten Reich*. Frankfurt, Germany: Suhrkamp.

Weingart, Peter, Jürgen Kroll, and Kurt Bayertz. 1992. *Rasse, Blut und Gene. Geschichte der Eugenik und Rassenhygiene in Deutschland*. Frankfurt, Germany: Suhrkamp Taschenbuch.

Weiss, Sheila F. 1994. Pedagogy, professionalism and politics: Biology instruction during the Third Reich, pp. 184–196. In: Renneberg, M. and M. Walker (Eds.), *Science, Technology, and National Socialism*. Cambridge, UK: Cambridge University Press.

Wenk, Peter. 2008/09. Biographisches zu Hermann Weber (21.11.1899–18.11.1956). *Entomologia Generalis*, 31: 109–112.

Wettstein, Richard von. 1901. *Handbuch der systematischen Botanik*. Wien, Austria: Franz Deuticke.

Weyl, Hermann. 1921. *Raum. Zeit. Materie. Vorlesungen über allgemeine Relativitätstheorie*, vierte Auflage. Berlin, Germany: Springer.

Weyl, Hermann. 1923. *Raum. Zeit. Materie. Vorlesungen über allgemeine Relativitätstheorie*, fünfte Auflage, Berlin, Germany: Springer.

Whitehead, Alfred N. 1920. *The Concept of Nature. Tarner Lectures delivered in Trinity College November 1919*. Cambridge, UK: Cambridge University Press.

Wiegand, Albert. 1874–77. *Der Darwinismus und die Naturforschung Newtons und Cuviers. Beiträge zur Methodik der Naturforschung und die Speciesfrage*; three volumes. Braunschweig, Germany: Vieweg und Sohn.

Wiley, Edward O. 2008. Emil Hans Willi Hennig. http://www.encyclopedia.com/topic/Emil_Hans_Willi_Hennig.aspx.

Williams, David M. 2001. Homologues and homology, phenetics and cladistics: 150 years of progress, pp. 191–224. In: Williams, D.M. and P.L. Forey (Eds.), *Milestones in Systematics*. Boca Raton, FL: CRC Press.

Williams, David M. 2006. Otto Kleinschmidt (1870–1954), biogeography and the 'origin' of species: From Formenkreis to progression rule. *Biogeographía*, 1: 3–10.

Williams David. M. and Malte C. Ebach. 2008. *Foundations of Systematics and Biogeography*. Berlin, Germany: Springer.

Williams, Patricia A. 1992. Confusion in cladism. *Synthese*, 91: 135–152.

Winsor, Mary P. 1976. *Starfish, Jellyfish, and the Order of Life, Issues in Nineteenth—Century Science*. New Haven, CT: Yale University Press.

Wolf, Karl Lothar and R. Ramsauer. 1935/36. Zur Geschichte der Naturanschauung in Deutschland, I. *Zeitschrift für die gesamte Naturwissenschaft*, 1: 129–149.

Wunderlich, Klaus. 1991. Erwin Stresemann—Ein Leben für die Wissenschaft. *Mitteilungen aus dem Zoologischen Museum in Berlin*, 67 (Supplementheft, Annalen für Ornithologie, 15): 7–14.

Wüst, Walther. 1939. Die Arbeit des Ahnenerbe. *Der Biologe*, 8: 241–245.

Zachos, Frank and Uwe Hoßfeld. 2001. Adolf Remane (1898–1976): Biographie und ausgewählte evolutionsbiologische Aspekte in seinem Werk, pp. 313–358. In: Hoßfeld, U. and R. Brömer (Eds.), *Darwinismus und/als Ideologie*. Berlin, Germany: VWB—Verlag für Wissenschaft und Bildung.

Zachos, Frank and Uwe Hoβfeld. 2006. Adolf Remane (1898–1976) and his views on systematics, homology and the modern synthesis. *Theory in Biosciences*, 124: 335–348.

Zangerl, Rainer and Gerard R. Case. 1976. *Obelodus aculeatus* (Cope), an anacanthous shark from Pennsylvanian Black Shales of North America. *Palaeontographica*, A154: 107–157.

Zeller, Thomas. 2005. Modeling the landscape of Nazi environmentalism, pp. 147–170. In: Brüggemeier, F.-J., M. Cioc, and T. Zeller (Eds.), *How Green Were the Nazis? Nature, Environment, and Nation in the Third Reich*. Athens, OH: Ohio University Press.

Ziehen, Theodor. 1919. Haeckel als Philosoph. *Die Naturwissenschaften*, 7: 958–961.

Ziehen, Theodor. 1934. *Erkenntnistheorie*. Zweite Auflage. Erster Teil. Allgemeine Grundlegung der Erkenntnistheorie. Spezielle Erkenntnistheorie der Empfindungstatsachen einschliesslich Raumtheorie. Jena, Germany: Gustav Fischer.

Ziehen, Theodor. 1939. *Erkenntnistheorie*. Zweite Auflage. Zweiter Teil. Zeittheorie. Wirklichkeitsproblem. Erkenntnistheorie der anorganischen Natur (erkenntnistheoretische Grundlagen der Physik). Kausalität. Jena, Germany: Gustav Fischer.

Zimmermann, Walter. 1930. *Die Phylogenie der Pflanzen. Ein Überblick über Tatsachen und Probleme*. Jena, Germany: Gustav Fischer.

Zimmermann, Walter. 1933. Paläobotanische und phylogenetische Beiträge, I–V. *Palaeobiologica*, 5: 321–348.

Zimmermann, Walter. 1935. Heidenhain, M.: Die Spaltungsgesetze der Blätter. Beitrag XVI zur synthetischen Morphologie. Jena: Gustav Fischer 1932. XII, 424 S., 221 Abb. und 11 Tafeln. Preis geh. RM. 30.-, geb. RM. 32,-. *Der Biologe*, 4: 26.

Zimmermann, Walter. 1936. System und Geschichte der Lebewesen in ihrer Bedeutung für die Forschung an menschlichen Rassen. *Rasse. Monatsschrift der nordischen Bewegung*, 3: 417–431.

Zimmermann, Walter. 1937. Arbeitsweise der botanischen Phylogenetik und anderer Gruppierungswissenschaften, pp. 941–1053. In: Abderhalden, E. (Ed.), *Handbuch der biologischen Arbeitsmethoden*, 3. Abteilung, Teil IX. Berlin: Urban und Schwarzenberg.

Zimmermann, Walter. 1937/38. Strenge Objekt/Subjekt-Scheidung als Voraussetzung wissenschaftlicher Biologie. *Erkenntnis*, 7: 1–44.

Zimmermann, Walter. 1938. *Vererbung "erworbener Eigenschaften" und Auslese*. Jena, Germany: Gustav Fischer.

Zimmermann, Walter. 1941. Über phylogenetische Methoden. *Der Biologe*, 10: 47–49.

Zimmermann, Walter. 1943. Die Methoden der Phylogenetik, pp. 20–56. In: Heberer, G. (Ed.), *Die Evolution der Organismen. Ergebnisse und Probleme der Abstammungslehre*. Jena, Germany: Gustav Fischer.

Zimmermann, Walter. 1953. *Evolution. Die Geschichte ihrer Probleme und Erkenntnisse*. Munich, Germany: Karl Albert Freiburg.

Zimmermann, Walter. 1954/1959. Methoden der Phylogenetik, pp. 25–102. In: Heberer, G. (Ed.), *Die Evolution der Organismen. Ergebnisse und Probleme der Abstammungslehre*, 2. erweiterte Ausgabe. Stuttgart, Germany: Gustav Fischer.

Zimmermann, Walter. 1959. *Die Phylogenie der Pflanzen. Ein Überblick über Tatsachen und Probleme*. 2. Völlig neu bearbeitete Auflage. Jena, Germany: Gustav Fischer.

Zündorf, Werner. 1938. Die biologische Arbeit der Tübinger Fachgruppe Naturwissenschaft der Tübinger Studentenführung. *Der Biologe*, 7: 311.

Zündorf, Werner. 1938/39. W. Zimmermann: Vererbung erworbener Eigenschaften und Auslese. Jena, Germany: Gustav Fischer, 1938. *Zeitschrift für die gesamte Naturwissenschaft*, 4: 323–324.

Zündorf, Werner. 1939a. Ernst Haeckel's Stammbaum der Pflanzen. *Der Biologe*, 8: 273–279.

Zündorf, Werner. 1939b. Der Lamarckismus in der heutigen Biologie. *Archiv für Rassen- und Gesellschaftsbiologie*, 33: 281–303.

Zündorf, Werner. 1940. Phylogenetische oder Idealistische Morphologie? *Der Biologe*, 9: 10–24.

Zündorf, Werner. 1942. Nochmals: Phylogenetik und Typologie? Entgegnung auf *E. Bergdolt*: 'Über Formwandlungen—Zugleich eine Kritik von Artbildungstheorien' (Der Biologe 9, 398–407, 1940). *Der Biologe*, 11: 125–129.

Zündorf, Werner. 1943. Idealistische Morphologie und Stammeslehre, pp. 86–104. In: Heberer, G. (Ed.), *Die Evolution der Organismen. Ergebnisse und Probleme der Stammesgeschichte.* Jena, Germany: Gustav Fischer.

Index

Note: Page numbers followed by f refer to figures.

Species and Systematics

SPECIES: A HISTORY OF THE IDEA
John S. Wilkins

COMPARATIVE BIOGEOGRAPHY: DISCOVERING AND
CLASSIFYING BIO-GEOGRAPHICAL PATTERNS OF A
DYNAMIC EARTH
Lynee R. Parenti and Malte C. Ebach

BEYOND CLADISTICS
Edited by David M. Williams and Sandra Knapp

MOLECULAR PANBIOGEOGRAPHY ON THE TROPICS
Michael Heads

THE EVOLUTION OF PHYLOGENETIC SYSTEMATICS
Edited by Andrew Hamilton

EVOLUTION BY NATURAL SELECTION: CONFIDENCE,
EVIDENCE AND THE GAP
Michaelis Michael

PHYLOGENETIC SYSTEMATICS: HAECKEL TO HENNIG
Olivier Rieppel